Pg 40 - Question. Final — Also in lecture 10

INTRODUCTION TO ROBOTICS

ANALYSIS, CONTROL, APPLICATIONS

Second Edition

Saeed Benjamin Niku, Ph.D., P.E.

Professor
Mechanical Engineering Department
California Polytechnic State University
San Luis Obispo

WILEY

JOHN WILEY & SONS, INC.

VP & Publisher	Don Fowley
Executive Editor	Linda Ratts
Editorial Assistant	Renata Marchione
Marketing Manager	Christopher Ruel
Media Editor	Lauren Sapira
Production Manager	Janis Soo
Assistant Production Editor	Yee Lyn Song
Cover Designer	RDC Publishing Group Sdn Bhd
Cover Photo Credit	© Alexey Dudoladov/iStockphoto

This book was set in 11/12 Bembo by Thomson Digital, and printed and bound by Courier Westford. The cover was printed by Courier Westford.

This book is printed on acid free paper. ∞
This book was previously published by: Pearson Education, Inc.

Library of Congress Cataloging-in-Publication Data
Niku, Saeed B. (Saeed Benjamin)
    An introduction to robotics : analysis, control, applications / Saeed Niku.—2nd ed.
        p.   cm.
    Includes index.
    ISBN 978-0-470-60446-5 (hardback)
    1. Robotics.   I. Title.   II. Title: Robotics analysis, control, applications.
    TJ211.N547 2010
    629.8'92—dc22

                                                          2010024723

Printed in the United States of America

10 9 8 7 6 5 4 3 2 1

Dedicated to
Shohreh, Adam, and Alan Niku
and to
Sara Niku and the memory of Saleh Niku

Preface

This is the second edition of the Introduction to Robotics textbook. As such, it has all the features and the material covered in the first edition, but also features more examples, more homework, new projects, more detailed material in all chapters, and as a new feature, it also includes a new chapter on automatic controls and control of robots as well as information about downloading a commercially available software system called SimulationX^{TM}.

What one of my students once said years ago still stands: "In the life of every product, there comes a time when you have to shoot the designer and go into production." Therefore, although no textbook is ever perfect, each has unique features that make it stand tall. So is this textbook. The intention behind writing this book was, and is, to cover most subjects that an undergraduate engineering student or a practicing engineer may need to know to be familiar with robotics, to be able to understand robots, design a robot, and integrate a robot in appropriate applications. As such, it covers all necessary fundamentals of robotics, robot components and subsystems, and applications.

The book is intended for senior or introductory graduate courses in robotics as well as for practicing engineers who would like to learn about robotics. Although the book covers a fair amount of mechanics and kinematics, it also covers microprocessor applications, control systems, vision systems, sensors, and actuators. Therefore, it can easily be used by mechanical engineers, electronic and electrical engineers, computer engineers, and engineering technologists. With the new chapter about control theory, even if the student has not had a controls course, he or she can learn enough material to be able to understand robotic control and design.

The book is comprised of 10 chapters. Chapter 1 covers introductory subjects that familiarize the reader with the necessary background information. This includes some historical information, robot components, robot characteristics, robot languages, and robotic applications. Chapter 2 explores the forward and inverse kinematics of robots, including frame representations, transformations, position and orientation analysis, as well as the Denavit-Hartenberg representation of robot kinematics. Chapter 3 continues with differential motions and velocity analysis of robots and frames. Chapter 4 presents an analysis of robot dynamics and forces. Lagrangian mechanics is used as the primary method of analysis and development for this chapter. Chapter 5 discusses methods of path and trajectory planning, both in joint-space and in Cartesian-space. Chapter 6 covers fundamentals of control engineering, including analysis and design tools. Among other things, it discusses root locus, proportional, derivative, and integral control as well as electromechanical system modeling. Chapter 6 also includes an introduction to multi-input-multi-output (MIMO) systems, digital systems, and nonlinear systems. However, the assumption is that students will need additional instruction to be proficient in actually designing systems. One chapter on this subject cannot be adequate, but can nicely serve as an introduction for majors in which a separate course in control engineering is not offered. Chapter 7 covers actuators, including hydraulic devices, electric motors such as DC servomotors and stepper motors, pneumatic devices, as well as many other novel actuators. It also covers microprocessor control of these actuators. Although this book is not a complete mechatronics book, it does cover a fair amount of mechatronics. Except for the design of a microprocessor, many aspects of mechatronic applications are covered in this chapter. Chapter 8 is a discussion of sensors used in robotics and robotic applications. Chapter 9 covers vision systems, including many different techniques for image processing and image analysis. Chapter 10 discusses the basic principles of fuzzy logic and its applications in microprocessor control and robotics. This coverage is not intended to be a complete and thorough analysis of fuzzy logic, but an introduction. It is believed that students and engineers who find it interesting will continue on their own. Appendix A is a quick review of matrix algebra and some other mathematical facts that are needed throughout this book. Appendix B covers image acquisition. Appendix C presents the application of MATLAB in control engineering. Appendix D includes references to commercial software that can be used to model and simulate robots and their dynamics. The student version of this program can be downloaded for free. Consequently, if robotic simulation is to be covered, the program and associated tutorials may be used without additional cost to students.

Most of the material in this book is generally covered in a four-unit, 10-week course at Cal Poly, with three one-hour lectures and one three-hour lab. However, it is easily possible to cover the entire course in a semester-long course as well. The following breakdown can be used as a model for setting up a course in robotics in a quarter system. In this case, certain subjects must be eliminated or shortened, as shown:

Introductory material and review: 3 lectures
Kinematics of position: 7 lectures
Differential motions: 4 lectures
Robot dynamics and force control: 2 lectures
Path and trajectory planning: 1 lecture
Actuators: 3 lectures
Sensors: 3 lectures
Vision systems: 5 lectures
Fuzzy logic: 1 lectures
Exam: 1 lecture

Alternately, for a 14-week long semester course with three lectures per week, the course may be set up as follows:

Introductory material and review: 3 lectures
Kinematics of position: 7 lectures
Differential motions: 5 lectures
Robot dynamics and force control: 5 lectures
Path and trajectory planning: 3 lectures
Robot control and modeling: 5 lectures
Actuators: 5 lectures
Sensors: 2 lectures
Vision systems: 5 lecture
Fuzzy logic: 1 lectures
Exam: 1 lecture

The book also features design projects that start in Chapter 2 and continue throughout the book. At the end of each chapter, the student is directed to continue with the design projects in reference to the present subject. Therefore, by the end of the book, complete systems may be designed.

I would like to thank all the people who, in one way or another, have helped me. This includes my colleagues, including Bill Murray, Charles Birdsong, Lynne Slivovsky, and John Ridgely, all the countless individuals who did the research, development, and hard work that came before my time and which enabled me to learn the subject myself, all the users and students and anonymous reviewers who made countless suggestions to improve the first draft, including Thomas Cavicchi, Ed Foley, and the students who helped with the design and development of projects at Cal Poly, including the Robotics Club. I also thank Mike McDonald, the acquisition editor at John Wiley and Sons, who was instrumental in getting the second edition published, Renata Marchione, Don Fowley, Linda Ratts, and Yee Lyn Song for their assistance throughout, and the editors and the artists who made the book look as it does. I also would like to thank the staff at Prentice Hall who published the first edition. Finally, I thank my family, Shohreh, Adam, and Alan, who let me work on this manuscript for long hours instead of spending time with them. Their patience is much appreciated. To all of you, my sincere thanks.

I hope that you will enjoy reading the book and, more importantly, that you will learn the subject. The joy of robotics comes from learning it.

Saeed Benjamin Niku, Ph.D., P.E.
San Luis Obispo, California
2010

Brief Contents

Chapter 1   Fundamentals   1

Chapter 2   Kinematics of Robots: Position Analysis   33

Chapter 3   Differential Motions and Velocities   114

Chapter 4   Dynamic Analysis and Forces   147

Chapter 5   Trajectory Planning   178

Chapter 6   Motion Control Systems   203

Chapter 7   Actuators and Drive Systems   266

Chapter 8   Sensors   319

Chapter 9   Image Processing and Analysis with Vision Systems   350

Chapter 10   Fuzzy Logic Control   423

Appendix A   Review of Matrix Algebra and Trigonometry   443

Appendix B   Image Acquisition Systems   450

Appendix C   Root Locus and Bode Diagram with MATLAB™   454

Appendix D   Simulation of Robots with Commercial Software   458

Index   459

Contents

Chapter 1   Fundamentals   1

1.1   Introduction   1

1.2   What Is a Robot?   2

1.3   Classification of Robots   3

1.4   What Is Robotics?   4

1.5   History of Robotics   4

1.6   Advantages and Disadvantages of Robots   6

1.7   Robot Components   6

1.8   Robot Degrees of Freedom   8

1.9   Robot Joints   11

1.10   Robot Coordinates   11

1.11   Robot Reference Frames   12

1.12   Programming Modes   14

1.13   Robot Characteristics   14

1.14   Robot Workspace   15

1.15   Robot Languages   16

1.16   Robot Applications   19

1.17   Other Robots and Applications   26

1.18   Social Issues   29

Summary   29

References   30

Problems   31

Chapter 2   Kinematics of Robots: Position Analysis   33

2.1   Introduction   33

2.2   Robots as Mechanisms   34

2.3   Conventions   35

2.4   Matrix Representation   36

2.4.1   Representation of a Point in Space   36

2.4.2   Representation of a Vector in Space   36

2.4.3   Representation of a Frame at the Origin of a Fixed Reference Frame   38

2.4.4   Representation of a Frame Relative to a Fixed Reference Frame   40

2.4.5   Representation of a Rigid Body   41

2.5   Homogeneous Transformation Matrices   44

2.6   Representation of Transformations   45

2.6.1   Representation of a Pure Translation   45

2.6.2   Representation of a Pure Rotation about an Axis   46

2.6.3   Representation of Combined Transformations   49

2.6.4   Transformations Relative to the Rotating Frame   52

2.7   Inverse of Transformation Matrices   54

2.8   Forward and Inverse Kinematics of Robots   59

2.9   Forward and Inverse Kinematic Equations: Position   60

2.9.1   Cartesian (Gantry, Rectangular) Coordinates   60

2.9.2   Cylindrical Coordinates   61

2.9.3   Spherical Coordinates   64

2.9.4   Articulated Coordinates   66

2.10   Forward and Inverse Kinematic Equations: Orientation   66

2.10.1   Roll, Pitch, Yaw (RPY) Angles   66

2.10.2 Euler Angles   70

2.10.3 Articulated Joints   72

2.11 Forward and Inverse Kinematic Equations: Position and Orientation   72

2.12 Denavit-Hartenberg Representation of Forward Kinematic Equations of Robots   73

2.13 The Inverse Kinematic Solution of Robots   87

2.13.1 General Solution for Articulated Robot Arms   89

2.14 Inverse Kinematic Programming of Robots   93

2.15 Degeneracy and Dexterity   95

2.15.1 Degeneracy   95

2.15.2 Dexterity   96

2.16 The Fundamental Problem with the Denavit-Hartenberg Representation   96

2.17 Design Projects   99

2.17.1 A 3-DOF Robot   99

2.17.2 A 3-DOF Mobile Robot   101

Summary   102

References   102

Problems   103

Chapter 3   Differential Motions and Velocities   114

3.1 Introduction   114

3.2 Differential Relationships   114

3.3 Jacobian   116

3.4 Differential versus Large-Scale Motions   118

3.5 Differential Motions of a Frame versus a Robot   119

3.6 Differential Motions of a Frame   120

3.6.1 Differential Translations   120

3.6.2 Differential Rotations about the Reference Axes   120

3.6.3 Differential Rotation about a General Axis q 122

3.6.4 Differential Transformations of a Frame   123

3.7 Interpretation of the Differential Change   124

3.8 Differential Changes between Frames   125

3.9 Differential Motions of a Robot and its Hand Frame   127

3.10 Calculation of the Jacobian   128

3.11 How to Relate the Jacobian and the Differential Operator   131

3.12 Inverse Jacobian   134

3.13 Design Projects   141

3.13.1 The 3-DOF Robot   141

3.13.2 The 3-DOF Mobile Robot   142

Summary   142

References   143

Problems   143

Chapter 4   Dynamic Analysis and Forces   147

4.1 Introduction   147

4.2 Lagrangian Mechanics: A Short Overview   148

4.3 Effective Moments of Inertia   158

4.4 Dynamic Equations for Multiple-DOF Robots   158

4.4.1 Kinetic Energy   158

4.4.2 Potential Energy   163

4.4.3 The Lagrangian   164

4.4.4 Robot's Equations of Motion   164

4.5 Static Force Analysis of Robots   170

4.6 Transformation of Forces and Moments between Coordinate Frames   172

4.7 Design Project   175

Summary   175

References   175

Problems   176

Chapter 5   Trajectory Planning   178

5.1   Introduction   178

5.2   Path versus Trajectory   178

5.3   Joint-Space versus Cartesian-Space Descriptions   179

5.4   Basics of Trajectory Planning   180

5.5   Joint-Space Trajectory Planning   184

   5.5.1   Third-Order Polynomial Trajectory Planning   184

   5.5.2   Fifth-Order Polynomial Trajectory Planning   187

   5.5.3   Linear Segments with Parabolic Blends   188

   5.5.4   Linear Segments with Parabolic Blends and Via Points   191

   5.5.5   Higher-Order Trajectories   191

   5.5.6   Other Trajectories   195

5.6   Cartesian-Space Trajectories   195

5.7   Continuous Trajectory Recording   200

5.8   Design Project   200

Summary   201

References   201

Problems   202

Chapter 6   Motion Control Systems   203

6.1   Introduction   203

6.2   Basic Components and Terminology   204

6.3   Block Diagrams   204

6.4   System Dynamics   205

6.5   Laplace Transform   208

6.6   Inverse Laplace Transform   211

   6.6.1   Partial Fraction Expansion when $F(s)$ Involves Only Distinct Poles   212

   6.6.2   Partial Fraction Expansion when $F(s)$ Involves Repeated Poles   213

   6.6.3   Partial Fraction Expansion when $F(s)$ Involves Complex Conjugate Poles   214

6.7   Transfer Function   216

6.8   Block Diagram Algebra   219

6.9   Characteristics of First-Order Transfer Functions   221

6.10   Characteristics of Second-Order Transfer Functions   223

6.11   Characteristic Equation: Pole/Zero Mapping   225

6.12   Steady-State Error   228

6.13   Root Locus Method   230

6.14   Proportional Controllers   235

6.15   Proportional-plus-Integral Controllers   239

6.16   Proportional-plus-Derivative Controllers   241

6.17   Proportional-Integral-Derivative Controller (PID)   244

6.18   Lead and Lag Compensators   246

6.19   The Bode Diagram and Frequency Domain Analysis   247

6.20   Open-Loop versus Closed-Loop Applications   247

6.21   Multiple-Input and Multiple-Output Systems   249

6.22   State-Space Control Methodology   250

6.23   Digital Control   254

6.24   Nonlinear Control Systems   256

6.25   Electromechanical Systems Dynamics: Robot Actuation and Control   257

6.26   Design Projects   262

Summary   263

References   263

Problems   263

Chapter 7   Actuators and Drive Systems   266

7.1   Introduction   266

7.2   Characteristics of Actuating Systems   267

7.2.1   Nominal Characteristics— Weight, Power to Weight Ratio, Operating Pressure, Voltage, and Others   267

7.2.2   Stiffness versus Compliance   267

7.2.3   Use of Reduction Gears   268

7.3   Comparison of Actuating Systems   271

7.4   Hydraulic Actuators   272

7.5   Pneumatic Devices   278

7.6   Electric Motors   279

7.6.1   Fundamental Differences between AC and DC-Type Motors   280

7.6.2   DC Motors   283

7.6.3   AC Motors   285

7.6.4   Brushless DC Motors   286

7.6.5   Direct Drive Electric Motors   286

7.6.6   Servomotors   287

7.6.7   Stepper Motors   288

7.7   Microprocessor Control of Electric Motors   303

7.7.1   Pulse Width Modulation   304

7.7.2   Direction Control of DC Motors with an H-Bridge   306

7.8   Magnetostrictive Actuators   307

7.9   Shape-Memory Type Metals   307

7.10   Electroactive Polymer Actuators (EAP)   308

7.11   Speed Reduction   309

7.12   Other Systems   311

7.13   Design Projects   312

7.13.1   Design Project 1   312

7.13.2   Design Project 2   312

7.13.3   Design Project 3   314

7.13.4   Design Project 4   314

Summary   315

References   316

Problems   317

Chapter 8   Sensors   319

8.1   Introduction   319

8.2   Sensor Characteristics   319

8.3   Sensor Utilization   322

8.4   Position Sensors   323

8.4.1   Potentiometers   323

8.4.2   Encoders   324

8.4.3   Linear Variable Differential Transformers (LVDT)   327

8.4.4   Resolvers   328

8.4.5   (Linear) Magnetostrictive Displacement Transducers (LMDT or MDT)   328

8.4.6   Hall-effect Sensors   329

8.4.7   Other Devices   329

8.5   Velocity Sensors   330

8.5.1   Encoders   330

8.5.2   Tachometers   330

8.5.3   Differentiation of Position Signal   331

8.6   Acceleration Sensors   331

8.7   Force and Pressure Sensors   331

8.7.1   Piezoelectric   331

8.7.2   Force Sensing Resistor   332

8.7.3   Strain Gauge   332

8.7.4   Antistatic Foam   333

8.8   Torque Sensors   333

8.9   Microswitches   334

8.10   Visible Light and Infrared Sensors   335

8.11   Touch and Tactile Sensors   335

8.12   Proximity Sensors   336

    8.12.1   Magnetic Proximity Sensors   337

    8.12.2   Optical Proximity Sensors   337

    8.12.3   Ultrasonic Proximity Sensors   338

    8.12.4   Inductive Proximity Sensors   338

    8.12.5   Capacitive Proximity Sensors   338

    8.12.6   Eddy Current Proximity Sensors   339

8.13   Range Finders   339

    8.13.1   Ultrasonic Range Finders   340

    8.13.2   Light-Based Range Finders   341

    8.13.3   Global Positioning System (GPS)   342

8.14   Sniff Sensors   343

8.15   Taste Sensors   343

8.16   Vision Systems   343

8.17   Voice Recognition Devices   343

8.18   Voice Synthesizers   344

8.19   Remote Center Compliance (RCC) Device   344

8.20   Design Project   348

    Summary   348

    References   348

Final Test

Chapter 9   Image Processing and Analysis with Vision Systems   350

9.1   Introduction   350

9.2   Basic Concepts   350

    9.2.1   Image Processing versus Image Analysis   350

    9.2.2   Two- and Three-Dimensional Image Types   351

    9.2.3   The Nature of an Image   351

    9.2.4   Acquisition of Images   352

    9.2.5   Digital Images   352

    9.2.6   Frequency Domain versus Spatial Domain   354

9.3   Fourier Transform and Frequency Content of a Signal   354

9.4   Frequency Content of an Image; Noise, Edges   357

9.5   Resolution and Quantization   358

9.6   Sampling Theorem   360

9.7   Image-Processing Techniques   363

9.8   Histogram of Images   364

9.9   Thresholding   365

9.10   Spatial Domain Operations: Convolution Mask   368

9.11   Connectivity   372

9.12   Noise Reduction   374

    9.12.1   Neighborhood Averaging with Convolution Masks   374

    9.12.2   Image Averaging   375

    9.12.3   Frequency Domain   376

    9.12.4   Median Filters   377

9.13   Edge Detection   377

9.14   Sharpening an Image   383

9.15   Hough Transform   385

9.16   Segmentation   388

9.17   Segmentation by Region Growing and Region Splitting   389

9.18   Binary Morphology Operations   391

    9.18.1   Thickening Operation   392

    9.18.2   Dilation   393

    9.18.3   Erosion   393

    9.18.4   Skeletonization   394

    9.18.5   Open Operation   395

    9.18.6   Close Operation   395

    9.18.7   Fill Operation   396

9.19   Gray Morphology Operations   396

    9.19.1   Erosion   396

    9.19.2   Dilation   396

9.20   Image Analysis   396

9.21   Object Recognition by Features   397

    9.21.1   Basic Features Used for Object Identification   397

    9.21.2   Moments   398

    9.21.3   Template Matching   404

    9.21.4   Discrete Fourier Descriptors   405

    9.21.5   Computed Tomography (CT)   405

9.22   Depth Measurement with Vision Systems   406

    9.22.1   Scene Analysis versus Mapping   406

    9.22.2   Range Detection and Depth Analysis   406

    9.22.3   Stereo Imaging   406

    9.22.4   Scene Analysis with Shading and Sizes   408

9.23   Specialized Lighting   408

9.24   Image Data Compression   409

    9.24.1   Intraframe Spatial Domain Techniques   409

    9.24.2   Interframe Coding   410

    9.24.3   Compression Techniques   411

9.25   Color Images   411

9.26   Heuristics   412

9.27   Applications of Vision Systems   412

9.28   Design Project   413

    Summary   414

    References   414

    Problems   415

Chapter 10 Fuzzy Logic Control   423

10.1   Introduction   423

10.2   Fuzzy Control: What Is Needed   425

10.3   Crisp Values versus Fuzzy Values   425

10.4   Fuzzy Sets: Degrees of Membership and Truth   426

10.5   Fuzzification   427

10.6   Fuzzy Inference Rule Base   429

10.7   Defuzzification   430

    10.7.1   Center of Gravity Method   431

    10.7.2   Mamdani's Inference Method   431

10.8   Simulation of Fuzzy Logic Controller   435

10.9   Applications of Fuzzy Logic in Robotics   437

10.10   Design Project   440

    Summary   440

    References   440

    Problems   441

Appendix A   Review of Matrix Algebra and Trigonometry   443

A.1   Matrix Algebra and Notation: A Review   443

A.2   Calculation of An Angle from its Sine, Cosine, or Tangent   448

    Problems   449

Appendix B   Image Acquisition Systems   450

B.1   Vidicon Camera   450

B.2   Digital Camera   452

    References   453

Appendix C   Root Locus and Bode Diagram with MATLAB™   454

C.1   Root Locus   454

C.2   Bode Diagram   457

Appendix D   Simulation of Robots with Commercial Software   458

Index   459

CHAPTER 1

Fundamentals

1.1 Introduction

Robotics, in different forms, has been on humans' minds since the time we could build things. You may have seen machines that artisans made that try to mimic human motions and behavior. Examples include the statues in Venice's San Marcos clock tower that hit the clock on the hour and figurines that tell a story in the fifteenth-century Astronomical Clock on the side of the Old Town Hall Tower in Prague (Figure 1.1). Toys, from simple types to sophisticated machines with repeating movements, are other examples. In Hollywood, movies have even portrayed robots and humanoids as superior to humans.

Although in principle humanoids are robots and are designed and governed by the same basics, in this book, we will primarily study industrial manipulator type robots. This book covers some basic introductory material that familiarizes you with the subject; it presents an analysis of the mechanics of robots including kinematics, dynamics, and trajectory planning; and it discusses the elements used in robots and in robotics, such as actuators, sensors, vision systems, and so on. Robot rovers are no different, although they usually have fewer degrees of freedom and generally move in a plane. Exoskeletal and humanoid robots, walking machines, and robots that mimic animals and insects have many degrees of freedom (DOF) and may possess unique capabilities. However, the same principles we learn about manipulators apply to robot rovers too, whether kinematics, differential motions, dynamics, or control.

Robots are very powerful elements of today's industry. They are capable of performing many different tasks and operations, are accurate, and do not require common safety and comfort elements humans need. However, it takes much effort and many resources to make a robot function properly. Most companies of the mid-1980s that made robots are gone, and with few exceptions, only companies that make real industrial robots have remained in the market (such as Adept, Staubli, Fanuc, Kuka, Epson, Motoman, Denso, Fuji, and IS Robotics as well as specialty robotic companies such as Mako Surgical Corp. and Intuitive Surgical). Early industrialist predictions about the possible number of robots

1

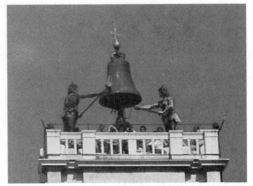

Figure 1.1   Centuries-old figurines and statues that mimic human motions.

in industry never materialized because high expectations could not be satisfied with the present robots. As a result, although there are many thousands of robots in industry working tirelessly and satisfactorily for the intended jobs, robots have not overwhelmingly replaced workers. They are used where they are useful. Like humans, robots can do certain things, but not others. As long as they are designed properly for the intended purposes, they are very useful and continue to be used.

The subject of robotics covers many different areas. Robots alone are hardly ever useful. They are used together with other devices, peripherals, and other manufacturing machines. They are generally integrated into a system, which as a whole, is designed to perform a task or do an operation. In this book, we will refer to some of these other devices and systems used with robots.

1.2 What Is a Robot?

If you compare a conventional robot manipulator with a crane attached to, say, a utility or towing vehicle, you will notice that the robot manipulator is very similar to the crane. Both possess a number of links attached serially to each other with joints, where each joint can be moved by some type of actuator. In both systems, the "hand" of the manipulator can be moved in space and placed in any desired location within the workspace of the system. Each one can carry a certain load and is controlled by a central controller that controls the actuators. However, one is called a robot and one is called a manipulator (or, in this case, a crane). Similarly, material handling manipulators that move heavy objects in manufacturing plants look just like robots, but they are not robots. The fundamental difference between the two is that the crane and the manipulator are controlled by a human who operates and controls the actuators, whereas the robot manipulator is controlled by a computer that runs

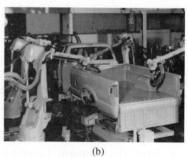

(a)   (b)

Figure 1.2   (a) Dalmec PM human-operated manipulator.   (Printed with permission from Dalmec S.p.A.) (b) Fanuc S-500 robots performing seam-sealing on a truck.   (Reprinted with permission from Fanuc Robotics, North America, Inc.) Both have similar construction and elements, but only the robot is controlled by a computer whereas the manipulator is controlled by an operator.

a program (Figure 1.2). This difference between the two determines whether a device is a simple manipulator or a robot. In general, robots are designed and meant to be controlled by a computer or similar device. The motions of the robot are controlled through a controller under the supervision of the computer, which is running some type of a program. Therefore, if the program is changed, the actions of the robot will change accordingly. The intention is to have a device that can perform many different tasks; consequently, it is very flexible in what it can do without having to be redesigned. Therefore, the robot is designed to be able to perform many tasks based on the running program(s) simply by changing the program. The simple manipulator (or the crane) cannot do this without an operator running it all the time.

Different countries have different standards for what they consider a robot. In American standards, a device must be easily reprogrammable to be considered a robot. Therefore, manual handling devices (devices that have multiple degrees of freedom and are actuated by an operator) or fixed sequence robots (devices controlled by hard stops to control actuator motions on a fixed sequence that are difficult to change) are not considered robots.

1.3 Classification of Robots

The following is the classification of robots according to the Japanese Industrial Robot Association (JIRA):

- Class 1: *Manual Handling Device*: a device with multiple degrees of freedom, actuated by an operator

- Class 2: *Fixed Sequence Robot*: a device that performs the successive stages of a task according to a predetermined, unchanging method, which is hard to modify
- Class 3: *Variable Sequence Robot*: same as in class 2, but easy to modify
- Class 4: *Playback Robot*: a human operator performs the task manually by leading the robot, which records the motions for later playback; the robot repeats the same motions according to the recorded information
- Class 5: *Numerical Control Robot*: the operator supplies the robot with a movement program rather than teaching it the task manually
- Class 6: *Intelligent Robot*: a robot with the means to understand its environment and the ability to successfully complete a task despite changes in the surrounding conditions under which it is to be performed

The Robotics Institute of America (RIA) only considers classes 3–6 of the above as robots. The Association Francaise de Robotique (AFR) has the following classification:

- Type A: handling devices with manual control to telerobotics
- Type B: automatic handling devices with predetermined cycles
- Type C: programmable, servo controlled robots with continuous or point-to-point trajectories
- Type D: same as C but with capability to acquire information from its environment

1.4 What Is Robotics?

Robotics is the art, knowledge base, and the know-how of designing, applying, and using robots in human endeavors. Robotic systems consist of not just robots, but also other devices and systems used together with the robots. Robots may be used in manufacturing environments, in underwater and space exploration, for aiding the disabled, or even for fun. In any capacity, robots can be useful, but they need to be programmed and controlled. Robotics is an interdisciplinary subject that benefits from mechanical engineering, electrical and electronic engineering, computer science, cognitive sciences, biology, and many other disciplines.

1.5 History of Robotics

Disregarding the early machines that were made to mimic humans and their actions and concentrating on the recent history, one can see a close relationship between the state of industry, the revolution in numeric and computer control of machinery, space exploration, and the vivid imagination of creative people. Starting with Karel Capek and his book, *Rossum's Universal Robots*,[1] and later, movies like *Flash Gordon, Metropolis, Lost in Space, The Day The Earth Stood Still*, and *The Forbidden Planet*,[2] the stage was set for a machine to be built to do a human's job (and, of course, R2D2, C3PO, Robocop, and others continued the trend).

Capek dreamed of a scenario where a bioprocess could create human-like machines, devoid of emotions and souls, who were strong, obeyed their masters, and could be produced quickly and cheaply. Soon, the market grew tremendously when all major countries wanted to "equip" their armies with hundreds of thousands of slave robotic soldiers, who would fight with dedication, but whose death would not matter. Eventually, the robots decided that they were actually superior to the humans, took over the whole world, and killed everyone. In this story, the word "rabota," or worker, was coined and is used even today. After World War II, automatic machines were designed to increase productivity, and machine-tool manufacturers made numerically controlled (NC) machines to enable manufacturers to produce better products. At the same time, multi-degree-of-freedom manipulators were developed for work on nuclear materials. Integration between the NC capability of machine tools and the manipulators created a simple robot. The first robots were controlled by strips of paper with holes, which electric eyes could detect and which controlled the robot's movements. As industry improved, the strip of paper gave way to magnetic tapes, to memory devices, and personal computers. The following is a summary of events that have marked changes in the direction of this industry.

1922	Czech author Karel Capek wrote a story called *Rossum's Universal Robots* and introduced the word rabota (worker).
1946	George Devol developed the magnetic controller, a playback device. Eckert and Mauchley built the ENIAC computer at the University of Pennsylvania.
1952	The first numerically controlled machine was built at MIT.
1954	George Devol developed the first programmable robot.
1955	Denavit and Hartenberg developed homogeneous transformation matrices.
1961	U.S. patent 2,988,237 was issued to George Devol for "Programmed Article Transfer," a basis for Unimate^{TM} robots.
1962	Unimation^{TM} was formed, the first industrial robots appeared, and GM installed its first robot from Unimation^{TM}.
1967	Unimate^{TM} introduced MarkII^{TM} robot. The first robot was imported to Japan for paint spraying applications.
1968	An intelligent robot called Shakey was built at the Stanford Research Institute (SRI).
1972	IBM worked on a rectangular coordinate robot for internal use. It eventually developed the IBM 7565 for sale.
1973	Cincinnati Milacron^{TM} introduced T3 model robot which became very popular in industry.
1978	The first PUMA robot was shipped to GM by Unimation^{TM}.
1982	GM and Fanuc of Japan signed an agreement to build GMFanuc robots.
1983	Robotics became a very popular subject, both in industry as well as academia. Many programs in the nation started teaching robotic courses.
1983	Unimation^{TM} was sold to Westinghouse Corporation, who subsequently sold it to the Staubli of Switzerland in 1988.
1986	Honda introduced its first humanoid robot called H0. First Asimo was introduced in 2000.
2005	Between January and March, over 5,300 robots were ordered by the North American manufacturing companies at a value of $302 million.

1.6 Advantages and Disadvantages of Robots

- Robotics and automation can, in many situations, increase productivity, safety, efficiency, quality, and consistency of products.
- Robots can work in hazardous environments such as radiation, darkness, hot and cold, ocean bottoms, space, and so on without the need for life support, comfort, or concern for safety.
- Robots need no environmental comfort like lighting, air conditioning, ventilation, and noise protection.
- Robots work continuously without tiring or fatigue or boredom. They do not get mad, do not have hangovers, and need no medical insurance or vacation.
- Robots have repeatable precision at all times unless something happens to them or unless they wear out.
- Robots can be much more accurate than humans. Typical linear accuracies are a few ten-thousandths of an inch. New wafer-handling robots have micro-inch accuracies.
- Robots and their accessories and sensors can have capabilities beyond those of humans.
- Robots can process multiple stimuli or tasks simultaneously. Humans can only process one active stimulus.
- Robots replace human workers, causing economic hardship, worker dissatisfaction and resentment, and the need for retraining the replaced workforce.
- Robots lack capability to respond in emergencies, unless the situation is predicted and the response is included in the system. Safety measures are needed to ensure that they do not injure operators and other machines that are working with them.[3] This includes:
  - Inappropriate or wrong responses
  - Lack of decision-making power
  - Loss of power
  - Damage to the robot and other devices
  - Injuries to humans
- Robots, although superior in certain senses, have limited capabilities in:
  - Cognition, creativity, decision-making, and understanding
  - Degrees of freedom and dexterity
  - Sensors and vision systems
  - Real-time response
- Robots are costly due to:
  - Initial cost of equipment and installation
  - Need for peripherals
  - Need for training
  - Need for programming

1.7 Robot Components

A robot, as a system, consists of the following elements, which are integrated together to form a whole:

Manipulator or the rover: This is the main body of the robot which consists of the links, the joints, and other structural elements of the robot. Without other elements, the manipulator alone is not a robot (Figure 1.3).

Figure 1.3 A Fanuc M–410iWW palletizing robotic manipulator with its end effector. (Reprinted by permission from Fanuc Robotics, North America, Inc.)

End effector: This part is connected to the last joint (hand) of a manipulator that generally handles objects, makes connections to other machines, or performs the required tasks (Figure 1.3). Robot manufacturers generally do not design or sell end effectors. In most cases, all they supply is a simple gripper. Generally, the hand of a robot has provisions for connecting specialty end effectors specifically designed for a purpose. This is the job of a company's engineers or outside consultants to design and install the end effector on the robot, and to make it work for the given situation. A welding torch, a paint spray gun, a glue laying device, or a parts handler are but a few possibilities. In most cases, the action of the end effector is either controlled by the robot's controller, or the controller communicates with the end effector's controlling device (such as a PLC).

Actuators: Actuators are the "muscles" of the manipulators. The controller sends signals to the actuators, which, in turn, move the robot joints and links. Common types are servomotors, stepper motors, pneumatic actuators, and hydraulic actuators. Other novel actuators are used in specific situations (this will be discussed later in Chapter 7). Actuators are under the control of the controller.

Sensors: Sensors are used to collect information about the internal state of the robot or to communicate with the outside environment. As in humans, the robot controller needs to know the location of each link of the robot in order to know the robot's configuration. When you wake up in the morning, even without opening your eyes, or when it is completely dark, you still know where your arms and legs are. This is because feedback sensors in your central nervous system embedded in muscle tendons send information to the brain. The brain uses this information to determine the length of your muscles and, consequently, the state of your arms, legs, and so on. The same is true for robots, where sensors integrated into the robot send information about each joint or link to the controller that determines the configuration of the robot. Still similar to your major senses of sight, touch, hearing, taste, and speech, robots are equipped with external sensory devices such as a vision system, touch and tactile sensors, speech synthesizer, and the like that enable the robot to communicate with the outside world.

Controller: The controller is rather similar to your cerebellum; although it does not have the power of the brain, it still controls your motions. The controller receives its data from the computer (the brain of the system), controls the motions of the actuators, and coordinates the motions with the sensory feedback information. Suppose that in order for the robot to pick up a part from a bin, it is necessary that its first joint be at 35°. If the joint is not already at this magnitude, the controller will send a signal to the actuator—a current to an electric motor, air to a pneumatic cylinder, or a signal to a hydraulic servo valve— causing it to move. It will then measure the change in the joint angle through the feedback sensor attached to the joint (a potentiometer, an encoder, etc.). When the joint reaches the desired value, the signal is stopped. In more sophisticated robots, the velocity and the force exerted by the robot are also controlled by the controller.

Processor: The processor is the brain of the robot. It calculates the motions of the robot's joints, determines how much and how fast each joint must move to achieve the desired location and speeds, and oversees the coordinated actions of the controller and the sensors. The processor is generally a computer, which works like all other computers, but is dedicated to this purpose. It requires an operating system, programs, peripheral equipment like a monitor, and has the same limitations and capabilities. In some systems, the controller and the processor are integrated together into one unit. In others, they are separate units, and in some, although the controller is provided by the manufacturer, the processor is not; they expect the user to provide his or her processor.

Software: Three groups of software programs are used in a robot. One is the operating system that operates the processor. The second is the robotic software that calculates the necessary motions of each joint based on the kinematic equations of the robot. This information is sent to the controller. This software may be at many different levels, from machine language to sophisticated languages used by modern robots. The third group is the collection of application-oriented routines and programs developed to use the robot or its peripherals for specific tasks such as assembly, machine loading, material handling, and vision routines.

1.8 Robot Degrees of Freedom

As you may remember from your engineering mechanics courses, in order to locate a point in space, one needs to specify three coordinates (such as the x-, y-, z-coordinates along the three Cartesian axes). Three coordinates are necessary and enough to completely define the location of the point. Although different coordinate systems may be used to express this information, they are always necessary. However, neither two nor four will be possible; two is inadequate to locate a point in space, and four is impossible. There is simply too much information. Similarly, if you consider a three-dimensional device that has 3 degrees of freedom within the workspace of the device, you should be able to place the device at any desired location. For example, a gantry (x, y, z) crane can place a ball at any location within its workspace as specified by the operator.

Similarly, to locate a rigid body (a three-dimensional object rather than a point) in space, we need to specify the location of a selected point on it; therefore, it requires three pieces of information to be located as desired. However, although the location of the object is specified, there are infinite possible ways to orientate the object about the selected point. To fully specify the object in space, in addition to the location of a selected point on it, we need to specify the orientation of the object as well. This means that six pieces of information are needed to fully specify the location and orientation of a rigid body. By the same token, there need to be 6 degrees of freedom available to fully place the object in space and orientate it as desired.

For this reason, robots need to have 6 degrees of freedom to freely place and orientate objects within their workspace. A robot that has 6 degrees of freedom can be requested to place objects at any desired location and orientation. If a robot has fewer degrees of freedom, we cannot arbitrarily specify any location and orientation for the robot; it can only go to places and to orientations that the fewer joints allow. To demonstrate this, consider a robot with 3 degrees of freedom, where it can only move along the x-, y-, and z-axes. In this case, no orientation can be specified; all the robot can do is to pick up the part and move it in space parallel to the reference axes. The orientation always remains the same. Now consider another robot with 5 degrees of freedom, capable of rotating about the three axes, but only moving along the x- and y-axes. Although you may specify any orientation desired, the positioning of the part is only possible along the x- and y-, but not z-axes. The same is true for any other robot configurations.

A system with 7 degrees of freedom would not have a unique solution. This means that if a robot has 7 degrees of freedom, there are infinite ways it can position a part and orientate it at the desired location. In order for the controller to know what to do, there must be some additional decision-making routine that allows it to pick only one of the infinite solutions. As an example, we may use an optimization routine to pick the fastest or the shortest path to the desired destination. Then the computer has to check all solutions to find the shortest or fastest response and perform it. Due to this additional requirement, which can take much computing power and time, no 7-degree of freedom robot is used in industry. A similar issue arises when a manipulator robot is mounted on a moving base such as a mobile platform or a conveyor belt (Figure 1.4). In either case, the robot has an additional degree of freedom, which, based on the above discussion, is impossible to control. The robot can be at a desired location and orientation from infinite distinct positions on the conveyor belt or the mobile platform. However, in this case, although there are too many degrees of freedom, the additional degrees of freedom are known and there is no need to solve for them. In other words, generally, when a robot is

Figure 1.4   A Fanuc P-15 robot.   (Reprinted with permission from Fanuc Robotics, North America, Inc.)

mounted on a conveyor belt or is otherwise mobile, the location of the base of the robot relative to the belt or other reference frame is known. Since this location does not need to be defined by the controller, the remaining number of degrees of freedom is still six, and consequently, unique. So long as the location of the base of the robot on the belt or the location of the mobile platform is known (or selected by the user), there is no need to find it by solving the set of equations of robot motions, and as a result, the system can be solved.

Can you determine how many degrees of freedom the human arm has? This should exclude the hand (palm and the fingers), but should include the wrist. Before you go on, try to see if you can determine it.

The human arm has three joint clusters: the shoulder, the elbow, and the wrist. The shoulder has 3 degrees of freedom, since the upper arm (humerus) can rotate in the sagittal plane, which is parallel to the mid-plane of the body; the coronal plane (a plane from shoulder to shoulder); and about the humerus (please verify this by rotating your arm about the three different axes). The elbow has just 1 degree of freedom; it can only flex and extend about the elbow joint. The wrist also has 3 degrees of freedom. It can abduct and adduct, flex and extend, and, since the radius bone can roll over the ulna, it can rotate longitudinally (pronate and supinate). Consequently, the human arm has a total of 7 degrees of freedom, even if the ranges of some movements are small. Since a 7-DOF system does not have a unique solution, how do you think we can use our arms?

Please note that the end effector of the robot is never considered as one of the degrees of freedom. All robots have this additional capability, which may appear to be similar to a degree of freedom. However, none of the movements in the end effector are counted toward the robot's degrees of freedom.

There are cases where a joint may have the ability to move, but its movement is not fully controlled. For example, consider a linear joint actuated by a pneumatic cylinder, where the arm is fully extended or fully retracted, but no controlled position can be achieved between the two extremes. In this case, the convention is to assign only a ½-degree of freedom to the joint. This means that the joint can only be at specified locations within its limits of movement. Another possibility for a ½-degree of freedom is to assign only particular values to the joint. For example, suppose a joint is made to be only at 0, 30, 60, and 90 degrees. Then, as before, the joint is limited to only a few possibilities, and therefore, has a partial degree of freedom.

Many industrial robots possess fewer than 6 degrees of freedom. Robots with 3.5, 4, and 5 degrees of freedom are in fact very common. So long as there is no need for the additional degrees of freedom, these robots perform very well. For example, suppose you intend to insert electronic components into a circuit board. The circuit board is always laid flat on a known work surface, and consequently, its height (z value) relative to the base of the robot is known. Therefore, there is only a need for 2 degrees of freedom along the x- and y-axes to specify any location on the board for insertion. Additionally, suppose that the components are to be inserted in any direction on the board, but the board is always flat. In that case, there is a need for 1 degree of freedom to rotate about the vertical axis (z) in order to orientate the component above the surface. Since there is also need for a ½-degree of freedom to fully extend the end effector to insert the part or to fully retract it to lift the robot before moving, only 3.5 degrees of freedom are needed: two to move over the board, one to rotate the component, and 1/2 to insert or retract. Insertion robots are very common and are extensively used in electronic industry. Their advantage is that they are simple to program, less expensive, smaller, and faster. Their disadvantage is that, although they may be programmed to insert components on any size board in any direction, they cannot perform other jobs. They are limited to what 3.5 degrees of freedom can achieve, but they can perform a variety of functions within this design limit.

1.9 Robot Joints

Robots may have different types of joints, such as linear, rotary, sliding, or spherical. Spherical joints are common in many systems but they possess multiple degrees of freedom, and therefore, are difficult to control. Consequently, they are not common in robotics except in research.[4] Most robots have either a linear (prismatic) joint or a rotary (revolute) joint. Prismatic joints are linear; there is no rotation involved. They are either hydraulic or pneumatic cylinders or linear electric actuators. These joints are used in gantry, cylindrical, or spherical robot variations. Revolute joints are rotary, and although hydraulic and pneumatic rotary joints are common, most rotary joints are electrically driven, either by stepper motors or, more commonly, by servomotors.

1.10 Robot Coordinates

Robot configurations generally follow the coordinate frames with which they are defined, as shown in Figure 1.5. Prismatic joints are denoted by P, revolute joints are denoted by R, and spherical joints are denoted by S. Robot configurations are

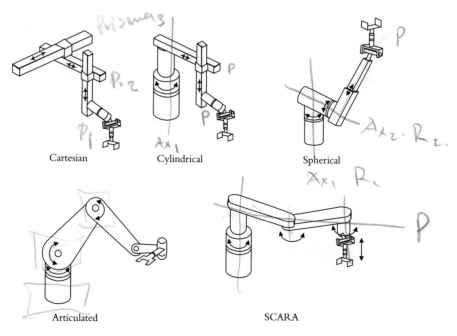

Cartesian   Cylindrical   Spherical

Articulated   SCARA

Figure 1.5 Some possible robot coordinate frames.

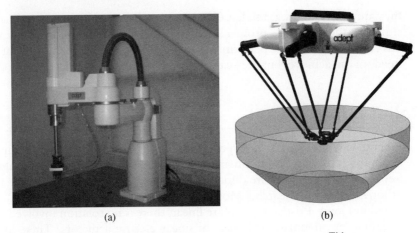

Figure 1.6   (a) An Adept SCARA robot. (b) The Adept Quattro™ s650H robot.   (Printed with permission from Adept Technology, Inc.)

specified by a succession of P, R, or S designations. For example, a robot with three prismatic and three revolute joints is specified by 3P3R. The following configurations are common for positioning the hand of the robot:

Cartesian/rectangular/gantry (3P): These robots are made of three linear joints that position the end effector, which are usually followed by additional revolute joints that orientate the end effector.

Cylindrical (PRP): Cylindrical coordinate robots have two prismatic joints and one revolute joint for positioning the part, plus revolute joints for orientating the part.

Spherical (P2R): Spherical coordinate robots follow a spherical coordinate system, which has one prismatic and two revolute joints for positioning the part, plus additional revolute joints for orientation.

Articulated/anthropomorphic (3R): An articulated robot's joints are all revolute, similar to a human's arm. They are the most common configuration for industrial robots.

Selective Compliance Assembly Robot Arm (SCARA): SCARA robots have two (or three) revolute joints that are parallel and allow the robot to move in a horizontal plane, plus an additional prismatic joint that moves vertically (Figure 1.6). SCARA robots are very common in assembly operations. Their specific characteristic is that they are more compliant in the $x-y$ plane but are very stiff along the z-axis, therefore providing selective compliance. This is an important issue in assembly, and will be discussed in Chapter 8.

1.11 Robot Reference Frames

Robots may be moved relative to different coordinate frames. In each type of coordinate frame, the motions will be different. Robot motions are usually accomplished in the following three coordinate frames (Figure 1.7):

World Reference Frame: This is a universal coordinate frame, as defined by the x-, y-, and z-axes. In this case, the joints of the robot move simultaneously in a coordinated

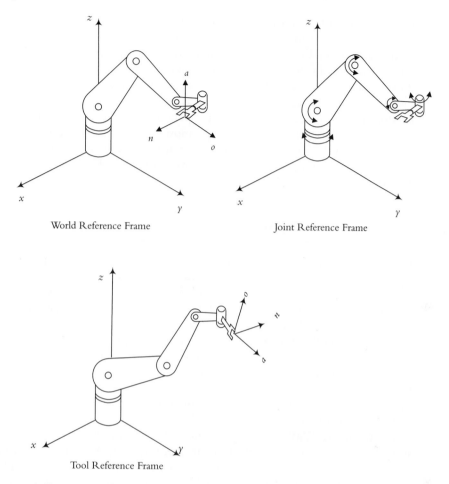

Figure 1.7 A robot's World, Joint, and Tool reference frames. Most robots may be programmed to move relative to any of these reference frames.

manner to create motions along the three major axes. In this frame, no matter where the arm, a positive movement along the x-axis is always in the plus direction of the x-axis, etc. The World reference frame is used to define the motions of the robot relative to other objects, define other parts and machines with which the robot communicates, and define motion trajectories.

   Joint Reference Frame: This is used to specify movements of individual joints of the robot. In this case, each joint is accessed and moved individually; therefore, only one joint moves at a time. Depending on the type of joint used (prismatic, revolute, or spherical), the motion of the robot hand will be different. For instance, if a revolute joint is moved, the hand will move on a circle defined by the joint axis.

   Tool Reference Frame: This specifies movements of the robot's hand relative to a frame attached to the hand, and consequently, all motions are relative to this local n,o,a– frame. Unlike the universal World frame, the local Tool frame moves with the robot. Suppose the hand is pointed as shown in Figure 1.7. Moving the hand relative to the positive *n*-axis of the local Tool frame will move the hand along the *n*-axis of the Tool frame. If the arm were pointed elsewhere, the same motion along the local *n*-axis of the Tool frame would be completely different from the first motion. The same +*n*-axis movement would be upward if the *n*-axis were pointed upward, and it would be downward if the *n*-axis were pointed downward. As a result, the Tool reference frame is a moving frame that changes continuously as the robot moves; therefore, the ensuing motions relative to it are also different depending on where the arm is and what direction the tool frame has. All joints of the robot must move simultaneously to create coordinated motions about the Tool frame. The Tool reference frame is an extremely useful frame in robotic programming where the robot is to approach and depart from other objects or to assemble parts.

1.12 Programming Modes

Robots may be programmed in a number of different modes, depending on the robot and how sophisticated it is. The following programming modes are common:

Physical Set-up: In this mode, an operator sets up switches and hard stops that control the motions of the robot. This mode is usually used along with other devices such as Programmable Logic Controllers (PLC).

Lead Through or Teach Mode: In this mode, the robot's joints are moved with a teach pendant. When the desired location and orientation is achieved, the location is entered (taught) into the controller. During playback, the controller moves the joints to the same locations and orientations. This mode is usually point-to-point; as such, the motion between points is not specified or controlled. Only the points that are taught are guaranteed to reach.

Continuous Walk-Through Mode: In this mode, all robot joints are moved simultaneously, while the motion is continuously sampled and recorded by the controller. During playback, the exact motion that was recorded is executed. The motions are taught by an operator, either through a model, by physically moving the end-effector, or by "wearing" the robot arm and moving it through its workspace. Painting robots, for example, may be programmed by skilled painters through this mode.

Software Mode: In this mode of programming the robot, a program is written offline or online and is executed by the controller to control the motions. The programming mode is the most sophisticated and versatile mode and can include sensory information, conditional statements (such as if . . . then statements), and branching. However, it requires a working knowledge of the programming syntax of the robot before any program is written. Most industrial robots can be programmed in more than one mode.

1.13 Robot Characteristics

The following definitions are used to characterize robot specifications:

Payload: Payload is the weight a robot can carry and still remain within its other specifications. As an example, a robot's maximum load capacity may be much larger than its specified payload, but at these levels, it may become less accurate, may not follow its intended trajectory accurately, or may have excessive deflections. The payload of robots compared to their own weight is usually very small. For example, Fanuc Robotics LR Mate^{TM} robot has a mechanical weight of 86 lb and a payload of 6.6 lb, and the M–16i ^{TM} robot has a mechanical weight of 594 lb and a payload of 35 lb.

Reach: Reach is the maximum distance a robot can reach within its work envelope. As will be seen later, many points within the work envelope of the robot may be reached with any desired orientation (called dexterous). However, for other points close to the limit of robot's reach capability, orientation cannot be specified as desired (called nondexterous point). Reach is a function of the robot's joints and lengths and its configuration. This is an important specification for industrial robots and must be considered before a robot is selected and installed.

Precision (validity): Precision is defined as how accurately a specified point can be reached. This is a function of the resolution of the actuators as well as the robot's feedback devices. Most industrial robots can have precision in the range of 0.001 inches or better. The precision is a function of how many positions and orientations were used to test the robot, with what load, and at what speed. When the precision is an important specification, it is crucial to investigate these issues.

Repeatability (variability): Repeatability is how accurately the same position can be reached if the motion is repeated many times. Suppose a robot is driven to the same point 100 times. Since many factors may affect the accuracy of the position, the robot may not reach the same point every time but will be within a certain radius from the desired point. The radius of a circle formed by the repeated motions is called repeatability. Repeatability is much more important than precision. If a robot is not precise, it will generally show a consistent error, which can be predicted, and therefore, corrected through programming. For example, suppose a robot is consistently off by 0.05 inches to the right. In that case, all desired points can be specified at 0.05 inches to the left and thereby eliminate the error. However, if the error is random, it cannot be predicted and consequently cannot be eliminated. Repeatability defines the extent of this random error. Repeatability is usually specified for a certain number of runs. Larger numbers of tests yield larger (bad for manufacturers) results, but more realistic (good for the users) results. Manufacturers must specify repeatability in conjunction with the number of tests, the applied payload during the tests, and the orientation of the arm. For example, the repeatability of an arm in a vertical direction will be different from when the arm is tested in a horizontal configuration. Most industrial robots have repeatability in the 0.001 inch range. It is crucial to find out about the details of repeatability if it is an important specification for the application.

1.14 Robot Workspace

Depending on their configuration and the size of their links and wrist joints, robots can reach a collection of points around them that constitute a workspace. The shape of the workspace for each robot is uniquely related to its design. The workspace may be found mathematically by writing equations that define the robot's links and joints and that include their limitations such as ranges of motions for each joint.[5] Alternately, the workspace may be found empirically by virtually moving each joint through its range of motions, combining all the space it can reach, and subtracting what it cannot reach.

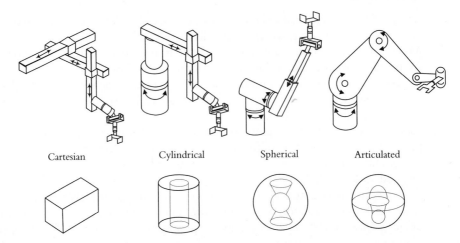

Cartesian   Cylindrical   Spherical   Articulated

Figure 1.8   Typical approximate workspaces for common robot configurations.

Figure 1.8 shows the approximate workspace for some common configurations. When a robot is considered for a particular application, its workspace must be studied to ensure that the robot will be able to reach the desired points. For accurate workspace determination, refer to manufacturers' data sheets.

1.15 Robot Languages

There are perhaps as many robot languages as there are robot manufacturers. Each manufacturer designs its own robotic language; therefore, in order to use any particular robot, its brand of the programming language must be learned. Many robot languages are based on some other common language such as Cobol, Basic, C, and Fortran. Other languages are stand-alone and do not relate to any other common language.

Robotic languages are at different levels of sophistication, depending on their design and application. This ranges from machine level to a proposed human intelligence level.[6-9] High-level languages are either interpreter-based or compiler-based.

Interpreter-based languages execute one line of the program at a time. Each line of the program has a line number. The interpreter interprets the line every time it is encountered (it converts the robot program to a machine language program that the processor can understand and execute) and executes each line sequentially. The execution continues until the last line is encountered or until an error is detected, at which time execution stops. The advantage of an interpreter-based language is in its ability to continue execution until an error is detected, which allows the user to run and debug the program, portion by portion. As a result, debugging programs is much faster and easier. However, because each line is interpreted every time, execution is slower and not very efficient. Many robot languages such as Unimation[TM] VAL[®], Adept's V[+®], and IBM's AML[®] (A Manufacturing Language) are interpreter based.[9,10]

Compiler-based languages use a compiler to translate the whole program into machine language (which creates an object code) before it is executed. Since the processor executes the object code, these programs are much faster and more efficient. However, since the whole program must first be compiled, it will be impossible to run any part of the program if there are any syntax errors present, even before the logic of the program is tested. As a result, debugging these programs is much more difficult. Certain languages such as AL© are more flexible. They allow the user to debug the program in interpreter mode, while the actual execution is in compiler mode. The following is a general description of different levels of robotic languages.[7]

Micro-Computer Machine Language Level: In this level, the programs are written in machine language. This level of programming is the most basic and is very efficient, but it is difficult to understand and difficult for others to follow. All languages will eventually be interpreted or compiled to this level. However, in the case of higher level programs, the user writes the programs in a higher level language that is easier to follow and understand.

Point-to-Point Level: In this level (such as in Funky and Cincinnati Milacron's T3), the coordinates of the points are entered sequentially, and the robot follows the points as specified. This is a very primitive and simple type of program, and it is easy to use, but not very powerful. It also lacks branching, sensory information, and conditional statements.

Primitive Motion Level: In these languages, it is possible to develop more sophisticated programs, including sensory information, branching, and conditional statements (such as VAL by Unimation, V^+ by Adept, and so on). Most languages in this level are interpreter-based.

Structured Programming Level: Most languages in this level are compiler-based, are powerful, and allow more sophisticated programming. However, they are also more difficult to learn.

Task-Oriented Level: There are no actual languages in existence in this level—yet. Autopass, proposed by IBM in the 1980s, never materialized. Autopass was supposed to be task-oriented. This means that instead of programming a robot to perform a task by programming each and every step necessary to complete it, the user was to only mention the task, while the controller would create the necessary sequence. Imagine that a robot is to sort three boxes by size. In all existing languages, the programmer will have to tell the robot exactly what to do; therefore, every step must be programmed. The robot must be told how to go to the largest box, how to pick up the box, where to place it, where to go to the next box, and so on. In Autopass, the user would only indicate "sort," while the robot controller would create this sequence automatically. This never happened.

Example 1.1

The following is an example of a program written in V^+, which is used with Adept robots, is interpreter-based, and allows for branching, sensory input and output communication, straight-line movements, and many other features. As an example, the user may define a distance "height" along the z-axis of the end effector, which can be used with commands called APPRO (for approach) and DEPART in order to approach an object or depart from an object without collision. A command called MOVE will allow the robot to move from its present location to the next specified location. However, MOVES will do the same in a straight line. The difference is discussed in detail in Chapter 5. In the following listing, a number of different commands are described in order to show some of the capabilities of V^+.

1	PROGRAM TEST	Declaration of the program name.
2	SPEED 30 ALWAYS	Sets the speed of the robot.
3	height=50	Specifies a distance for the lift-off and set-down points along the z-axis of the end effector.
4	MOVES p1	Moves the robot in straight line to point p1.
5	MOVE p2	Moves the robot to a second point p2 in joint interpolated motion.
6	REACTI 1001	Stops the robot immediately if an input signal to port 1 goes high (is closed).
7	BREAK	Stops execution until the previous motion is finished.
8	DELAY 2	Delays execution for 2 seconds.
9	IF SIG(1001) GOTO 100	Checks input port 1. If it is high (closed), execution continues at line 100. Otherwise, execution continues with the next line.
10	OPEN	Opens the gripper.
11	MOVE p5	Moves to point p5.
12	SIGNAL 2	Turns on output port 2.
13	APPRO p6, height	Moves the robot toward p6, but away from it a distance specified as ''height,'' along the z-axis of the gripper (Tool frame). This is called a lift-off point.
14	MOVE p6	Moves to the object at point p6.
15	CLOSEI	Closes the gripper and waits until it closes.
16	DEPART height	Moves up along the z-axis of the gripper (Tool frame) a distance specified by ''height.''
17	100 MOVE p1	Moves the robot to point p1.
18	TYPE ''all done.''	Writes the message to the monitor.
19	END	

Example 1.2

The following is an example of a program written in IBM's AML (A Manufacturing Language). AML is no longer common. However, the example is provided to show how one language may differ from another in its features and syntax. The program is written for a gantry 3P3R robot, with three prismatic linear positioning joints, three revolute orientation joints, and a gripper. Joints may be referred to by numbers <1, 2, 3, 4, 5, 6, 7>, where 1, 2, 3 indicate the prismatic joints, 4, 5, 6 indicate the revolute joints, and 7 indicates the gripper. The joints may also be referred to by index letters JX, JY, JZ for motions along the x-, y-, z-axes, JR, JP, JY, for rotations about the Roll, Pitch, and Yaw axes (used for orientation), and JG, for gripper. Please note that since this robot is gantry, the path the robot takes is different from a revolute robot's path. Therefore, the way it is programmed is also different. Instead of specifying a point, joint movements are specified, although all simultaneously.

There are two types of movements allowed in AML. MOVE commands are absolute. This means the robot will move along the specified joint to the specified value. DMOVE commands are differential. This means the joint will move the specified amount from wherever it is. Therefore, MOVE (1, 10) means the robot will move along the *x*-axis to 10 inches from the origin of the reference frame, whereas DMOVE (1, 10) means the robot will move 10 inches along the *x*-axis from its current position. There is a large set of commands in AML, allowing the user to write sophisticated programs.

The following simple program will direct the robot to pick an object from one location and place it at another. This is written to show you how a robotic program may be structured:

```
10   SUBR(PICK-PLACE);                 Subroutine's name.
20   PT1: NEW <4., -24, 2, 0, 0, -13>;  Declaration of a location.
30   PT2: NEW <-2, 13, 2, 135, -90, -33>;
40   PT3: NEW <-2, 13, 2, 150, -90, -33, 1>;
50   SPEED (0.2);                       Specifies velocity of the
                                        robot (20% of full speed).
60   MOVE (ARM, 0.0);                   Moves the robot (ARM) to
                                        its reset position at the
                                        origin of the reference
                                        frame.
70   MOVE (<1,2,3,4,5,6>,PT1);          Moves the arm to point-1
                                        above the object.
80   MOVE (7,3);                        Opens the gripper to 3
                                        inches.
90   DMOVE (3, -1);                     Moves the arm down 1 inch
                                        along the z-axis.
100  DMOVE (7, -1.5);                   Closes the gripper by 1.5
                                        inches
110  DMOVE (3, 1);                      Moves up 1 inch along the
                                        z-axis to lift the object.
120  MOVE (<JX, JY, JZ, JR, JP, JY>, PT2);  Moves the arm to point-2.
130  DMOVE (JZ, -3);                    Moves the arm down 3 inches
                                        along z-axis to place the
                                        object.
140  MOVE (JG,3);                       Opens the gripper to 3
                                        inches.
150  DMOVE (JZ, 11);                    Moves the arm up 11 inches
                                        along the z-axis.
160  MOVE (ARM, PT3);                   Moves the arm to point-3.
170  END;
```

1.16 Robot Applications

Robots are best suited to work in environments and on tasks where humans are not. Robots have already been used in many industries and for many purposes. They have excelled when they can perform better than humans or at lower costs. For example, a

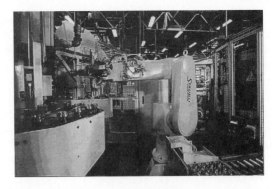

Figure 1.9 A Staubli robot, loading and unloading components into and from a machining center. (Reprinted with permission from Staubli Robotics.)

welding robot can probably weld better than a human welder because it can move more uniformly and more consistently. In addition, robots do not need protective goggles, protective clothing, ventilation, or many other necessities that their human counterparts would need. As a result, robots can be more productive and better suited for the job, as long as the welding job is set up for the robot for automatic operations, nothing happens to the set-up, and the welding job is not too complicated. Similarly, a robot exploring the ocean floor requires far less attention than a human diver, can stay underwater for long periods of time, can go to very large depths and still survive the pressure, and does not require oxygen.

The following is a list of some robotic applications. The list is not complete by any stretch of the imagination. There are many other uses as well, and other applications find their way into the industry and society all the time:

Machine loading, where robots supply other machines with parts, or remove the processed parts from other machines (Figure 1.9). In this type of work, the robot may not even perform any operation on the part, but rather it facilitates material and parts handling and loading other machines within the context of a task.

Pick and place operations, where the robot picks up parts and places them elsewhere (Figure 1.10). This may include palletizing, placement of cartridges, simple assembly where two parts are put together (such as placing tablets into a bottle), placing parts in an oven and removing the treated parts from the oven, or other similar routines.

Figure 1.10 Staubli robots placing dishwasher tubs into welding stations. (Reprinted with permission from Staubli Robotics.)

Figure 1.11 An AM120 Fanuc robot. (Reprinted with permission from Fanuc Robotics, North America, Inc.)

Welding, where the robot, along with proper set-ups and a welding end effector, is used to weld parts together. This is one of the most common applications of robots in the auto industry. Due to their consistent movements, robotic welds are very uniform and accurate. Welding robots are usually large and powerful (Figure 1.11).

Painting is another very common application of robots, especially in the automobile industry. Since maintaining a ventilated but clean room suitable for humans is difficult to achieve, and because compared to humans, robotic operations are more consistent, painting robots are very well-suited for their job (Figure 1.12).

Inspection of parts, circuit boards, and other similar products is also a very common application for robots. In general, robots are one component of an inspection system that may include a vision system, x-ray device, ultrasonic detector, or other similar devices (Figure 1.13). In one application, a robot was equipped with an ultrasonic crack detector, was given the CAD data about the shape of an airplane fuselage and wings, and was used to follow the airplane's body contours and check each joint, weld, or rivet. In a similar

Figure 1.12 A P200-EPS Fanuc robot painting automobile bodies. (Reprinted with permission from Fanuc Robotics, North America, Inc.)

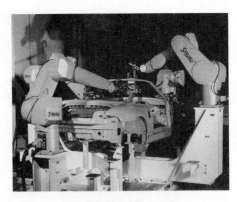

Figure 1.13 Staubli RX FRAMS (Flexible Robotic Absolute Measuring System) robots in a BMW manufacturing facility. (Reprinted with permission from Staubli Robotics.)

application, a robot would search for and find the location of each rivet, detect and mark the rivets with fatigue cracks, would drill them out, and move on. The technicians would insert and install new rivets. Robots have also been extensively used for circuit board and chip inspection. In most applications like this, including part identification, the characteristics of the part (such as the circuit diagram of a board, the nameplate of a part, and so on) are stored in the system in a data library. The system uses this information to match the part with the stored data. Based on the result of the inspection, the part is either accepted or rejected.

Sampling with robots is used in the agriculture industry as well as in many other industries. Sampling can be similar to pick and place and inspection, except that it is performed only on a certain number of products.

Assembly tasks usually involve many operations. For example, the parts must be located and identified, they must be carried in a particular order with many obstacles around the set-up, they must be fit together, and then assembled. Many of the fitting and assembling tasks are complicated and may require pushing, turning, bending, wiggling, pressing, snapping the tabs to connect the parts, and other operations. Slight variations in parts and their dimensions due to larger tolerances also complicate the process since the robot has to know the difference between variations in parts and wrong parts.

Manufacturing by robots may include many different operations such as material removal (Figure 1.14), drilling, de-burring, laying glue, cutting, and so on. It also includes insertion of parts such as electronic components into circuit boards, installation of boards into electronic devices, and other similar operations. Insertion robots are very common and are extensively used in the electronic industry.

Medical applications are also becoming increasingly common. As an example, Curexo Technology Corporation's Robodoc® was designed to assist a surgeon in total joint replacement operations. Since many of the functions performed during this procedure—such as cutting the head of the bone, drilling a hole in the bone's body, reaming the hole for precise dimension, and installation of the manufactured implant joint—can be performed with better precision by a robot, the mechanical parts of the operation are assigned to the robot. This is also important because the orientation and the shape of the bone can be determined by a CAT scan and downloaded to the robot controller, where it is used to direct the motions of the robot for a best fit with the implant.

Figure 1.14 A Fanuc LR Mate 200i robot is used in a material removal operation on a piece of jewelry. (Reprinted with permission from Fanuc Robotics, North America, Inc.)

Other surgical robots such as Mako Surgical Corporation's robot system and Intuitive Surgical's da Vinci system are used in a variety of surgical procedures, including orthopedic and internal surgery operations. For instance, da Vinci possesses four arms, three that hold instruments (one more than a surgeon could) and one to hold a 3D imaging scope that displays the surgical area to a surgeon behind a monitor (Figure 1.15). The surgeon directs the robotic movements with the help of haptic guidance systems, even remotely.[11] Similarly, other robots have been used to assist surgeons during microsurgery, including operations on heart valves in Paris and Liepzig.[12]

Assisting disabled individuals has also been tried with interesting results. Much can be done to help the disabled in their daily lives. In one study, a small tabletop robot was programmed to communicate with a disabled person and to perform simple tasks such as placing a food plate into the microwave oven or placing it in front of the disabled person

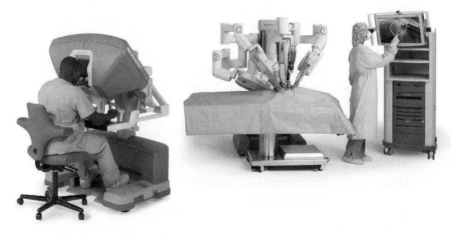

Figure 1.15 The da Vinci surgical system. (Image courtesy of Intuitive Surgical, Inc. (2010).)

Figure 1.16 Finger-spelling hand for communication with blind-and-deaf individuals. (Supported by the Smith Kettlewell Eye Research Institute, San Francisco).

to eat.[13] Many other tasks were also programmed in the same fashion. The finger-spelling hand (Figure 1.16), designed for communication with the blind-and-deaf individuals, is capable of making gestures that spell all letters of the alphabet. With its 17 servomotors, the hand is mounted on an arm and can be held by one hand while the other hand reads the letters. The letters typed in a computer are coded and sent to the hand.[14]

Hazardous environments are well-suited for robotics. Because of their inherent danger, in these environments humans must be well-protected. However, robots can access, traverse, maintain, and explore these areas without the same level of concern. Servicing a radioactive environment, for example, is easier with a robot than a human. In 1993, an eight-legged robot called Dante was to reach the lava lake of constantly erupting volcano Mount Erebus in Antarctica and study its gases.[15] A variety of mine-detecting robots have also been put to use with the idea that a robot may be expendable whereas a human is not. One such robot uses vibrating ultrasonic pods to identify underground mines, therefore eliminating the need for human searches.[16] Another minesweeping robot, running on two all-terrain spiral tubes, has a bare-bones design and is meant to be expendable. Therefore, it is used to move in suspected areas and explode the mine.[17] A snakelike robot with articulating serial sections can maneuver through tight spaces. It is made of a series of sections, each composed of a pair of plates with fixed-length struts holding them together while linear actuators move one plate relative to the other, creating a snakelike motion.[18] Similarly, a crustacean-looking "lobster" robot was developed for searching ocean bottoms for mines and other weapons.[19] Another robot called Talon by QinetiQ™, designed for dangerous duty, can run alongside a soldier, cross broken terrain, and clear mines.[20]

Underwater, space, and inaccessible locations can also be serviced or explored by robots. So far, it is still impractical to send a human to other planets, even Mars, but there have been a number of rovers (Figure 1.17) that have already landed and explored it.[21] The same is true for other space and underwater applications.[22–24] Until recently for example, very few sunken ships were explored in deep oceans because no one could

Figure 1.17 NASA Mars rovers in a lab test area, showing the FIDO rover (left) next to models of the Sojourner and MER rovers (center and right, respectively). (Courtesy NASA/JPL-Caltech.)

access those depths. Many crashed airplanes as well as sunken ships and submarines are nowadays recovered quickly by underwater robots.

In an attempt to clean the sludge from inside of a steam generator blowdown pipe, a teleoperated robot called Cecil was designed to crawl down the pipe and wash away the sludge with a stream of water at 5000 psi.[25] Figure 1.18 shows The Arm, a 6-DOF, bilateral force-feedback manipulator, used primarily on manned submersibles and remotely operated vehicles. The Arm is controlled via a remote master that also "feels" everything the slave arm feels. The system can also perform preprogrammed motions through a "teach and repeat" system.

NASA has developed a Robonaut, a humanoid anthropomorphic robot that functions as an astronaut. It has two five-fingered, tool-handling end effectors, modular robotic

Figure 1.18 The Arm, a 6-DOF, bilateral force-feedback manipulator, used primarily on manned submersibles and remotely operated vehicles. (Reprinted with permission from Western Space and Marine, Inc.)

components, and telepresence capability.[26] And finally, in another application, a tele-robot was used for microsurgery.[27] In this case, the location of the telerobot is of secondary concern. The primary intention is to have the telerobot repeat the surgeon's hand movements at a smaller scale for reduced tremor during microsurgery.

1.17 Other Robots and Applications

Since the first edition of this book was published, new robots and issues have appeared. Such is the nature of this active subject. Therefore, you should expect that there will be applications and robots that are not included in this edition either. However, the following are just a sample of some systems that show a trend and future possibilities.

Roomba[TM], a robot vacuum cleaner, commercially available for years, autonomously and randomly moves throughout an area and vacuums the dust. It also finds its own docking station to recharge. All its intelligence is based on a few simple rules: randomly move around, turn left or right when hitting an obstacle, back up and turn around when in a corner, and find the docking station.[28]

Robots such as Honda's *ASIMO*, Bluebotic's *Gilbert*, Nestle's *Nesbot*, Anybots's *Monty*, and many others are intelligent humanoid robots with humanlike features and behavior. ASIMO walks, runs, goes up and down staircases, and interacts with people. Nesbot brings coffee to workers who have ordered it online.[28] Monty loads a dishwasher and does other chores, while Robomower mows your lawn while you read.[29] Figure 1.19 shows a picture of Nao robot.[30] Like others, Nao is a fully programmable robot that can behave autonomously–it communicates with humans and it walks, dances, and performs tasks.

A number of different robots have also been designed and used for emergency services during natural and human-caused disasters. These robots, equipped with special sensors, are capable of looking for live humans and animals buried under rubble and reporting their locations to rescuers. Similar robots are also used for diffusing bombs and other explosive devices. SDA10 dual-arm robot by Motoman, Inc. (Figure 1.20), has 15 axes of motion. The two arms can move independently or in a coordinated manner. It can transfer a part from one gripper to the other without the need to set it down.

Figure 1.19   Nao humanoid robot. (Reprinted with permission from Aldebaran Robotics (Picture by C. De Torquat).)

Figure 1.20 SDA10 dual arm robot. (Reprinted with permission from Motoman, Inc.)

Exoskeletal assistive devices, although not robots, follow the same logic and allow a human to carry large loads for extended periods of time. In fact, these devices can conceivably be used to aid the disabled in different forms, including helping a wheel-chair-bound person walk. One lightweight exoskeleton device called Human Universal Load Carrier (HULC) assists people in carrying heavy loads (as of the date of this writing, over 200 lbs. for 10 hours). As shown in Figure 1.21, the skeleton is worn by the person who directs the motions of the frame, but the frame carries the weight and is actuated by a battery driven hydraulic pump.[31,32]

Figure 1.21 The Human Universal Load Carrier (HULC) is an un-tethered, hydraulic-powered anthropomorphic exoskeleton that provides users with the ability to carry loads of up to 200 lb for extended periods of time and over all terrains. (Reprinted by permission from Berkeley Bionics.)

As mentioned earlier, in addition to humanoid robots, other life-form robots—such as insects, creatures, animals, or fish—have become very popular. Some of these are designed and studied for their robotic aspects, others for specific applications, and yet others for the study of the animal's behavior. For example, in one study, small robotic roaches were doused with the roach's sex hormone. Therefore, real roaches that cannot see very well, follow the robotic roach. Normally, they tend to congregate in dark places; however, researchers could alter the behavior of roaches by sending the robotic roach to unexpectedly lighter places. The real roaches would follow the robotic roaches against their instinct.[33]

Other insect robots have been designed for applications that range from pure research to entertainment and from civilian to military uses. They mimic six-legged insects, eight-legged insects, flying insects, swarms of insects, and more.[34-36] Developed at the Ben–Gurion University and inspired by Spider-Man, Spiderbot is an insect robot that launches four (magnetic) grapplers that stick to the ceiling, by which it pulls itself up. It then releases the magnets one at a time, retracts them, and launches again to a new point.[37] Other application-oriented life forms include wormlike robots, snakelike robots, robots that swim like fish, a lobsterlike robot, birdlike robots, "dinosaur" robots, as well as unidentified life forms.[38-45] Figure 1.22 shows BigDog, another life-form robot with both civilian and military applications. BigDog is a load-carrying quadruped robot that walks, runs, and climbs hills with an on-board gasoline engine and hydraulic pump to power it. Not only does it carry a load in rough terrain, it can balance itself even if pushed.

The Grand Challenge, a 2007 contest sponsored by the Pentagon through the Defense Advanced Research Projects Agency (DARPA), involved designing a vehicle that could autonomously navigate a 60-mile course in a simulated city in less than 6 hours. The vehicles use GPS navigation, sophisticated vision systems, and automatic control algorithms that provide the ability to navigate and avoid collisions throughout the ordeal. No doubt these vehicles will be increasingly capable in the near future in order to meet a congressional

Figure 1.22   BigDog robot. (BigDog image courtesy of Boston Dynamics © 2009.)

directive that requires the military to replace up to one-third of its logistics vehicles with these "robots" by the mid 2010s.[46]

Animatronics refers to the design and development of systems used in animated robotic figures and machines that look and behave like humans and animals. Examples include animatronic lips, eyes, and hands.[47,48] As more sophisticated animatronic components become available, the action figures they replace become increasingly real.

Another area somewhat related to robotics and its applications is Micro-Electro-Mechanical-Systems (MEMS). These are micro-level devices designed to perform functions within a system that include medical, mechanical, electrical, and physical tasks. As an example, a micro-level robotic device may be sent through major veins to the heart for exploratory or surgical functions, a MEMS sensor may be used to measure the levels of various elements in blood, or a MEMS actuator may be used to deploy automobile airbags in a collision.

1.18 Social Issues

We must always remember the social consequences of using robots. Although there are many applications of robots where they are used because no workers can do the same job, there are many other applications in which a robot replaces a human worker. The worker who is replaced by a robot will lose his or her income. If the trend continues without consideration, it is conceivable that most products will be made by robots, without the need for any human workers. The result will be fewer workers with jobs who have the money to buy the products the robots make. Of importance is the issue of social problems that arise as increasingly more workers are out of jobs as well as its social and economic consequences. One of the important points of negotiations between the automobile manufacturers and the United Auto Workers (UAW) is how many human jobs may be replaced by robots, and at what rate.

Although no solution is presented in this book, many references are available for further study of the problem.[49,50] However, as engineers who strive to make better products at lower costs and who may consider using robots to replace human workers, we must always remember the consequences of this choice. Our academic and professional interest in robotics must always be intertwined with its social and economic considerations.

Summary

Many people who are interested in robotics have background information about robots and may even have interacted with robots too. However, it is necessary that certain ideas are understood by everyone. In this chapter, we discussed some fundamental ideas about robotics that enable us to better understand what they are for, how they can be used, and what they can do. Robots can be used for many purposes, including industrial applications, entertainment, and other specific and unique applications such as in space and underwater exploration and in hazardous environments. Obviously, as time goes by, robots will be used for other unique applications. The remainder of this book will discuss the kinematics and kinetics of robots, their components such as actuators, sensors and vision systems, and robot applications.

References

1. Capek, Karel, "Rossum's Universal Robots," translated by Paul Selver, Doubleday, NY, 1923.
2. Valenti, Michael, "A Robot Is Born," *Mechanical Engineering*, June 1996, pp. 50–57.
3. "Robot Safety," Bonney, M. C., Y. F. Yong, Editors, IFS Publications Ltd., UK, 1985.
4. Stein, David, Gregory S. Chirikjian, "Experiments in the Commutation and Motion Planning of a Spherical Stepper Motor," Proceedings of DETC'00, ASME 2000 Design Engineering Technical Conferences and Computers and Information in Engineering Conference, Baltimore, Maryland, September 2000, pp. 1–7.
5. Wiitala, Jared M., B. J., Rister, J. P. Schmiedler, "A More Flexible Robotic Wrist," *Mechanical Engineering*, July 1997, pp. 78–80.
6. W. A. Gruver, B. I. Soroka, J. J. Craig, and T. L. Turner, "Industrial Robot Programming Languages: A Comparative Evaluation." *IEEE Transactions on Systems, Man, and Cybernetics* SMC–14(4), July/August 1984.
7. Bonner, Susan, K. G. Shin, "A Comprehensive Study of Robot Languages," *IEEE Computer*, December 1982, pp. 82–96.
8. Kusiak, Andrew, "Programming, Off-Line Languages,"International Encyclopedia of Robotics: Applications and Automation, Richard C. Dorf, Editor, John Wiley and Sons, NY, 1988, pp. 1235–1250.
9. Gruver, William, B. I. Soroka, "Programming, High Level Languages," International Encyclopedia of Robotics: Applications and Automation, Richard C. Dorf, Editor, John Wiley and Sons, NY, 1988, pp. 1203–1234.
10. VAL-II Programming Manual, Version 4, Unimation, Inc., Pittsburgh, 1988.
11. McGuinn, Jack, Senior Editor, "These Bots are Cutting-Edge," *Power Transmission Engineering*, October 2008, pp. 32–36.
12. Salisbury, Kenneth, Jr., "The Heart of Microsurgery," *Mechanical Engineering*, December 1998, pp. 46–51.
13. "Stanford Rehabilitation Center," Stanford University, California.
14. Garcia, Mario, Saeed Niku,"Finger-Spelling Hand," California Polytechnic State University masters thesis, San Luis Obispo, 2009.
15. Leary, Warren, "Robot Named Dante to Explore Inferno of Antarctic Volcano," *The New York Times*, December 8, 1992, p. B7.
16. "Ultrasonic Detector and Identifier of Land Mines," *NASA Motion Control Tech Briefs*, 2001, p. 8b.
17. Chalmers, Peggy, "Lobster Special," *Mechanical Engineering*, September 2000, pp. 82–84.
18. "Snakelike Robots Would Maneuver in Tight Spaces," *NASA Tech Briefs*, August 1998, pp. 36–37.
19. "Lobster Special," *Mechanical Engineering*, September 2000, pp. 82–84.
20. "Robot Population Explosion," *Mechanical Engineering*, February 2009, p. 64.
21. http://www.jpl.nasa.gov/pictures/
22. Wernli, Robert L., "Robotics Undersea," *Mechanical Engineering*, August 1982, pp. 24–31.
23. Asker, James, "Canada Gives Station Partners a Hand—And an Arm," *Aviation Week & Space Technology*, December 1997, pp. 71–73.
24. Puttre, Michael, "Space-Age Robots Come Down to Earth," *Mechanical Engineering*, January 1995, pp. 88–89.
25. Trovato, Stephen A., "Robot Hunts Sludge and Hoses It Away," *Mechanical Engineering*, May 1988, pp. 66–69.
26. "The Future of Robotics Is Now," *NASA Tech Briefs*, October 2002, p. 27.
27. "Telerobot Control for Microsurgery," *NASA Tech Briefs*, October 1997, p. 46.
28. "From Simple Rules, Complex Behavior," *Mechanical Engineering*, July 2009, pp. 22–27.
29. Drummond, Mike, "Rise of the Machines," *Inventors Digest*, February 2008, pp. 16–23.
30. http://www.aldebaran-robotics.com/eng/PressFiles.php
31. www.berkeleybionics.com
32. "Lightweight Exoskeleton with Controllable Actuators," *NASA Tech Briefs*, October 2004, pp. 54–55.

33. Chang, Kenneth, John Scwartz, "Led by Robots, Roaches Abandon Instincts," *The New York Times*, November 15, 2007.

34. Yeaple, Judith A., "Robot Insects," *Popular Science*, March 1991, pp. 52–55 and 86.

35. Freedman, David, "Invasion of the Insect Robots," *Discover*, March 1991, pp. 42–50.

36. Thakoor, Sarita, B. Kennedy, A. Thakoor, "Insectile and Vemiform Exploratory Robots," *NASA Tech Briefs*, November 1999, pp. 61–63.

37. http://www.youtube.com/watch?v=uBikHgnt16E

38. Terry, Bryan, "The Robo-Snake," Senior Project report, Cal Poly, San Luis Obispo, California, June 2000.

39. O'Conner, Leo, "Robotic Fish Gotta Swim, Too," *Mechanical Engineering*, January 1995, p. 122.

40. Lipson, Hod, J. B. Pollack, "The Golem Project: Automatic Design and Manufacture of Robotic Lifeforms," http://golem03.cs-i.brandies.edu/index.html.

41. Corrado, Joseph K., "Military Robots," *Design News*, October 83, pp. 45–66.

42. "Low-Cost Minesweeping," *Mechanical Engineering*, April 1996, p. 66.

43. IS Robotics, Somerville, Massachusetts.

44. "Biomorphic Gliders," *NASA Tech Briefs*, April 2001, pp. 65–66.

45. Baumgartner, Henry, "When Bugs Are the Machines," *Mechanical Engineering*, April 2001, p. 108.

46. Markoff, John, "Crashes and Traffic Jams in Military Test of Robotic Vehicles," *The New York Times*, November 5, 2007.

47. Jones, Adam, S. B., Niku, "Animatronic Lips with Speech Synthesis (AliSS)," Proceedings of the 8th Mechatronics Forum and International Conference, University of Twente, The Netherlands, June 2002.

48. Sanders, John K., S. B. Shooter, "The Design and Development of an Animatronic Eye," Proceedings of DETC 98/MECH: 25th ASME Biennial Mechanisms Conference, September 1998.

49. Coates, V. T., "The Potential Impacts of Robotics," Paper number 83-WA/TS-9, American Society of Mechanical Engineers, 1983.

50. Albus, James, "Brains, Behavior, and Robotics," Byte Books, McGraw-Hill, 1981.

Problems

1.1. Draw the approximate workspace for the following robot. Assume the dimensions of the base and other parts of the structure of the robot are as shown.

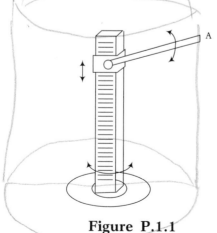

Figure P.1.1

1.2. Draw the approximate workspace for the following robot. Assume the dimensions of the base and other parts of the structure of the robot are as shown.

A

Figure P.1.2

1.3. Draw the approximate workspace for the following robot. Assume the dimensions of the base and other parts of the structure of the robot are as shown.

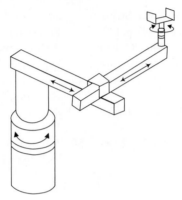

Figure P.1.3

CHAPTER 2

Kinematics of Robots: Position Analysis

2.1 Introduction

In this chapter, we will study forward and inverse kinematics of robots. With forward kinematic equations, we can determine where the robot's end (hand) will be if all joint variables are known. Inverse kinematics enables us to calculate what each joint variable must be in order to locate the hand at a particular point and a particular orientation. Using matrices, we will first establish a method of describing objects, locations, orientations, and movements. Then we will study the forward and inverse kinematics of different robot configurations such as Cartesian, cylindrical, and spherical coordinates. Finally, we will use the Denavit–Hartenberg representation to derive forward and inverse kinematic equations of all possible configurations of robots—regardless of number of joints, order of joints, and presence (or lack) of offsets and twists.

It is important to realize that in practice, manipulator-type robots are delivered with no end effector. In most cases, there may be a gripper attached to the robot; however, depending on the actual application, different end effectors are attached to the robot by the user. Obviously, the end effector's size and length determine where the end of the robot will be. For a short end effector, the end will be at a different location compared to a long end effector. In this chapter, we will assume that the end of the robot is a plate to which the end effector can be attached, as necessary. We will call this the "hand" or the "end plate" of the robot. If necessary, we can always add the length of the end effector to the robot for determining the location and orientation of the end effector. It should be mentioned here that a real robot manipulator, for which the length of the end effector is not defined, will calculate its joint values based on the end plate location and orientation, which may be different from the position and orientation perceived by the user.

2.2 Robots as Mechanisms

Manipulator-type robots are multi-degree-of-freedom (DOF), three-dimensional, open loop, chain mechanisms, and are discussed in this section.

Multi-degree-of-freedom means that robots possess many joints, allowing them to move freely within their envelope. In a 1-DOF system, when the variable is set to a particular value, the mechanism is totally set and all its other variables are known. For example, in the 1-DOF 4-bar mechanism of Figure 2.1, when the crank is set to 120°, the angles of the coupler link and the rocker arm are also known, whereas in a multi-DOF mechanism, all input variables must be individually defined in order to know the remaining parameters. Robots are multi-DOF machines, where each joint variable must be known in order to determine the location of the robot's hand.

Robots are three-dimensional machines if they are to move in space. Although it is possible to have a two-dimensional multi-DOF robot, they are not common (or useful).

Robots are open-loop mechanisms. Unlike mechanisms that are closed-loop (e.g., 4-bar mechanisms), even if all joint variables are set to particular values, there is no guarantee that the hand will be at the given location. This is because deflections in any joint or link will change the location of all subsequent links without feedback. For example, in the 4-bar mechanism of Figure 2.2, when link AB deflects as a result of load F, link BO_2 will also move; therefore, the deflection can be detected. In an open-loop system such as the robot, the deflections will move all succeeding members without any

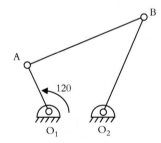

Figure 2.1   A 1-DOF closed-loop 4-bar mechanism.

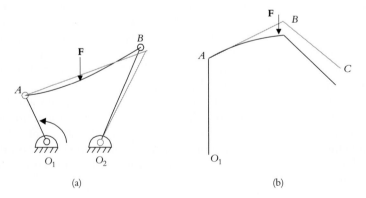

Figure 2.2   Closed-loop (a) versus open-loop (b) mechanisms.

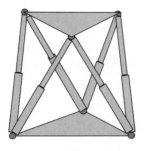 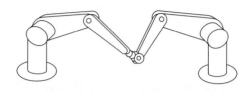

Figure 2.3 Possible parallel manipulator configurations.

feedback. Therefore, in open-loop systems, either all joint and link parameters must continuously be measured, or the end of the system must be monitored; otherwise, the kinematic position of the machine is not completely known. This difference can be expressed by comparing the vector equations describing the relationship between different links of the two mechanisms as follows:

$$\text{For the 4-bar mechanism:} \quad \overline{O_1A} + \overline{AB} = \overline{O_1O_2} + \overline{O_2B} \tag{2.1}$$

$$\text{For the robot:} \quad \overline{O_1A} + \overline{AB} + \overline{BC} = \overline{O_1C} \tag{2.2}$$

As you can see, if there is a deflection in link AB, link O_2B will move accordingly. However, the two sides of Equation (2.1) have changed corresponding to the changes in the links. On the other hand, if link AB of the robot deflects, all subsequent links will move too; however, unless O_1C is measured by other means, the change will not be known. To remedy this problem in open loop robots, either the position of the hand is constantly measured with devices such as a camera, the robot is made into a closed loop system with external means such as the use of secondary arms or laser beams,[1,2,3] or as standard practice, the robot links and joints are made excessively strong to eliminate all deflections. This will render the robot very heavy, massive, and slow, and its specified payload will be very low compared to what it can actually carry.

Alternatives, also called *parallel manipulators*, are based on closed-loop parallel architecture (Figure 2.3). The tradeoff is much-reduced range of motions and workspace.

2.3 Conventions

Throughout this book, we will use the following conventions for describing vectors, frames, transformations, and so on:

Vectors	**i, j, k, x, y, z, n, o, a, p**
Vector components	$n_x, n_y, n_z, a_x, a_y, a_z$
Frames	$F_{xyz}, F_{noa}, xyz, noa, F_{camera}$
Transformations	$T_1, T_2, {}^uT, {}^BP, {}^UT_R$ (transformation of robot relative to the Universe, where Universe is a fixed frame)

2.4 Matrix Representation

Matrices can be used to represent points, vectors, frames, translations, rotations, transformations, as well as objects and other kinematic elements. We will use this representation throughout the book.

2.4.1 Representation of a Point in Space

A point P in space (Figure 2.4) can be represented by its three coordinates relative to a
reference frame as:

$$P = a_x\mathbf{i} + b_y\mathbf{j} + c_z\mathbf{k} \tag{2.3}$$

where a_x, b_y, and c_z are the three coordinates of the point represented in the reference
frame. Obviously, other coordinate representations can also be used to describe the
location of a point in space.

2.4.2 Representation of a Vector in Space

A vector can be represented by three coordinates of its tail and its head. If the vector starts
at point A and ends at point B, then it can be represented by $\mathbf{P}_{AB} = (B_x - A_x)\mathbf{i} + (B_y - A_y)\mathbf{j} + (B_z - A_z)\mathbf{k}$. Specifically, if the vector starts at the origin (Figure 2.5),

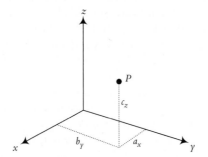

Figure 2.4   Representation of a point in space.

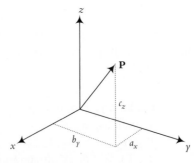

Figure 2.5   Representation of a vector in space.

then:

$$\mathbf{P} = a_x\mathbf{i} + b_y\mathbf{j} + c_z\mathbf{k} \tag{2.4}$$

where a_x, b_y, and c_z are the three components of the vector in the reference frame. In fact, point P in the previous section is in reality represented by a vector connected to it at point P and expressed by the three components of the vector.

The three components of the vector can also be written in matrix form, as in Equation (2.5). This format will be used throughout this book to represent all kinematic elements:

$$\mathbf{P} = \begin{bmatrix} a_x \\ b_y \\ c_z \end{bmatrix} \tag{2.5}$$

This representation can be slightly modified to also include a scale factor w such that if P_x, P_y, and P_z are divided by w, they will yield a_x, b_y, and c_z. Therefore the vector can be written as:

$$\mathbf{P} = \begin{bmatrix} P_x \\ P_y \\ P_z \\ w \end{bmatrix} \quad \text{where } a_x = \frac{P_x}{w}, b_y = \frac{P_y}{w}, \text{ etc.} \tag{2.6}$$

w may be any number and, as it changes, it can change the overall size of the vector. This is similar to the zooming function in computer graphics. As the value of w changes, the size of the vector changes accordingly. If w is bigger than 1, all vector components enlarge; if w is smaller than 1, all vector components become smaller.

When w is 1, the size of these components remains unchanged. However, if $w = 0$, then a_x, b_y, and c_z will be infinity. In this case, P_x, P_y, and P_z (as well as a_x, b_y, and c_z) will represent a vector whose length is infinite but nonetheless is in the direction represented by the vector. This means that a *direction vector* can be represented by a scale factor of $w = 0$, where the length is not important, but the direction is represented by the three components of the vector. This will be used throughout the book to represent direction vectors.

In computer graphics applications, the addition of a scale factor allows the user to zoom in or out simply by changing this value. Since the scale factor increases or decreases all vector dimensions accordingly, the size of a vector (or drawing) can be easily changed without the need to redraw it. However, our reason for this inclusion is different, and it will become apparent shortly.

Example 2.1

A vector is described as $\mathbf{P} = 3\mathbf{i} + 5\mathbf{j} + 2\mathbf{k}$. Express the vector in matrix form:

(a) With a scale factor of 2.
(b) If it were to describe a direction as a unit vector.

Solution: The vector can be expressed in matrix form with a scale factor of 2 as well as 0 for direction as:

$$\mathbf{P} = \begin{bmatrix} 6 \\ 10 \\ 4 \\ 2 \end{bmatrix} \quad \text{and} \quad \mathbf{P} = \begin{bmatrix} 3 \\ 5 \\ 2 \\ 0 \end{bmatrix}$$

However, in order to make the vector into a unit vector, we normalize the length to be equal to 1. To do this, each component of the vector is divided by the square root of the sum of the squares of the three components:

$$\lambda = \sqrt{P_x^2 + P_y^2 + P_z^2} = 6.16 \text{ and } P_x = {}^3/_{6.16} = 0.487, \text{ etc. Therefore,}$$

$$\mathbf{P}_{unit} = \begin{bmatrix} 0.487 \\ 0.811 \\ 0.324 \\ 0 \end{bmatrix}$$

Note that $\sqrt{0.487^2 + 0.811^2 + 0.324^2} = 1$. ∎

Example 2.2

A vector **p** is 5 units long and is in the direction of a unit vector **q** described below. Express the vector in matrix form.

$$\mathbf{q}_{unit} = \begin{bmatrix} 0.371 \\ 0.557 \\ q_z \\ 0 \end{bmatrix}$$

Solution: The unit vector's length must be 1. Therefore,

$$\lambda = \sqrt{q_x^2 + q_y^2 + q_z^2} = \sqrt{0.138 + 0.310 + q_z^2} = 1 \quad \rightarrow \quad q_z = 0.743$$

$$\mathbf{q}_{unit} = \begin{bmatrix} 0.371 \\ 0.557 \\ 0.743 \\ 0 \end{bmatrix} \quad \text{and} \quad \mathbf{p} = \mathbf{q}_{unit} \times 5 = \begin{bmatrix} 1.855 \\ 2.785 \\ 3.715 \\ 1 \end{bmatrix}$$

∎

2.4.3 Representation of a Frame at the Origin of a Fixed Reference Frame

A frame is generally represented by three mutually orthogonal axes (such as x, y, and z). Since we may have more than one frame at any given time, we will use axes x, y, and z to represent the fixed Universe reference frame $F_{x,y,z}$ and a set of axes n, o, and a to represent another (moving) frame $F_{n,o,a}$ relative to the reference frame. This way, there should be no confusion about which frame is referenced.

The letters n, o, and a are derived from the words *normal*, *orientation*, and *approach*. Referring to Figure 2.6, it should be clear that in order to avoid hitting the part while trying to pick it up, the robot would have to approach it along the z-axis of the gripper. In robotic nomenclature, this axis is called *approach-axis* and is referred to as the a-axis. The orientation with which the gripper frame approaches the part is called *orientation-axis*, and it is referred to as the o-axis. Since the x-axis is normal to both, it is referred to as n-axis. Throughout this book, we will refer to a moving frame as $F_{n,o,a}$ with *normal*, *orientation*, and *approach* axes.

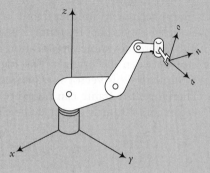

Figure 2.6 The normal-, orientation-, and approach-axis of a moving frame.

Each direction of each axis of a frame $F_{n,o,a}$ located at the origin of a reference frame $F_{x,y,z}$ (Figure 2.7) is represented by its three directional cosines relative to the reference frame as in section 2.4.2. Consequently, the three axes of the frame can be represented by three vectors in matrix form as:

$$F = \begin{bmatrix} n_x & o_x & a_x \\ n_y & o_y & a_y \\ n_z & o_z & a_z \end{bmatrix} \qquad (2.7)$$

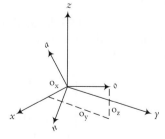

Figure 2.7 Representation of a frame at the origin of the reference frame.

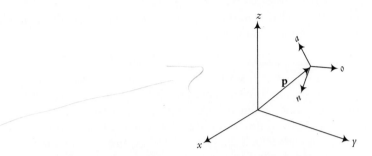

Figure 2.8 Representation of a frame in a frame.

2.4.4 Representation of a Frame Relative to a Fixed Reference Frame

To fully describe a frame relative to another frame, both the location of its origin and the directions of its axes must be specified. If a frame is not at the origin (or, in fact, even if it is at the origin) of the reference frame, its location relative to the reference frame is described by a vector between the origin of the frame and the origin of the reference frame (Figure 2.8). Similarly, this vector is expressed by its components relative to the reference frame. Therefore, the frame can be expressed by three vectors describing its directional unit vectors and a fourth vector describing its location as:

Test Question ⟶

$$F = \begin{bmatrix} n_x & o_x & a_x & p_x \\ n_y & o_y & a_y & p_y \\ n_z & o_z & a_z & p_z \\ 0 & 0 & 0 & 1 \end{bmatrix} \tag{2.8}$$

As shown in Equation (2.8), the first three vectors are directional vectors with $w = 0$, representing the directions of the three unit vectors of the frame $F_{n,o,a}$, while the fourth vector with $w = 1$ represents the location of the origin of the frame relative to the reference frame. Unlike the unit vectors, the length of vector p is important. Consequently, we use a scale factor of 1.

A frame may also be represented by a 3×4 matrix without the scale factors, but it is not common. Adding the fourth row of scale factors to the matrix makes it a 4×4 or *homogeneous* matrix.

Example 2.3

The frame F shown in Figure 2.9 is located at 3,5,7 units, with its n-axis parallel to x, its o-axis at 45° relative to the y-axis, and its a-axis at 45° relative to the z-axis. The frame can be described by:

$$F = \begin{bmatrix} 1 & 0 & 0 & 3 \\ 0 & 0.707 & -0.707 & 5 \\ 0 & 0.707 & 0.707 & 7 \\ 0 & 0 & 0 & 1 \end{bmatrix}$$

∎

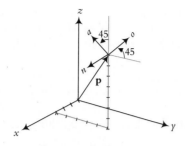

Figure 2.9 An example of representation of a frame.

2.4.5 Representation of a Rigid Body

An object can be represented in space by attaching a frame to it and representing the frame. Since the object is permanently attached to this frame, its position and orientation relative to the frame is always known. As a result, so long as the frame can be described in space, the object's location and orientation relative to the fixed frame will be known (Figure 2.10). As before, a frame can be represented by a matrix, where the origin of the frame and the three vectors representing its orientation relative to the reference frame are expressed. Therefore,

$$F_{object} = \begin{bmatrix} n_x & o_x & a_x & p_x \\ n_y & o_y & a_y & p_y \\ n_z & o_z & a_z & p_z \\ 0 & 0 & 0 & 1 \end{bmatrix} \qquad (2.9)$$

As we discussed in Chapter 1, a point in space has only three degrees of freedom; it can only move along the three reference axes. However, a rigid body in space has six degrees of freedom, meaning that not only can it move along x-, y-, and z-axes, it can also rotate about these three axes. Consequently, all that is needed to completely define an object in space is six pieces of information describing the location of the origin of the object in the reference frame and its orientation about the three axes. However, as can be seen in Equation (2.9), twelve pieces of information are given: nine for orientation, and three for position (this excludes the scale factors on the last row of the matrix because they do not add to this information). Obviously, there must be some constraints present in this

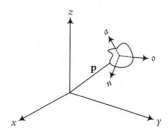

Figure 2.10 Representation of an object in space.

representation to limit the above to six. Therefore, we need 6 constraint equations to reduce the above from twelve to six. The constraints come from the known characteristics of a frame that have not been used yet, that:

- the three unit vectors **n, o, a** are mutually perpendicular, and
- each unit vector's length, represented by its directional cosines, must be equal to 1.

These constraints translate into the following six constraint equations:

1. $\mathbf{n} \cdot \mathbf{o} = 0$ (the dot-product of **n** and **o** vectors must be zero)
2. $\mathbf{n} \cdot \mathbf{a} = 0$
3. $\mathbf{a} \cdot \mathbf{o} = 0$
4. $|\mathbf{n}| = 1$ (the magnitude of the length of the vector must be 1)   (2.10)
5. $|\mathbf{o}| = 1$
6. $|\mathbf{a}| = 1$

As a result, the values representing a frame in a matrix must be such that the above equations remain true. Otherwise, the frame will not be correct. Alternatively, the first three equations in Equation (2.10) can be replaced by a cross product of the three vectors as:

$$\mathbf{n} \times \mathbf{o} = \mathbf{a} \qquad (2.11)$$

Since Equation (2.11) includes the correct right-hand–rule relationship too, it is recommended that this equation be used to determine the correct relationship between the three vectors.

Example 2.4

For the following frame, find the values of the missing elements and complete the matrix representation of the frame:

$$F = \begin{bmatrix} ? & 0 & ? & 5 \\ 0.707 & ? & ? & 3 \\ ? & ? & 0 & 2 \\ 0 & 0 & 0 & 1 \end{bmatrix}$$

Solution: Obviously, the 5,3,2 values representing the position of the origin of the frame do not affect the constraint equations. Please notice that only 3 values for directional vectors are given. This is all that is needed. Using Equation (2.10), we will get:

$$n_x o_x + n_y o_y + n_z o_z = 0 \quad \text{or} \quad n_x(0) + 0.707(o_y) + n_z(o_z) = 0$$
$$n_x a_x + n_y a_y + n_z a_z = 0 \quad \text{or} \quad n_x(a_x) + 0.707(a_y) + n_z(0) = 0$$
$$a_x o_x + a_y o_y + a_z o_z = 0 \quad \text{or} \quad a_x(0) + a_y(o_y) + 0(o_z) = 0$$
$$n_x^2 + n_y^2 + n_z^2 = 1 \quad \text{or} \quad n_x^2 + 0.707^2 + n_z^2 = 1$$
$$o_x^2 + o_y^2 + o_z^2 = 1 \quad \text{or} \quad 0^2 + o_y^2 + o_z^2 = 1$$
$$a_x^2 + a_y^2 + a_z^2 = 1 \quad \text{or} \quad a_x^2 + a_y^2 + 0^2 = 1$$

Simplifying these equations yields:

$$0.707\, o_y + n_z o_z = 0$$

$$n_x a_x + 0.707\, a_y = 0$$

$$a_y o_y = 0$$

$$n_x^2 + n_z^2 = 0.5$$

$$o_y^2 + o_z^2 = 1$$

$$a_x^2 + a_y^2 = 1$$

Solving these six equations will yield $n_x = \pm 0.707$, $n_z = 0$, $o_y = 0$, $o_z = 1$, $a_x = \pm 0.707$, and $a_y = -0.707$. Notice that both n_x and a_x must have the same sign. The reason for multiple solutions is that with the given parameters, it is possible to have two sets of mutually perpendicular vectors in opposite directions. The final matrix will be:

$$F_1 = \begin{bmatrix} 0.707 & 0 & 0.707 & 5 \\ 0.707 & 0 & -0.707 & 3 \\ 0 & 1 & 0 & 2 \\ 0 & 0 & 0 & 1 \end{bmatrix} \quad \text{or} \quad F_2 = \begin{bmatrix} -0.707 & 0 & -0.707 & 5 \\ 0.707 & 0 & -0.707 & 3 \\ 0 & 1 & 0 & 2 \\ 0 & 0 & 0 & 1 \end{bmatrix}$$

As you can see, both matrices satisfy all the requirements set by the constraint equations. It is important to realize that the values representing the three direction vectors are not arbitrary but bound by these equations. Therefore, you may not randomly use any desired values in the matrix.

The same problem may be solved using $\mathbf{n} \times \mathbf{o} = \mathbf{a}$, or:

$$\begin{vmatrix} \mathbf{i} & \mathbf{j} & \mathbf{k} \\ n_x & n_y & n_z \\ o_x & o_y & o_z \end{vmatrix} = a_x \mathbf{i} + a_y \mathbf{j} + a_z \mathbf{k}$$

or $\quad \mathbf{i}(n_y o_z - n_z o_y) - \mathbf{j}(n_x o_z - n_z o_x) + \mathbf{k}(n_x o_y - n_y o_x) = a_x \mathbf{i} + a_y \mathbf{j} + a_z \mathbf{k}$ $\quad$ (2.12)

Substituting the values into this equation yields:

$$\mathbf{i}(0.707 o_z - n_z o_y) - \mathbf{j}(n_x o_z) + \mathbf{k}(n_x o_y) = a_x \mathbf{i} + a_y \mathbf{j} + 0\mathbf{k}$$

Solving the three simultaneous equations will result in:

$$0.707\, o_z - n_z o_y = a_x$$

$$-n_x o_z = a_y$$

$$n_x o_y = 0$$

which replace the three equations for the dot products. Together with the three unit-vector length constraint equations, there will be six equations. However, as you will see, only one of the two solutions (F_1) obtained in the first part will satisfy these equations. This is because the dot-product equations are scalar, and therefore, are the same whether the unit vectors are right-handed or left-handed frames, whereas the cross-product equations do indicate the correct right-handed frame configuration. Consequently, it is recommended that the cross-product equation be used. ■

Example 2.5

Find the missing elements of the following frame representation:

$$F = \begin{bmatrix} ? & 0 & ? & 3 \\ 0.5 & ? & ? & 9 \\ 0 & ? & ? & 7 \\ 0 & 0 & 0 & 1 \end{bmatrix}$$

Solution:

$$n_x^2 + n_y^2 + n_z^2 = 1 \quad \rightarrow \quad n_x^2 + 0.25 = 1 \quad \rightarrow \quad n_x = 0.866$$

$$\mathbf{n} \cdot \mathbf{o} = 0 \quad \rightarrow \quad (0.866)(0) + (0.5)(o_y) + (0)(o_z) = 0 \quad \rightarrow \quad o_y = 0$$

$$|\mathbf{o}| = 1 \quad \rightarrow \quad o_z = 1$$

$$\mathbf{n} \times \mathbf{o} = \mathbf{a} \quad \rightarrow \quad \mathbf{i}(0.5) - \mathbf{j}(0.866) + \mathbf{k}(0) = a_x\mathbf{i} + a_y\mathbf{j} + a_z\mathbf{k}$$

$$a_x = 0.5$$

$$a_y = -0.866$$

$$a_z = 0$$

■

2.5 Homogeneous Transformation Matrices

For a variety of reasons, it is desirable to keep matrices in square form, either 3×3 or 4×4. First, as we will see later, it is much easier to calculate the inverse of square matrices than rectangular matrices. Second, in order to multiply two matrices, their dimensions must match, such that the number of columns of the first matrix must be the same as the number of rows of the second matrix, as in $(m \times n)$ and $(n \times p)$, which results in a matrix of $(m \times p)$ dimensions. If two matrices, A and B, are square with $(m \times m)$ and $(m \times m)$ dimensions, we may multiply A by B, or B by A, both resulting in the same $(m \times m)$ dimensions. However, if the two matrices are not square, with $(m \times n)$ and $(n \times p)$ dimensions respectively, A can be multiplied by B, but B may not be multiplied by A, and the result of AB has a dimension different from A and B. Since we will have to multiply many matrices together, in different orders, to find the equations of motion of the robots, we want to have square matrices.

In order to keep representation matrices square, if we represent both orientation and position in the same matrix, we will add the scale factors to the matrix to make it 4×4. If we represent the orientation alone, we may either drop the scale factors and use 3×3 matrices, or add a fourth column with zeros for position in order to keep the matrix square. Matrices of this form are called homogeneous matrices, and we refer to them as:

$$F = \begin{bmatrix} n_x & o_x & a_x & p_x \\ n_y & o_y & a_y & p_y \\ n_z & o_z & a_z & p_z \\ 0 & 0 & 0 & 1 \end{bmatrix} \tag{2.13}$$

2.6 Representation of Transformations

A transformation is defined as making a movement in space. When a frame (a vector, an object, or a moving frame) moves in space relative to a fixed reference frame, we represent this motion in a form similar to a frame representation. This is because a transformation is a change in the state of a frame (representing the change in its location and orientation); therefore, it can be represented like a frame. A transformation may be in one of the following forms:

- A pure translation
- A pure rotation about an axis
- A combination of translations and/or rotations

In order to see how these can be represented, we will study each one separately.

2.6.1 Representation of a Pure Translation

If a frame (that may also be representing an object) moves in space without any change in its orientation, the transformation is a pure translation. In this case, the directional unit vectors remain in the same direction, and therefore, do not change. The only thing that changes is the location of the origin of the frame relative to the reference frame, as shown in Figure 2.11. The new location of the frame relative to the fixed reference frame can be found by adding the vector representing the translation to the vector representing the original location of the origin of the frame. In matrix form, the new frame representation may be found by pre-multiplying the frame with a matrix representing the transformation. Since the directional vectors do not change in a pure translation, the transformation T will simply be:

$$T = \begin{bmatrix} 1 & 0 & 0 & d_x \\ 0 & 1 & 0 & d_y \\ 0 & 0 & 1 & d_z \\ 0 & 0 & 0 & 1 \end{bmatrix} \qquad (2.14)$$

where d_x, d_y, and d_z are the three components of a pure translation vector $\mathbf{d}$ relative to the x-, y-, and z-axes of the reference frame. The first three columns represent no rotational movement (equivalent of a 1), while the last column represents the translation. The new

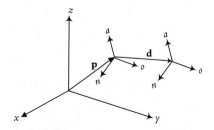

Figure 2.11   Representation of a pure translation in space.

location of the frame will be:

$$F_{new} = \begin{bmatrix} 1 & 0 & 0 & d_x \\ 0 & 1 & 0 & d_y \\ 0 & 0 & 1 & d_z \\ 0 & 0 & 0 & 1 \end{bmatrix} \times \begin{bmatrix} n_x & o_x & a_x & p_x \\ n_y & o_y & a_y & p_y \\ n_z & o_z & a_z & p_z \\ 0 & 0 & 0 & 1 \end{bmatrix} = \begin{bmatrix} n_x & o_x & a_x & p_x + d_x \\ n_y & o_y & a_y & p_y + d_y \\ n_z & o_z & a_z & p_z + d_z \\ 0 & 0 & 0 & 1 \end{bmatrix} \quad (2.15)$$

This equation is also symbolically written as:

$$F_{new} = Trans(d_x, d_y, d_z) \times F_{old} \quad (2.16)$$

First, as you can see, pre-multiplying the frame matrix by the transformation matrix will yield the new location of the frame. Second, notice that the directional vectors remain the same after a pure translation, but the new location of the frame is at **d+p**. Third, notice how homogeneous transformation matrices facilitate the multiplication of matrices, resulting in the same dimensions as before.

Example 2.6

A frame *F* has been moved 10 units along the *y*-axis and 5 units along the *z*-axis of the reference frame. Find the new location of the frame.

$$F = \begin{bmatrix} 0.527 & -0.574 & 0.628 & 5 \\ 0.369 & 0.819 & 0.439 & 3 \\ -0.766 & 0 & 0.643 & 8 \\ 0 & 0 & 0 & 1 \end{bmatrix}$$

Solution: Using Equation (2.15) or (2.16), we get:

$$F_{new} = Trans(d_x, d_y, d_z) \times F_{old} = Trans(0, 10, 5) \times F_{old}$$

and

$$F_{new} = \begin{bmatrix} 1 & 0 & 0 & 0 \\ 0 & 1 & 0 & 10 \\ 0 & 0 & 1 & 5 \\ 0 & 0 & 0 & 1 \end{bmatrix} \times \begin{bmatrix} 0.527 & -0.574 & 0.628 & 5 \\ 0.369 & 0.819 & 0.439 & 3 \\ -0.766 & 0 & 0.643 & 8 \\ 0 & 0 & 0 & 1 \end{bmatrix}$$

$$= \begin{bmatrix} 0.527 & -0.574 & 0.628 & 5 \\ 0.369 & 0.819 & 0.439 & 13 \\ -0.766 & 0 & 0.643 & 13 \\ 0 & 0 & 0 & 1 \end{bmatrix} \quad \blacksquare$$

2.6.2 Representation of a Pure Rotation about an Axis

To simplify the derivation of rotations about an axis, let's first assume that the frame is at the origin of the reference frame and is parallel to it. We will later expand the results to other rotations as well as combinations of rotations.

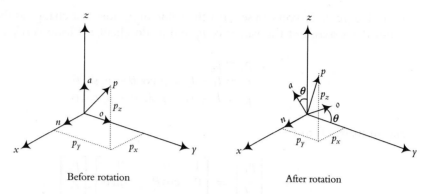

Before rotation   After rotation

Figure 2.12 Coordinates of a point in a rotating frame before and after rotation.

Let's assume that a frame F_{noa}, located at the origin of the reference frame F_{xyz}, rotates an angle of θ about the x-axis of the reference frame. Let's also assume that attached to the rotating frame F_{noa}, is a point p, with coordinates p_x, p_y, and p_z relative to the reference frame and p_n, p_o, and p_a relative to the moving frame. As the frame rotates about the x-axis, point p attached to the frame will also rotate with it. Before rotation, the coordinates of the point in both frames are the same (remember that the two frames are at the same location and are parallel to each other). After rotation, the p_n, p_o, and p_a coordinates of the point remain the same in the rotating frame F_{noa}, but p_x, p_y, and p_z will be different in the F_{xyz} frame (Figure 2.12). We want to find the new coordinates of the point relative to the fixed reference frame after the moving frame has rotated.

Now let's look at the same coordinates in 2-D as if we were standing on the x-axis. The coordinates of point p are shown before and after rotation in Figure 2.13. The coordinates of point p relative to the reference frame are p_x, p_y, and p_z, while its coordinates relative to the rotating frame (to which the point is attached) remain as p_n, p_o, and p_a.

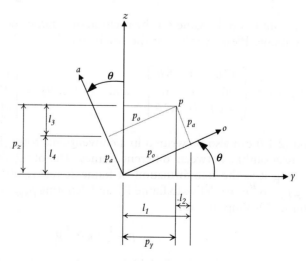

Figure 2.13 Coordinates of a point relative to the reference frame and rotating frame as viewed from the x-axis.

From Figure 2.13, you can see that the value of p_x does not change as the frame rotates about the *x*-axis, but the values of p_y and p_z do change. Please verify that:

$$p_x = p_n$$
$$p_y = l_1 - l_2 = p_o \cos \theta - p_a \sin \theta \qquad (2.17)$$
$$p_z = l_3 + l_4 = p_o \sin \theta + p_a \cos \theta$$

and in matrix form:

$$\begin{bmatrix} p_x \\ p_y \\ p_z \end{bmatrix} = \begin{bmatrix} 1 & 0 & 0 \\ 0 & \cos\theta & -\sin\theta \\ 0 & \sin\theta & \cos\theta \end{bmatrix} \begin{bmatrix} p_n \\ p_o \\ p_a \end{bmatrix} \qquad (2.18)$$

This means that the coordinates of the point *p* (or vector **p**) in the rotated frame must be pre-multiplied by the rotation matrix, as shown, to get the coordinates in the reference frame. This rotation matrix is only for a pure rotation about the *x*-axis of the reference frame and is denoted as:

$$p_{xyz} = Rot(x, \theta) \times p_{noa} \qquad (2.19)$$

Notice that the first column of the rotation matrix in Equation (2.18)—which expresses the location relative to the *x*-axis—has 1,0,0 values, indicating that the coordinate along the *x*-axis has not changed.

To simplify writing these matrices, it is customary to designate $C\theta$ to denote $\cos \theta$ and $S\theta$ to denote $\sin \theta$. Therefore, the rotation matrix may be also written as:

$$Rot(x, \theta) = \begin{bmatrix} 1 & 0 & 0 \\ 0 & C\theta & -S\theta \\ 0 & S\theta & C\theta \end{bmatrix} \qquad (2.20)$$

You may want to do the same for the rotation of a frame about the *y*- and *z*-axes of the reference frame. Please verify that the results will be:

$$Rot(y, \theta) = \begin{bmatrix} C\theta & 0 & S\theta \\ 0 & 1 & 0 \\ -S\theta & 0 & C\theta \end{bmatrix} \quad \text{and} \quad Rot(z, \theta) = \begin{bmatrix} C\theta & -S\theta & 0 \\ S\theta & C\theta & 0 \\ 0 & 0 & 1 \end{bmatrix} \qquad (2.21)$$

Equation (2.19) can also be written in a conventional form that assists in easily following the relationship between different frames. Denoting the transformation as $^U T_R$ (and reading it as the transformation of frame *R* relative to frame *U* (for Universe)), denoting p_{noa} as $^R p$ (*p* relative to frame *R*), and denoting p_{xyz} as $^U p$ (*p* relative to frame *U*), Equation (2.19) simplifies to:

$$^U p = {^U T_R} \times {^R p} \qquad (2.22)$$

As you see, canceling the *R*s will yield the coordinates of point *p* relative to *U*. The same notation will be used throughout this book to relate to multiple transformations.

Example 2.7

A point $p(2,3,4)^T$ is attached to a rotating frame. The frame rotates $90°$ about the x-axis of the reference frame. Find the coordinates of the point relative to the reference frame after the rotation, and verify the result graphically.

Solution: Of course, since the point is attached to the rotating frame, the coordinates of the point relative to the rotating frame remain the same after the rotation. The coordinates of the point relative to the reference frame will be:

$$
\begin{bmatrix} p_x \\ p_y \\ p_z \end{bmatrix} = \begin{bmatrix} 1 & 0 & 0 \\ 0 & C\theta & -S\theta \\ 0 & S\theta & C\theta \end{bmatrix} \times \begin{bmatrix} p_n \\ p_o \\ p_a \end{bmatrix} = \begin{bmatrix} 1 & 0 & 0 \\ 0 & 0 & -1 \\ 0 & 1 & 0 \end{bmatrix} \times \begin{bmatrix} 2 \\ 3 \\ 4 \end{bmatrix} = \begin{bmatrix} 2 \\ -4 \\ 3 \end{bmatrix}
$$

As shown in Figure 2.14, the coordinates of point p relative to the reference frame after rotation are 2, -4, 3, as obtained by the above transformation.

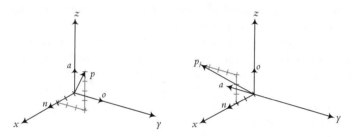

Figure 2.14   Rotation of a frame relative to the x-axis of the reference frame.   ∎

2.6.3 Representation of Combined Transformations

Combined transformations consist of a number of successive translations and rotations about the fixed reference frame axes or the moving current frame axes. Any transformation can be resolved into a set of translations and rotations in a particular order. For example, we may rotate a frame about the x-axis, then translate about the x-, y-, and z-axes, then rotate about the y-axis in order to accomplish the desired transformation. As we will see later, this order is very important, such that if the order of two successive transformations changes, the result may be completely different.

To see how combined transformations are handled, let's assume that a frame F_{noa} is subjected to the following three successive transformations relative to the reference frame F_{xyz}:

1. Rotation of α degrees about the x-axis,
2. Followed by a translation of $[l_1, l_2, l_3]$ (relative to the x-, y-, and z-axes respectively),
3. Followed by a rotation of β degrees about the y-axis.

Also, let's say that a point p_{noa} is attached to the rotating frame at the origin of the reference frame. As the frame F_{noa} rotates or translates relative to the reference frame, point p within the frame moves as well, and the coordinates of the point relative to the reference frame change. After the first transformation, as we saw in the previous section, the coordinates of point p relative to the reference frame can be calculated by:

$$p_{1,xyz} = Rot(x,\alpha) \times p_{noa} \qquad (2.23)$$

where $p_{1,xyz}$ is the coordinates of the point after the first transformation relative to the reference frame. The coordinates of the point relative to the reference frame at the conclusion of the second transformation will be:

$$p_{2,xyz} = Trans(l_1, l_2, l_3) \times p_{1,xyz} = Trans(l_1, l_2, l_3) \times Rot(x,\alpha) \times p_{noa} \qquad (2.24)$$

Similarly, after the third transformation, the coordinates of the point relative to the reference frame will be:

$$p_{xyz} = p_{3,xyz} = Rot(y,\beta) \times p_{2,xyz} = Rot(y,\beta) \times Trans(l_1, l_2, l_3) \times Rot(x,\alpha) \times p_{noa}$$

As you can see, the coordinates of the point relative to the reference frame at the conclusion of each transformation is found by pre-multiplying the coordinates of the point by each transformation matrix. Of course, as shown in Appendix A, the order of matrices cannot be changed, therefore this order is very important. You will also notice that for each transformation relative to the reference frame, the matrix is pre-multiplied. Consequently, the order of matrices *written* is the opposite of the order of transformations *performed*.

Example 2.8

A point $p(7,3,1)^T$ is attached to a frame F_{noa} and is subjected to the following transformations. Find the coordinates of the point relative to the reference frame at the conclusion of transformations.

1. Rotation of $90°$ about the z-axis,
2. Followed by a rotation of $90°$ about the y-axis,
3. Followed by a translation of $[4,-3,7]$.

Solution: The matrix equation representing the transformation is:

$$p_{xyz} = Trans(4,-3,7)Rot(y,90)Rot(z,90)p_{noa}$$

$$= \begin{bmatrix} 1 & 0 & 0 & 4 \\ 0 & 1 & 0 & -3 \\ 0 & 0 & 1 & 7 \\ 0 & 0 & 0 & 1 \end{bmatrix} \times \begin{bmatrix} 0 & 0 & 1 & 0 \\ 0 & 1 & 0 & 0 \\ -1 & 0 & 0 & 0 \\ 0 & 0 & 0 & 1 \end{bmatrix} \times \begin{bmatrix} 0 & -1 & 0 & 0 \\ 1 & 0 & 0 & 0 \\ 0 & 0 & 1 & 0 \\ 0 & 0 & 0 & 1 \end{bmatrix} \times \begin{bmatrix} 7 \\ 3 \\ 1 \\ 1 \end{bmatrix} = \begin{bmatrix} 5 \\ 4 \\ 10 \\ 1 \end{bmatrix}$$

As you can see, the first transformation of $90°$ about the z-axis rotates the F_{noa} frame as shown in Figure 2.15, followed by the second rotation about the y-axis, followed by the translation relative to the reference frame F_{xyz}. The point p in the frame can then be found relative to the F_{noa} as shown. The final coordinates of the point can be

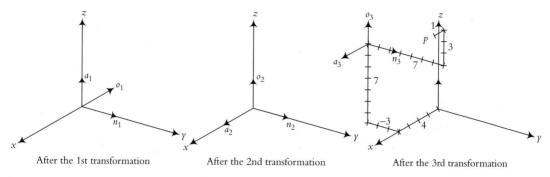

Figure 2.15 Effects of three successive transformations.

traced on the x-, y-, z-axes to be $4 + 1 = 5$, $-3 + 7 = 4$, and $7 + 3 = 10$. Be sure to follow this graphically. ∎

Example 2.9

In this case, assume the same point $p(7,3,1)^{\mathrm{T}}$, attached to F_{noa}, is subjected to the same transformations, but the transformations are performed in a different order, as shown. Find the coordinates of the point relative to the reference frame at the conclusion of transformations.

1. A rotation of $90°$ about the z-axis,
2. Followed by a translation of $[4,-3,7]$,
3. Followed by a rotation of $90°$ about the y-axis.

Solution: The matrix equation representing the transformation is:

$$P_{xyz} = Rot(y, 90)\, Trans(4,-3,7)\, Rot(z, 90) p_{noa}$$

$$= \begin{bmatrix} 0 & 0 & 1 & 0 \\ 0 & 1 & 0 & 0 \\ -1 & 0 & 0 & 0 \\ 0 & 0 & 0 & 1 \end{bmatrix} \times \begin{bmatrix} 1 & 0 & 0 & 4 \\ 0 & 1 & 0 & -3 \\ 0 & 0 & 1 & 7 \\ 0 & 0 & 0 & 1 \end{bmatrix} \times \begin{bmatrix} 0 & -1 & 0 & 0 \\ 1 & 0 & 0 & 0 \\ 0 & 0 & 1 & 0 \\ 0 & 0 & 0 & 1 \end{bmatrix} \times \begin{bmatrix} 7 \\ 3 \\ 1 \\ 1 \end{bmatrix} = \begin{bmatrix} 8 \\ 4 \\ -1 \\ 1 \end{bmatrix}$$

As you can see, although the transformations are exactly the same as in Example 2.8, since the order of transformations is changed, the final coordinates of the point are completely different. This can clearly be demonstrated graphically as in Figure 2.16. In this case, you can see that although the first transformation creates exactly the same change in the frame, the second transformation's result is very different because the translation relative to the reference frame axes will move the rotating frame F_{noa} outwardly. As a result of the third transformation, this frame will rotate about the y-axis, therefore rotating downwardly. The location of point p, attached to the frame is also shown. Please verify that the coordinates of this point

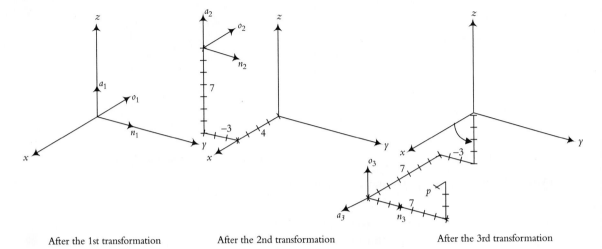

After the 1st transformation   After the 2nd transformation   After the 3rd transformation

Figure 2.16   Changing the order of transformations will change the final result.

relative to the reference frame are $7+1=8$, $-3+7=4$, and $-4+3=-1$, which is the same as the analytical result. ∎

2.6.4 Transformations Relative to the Rotating Frame

All transformations we have discussed so far have been relative to the fixed reference frame. This means that all translations, rotations, and distances (except for the location of a point relative to the moving frame) have been measured relative to the reference frame axes. However, it is possible to make transformations relative to the axes of a moving or current frame. This means that, for example, a rotation of 90° may be made relative to the n-axis of the moving frame (also referred to as the current frame), and not the x-axis of the reference frame. To calculate the changes in the coordinates of a point attached to the current frame relative to the reference frame, the transformation matrix is post-multiplied instead. Note that since the position of a point or an object attached to a moving frame is always measured relative to that moving frame, the position matrix describing the point or object is also always post-multiplied.

Example 2.10

Assume that the same point as in Example 2.9 is now subjected to the same transformations, but all relative to the current moving frame, as listed below. Find the coordinates of the point relative to the reference frame after transformations are completed.

1. A rotation of 90° about the a-axis,
2. Then a translation of $[4,-3,7]$ along n-, o-, a-axes
3. Followed by a rotation of 90° about the o-axis.

Solution: In this case, since the transformations are made relative to the current frame, each transformation matrix is post-multiplied. As a result, the equation

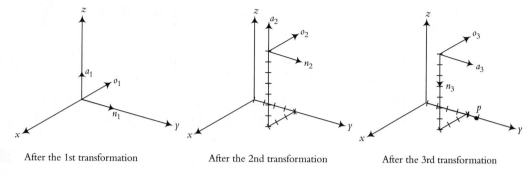

Figure 2.17 Transformations relative to the current frames.

representing the coordinates is:

$$p_{xyz} = Rot(a, 90)\,Trans(4, -3, 7)\,Rot(o, 90)\,p_{noa}$$

$$= \begin{bmatrix} 0 & -1 & 0 & 0 \\ 1 & 0 & 0 & 0 \\ 0 & 0 & 1 & 0 \\ 0 & 0 & 0 & 1 \end{bmatrix} \times \begin{bmatrix} 1 & 0 & 0 & 4 \\ 0 & 1 & 0 & -3 \\ 0 & 0 & 1 & 7 \\ 0 & 0 & 0 & 1 \end{bmatrix} \times \begin{bmatrix} 0 & 0 & 1 & 0 \\ 0 & 1 & 0 & 0 \\ -1 & 0 & 0 & 0 \\ 0 & 0 & 0 & 1 \end{bmatrix} \times \begin{bmatrix} 7 \\ 3 \\ 1 \\ 1 \end{bmatrix} = \begin{bmatrix} 0 \\ 5 \\ 0 \\ 1 \end{bmatrix}$$

As expected, the result is completely different from the other cases, both because the transformations are made relative to the current frame, and because the order of the matrices is now different. Figure 2.17 shows the results graphically. Notice how the transformations are accomplished relative to the current frames.

Notice how the 7,3,1 coordinates of point p in the current frame will result in 0,5,0 coordinates relative to the reference frame. ∎

Example 2.11

A frame B was rotated about the x-axis 90°, then it was translated about the current a-axis 3 inches before it was rotated about the z-axis 90°. Finally, it was translated about current o–axis 5 inches.

(a) Write an equation that describes the motions.
(b) Find the final location of a point $p(1,5,4)^T$ attached to the frame relative to the reference frame.

Solution: In this case, motions alternate relative to the reference frame and current frame.

(a) Pre- or post-multiplying each motion's matrix accordingly, we will get:

$$^U T_B = Rot(z, 90)\,Rot(x, 90)\,Trans(0, 0, 3)\,Trans(0, 5, 0)$$

(b) Substituting the matrices and multiplying them, we will get:

$$^Up = {}^UT_B \times {}^Bp$$

$$= \begin{bmatrix} 0 & -1 & 0 & 0 \\ 1 & 0 & 0 & 0 \\ 0 & 0 & 1 & 0 \\ 0 & 0 & 0 & 1 \end{bmatrix} \begin{bmatrix} 1 & 0 & 0 & 0 \\ 0 & 0 & -1 & 0 \\ 0 & 1 & 0 & 0 \\ 0 & 0 & 0 & 1 \end{bmatrix} \begin{bmatrix} 1 & 0 & 0 & 0 \\ 0 & 1 & 0 & 0 \\ 0 & 0 & 1 & 3 \\ 0 & 0 & 0 & 1 \end{bmatrix} \begin{bmatrix} 1 & 0 & 0 & 0 \\ 0 & 1 & 0 & 5 \\ 0 & 0 & 1 & 0 \\ 0 & 0 & 0 & 1 \end{bmatrix} \begin{bmatrix} 1 \\ 5 \\ 4 \\ 1 \end{bmatrix} = \begin{bmatrix} 7 \\ 1 \\ 10 \\ 1 \end{bmatrix}$$

■

Example 2.12

A frame F was rotated about the y-axis $90°$, followed by a rotation about the o-axis of $30°$, followed by a translation of 5 units along the n-axis, and finally, a translation of 4 units along the x-axis. Find the total transformation matrix.

Solution: The following set of matrices, written in the proper order to represent transformations relative to the reference frame or the current frame describes the total transformation:

$$T = Trans(4,0,0)Rot(y,90)Rot(o,30)Trans(5,0,0)$$

$$= \begin{bmatrix} 1 & 0 & 0 & 4 \\ 0 & 1 & 0 & 0 \\ 0 & 0 & 1 & 0 \\ 0 & 0 & 0 & 1 \end{bmatrix} \times \begin{bmatrix} 0 & 0 & 1 & 0 \\ 0 & 1 & 0 & 0 \\ -1 & 0 & 0 & 0 \\ 0 & 0 & 0 & 1 \end{bmatrix} \times \begin{bmatrix} 0.866 & 0 & 0.5 & 0 \\ 0 & 1 & 0 & 0 \\ -0.5 & 0 & 0.866 & 0 \\ 0 & 0 & 0 & 1 \end{bmatrix} \times \begin{bmatrix} 1 & 0 & 0 & 5 \\ 0 & 1 & 0 & 0 \\ 0 & 0 & 1 & 0 \\ 0 & 0 & 0 & 1 \end{bmatrix}$$

$$= \begin{bmatrix} -0.5 & 0 & 0.866 & 1.5 \\ 0 & 1 & 0 & 0 \\ -0.866 & 0 & -0.5 & -4.33 \\ 0 & 0 & 0 & 1 \end{bmatrix}$$

Please verify graphically that this is true.

■

2.7 Inverse of Transformation Matrices

As mentioned earlier, there are many situations where the inverse of a matrix will be needed in robotic analysis. One situation where transformation matrices may be involved can be seen in the following example. Suppose the robot in Figure 2.18 is to be moved toward part P in order to drill a hole in the part. The robot's base position relative to the reference frame U is described by a frame R, the robot's hand is described by frame H, and the end effector (let's say the end of the drill bit that will be used to drill the hole) is described by frame E. The part's position is also described by frame P. The location of the point where the hole will be drilled can be related to the reference frame U through two independent paths: one through the part, one through the robot. Therefore, the

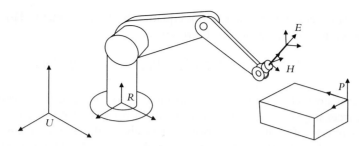

Figure 2.18 The Universe, robot, hand, part, and end effector frames.

following equation can be written:

$$^U T_E = {}^U T_R\, {}^R T_H\, {}^H T_E = {}^U T_P\, {}^P T_E \tag{2.25}$$

The location of point E on the part can be achieved by moving from U to P and from P to E, or it can alternately be achieved by a transformation from U to R, from R to H, and from H to E.

In reality, the transformation of frame R relative to the Universe frame $(^U T_R)$ is known since the location of the robot's base must be known in any set-up. For example, if a robot is installed in a work cell, the location of the robot's base will be known since it is bolted to a table. Even if the robot is mobile or attached to a conveyor belt, its location at any instant is known because a controller must be following the position of the robot's base at all times. The $^H T_E$, or the transformation of the end effector relative to the robot's hand, is also known since any tool used at the end effector is a known tool and its dimensions and configuration is known. $^U T_P$, or the transformation of the part relative to the universe, is also known since we must know where the part is located if we are to drill a hole in it. This location is known by putting the part in a jig, through the use of a camera and vision system, through the use of a conveyor belt and sensors, or other similar devices. $^P T_E$ is also known since we need to know where the hole is to be drilled on the part. Consequently, the only unknown transformation is $^R T_H$, or the transformation of the robot's hand relative to the robot's base. This means we need to find out what the robot's joint variables—the angle of the revolute joints and the length of the prismatic joints of the robot—must be in order to place the end effector at the hole for drilling. As you can see, it is necessary to calculate this transformation, which will tell us what needs to be accomplished. The transformation will later be used to actually solve for joint angles and link lengths.

To calculate this matrix, unlike in an algebraic equation, we cannot simply divide the right side by the left side of the equation. We need to pre- or post-multiply by inverses of appropriate matrices to eliminate them. As a result, we will have:

$$\left(^U T_R\right)^{-1}\left(^U T_R\, {}^R T_H\, {}^H T_E\right)\left(^H T_E\right)^{-1} = \left(^U T_R\right)^{-1}\left(^U T_P\, {}^P T_E\right)\left(^H T_E\right)^{-1} \tag{2.26}$$

or, since $\left(^U T_R\right)^{-1}\left(^U T_R\right) = I$ and $\left(^H T_E\right)\left(^H T_E\right)^{-1} = I$, the left side of Equation (2.26) simplifies to $^R T_H$ and we get:

$$^R T_H = {}^U T_R^{-1}\, {}^U T_P\, {}^P T_E\, {}^H T_E^{-1} \tag{2.27}$$

We can check the accuracy of this equation by realizing that $\left({}^{H}T_{E}\right)^{-1}$ is the same as ${}^{E}T_{H}$. Therefore, the equation can be rewritten as:

$$ {}^{R}T_{H} = {}^{U}T_{R}^{-1}\,{}^{U}T_{P}\,{}^{P}T_{E}\,{}^{H}T_{E}^{-1} = {}^{R}T_{U}\,{}^{U}T_{P}\,{}^{P}T_{E}\,{}^{E}T_{H} = {}^{R}T_{H} \qquad (2.28) $$

It is now clear that we need to be able to calculate the inverse of transformation matrices for kinematic analysis as well. In order to see what transpires, let's calculate the inverse of a simple rotation matrix about the x-axis. Please review the process for calculation of square matrices in Appendix A. The rotation matrix about the x-axis is:

$$ Rot(x,\theta) = \begin{bmatrix} 1 & 0 & 0 \\ 0 & C\theta & -S\theta \\ 0 & S\theta & C\theta \end{bmatrix} \qquad (2.29) $$

Recall that the following steps must be taken to calculate the inverse of a matrix:

- Calculate the determinant of the matrix.
- Transpose the matrix.
- Replace each element of the transposed matrix by its own minor (adjoint matrix).
- Divide the converted matrix by the determinant.

Applying the process to the rotation matrix, we will get:

$$ \det[Rot(x,\theta)] = 1(C^2\theta + S^2\theta) + 0 = 1 $$
$$ Rot(x,\theta)^T = \begin{bmatrix} 1 & 0 & 0 \\ 0 & C\theta & S\theta \\ 0 & -S\theta & C\theta \end{bmatrix} $$

Now calculate each minor. As an example, the minor for the 2,2 element will be $C\theta - 0 = C\theta$, the minor for 1,1 element will be $C^2\theta + S^2\theta = 1$, and so on. As you will notice, the minor for each element will be the same as the element itself. Therefore:

$$ \mathrm{Adj}[Rot(x,\theta)] = Rot(x,\theta)^T_{minor} = Rot(x,\theta)^T $$

Since the determinant of the original rotation matrix is 1, dividing the Adj[$Rot\,(x,\,\theta)$] matrix by the determinant will yield the same result. Consequently, the inverse of a rotation matrix about the x-axis is the same as its transpose, or:

$$ Rot(x,\theta)^{-1} = Rot(x,\theta)^T \qquad (2.30) $$

Of course, you would get the same result with the second method mentioned in Appendix A. A matrix with this characteristic is called a unitary matrix. It turns out that all rotation matrices are unitary matrices. Therefore, all we need to do to calculate the inverse of a rotation matrix is to transpose it. Please verify that rotation matrices about the

y- and z-axes are also unitary in nature. Beware that only rotation matrices are unitary; if a matrix is not a simple rotation matrix, it may not be unitary.

The preceding result is also true only for a simple 3×3 rotation matrix without representation of a location. For a homogenous 4×4 transformation matrix, it can be shown that the matrix inverse can be written by dividing the matrix into two portions; the rotation portion of the matrix can be simply transposed, as it is still unitary. The position portion of the homogeneous matrix is the negative of the dot product of the **p**-vector with each of the **n**-, **o**-, and **a**-vectors, as follows:

$$T = \begin{bmatrix} n_x & o_x & a_x & p_x \\ n_y & o_y & a_y & p_y \\ n_z & o_z & a_z & p_z \\ 0 & 0 & 0 & 1 \end{bmatrix} \quad \text{and} \quad T^{-1} = \begin{bmatrix} n_x & n_y & n_z & -\mathbf{p} \cdot \mathbf{n} \\ o_x & o_y & o_z & -\mathbf{p} \cdot \mathbf{o} \\ a_x & a_y & a_z & -\mathbf{p} \cdot \mathbf{a} \\ 0 & 0 & 0 & 1 \end{bmatrix} \tag{2.31}$$

As shown, the rotation portion of the matrix is simply transposed, the position portion is replaced by the negative of the dot products, and the last row (scale factors) is not affected. This is very helpful, since we will need to calculate inverses of transformation matrices, but direct calculation of 4×4 matrices is a lengthy process.

Example 2.13

Calculate the matrix representing $Rot(x, 40°)^{-1}$.

Solution: The matrix representing a $40°$ rotation about the x-axis is:

$$Rot(x, 40°) = \begin{bmatrix} 1 & 0 & 0 & 0 \\ 0 & 0.766 & -0.643 & 0 \\ 0 & 0.643 & 0.766 & 0 \\ 0 & 0 & 0 & 1 \end{bmatrix}$$

The inverse of this matrix is:

$$Rot(x, 40°)^{-1} = \begin{bmatrix} 1 & 0 & 0 & 0 \\ 0 & 0.766 & 0.643 & 0 \\ 0 & -0.643 & 0.766 & 0 \\ 0 & 0 & 0 & 1 \end{bmatrix}$$

As you can see, since the position vector of the matrix is zero, its dot product with the **n**-, **o**-, and **a**-vectors is also zero. ∎

Example 2.14

Calculate the inverse of the given transformation matrix:

$$T = \begin{bmatrix} 0.5 & 0 & 0.866 & 3 \\ 0.866 & 0 & -0.5 & 2 \\ 0 & 1 & 0 & 5 \\ 0 & 0 & 0 & 1 \end{bmatrix}$$

Solution: Based on the above, the inverse of the transformation will be:

$$T^{-1} = \begin{bmatrix} 0.5 & 0.866 & 0 & -(3 \times 0.5 + 2 \times 0.866 + 5 \times 0) \\ 0 & 0 & 1 & -(3 \times 0 + 2 \times 0 + 5 \times 1) \\ 0.866 & -0.5 & 0 & -(3 \times 0.866 + 2 \times -0.5 + 5 \times 0) \\ 0 & 0 & 0 & 1 \end{bmatrix}$$

$$= \begin{bmatrix} 0.5 & 0.866 & 0 & -3.23 \\ 0 & 0 & 1 & -5 \\ 0.866 & -0.5 & 0 & -1.598 \\ 0 & 0 & 0 & 1 \end{bmatrix}$$

You may want to verify that TT^{-1} will be an identity matrix. ∎

Example 2.15

In a robotic set-up, a camera is attached to the fifth link of a 6-DOF robot. It observes an object and determines its frame relative to the camera's frame. Using the following information, determine the necessary motion the end effector must make to get to the object:

$$^5T_{cam} = \begin{bmatrix} 0 & 0 & -1 & 3 \\ 0 & -1 & 0 & 0 \\ -1 & 0 & 0 & 5 \\ 0 & 0 & 0 & 1 \end{bmatrix} \qquad ^5T_H = \begin{bmatrix} 0 & -1 & 0 & 0 \\ 1 & 0 & 0 & 0 \\ 0 & 0 & 1 & 4 \\ 0 & 0 & 0 & 1 \end{bmatrix}$$

$$^{cam}T_{obj} = \begin{bmatrix} 0 & 0 & 1 & 2 \\ 1 & 0 & 0 & 2 \\ 0 & 1 & 0 & 4 \\ 0 & 0 & 0 & 1 \end{bmatrix} \qquad ^HT_E = \begin{bmatrix} 1 & 0 & 0 & 0 \\ 0 & 1 & 0 & 0 \\ 0 & 0 & 1 & 3 \\ 0 & 0 & 0 & 1 \end{bmatrix}$$

Solution: Referring to Equation (2.25), we can write a similar equation that relates the different transformations and frames together as:

$$^RT_5 \times {}^5T_H \times {}^HT_E \times {}^ET_{obj} = {}^RT_5 \times {}^5T_{cam} \times {}^{cam}T_{obj}$$

Since $^RT_5$ appears on both sides of the equation, we can simply neglect it. All other matrices, with the exception of $^ET_{obj}$, are known. Then:

$$^ET_{obj} = {}^HT_E^{-1} \times {}^5T_H^{-1} \times {}^5T_{cam} \times {}^{cam}T_{obj} = {}^ET_H \times {}^HT_5 \times {}^5T_{cam} \times {}^{cam}T_{obj}$$

where $^HT_E^{-1} = \begin{bmatrix} 1 & 0 & 0 & 0 \\ 0 & 1 & 0 & 0 \\ 0 & 0 & 1 & -3 \\ 0 & 0 & 0 & 1 \end{bmatrix} \qquad ^5T_H^{-1} = \begin{bmatrix} 0 & 1 & 0 & 0 \\ -1 & 0 & 0 & 0 \\ 0 & 0 & 1 & -4 \\ 0 & 0 & 0 & 1 \end{bmatrix}$

Substituting the matrices and the inverses in the above equation will result:

$$
{}^{E}T_{obj} = \begin{bmatrix} 1 & 0 & 0 & 0 \\ 0 & 1 & 0 & 0 \\ 0 & 0 & 1 & -3 \\ 0 & 0 & 0 & 1 \end{bmatrix} \begin{bmatrix} 0 & 1 & 0 & 0 \\ -1 & 0 & 0 & 0 \\ 0 & 0 & 1 & -4 \\ 0 & 0 & 0 & 1 \end{bmatrix} \begin{bmatrix} 0 & 0 & -1 & 3 \\ 0 & -1 & 0 & 0 \\ -1 & 0 & 0 & 5 \\ 0 & 0 & 0 & 1 \end{bmatrix} \begin{bmatrix} 0 & 0 & 1 & 2 \\ 1 & 0 & 0 & 2 \\ 0 & 1 & 0 & 4 \\ 0 & 0 & 0 & 1 \end{bmatrix}
$$

or

$$
{}^{E}T_{obj} = \begin{bmatrix} -1 & 0 & 0 & -2 \\ 0 & 1 & 0 & 1 \\ 0 & 0 & -1 & -4 \\ 0 & 0 & 0 & 1 \end{bmatrix} \qquad \blacksquare
$$

2.8 Forward and Inverse Kinematics of Robots

Suppose we have a robot whose configuration is known. This means that all the link lengths and joint angles of the robot are known. Calculating the position and orientation of the hand of the robot is called forward kinematic analysis. In other words, if all robot joint variables are known, using forward kinematic equations, we can calculate where the robot is at any instant. However, if we want to place the hand of the robot at a desired location and orientation, we need to know how much each link length or joint angle of the robot must be such that—at those values—the hand will be at the desired position and orientation. This is called inverse kinematic analysis. This means that instead of substituting the known robot variables in the forward kinematic equations of the robot, we need to find the inverse of these equations to enable us to find the necessary joint values to place the robot at the desired location and orientation. In reality, the inverse kinematic equations are more important since the robot controller will calculate the joint values using these equations and it will run the robot to the desired position and orientation. We will first develop the forward kinematic equations of robots; then, using these equations, we will calculate the inverse kinematic equations.

For forward kinematics, we will have to develop a set of equations that relate to the particular configuration of a robot (the way it is put together) such that by substituting the joint and link variables in these equations, we may calculate the position and orientation of the robot. These equations will then be used to derive the inverse kinematic equations.

You may recall from Chapter 1 that in order to position and orientate a rigid body in space, we attach a frame to the body and then describe the position of the origin of the frame and the orientation of its three axes. This requires a total of 6 DOF, or alternately, six pieces of information, to completely define the position and orientation of the body. Here too, if we want to define or find the position and orientation of the hand of the robot in space, we will attach a frame to it and define the position and orientation of the hand frame of the robot. The means by which the robot accomplishes this determines the forward kinematic equations. In other words, depending on the configuration of the links and joints of the robot, a particular set of equations will relate the hand frame of the robot to the reference frame. Figure 2.19 shows a hand frame, the reference frame, and their relative positions and orientations. The undefined connection between the two frames is related to the configuration of the robot. Of

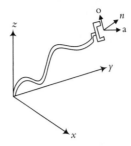

Figure 2.19   The hand frame of the robot relative to the reference frame.

course, there are many different possibilities for this configuration, and we will later see how we can develop the equations relating the two frames, depending on the robot configuration.

In order to simplify the process, we will analyze the position and orientation issues separately. First, we will develop the position equations, then we will do the same for orientation. Later, we will combine the two for a complete set of equations. Finally, we will see about the use of the Denavit-Hartenberg representation, which can model any robot configuration.

2.9 Forward and Inverse Kinematic Equations: Position

In this section, we will study the forward and inverse kinematic equations for position. As was mentioned earlier, the position of the origin of a frame attached to a rigid body has three degrees of freedom, and therefore, can be completely defined by three pieces of information. As a result, the position of the origin of the frame may be defined in any customary coordinates. As an example, we may position a point in space based on Cartesian coordinates, meaning there will be three linear movements relative to the x-, y-, and z-axes. Alternately, it may be accomplished through spherical coordinates, meaning there will be one linear motion and two rotary motions. The following possibilities will be discussed:

(a)   Cartesian (gantry, rectangular) coordinates
(b)   Cylindrical coordinates
(c)   Spherical coordinates
(d)   Articulated (anthropomorphic or all-revolute) coordinates

2.9.1 Cartesian (Gantry, Rectangular) Coordinates

In this case, there will be three linear movements along the x-, y-, and z-axes. In this type of robot, all actuators are linear (such as a hydraulic ram or a linear power screw), and the positioning of the hand of the robot is accomplished by moving the three linear joints along the three axes (Figure 2.20). A gantry robot is basically a Cartesian coordinate robot, except that the robot is usually attached to a rectangular frame upside down.

Of course, since there are no rotations, the transformation matrix representing this motion to point p is a simple translation matrix (shown next). Note that here we are only

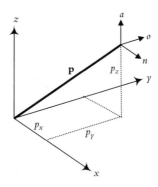

Figure 2.20   Cartesian coordinates.

concerned with the position of the origin of the frame—not its orientation. The transformation matrix representing the forward kinematic equation of the position of the hand of the robot in a Cartesian coordinate system will be:

$$^{R}T_{p} = T_{cart}(p_{x}, p_{y}, p_{z}) = \begin{bmatrix} 1 & 0 & 0 & p_{x} \\ 0 & 1 & 0 & p_{y} \\ 0 & 0 & 1 & p_{z} \\ 0 & 0 & 0 & 1 \end{bmatrix} \tag{2.32}$$

where $^{R}T_{p}$ is the transformation between the reference frame and the origin of the hand p, and $T_{cart}(p_{x}, p_{y}, p_{z})$ denotes Cartesian transformation matrix. For the inverse kinematic solution, simply set the desired position equal to p.

Example 2.16

It is desired to position the origin of the hand frame of a Cartesian robot at point $p = [3,4,7]^{T}$. Calculate the necessary Cartesian coordinate motions that need to be made.

Solution: Setting the forward kinematic equation, represented by the $^{R}T_{p}$ matrix of Equation (2.32), equal to the desired position will yield the following result:

$$^{R}T_{p} = \begin{bmatrix} 1 & 0 & 0 & p_{x} \\ 0 & 1 & 0 & p_{y} \\ 0 & 0 & 1 & p_{z} \\ 0 & 0 & 0 & 1 \end{bmatrix} = \begin{bmatrix} 1 & 0 & 0 & 3 \\ 0 & 1 & 0 & 4 \\ 0 & 0 & 1 & 7 \\ 0 & 0 & 0 & 1 \end{bmatrix} \quad \text{or} \quad p_{x} = 3, \ p_{y} = 4, \ p_{z} = 7$$

∎

2.9.2 Cylindrical Coordinates

A cylindrical coordinate system includes two linear translations and one rotation. The sequence is a translation of r along the x-axis, a rotation of α about the z-axis, and a

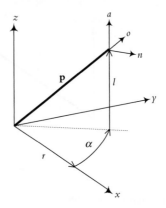

Figure 2.21   Cylindrical coordinates.

translation of l along the z-axis, as shown in Figure 2.21. Since these transformations are all relative to the Universe frame, the total transformation caused by these three transformations is found by pre-multiplying by each matrix, as follows:

$$^R T_p = T_{cyl}(r, \alpha, l) = Trans(0,0,l)Rot(z,\alpha)Trans(r,0,0) \qquad (2.33)$$

$$^R T_p = \begin{bmatrix} 1 & 0 & 0 & 0 \\ 0 & 1 & 0 & 0 \\ 0 & 0 & 1 & l \\ 0 & 0 & 0 & 1 \end{bmatrix} \times \begin{bmatrix} C\alpha & -S\alpha & 0 & 0 \\ S\alpha & C\alpha & 0 & 0 \\ 0 & 0 & 1 & 0 \\ 0 & 0 & 0 & 1 \end{bmatrix} \times \begin{bmatrix} 1 & 0 & 0 & r \\ 0 & 1 & 0 & 0 \\ 0 & 0 & 1 & 0 \\ 0 & 0 & 0 & 1 \end{bmatrix}$$

$$\qquad (2.34)$$

$$^R T_p = T_{cyl}(r, \alpha, l) = \begin{bmatrix} C\alpha & -S\alpha & 0 & rC\alpha \\ S\alpha & C\alpha & 0 & rS\alpha \\ 0 & 0 & 1 & l \\ 0 & 0 & 0 & 1 \end{bmatrix}$$

The first three columns represent the orientation of the frame after this series of transformations. However, at this point, we are only interested in the position of the origin of the frame, or the last column. Obviously, in cylindrical coordinate movements, due to the rotation of α about the z-axis, the orientation of the moving frame will change. This orientation change will be discussed later.

You may restore the original orientation of the frame by rotating the n,o,a frame about the a-axis an angle of $-\alpha$, which is equivalent of post-multiplying the cylindrical coordinate matrix by a rotation matrix of $Rot(a, -\alpha)$. As a result, the frame will be at the same location but will be parallel to the reference frame again, as follows:

$$T_{cyl} \times Rot(a, -\alpha) = \begin{bmatrix} C\alpha & -S\alpha & 0 & rC\alpha \\ S\alpha & C\alpha & 0 & rS\alpha \\ 0 & 0 & 1 & l \\ 0 & 0 & 0 & 1 \end{bmatrix} \times \begin{bmatrix} C(-\alpha) & -S(-\alpha) & 0 & 0 \\ S(-\alpha) & C(-\alpha) & 0 & 0 \\ 0 & 0 & 1 & 0 \\ 0 & 0 & 0 & 1 \end{bmatrix}$$

$$= \begin{bmatrix} 1 & 0 & 0 & rC\alpha \\ 0 & 1 & 0 & rS\alpha \\ 0 & 0 & 1 & l \\ 0 & 0 & 0 & 1 \end{bmatrix}$$

As you can see, the location of the origin of the moving frame has not changed, but it was restored back to being parallel to the reference frame. Notice that the last rotation was performed about the local a-axis in order to not cause any change in the location of the frame, but only in its orientation.

Example 2.17

Suppose we desire to place the origin of the hand frame of a cylindrical robot at $[3,4,7]^T$. Calculate the joint variables of the robot.

Solution: Setting the components of the location of the origin of the frame from the T_{cyl} matrix of Equation (2.34) to the desired values, we get:

$$l = 7$$

$$rC\alpha = 3 \quad \text{and} \quad rS\alpha = 4 \quad \text{and therefore, } \tan\alpha = {}^4/_3 \text{ and } \alpha = 53.1°$$

Substituting α into either equation will yield $r = 5$. The final answer is $r = 5$ units, $\alpha = 53.1°$, and $l = 7$ units. Note: As discussed in Appendix A, it is necessary to ensure that the angles calculated in robot kinematics are in correct quadrants. In this example, $rC\alpha$ and $rS\alpha$ are both positive and the length r is always positive, therefore $S\alpha$ and $C\alpha$ are also both positive. Consequently, the angle α is in quadrant 1 and is correctly 53.1°. ∎

Example 2.18

The position and restored orientation of a cylindrical robot are given. Find the matrix representing the original position and orientation of the robot before it was restored.

$$T = \begin{bmatrix} 1 & 0 & 0 & -2.394 \\ 0 & 1 & 0 & 6.578 \\ 0 & 0 & 1 & 9 \\ 0 & 0 & 0 & 1 \end{bmatrix}$$

Solution: Since r is always positive, it is clear that $S\alpha$ and $C\alpha$ are positive and negative, respectively. Therefore, α is in the second quadrant. From T, we get:

$$l = 9$$

$$\tan(\alpha) = \frac{6.578}{-2.394} = -2.748 \quad \rightarrow \quad \alpha = 180° - 70° = 110°$$

$$r\sin(\alpha) = 6.578 \rightarrow r = 7$$

Substituting these values into Equation (2.34), we will get the original orientation of the robot:

$$
^R T_p =
\begin{bmatrix}
C\alpha & -S\alpha & 0 & rC\alpha \\
S\alpha & C\alpha & 0 & rS\alpha \\
0 & 0 & 1 & l \\
0 & 0 & 0 & 1
\end{bmatrix}
=
\begin{bmatrix}
-0.342 & -0.9397 & 0 & -2.394 \\
0.9397 & -0.342 & 0 & 6.578 \\
0 & 0 & 1 & 9 \\
0 & 0 & 0 & 1
\end{bmatrix}
\blacksquare
$$

2.9.3 Spherical Coordinates

A spherical coordinate system consists of one linear motion and two rotations. The sequence is a translation of r along the z-axis, a rotation of β about the y-axis, and a rotation of γ about the z-axis as shown in Figure 2.22. Since these transformations are all relative to the Universe frame, the total transformation caused by these three transformations can be found by pre-multiplying by each matrix, as follows:

$$^R T_P = T_{sph}(r, \beta, \gamma) = Rot(z, \gamma)Rot(y, \beta)Trans(0, 0, r) \tag{2.35}$$

$$
^R T_p =
\begin{bmatrix}
C\gamma & -S\gamma & 0 & 0 \\
S\gamma & C\gamma & 0 & 0 \\
0 & 0 & 1 & 0 \\
0 & 0 & 0 & 1
\end{bmatrix}
\times
\begin{bmatrix}
C\beta & 0 & S\beta & 0 \\
0 & 1 & 0 & 0 \\
-S\beta & 0 & C\beta & 0 \\
0 & 0 & 0 & 1
\end{bmatrix}
\times
\begin{bmatrix}
1 & 0 & 0 & 0 \\
0 & 1 & 0 & 0 \\
0 & 0 & 1 & r \\
0 & 0 & 0 & 1
\end{bmatrix}
$$

$$\tag{2.36}$$

$$
^R T_P = T_{sph}(r, \beta, \gamma) =
\begin{bmatrix}
C\beta C\gamma & -S\gamma & S\beta C\gamma & rS\beta C\gamma \\
C\beta S\gamma & C\gamma & S\beta S\gamma & rS\beta S\gamma \\
-S\beta & 0 & C\beta & rC\beta \\
0 & 0 & 0 & 1
\end{bmatrix}
$$

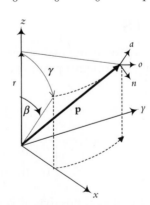

Figure 2.22   Spherical coordinates.

The first three columns represent the orientation of the frame after this series of transformations, while the last column is the position of the origin. We will discuss the orientation part of the matrix later. Note that spherical coordinates may be defined in other orders as well. Therefore, make sure correct equations are used.

Here, too, you may restore the original orientation of the final frame and make it parallel to the reference frame. This exercise is left for you to find the correct sequence of movements to get the right answer.

The inverse kinematic equations for spherical coordinates are more complicated than the simple Cartesian or cylindrical coordinates because the two angles β and γ are coupled. Let's see how this could be done through an example.

Example 2.19

Suppose we now want to place the origin of the hand of a spherical robot at $[3,4,7]^T$. Calculate the joint variables of the robot.

Solution: Setting the components of the location of the origin of the frame from T_{sph} matrix of Equation (2.36) to the desired values, we get:

$$rS\beta C\gamma = 3$$
$$rS\beta S\gamma = 4$$
$$rC\beta = 7$$

From the third equation, we determine that the $C\beta$ is positive, but there is no such information about $S\beta$. Therefore, because we do not know the actual sign of $S\beta$, there are two possible solutions. Later, we will have to check the final results to ensure they are correct.

$\tan\gamma = {}^4/_3$	$\rightarrow$ $\gamma = 53.1°$	or	$233.1°$
then	$S\gamma = 0.8$	or	-0.8
and	$C\gamma = 0.6$	or	-0.6
and	$rS\beta = {}^3/_{0.6} = 5$	or	-5
and since	$rC\beta = 7,$ $\rightarrow$ $\beta = 35.5°$	or	$-35.5°$
and	$r = 8.6$		

You may check both answers and verify that they both satisfy all position equations. If you also follow these angles about the given axes in 3-D, you will get to the same point physically. However, you must notice that only one set of answers will also satisfy the orientation equations. In other words, the two answers above will result in the same position, but at different orientations. Since we are not concerned with the orientation of the hand frame at this point, both position answers are correct. In fact, since we cannot specify any orientation for a 3-DOF robot anyway, we cannot determine which of the two answers relates to a desired orientation. ■

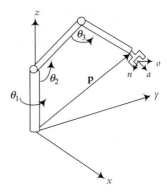

Figure 2.23   Articulated coordinates.

2.9.4 Articulated Coordinates

Articulated coordinates consist of three rotations, as shown in Figure 2.23. We will develop the matrix representation for this later, when we discuss the Denavit-Hartenberg representation.

2.10 Forward and Inverse Kinematic Equations: Orientation

Suppose the moving frame attached to the hand of the robot has already moved to a desired position—in Cartesian, cylindrical, spherical, or articulated coordinates—and is either parallel to the reference frame or is in an orientation other than what is desired. The next step will be to rotate the frame appropriately in order to achieve a desired orientation without changing its position. This can only be accomplished by rotating about the current frame axes; rotations about the reference frame axes will change the position. The appropriate sequence of rotations depends on the design of the wrist of the robot and the way the joints are assembled together. We will consider the following three common configurations:

(a)   Roll, Pitch, Yaw (RPY) angles
(b)   Euler angles
(c)   Articulated joints

2.10.1 Roll, Pitch, Yaw (RPY) Angles

This is a sequence of three rotations about current a-, o-, and n-axes respectively, which will orientate the hand of the robot to a desired orientation. The assumption here is that the current frame is parallel to the reference frame; therefore, its orientation is the same as the reference frame before the application of RPY. If the current moving frame is not parallel to the reference frame, then the final orientation of the robot's hand will be a combination of the previous orientation, post-multiplied by the RPY.

It is very important to realize that since we do not want to cause any change in the position of the origin of the moving frame (we have already placed it at the desired

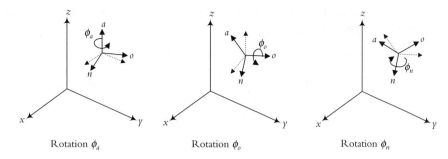

Rotation ϕ_a Rotation ϕ_o Rotation ϕ_n

Figure 2.24 RPY rotations about the current axes.

location and only want to rotate it to the desired orientation), the movements relating to RPY rotations are relative to the current moving axes. Otherwise, as we saw before, the position of the frame will change. Therefore, all matrices related to the orientation change due to RPY (as well as other rotations) will be post-multiplied. Referring to Figure 2.24, the RPY sequence of rotations consists of:

Rotation of ϕ_a about the a-axis (z-axis of the moving frame) called Roll,
Rotation of ϕ_o about the o-axis (y-axis of the moving frame) called Pitch,
Rotation of ϕ_n about the n-axis (x-axis of the moving frame) called Yaw.

The matrix representing the RPY orientation change will be:

$$\text{RPY}(\phi_a, \phi_o, \phi_n) = Rot(a, \phi_a)Rot(o, \phi_o)Rot(n, \phi_n)$$

$$= \begin{bmatrix} C\phi_a C\phi_o & C\phi_a S\phi_o S\phi_n - S\phi_a C\phi_n & C\phi_a S\phi_o C\phi_n + S\phi_a S\phi_n & 0 \\ S\phi_a C\phi_o & S\phi_a S\phi_o S\phi_n + C\phi_a C\phi_n & S\phi_a S\phi_o C\phi_n - C\phi_a S\phi_n & 0 \\ -S\phi_o & C\phi_o S\phi_n & C\phi_o C\phi_n & 0 \\ 0 & 0 & 0 & 1 \end{bmatrix} \quad (2.37)$$

This matrix represents the orientation change caused by the RPY alone. The location and the final orientation of the frame relative to the reference frame will be the product of the two matrices representing the position change and the RPY. For example, suppose that a robot is designed based on spherical coordinates and RPY. Then the robot may be represented by:

$$^R T_H = T_{sph}(r, \beta, \gamma) \times \text{RPY}(\phi_a, \phi_o, \phi_n)$$

The inverse kinematic solution for the RPY is more complicated than the spherical coordinates because here there are three coupled angles, where we need to have information about the sines and the cosines of all three angles individually to solve for the angles. To solve for these sines and cosines, we will have to de-couple these angles. To do this, we will pre-multiply both sides of Equation (2.37) by the inverse of $Rot(a, \phi_a)$:

$$Rot(a, \phi_a)^{-1} \text{RPY}(\phi_a, \phi_o, \phi_n) = Rot(o, \phi_o)Rot(n, \phi_n) \quad (2.38)$$

Assuming that the final desired orientation achieved by RPY is represented by the (n,o,a) matrix, we will have:

$$Rot(a, \phi_a)^{-1} \begin{bmatrix} n_x & o_x & a_x & 0 \\ n_y & o_y & a_y & 0 \\ n_z & o_z & a_z & 0 \\ 0 & 0 & 0 & 1 \end{bmatrix} = Rot(o, \phi_o)Rot(n, \phi_n) \qquad (2.39)$$

Multiplying the matrices, we will get:

$$\begin{bmatrix} n_x C\phi_a + n_y S\phi_a & o_x C\phi_a + o_y S\phi_a & a_x C\phi_a + a_y S\phi_a & 0 \\ n_y C\phi_a - n_x S\phi_a & o_y C\phi_a - o_x S\phi_a & a_y C\phi_a - a_x S\phi_a & 0 \\ n_z & o_z & a_z & 0 \\ 0 & 0 & 0 & 1 \end{bmatrix}$$

$$= \begin{bmatrix} C\phi_o & S\phi_o S\phi_n & S\phi_o C\phi_n & 0 \\ 0 & C\phi_n & -S\phi_n & 0 \\ -S\phi_o & C\phi_o S\phi_n & C\phi_o C\phi_n & 0 \\ 0 & 0 & 0 & 1 \end{bmatrix}$$

$$(2.40)$$

Remember that the n,o,a components in Equation (2.39) represent the final desired values normally given or known. The values of the RPY angles are the unknown variables. Equating the different elements of the right-hand and left-hand sides of Equation (2.40) will result in the following. Refer to Appendix A for an explanation of $ATAN2$ function.

From the 2,1 elements we get:

$$n_y C\phi_a - n_x S\phi_a = 0 \rightarrow \phi_a = ATAN2(n_y, n_x) \text{ and } \phi_a = ATAN2(-n_y, -n_x) \quad (2.41)$$

Note that since we do not know the signs of $\sin(\phi_a)$ or $\cos(\phi_a)$, two complementary solutions are possible. From the 3,1 and 1,1 elements we get:

$$S\phi_o = -n_z$$
$$C\phi_o = n_x C\phi_a + n_y S\phi_a \rightarrow \phi_o = ATAN2\left[-n_z, (n_x C\phi_a + n_y S\phi_a)\right]$$

$$(2.42)$$

And finally, from the 2,2 and 2,3 elements we get:

$$C\phi_n = o_y C\phi_a - o_x S\phi_a$$
$$S\phi_n = -a_y C\phi_a + a_x S\phi_a \rightarrow \phi_n = ATAN2\left[(-a_y C\phi_a + a_x S\phi_a), (o_y C\phi_a - o_x S\phi_a)\right]$$

$$(2.43)$$

Example 2.20

The desired final position and orientation of the hand of a Cartesian-RPY robot is given below. Find the necessary RPY angles and displacements.

$$^{R}T_{p} = \begin{bmatrix} n_x & o_x & a_x & p_x \\ n_y & o_y & a_y & p_y \\ n_z & o_z & a_z & p_z \\ 0 & 0 & 0 & 1 \end{bmatrix} = \begin{bmatrix} 0.354 & -0.674 & 0.649 & 4.33 \\ 0.505 & 0.722 & 0.475 & 2.50 \\ -0.788 & 0.160 & 0.595 & 8 \\ 0 & 0 & 0 & 1 \end{bmatrix}$$

Solution: From the above equations, we find two sets of answers:

$$\phi_a = ATAN2(n_y, n_x) = ATAN2(0.505, 0.354) = 55° \text{ or } 235°$$
$$\phi_o = ATAN2(-n_z, (n_x C\phi_a + n_y S\phi_a)) = ATAN2(0.788, 0.616) = 52° \text{ or } 128°$$
$$\phi_n = ATAN2((-a_y C\phi_a + a_x S\phi_a), (o_y C\phi_a - o_x S\phi_a))$$
$$= ATAN2(0.259, 0.966) = 15° \text{ or } 195°$$
$$p_x = 4.33 \quad p_y = 2.5 \quad p_z = 8 \text{ units.}$$

Example 2.21

For the same position and orientation as in Example 2.20, find all necessary joint variables if the robot is cylindrical-RPY.

Solution: In this case, we will use:

$$^{R}T_{p} = \begin{bmatrix} 0.354 & -0.674 & 0.649 & 4.33 \\ 0.505 & 0.722 & 0.475 & 2.50 \\ -0.788 & 0.160 & 0.595 & 8 \\ 0 & 0 & 0 & 1 \end{bmatrix} = T_{cyl}(r, \alpha, l) \times \text{RPY}(\phi_a, \phi_o, \phi_n)$$

The right-hand side of this equation now involves four coupled angles; as before, these must be de-coupled. However, since the rotation of α about the z-axis for the cylindrical coordinates does not affect the a-axis, it remains parallel to the z-axis. As a result, the rotation of ϕ_a about the a-axis for RPY will simply be added to α. This means that the 55° angle we found for ϕ_a is the summation of $\phi_a + \alpha$ (see Figure 2.25). Using the position information given, the solution of Example 2.20, and referring to Equation (2.34), we get:

$$rC\alpha = 4.33, \quad rS\alpha = 2.5 \rightarrow \alpha = 30°$$
$$\phi_a + \alpha = 55° \qquad\qquad \rightarrow \phi_a = 25°$$
$$S\alpha = 0.5 \qquad\qquad\quad \rightarrow r = 5$$
$$p_z = 8 \qquad\qquad\qquad \rightarrow l = 8$$

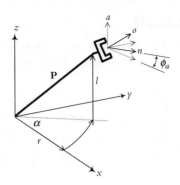

Figure 2.25 Cylindrical and RPY coordinates of Example 2.21.

As in Example 2.16:
$$\rightarrow \phi_o = 52°, \quad \phi_n = 15°$$

Of course, a similar solution may be found for the second set of answers. ∎

2.10.2 Euler Angles

Euler angles are very similar to RPY, except that the last rotation is also about the current a-axis (Figure 2.26). We still need to make all rotations relative to the current axes to prevent any change in the position of the robot. Therefore, the rotations representing the Euler angles will be:

Rotation of ϕ about the a-axis (z-axis of the moving frame) followed by,
Rotation of θ about the o-axis (y-axis of the moving frame) followed by,
Rotation of ψ about the a-axis (z-axis of the moving frame).

The matrix representing the Euler angles orientation change will be:

$$Euler(\phi, \theta, \psi) = Rot(a, \phi)Rot(o, \theta), Rot(a, \psi)$$

$$= \begin{bmatrix} C\phi C\theta C\psi - S\phi S\psi & -C\phi C\theta S\psi - S\phi C\psi & C\phi S\theta & 0 \\ S\phi C\theta C\psi + C\phi S\psi & -S\phi C\theta S\psi + C\phi C\psi & S\phi S\theta & 0 \\ -S\theta C\psi & S\theta S\psi & C\theta & 0 \\ 0 & 0 & 0 & 1 \end{bmatrix} \quad (2.44)$$

Rotation of ϕ about the a-axis	Rotation of θ about the o-axis	Rotation of ψ about the a-axis

Figure 2.26 Euler rotations about the current axes.

Once again, this matrix represents the orientation change caused by the Euler angles alone. The location and the final orientation of the frame relative to the reference frame will be the product of the two matrices representing the position change and the Euler angles.

The inverse kinematic solution for the Euler angles can be found in a manner very similar to RPY. We will pre-multiply the two sides of the Euler equation by $Rot^{-1}(a, \phi)$ to eliminate ϕ from one side. By equating the elements of the two sides to each other, we will find the following equations. Assuming the final desired orientation achieved by Euler angles is represented by the (n,o,a) matrix:

$$Rot^{-1}(a, \phi) \times \begin{bmatrix} n_x & o_x & a_x & 0 \\ n_y & o_y & a_y & 0 \\ n_z & o_z & a_z & 0 \\ 0 & 0 & 0 & 1 \end{bmatrix} = \begin{bmatrix} C\theta C\psi & -C\theta S\psi & S\theta & 0 \\ S\psi & C\psi & 0 & 0 \\ -S\theta C\psi & S\theta S\psi & C\theta & 0 \\ 0 & 0 & 0 & 1 \end{bmatrix} \quad (2.45)$$

or,

$$\begin{bmatrix} n_x C\phi + n_y S\phi & o_x C\phi + o_y S\phi & a_x C\phi + a_y S\phi & 0 \\ -n_x S\phi + n_y C\phi & -o_x S\phi + o_y C\phi & -a_x S\phi + a_y C\phi & 0 \\ n_z & o_z & a_z & 0 \\ 0 & 0 & 0 & 1 \end{bmatrix}$$

$$= \begin{bmatrix} C\theta C\psi & -C\theta S\psi & S\theta & 0 \\ S\psi & C\psi & 0 & 0 \\ -S\theta C\psi & S\theta S\psi & C\theta & 0 \\ 0 & 0 & 0 & 1 \end{bmatrix} \quad (2.46)$$

Remember that the n,o,a components in Equation (2.45) represent the final desired values that are normally given or known. The values of the Euler angles are the unknown variables. Equating the different elements of the right-hand and left-hand sides of Equation (2.46) will result in the following.

From the 2,3 elements we get:

$$- a_x S\phi + a_y C\phi = 0 \rightarrow \phi = ATAN2(a_y, a_x) \text{ or } \phi = ATAN2(-a_y, -a_x) \quad (2.47)$$

With ϕ evaluated, all the elements of the left-hand side of Equation (2.46) are known. From the 2,1 and 2,2 elements we get:

$$S\psi = -n_x S\phi + n_y C\phi$$
$$C\psi = -o_x S\phi + o_y C\phi \rightarrow \psi = ATAN2[(-n_x S\phi + n_y C\phi), (-o_x S\phi + o_y C\phi)] \quad (2.48)$$

And finally, from the 1,3 and 3,3 elements we get:

$$S\theta = a_x C\phi + a_y S\phi$$
$$C\theta = a_z \rightarrow \theta = ATAN2\left[\left(a_x C\phi + a_y S\phi\right), a_z\right)\right] \tag{2.49}$$

Example 2.22

The desired final orientation of the hand of a Cartesian-Euler robot is given. Find the necessary Euler angles.

$$^{R}T_{H} = \begin{bmatrix} n_x & o_x & a_x & p_x \\ n_y & o_y & a_y & p_y \\ n_z & o_z & a_z & p_z \\ 0 & 0 & 0 & 1 \end{bmatrix} = \begin{bmatrix} 0.579 & -0.548 & -0.604 & 5 \\ 0.540 & 0.813 & -0.220 & 7 \\ 0.611 & -0.199 & 0.766 & 3 \\ 0 & 0 & 0 & 1 \end{bmatrix}$$

Solution: From the above equations, we find:

$$\phi = ATAN2\left(a_y, a_x\right) = ATAN2(-0.220, -0.604) = 20° \text{ or } 200°$$

Realizing that both the sines and cosines of $20°$ and $200°$ can be used for the remainder,

$$\psi = ATAN2\left(-n_x S\phi + n_y C\phi, -o_x S\phi + o_y C\phi\right) = (0.31, 0.952) = 18° \text{ or } 198°$$
$$\theta = ATAN2\left(a_x C\phi + a_y S\phi, a_z\right) = ATAN2(-0.643, 0.766) = -40° \text{ or } 40° \quad\blacksquare$$

2.10.3 Articulated Joints

Articulated joints consist of three rotations other than the above. Similar to section 2.9.4., we will develop the matrix representing articulated joints later, when we discuss the Denavit-Hartenberg representation.

2.11 Forward and Inverse Kinematic Equations: Position and Orientation

The matrix representing the final location and orientation of the robot is a combination of the above, depending on which coordinates are used. If a robot is made of a Cartesian and an RPY set of joints, then the location and the final orientation of the frame relative to the reference frame will be the product of the two matrices representing the Cartesian position change and the RPY. The robot may be represented by:

$$^{R}T_{H} = T_{cart}(p_x, p_y, p_z) \times RPY(\phi_a, \phi_o, \phi_n) \tag{2.50}$$

If the robot is designed based on spherical coordinates for positioning and Euler angles for orientation, then the final answer will be the following equation, where the position is determined by the spherical coordinates, while the final orientation is affected by both the angles in the spherical coordinates as well as the Euler angles:

$$^{R}T_{H} = T_{sph}(r, \beta, \gamma) \times \text{Euler}(\phi, \theta, \psi) \qquad (2.51)$$

The forward and inverse kinematic solutions for these cases are not developed here, since many different combinations are possible. Instead, in complicated designs, the Denavit-Hartenberg representation is recommended. We will discuss this next.

2.12 Denavit-Hartenberg Representation of Forward Kinematic Equations of Robots

In 1955, Denavit and Hartenberg[4] published a paper in the *ASME Journal of Applied Mechanics* that was later used to represent and model robots and to derive their equations of motion. This technique has become the standard way of representing robots and modeling their motions, and therefore, is essential to learn. The Denavit-Hartenberg (D–H) model of representation is a very simple way of modeling robot links and joints that can be used for any robot configuration, regardless of its sequence or complexity. It can also be used to represent transformations in any coordinates we have already discussed, such as Cartesian, cylindrical, spherical, Euler, and RPY. Additionally, it can be used for representation of all-revolute articulated robots, SCARA robots, or any possible combinations of joints and links. Although the direct modeling of robots with the previous techniques are faster and more straightforward, the D-H representation has an added benefit; as we will see later, analysis of differential motions and Jacobians, dynamic analysis, force analysis, and others are based on the results obtained from D-H representation.[5–9]

Robots may be made of a succession of joints and links in any order. The joints may be either prismatic (linear) or revolute (rotational), move in different planes, and have offsets. The links may also be of any length, including zero; may be twisted and bent; and may be in any plane. Therefore, any general set of joints and links may create a robot. We need to be able to model and analyze any robot, whether or not it follows any of the preceding coordinates.

To do this, we assign a reference frame to each joint, and later define a general procedure to transform from one joint to the next (one frame to the next). If we combine all the transformations from the base to the first joint, from the first joint to the second joint, and so on, until we get to the last joint, we will have the robot's total transformation matrix. In the following sections, we will define the general procedure, based on the D-H representation, to assign reference frames to each joint. Then we will define how a transformation between any two successive frames may be accomplished. Finally, we will write the total transformation matrix for the robot.

Imagine that a robot may be made of a number of links and joints in any form. Figure 2.27 represents three successive joints and two links. Although these joints and links are not necessarily similar to any real robot joint or link, they are very general and can easily represent any joints in real robots. These joints may be revolute or prismatic, or both. Although in real robots it is customary to only have 1-DOF joints, the joints in Figure 2.27 represent 1- or 2-DOF joints.

Figure 2.27(a) shows three joints. Each joint may both rotate and/or translate. Let's assign joint number n to the first joint, $n + 1$ to the second joint, and $n + 2$ to the third

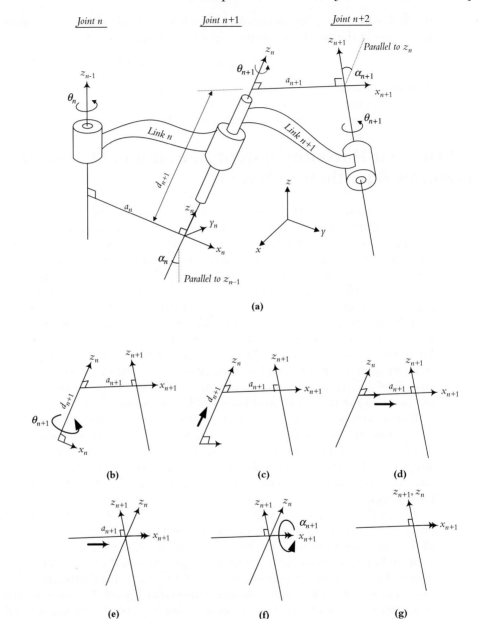

Figure 2.27 A Denavit-Hartenberg representation of a general purpose joint-link combination.

joint shown. There may be other joints before or after these. Each link is also assigned a link number as shown. Link n will be between joints n and $n+1$, and link $n+1$ is between joints $n+1$ and $n+2$.

To model the robot with the D-H representation, the first thing we need to do is assign a local reference frame for each and every joint. Therefore, for each joint, we will have to assign a z-axis and an x-axis. We normally do not need to assign a y-axis, since we always know that y-axes are mutually perpendicular to both x- and z-axes. In addition, the D-H representation does not use the y-axis at all. The following is the procedure for assigning a local reference frame to each joint.

- All joints, without exception, are represented by a z-axis. If the joint is revolute, the z-axis is in the direction of rotation as followed by the right-hand rule for rotations. If the joint is prismatic, the z-axis for the joint is along the direction of the linear movement. In each case, the index number for the z-axis of joint n (as well as the local reference frame for the joint) is $n - 1$. For example, the z-axis representing motions about joint number $n + 1$ is z_n. These simple rules will allow us to quickly assign z-axes to all joints. For revolute joints, the rotation about the z-axis (θ) will be the joint variable. For prismatic joints, the length of the link along the z-axis represented by d will be the joint variable.
- As you can see in Figure 2.27(a), in general, joints may not necessarily be parallel or intersecting. As a result, the z-axes may be skew lines. There is always one line mutually perpendicular to any two skew lines, called common normal, which is the shortest distance between them. We always assign the x-axis of the local reference frame in the direction of the common normal. Therefore, if a_n represents the common normal between z_{n-1} and z_n, the direction of x_n will be along a_n. Similarly, if the common normal between z_n and z_{n+1} is a_{n+1}, the direction of x_{n+1} will be along a_{n+1}. The common normal lines between successive joints are not necessarily intersecting or colinear. As a result, the origins of two successive frames may also not be at the same location. Based on the above, we can assign coordinate frames to all joints, with the following exceptions:

  - If two z-axes are parallel, there are an infinite number of common normals between them. We will pick the common normal that is colinear with the common normal of the previous joint. This will simplify the model.
  - If the z-axes of two successive joints are intersecting, there is no common normal between them (or it has a zero length). We will assign the x-axis along a line perpendicular to the plane formed by the two axes. This means that the common normal is a line perpendicular to the plane containing the two z-axes, which is the equivalent of picking the direction of the cross-product of the two z-axes. This also simplifies the model.

In Figure 2.27(a), θ represents a rotation about the z-axis, d represents the distance on the z-axis between two successive common normals (or joint offset), a represents the length of each common normal (the length of a link), and α represents the angle between two successive z-axes (also called joint twist angle). Commonly, only θ and d are joint variables.

The next step is to follow the necessary motions to transform from one reference frame to the next. Assuming we are at the local reference frame $x_n - z_n$, we will do the following four standard motions to get to the next local reference frame $x_{n+1} - z_{n+1}$:

1. Rotate about the z_n-axis an angle of θ_{n+1} (Figure 2.27(a) and (b)). This will make x_n and x_{n+1} parallel to each other. This is true because a_n and a_{n+1} are both perpendicular to z_n, and rotating z_n an angle of θ_{n+1} will make them parallel (and thus, coplanar).

2. Translate along the z_n-axis a distance of d_{n+1} to make x_n and x_{n+1} colinear (Figure 2.27(c)). Since x_n and x_{n+1} were already parallel and normal to z_n, moving along z_n will lay them over each other.
3. Translate along the (already rotated) x_n-axis a distance of a_{n+1} to bring the origins of x_n and x_{n+1} together (Figure 2.27(d) and (e)). At this point, the origins of the two reference frames will be at the same location.
4. Rotate z_n-axis about x_{n+1}-axis an angle of α_{n+1} to align z_n-axis with z_{n+1}-axis (Figure 2.27(f)). At this point, frames n and $n+1$ will be exactly the same (Figure 2.27(g)), and we have transformed from one to the next.

Continuing with exactly the same sequence of four movements between the $n+1$ and $n+2$ frames will transform one to the next, and by repeating this as necessary, we can transform between successive frames. Starting with the robot's reference frame, we can transform to the first joint, second joint and so on, until the end effector. Note that the above sequence of movements remains the same between any two frames.

The transformation $^n T_{n+1}$ (called A_{n+1}) between two successive frames representing the preceding four movements is the product of the four matrices representing them. Since all transformations are relative to the current frame (they are measured and performed relative to the axes of the current local frame), all matrices are post-multiplied. The result is:

$$^n T_{n+1} = A_{n+1} = Rot(z, \theta_{n+1}) \times Trans(0, 0, d_{n+1}) \times Trans(a_{n+1}, 0, 0) \times Rot(x, \alpha_{n+1})$$

$$= \begin{bmatrix} C\theta_{n+1} & -S\theta_{n+1} & 0 & 0 \\ S\theta_{n+1} & C\theta_{n+1} & 0 & 0 \\ 0 & 0 & 1 & 0 \\ 0 & 0 & 0 & 1 \end{bmatrix} \times \begin{bmatrix} 1 & 0 & 0 & 0 \\ 0 & 1 & 0 & 0 \\ 0 & 0 & 1 & d_{n+1} \\ 0 & 0 & 0 & 1 \end{bmatrix} \times \begin{bmatrix} 1 & 0 & 0 & a_{n+1} \\ 0 & 1 & 0 & 0 \\ 0 & 0 & 1 & 0 \\ 0 & 0 & 0 & 1 \end{bmatrix}$$

$$\times \begin{bmatrix} 1 & 0 & 0 & 0 \\ 0 & C\alpha_{n+1} & -S\alpha_{n+1} & 0 \\ 0 & S\alpha_{n+1} & C\alpha_{n+1} & 0 \\ 0 & 0 & 0 & 1 \end{bmatrix} \tag{2.52}$$

$$A_{n+1} = \begin{bmatrix} C\theta_{n+1} & -S\theta_{n+1} C\alpha_{n+1} & S\theta_{n+1} S\alpha_{n+1} & a_{n+1} C\theta_{n+1} \\ S\theta_{n+1} & C\theta_{n+1} C\alpha_{n+1} & -C\theta_{n+1} S\alpha_{n+1} & a_{n+1} S\theta_{n+1} \\ 0 & S\alpha_{n+1} & C\alpha_{n+1} & d_{n+1} \\ 0 & 0 & 0 & 1 \end{bmatrix} \tag{2.53}$$

Table 2.1   *D-H Parameters Table.*

#	θ	d	a	α
0–1				
1–2				
2–3				
3–4				
4–5				
5–6				

As an example, the transformation between joints 2 and 3 of a generic robot will simply be:

$$
{}^2T_3 = A_3 =
\begin{bmatrix}
C\theta_3 & -S\theta_3 C\alpha_3 & S\theta_3 S\alpha_3 & a_3 C\theta_3 \\
S\theta_3 & C\theta_3 C\alpha_3 & -C\theta_3 S\alpha_3 & a_3 S\theta_3 \\
0 & S\alpha_3 & C\alpha_3 & d_3 \\
0 & 0 & 0 & 1
\end{bmatrix}
\tag{2.54}
$$

At the base of the robot, we can start with the first joint and transform to the second joint, then to the third, and so on, until the hand of the robot and eventually the end effector. Calling each transformation an A_{n+1}, we will have a number of A matrices that represent the transformations. The total transformation between the base of the robot and the hand will be:

$$
{}^R T_H = {}^R T_1 \, {}^1 T_2 \, {}^2 T_3 \ldots {}^{n-1} T_n = A_1 A_2 A_3 \ldots A_n
\tag{2.55}
$$

where n is the joint number. For a 6-DOF robot, there will be six A matrices.

To facilitate the calculation of the A matrices, we will form a table of joint and link parameters, whereby the values representing each link and joint are determined from the schematic drawing of the robot and are substituted into each A matrix. Table 2.1 can be used for this purpose.

In the following examples, we will assign the necessary frames, fill out the parameters tables, and substitute the values into the A matrices. We will start with a simple robot, but will consider more difficult robots later.

Starting with a simple 2-axis robot and moving up to a robot with 6 axes, we will apply the D-H representation in the following examples to derive the forward kinematic equations for each one.

Example 2.23

For the simple 2-axis, planar robot of Figure 2.28, assign the necessary coordinate systems based on the D-H representation, fill out the parameters table, and derive the forward kinematic equations for the robot.

Solution: First, note that both joints rotate in the x-y plane and that a frame $x_H - z_H$ shows the end of the robot. We start by assigning the z-axes for the joints. z_0 will be assigned to joint 1, and z_1 will be assigned to joint 2. Figure 2.28 shows both

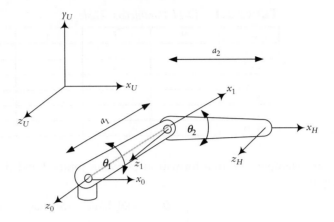

Figure 2.28   A simple 2-axis, articulated robot arm.

z-axes pointing out from the page (as are the z_U- and z_H-axes). Notice that the 0-frame is fixed and does not move. The robot moves relative to it.

Next, we need to assign the x-axes for each frame. Since the first frame (frame 0) is at the base of the robot, and therefore, there are no joints before it, the direction of x_0 is arbitrary. For convenience (only), we may choose to assign it in the same direction as the Universe x-axis. As we will see later, there is no problem if another direction is chosen; all it means is that if we were to specify $^U T_H$ instead of $^0 T_H$, we would have to include an additional fixed rotation to indicate that x_U- and x_0-axes are not parallel.

Since z_0 and z_1 are parallel, the common normal between them is in the direction between the two, and therefore, the x_1-axis is as shown.

Table 2.2 shows the parameters table for the robot. To identify the values, follow the four necessary transformations required to go from one frame to the next, according to the D-H convention:

1. Rotate about the z_0-axis an angle of θ_1 to make x_0 parallel to x_1.
2. Since x_0 and x_1 are in the same plane, translation d along the z_0-axis is zero.
3. Translate along the (already rotated) x_0-axis a distance of a_1.
4. Since z_0 and z_1-axes are parallel, the necessary rotation α about the x_1-axis is zero.

The same can be repeated for transforming between frames 1 and H.

Note that since there are two revolute joints, the two unknowns are also joint angles θ_1 and θ_2. The forward kinematic equation of the robot can be found by

Table 2.2   *D-H Parameters Table for Example 2.23.*

#	θ	d	a	α
0–1	θ_1	0	a_1	0
1–H	θ_2	0	a_2	0

substituting these parameters into the corresponding A matrices as follows:

$$A_1 = \begin{bmatrix} C_1 & -S_1 & 0 & a_1 C_1 \\ S_1 & C_1 & 0 & a_1 S_1 \\ 0 & 0 & 1 & 0 \\ 0 & 0 & 0 & 1 \end{bmatrix} \quad \text{and} \quad A_2 = \begin{bmatrix} C_2 & -S_2 & 0 & a_2 C_2 \\ S_2 & C_2 & 0 & a_2 S_2 \\ 0 & 0 & 1 & 0 \\ 0 & 0 & 0 & 1 \end{bmatrix}$$

$$^0T_H = A_1 \times A_2 = \begin{bmatrix} C_1 C_2 - S_1 S_2 & -C_1 S_2 - S_1 C_2 & 0 & a_2(C_1 C_2 - S_1 S_2) + a_1 C_1 \\ S_1 C_2 + C_1 S_2 & -S_1 S_2 + C_1 C_2 & 0 & a_2(S_1 C_2 + C_1 S_2) + a_1 S_1 \\ 0 & 0 & 1 & 0 \\ 0 & 0 & 0 & 1 \end{bmatrix}$$

Using functions $C_1 C_2 - S_1 S_2 = C(\theta_1 + \theta_2) = C_{12}$ and $S_1 C_2 + C_1 S_2 = S(\theta_1 + \theta_2)$ $= S_{12}$, the transformation simplifies to:

$$^0T_H = \begin{bmatrix} C_{12} & -S_{12} & 0 & a_2 C_{12} + a_1 C_1 \\ S_{12} & C_{12} & 0 & a_2 S_{12} + a_1 S_1 \\ 0 & 0 & 1 & 0 \\ 0 & 0 & 0 & 1 \end{bmatrix} \tag{2.56}$$

The forward kinematic solution allows us to find the location (and orientation) of the robot's end if values for θ_1, θ_2, a_1, and a_2 are specified. We will find the inverse kinematic solution for this robot later. ∎

Example 2.24

Assign the necessary frames to the robot of Figure 2.29 and derive the forward kinematic equation of the robot.

Solution: As you can see, this robot is very similar to the robot of Example 2.23, except that another joint is added to it. The same assignment of frames 0 and 1 are

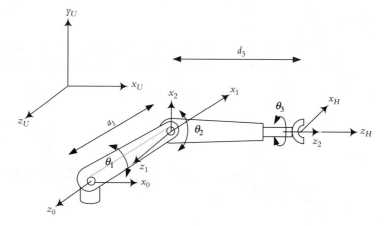

Figure 2.29 The 3-DOF robot of Example 2.24.

Table 2.3 *D-H Parameters Table for Example 2.24.*

#	θ	d	a	α
0–1	θ_1	0	a_1	0
1–2	$90 + \theta_2$	0	0	90
2–H	θ_3	d_3	0	0

applicable to this robot, but we need to add another frame for the new joint. Therefore, we will add a z_2-axis perpendicular to the joint, as shown. Since z_1 and z_2 axes intersect at joint 2, x_2-axis will be perpendicular to both at the same location, as shown.

Table 2.3 shows the parameters for the robot. Please follow the four required transformations between every two frames and make sure that you note the following:

- The direction of the H-frame is changed to represent the motions of the gripper.
- The physical length of link 2 is now a "d" and not an "a".
- Joint 3 is shown as a revolute joint. In this case, d_3 is fixed. However, the joint could have been a prismatic joint (in which case, d_3 would be a variable but θ_3 would be fixed), or both (in which case both θ_3 and d_3 would be variables).
- Remember that the rotations are measured with the right-hand rule. The curled fingers of your right hand, rotating in the direction of rotation, determine the direction of the axis of rotation along the thumb.
- Note that the rotation about z_1 is shown to be $90° + \theta_2$ and not θ_2. This is because even when θ_2 is zero, there is a $90°$ angle between x_1 and x_2 (see Figure 2.30). This is an extremely important factor in real life, when the reset position of the robot must be defined.

Noting that $\sin(90 + \theta) = \cos(\theta)$ and $\cos(90 + \theta) = -\sin(\theta)$, the matrices representing each joint transformation and the total transformation of the robot are:

$$A_1 = \begin{bmatrix} C_1 & -S_1 & 0 & a_1 C_1 \\ S_1 & C_1 & 0 & a_1 S_1 \\ 0 & 0 & 1 & 0 \\ 0 & 0 & 0 & 1 \end{bmatrix} \quad A_2 = \begin{bmatrix} -S_2 & 0 & C_2 & 0 \\ C_2 & 0 & S_2 & 0 \\ 0 & 1 & 0 & 0 \\ 0 & 0 & 0 & 1 \end{bmatrix} \quad A_3 = \begin{bmatrix} C_3 & -S_3 & 0 & 0 \\ S_3 & C_3 & 0 & 0 \\ 0 & 0 & 1 & d_3 \\ 0 & 0 & 0 & 1 \end{bmatrix}$$

$${}^0T_H = A_1 A_2 A_3$$

$$= \begin{bmatrix} (-C_1 S_2 - S_1 C_2)C_3 & -(-C_1 S_2 - S_1 C_2)S_3 & C_1 C_2 - S_1 S_2 & (C_1 C_2 - S_1 S_2)d_3 + a_1 C_1 \\ (C_1 C_2 - S_1 S_2)C_3 & -(C_1 C_2 - S_1 S_2)S_3 & C_1 S_2 + S_1 C_2 & (C_1 S_2 + S_1 C_2)d_3 + a_1 S_1 \\ S_3 & C_3 & 0 & 0 \\ 0 & 0 & 0 & 1 \end{bmatrix}$$

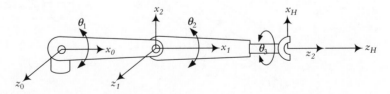

Figure 2.30 Robot of Example 2.24 in reset position.

Simplifying the matrix with $C_1 C_2 - S_1 S_2 = C_{12}$ and $S_1 C_2 + C_1 S_2 = S_{12}$, we get:

$$^0T_H = A_1 A_2 A_3 = \begin{bmatrix} -S_{12}C_3 & S_{12}S_3 & C_{12} & C_{12}d_3 + a_1 C_1 \\ C_{12}C_3 & -C_{12}S_3 & S_{12} & S_{12}d_3 + a_1 S_1 \\ S_3 & C_3 & 0 & 0 \\ 0 & 0 & 0 & 1 \end{bmatrix}$$

For $\begin{cases} \theta_1 = 0 \\ \theta_2 = 0, \\ \theta_3 = 0 \end{cases}$ $^0T_H = \begin{bmatrix} 0 & 0 & 1 & d_3 + a_1 \\ 1 & 0 & 0 & 0 \\ 0 & 1 & 0 & 0 \\ 0 & 0 & 0 & 1 \end{bmatrix}$, and

for $\begin{cases} \theta_1 = 90 \\ \theta_2 = 0 \ , \\ \theta_3 = 0 \end{cases}$ $^0T_H = \begin{bmatrix} -1 & 0 & 0 & 0 \\ 0 & 0 & 1 & d_3 + a_1 \\ 0 & 1 & 0 & 0 \\ 0 & 0 & 0 & 1 \end{bmatrix}$

Please verify that these values represent the robot correctly. ∎

Example 2.25

For the simple 6-DOF robot of Figure 2.31, assign the necessary coordinate frames based on the D-H representation, fill out the accompanying parameters table, and derive the forward kinematic equation of the robot.

Solution: As you will notice, when the number of joints increases, in this case to six, the analysis of the forward kinematics becomes more complicated. However, all principles apply the same as before. You will also notice that this 6-DOF robot is still simplified with no joint offsets or twist angles. In this example, for simplicity, we are assuming that joints 2, 3, and 4 are in the same plane, which will render their d_n values zero; otherwise, the presence of offsets will make the equations slightly more

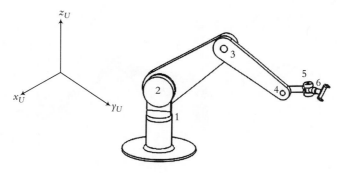

Figure 2.31 A simple 6-DOF articulate robot.

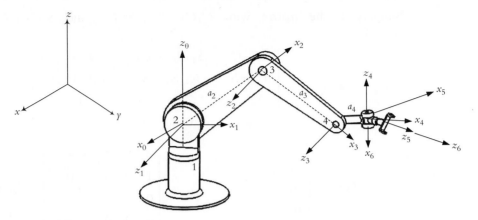

Figure 2.32   Reference frames for the simple 6-DOF articulate robot.

involved. Generally, offsets will change the position terms, but not orientation terms. To assign coordinate frames to the robot, we will first look for the joints (as shown). First, we will assign z-axes to each joint, followed by x-axes. Please follow the coordinates as shown in Figures 2.32 and 2.33. Figure 2.33 is a line drawing of the robot in Figure 2.31 for simplicity. Notice where the origin of each frame is, and why.

Start at joint 1. z_0 represents motions about the first joint. x_0 is chosen to be parallel to the reference frame x-axis. This is done only for convenience. x_0 is a fixed axis, representing the base of the robot, and does not move. The movement of the first joint *occurs around the $z_0 - x_0$ axes.* Next, z_1 is assigned at joint 2. x_1 will be normal to z_0 and z_1 because these two axes are intersecting. x_2 will be in the direction of the common normal between z_1 and z_2. x_3 is in the direction of the common normal between z_2 and z_3. Similarly, x_4 is in the direction of the common normal between z_3 and z_4. Finally, z_5 and z_6 are as shown, because they are parallel and colinear. z_5 represents the motions about joint 6, while z_6 represents the motions of the end effector. Although we normally do not include the end effector in the equations of motion, it is necessary to include the end effector frame because it will allow us to transform out of frame $z_5 - x_5$. Also important to notice is the location of the origins

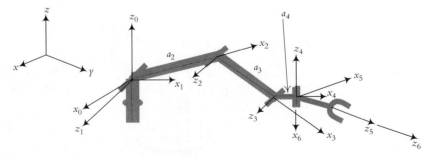

Figure 2.33   Line drawing of the reference frames for the simple 6-DOF articulate robot.

of the first and the last frames. This will determine the total transformation equation of the robot. You may be able to assign other (or different) intermediate coordinate frames between the first and the last, but as long as the first and the last frames are not changed, the total transformation of the robot will be the same. Notice that the origin of the first joint is *not* at the physical location of the joint. You can verify that this is correct, because physically, whether the actual joint is a little higher or lower will not make any difference in the robot's movements. Therefore, the origin can be as shown without regard to the physical location of the joint on the base. Note that we could have chosen to place the 0-frame at the base of the robot. In that case, the total transformation between the base and the end effector of the robot would have included the height of the robot too, whereas the way we have assigned the base frame, our measurements are relative to the present 0-frame. We can simply add the height to our equation later.

Next, we will follow the assigned coordinate frames to fill out the parameters of Table 2.4. Starting with $z_0 - x_0$, there will be a rotation of θ_1 to bring x_0 to x_1, a translation of zero along z_1 and zero along x_1 to align the x's together, and a rotation of $\alpha_1 = +90°$ to bring z_0 to z_1. Remember that the rotations are measured with the right-hand rule. The curled fingers of your right hand, rotating in the direction of rotation, determine the direction of the axis of rotation along the thumb. At this point, we will be at $z_1 - x_1$. Continue with the next joints the same way to fill out the table.

You *must* realize that like any other machine, robots do not stay in one configuration as shown in a drawing. You need to visualize the motions, even though the schematic is in 2-D. This means you must realize that the different links and joints of the robot move, as do the frames attached to them. If in this instant, due to the configuration in which the robot is drawn, the axes are shown to be in a particular position and orientation, they will be at other points and orientations as the robot moves. For example, x_3 is always in the direction of a_3 along the line between joints 3 and 4. As the lower arm of the robot rotates about joint 3, x_3 moves as well, but not x_2. However, x_2 will move as the upper arm rotates about joint 2. This must be kept in mind as we determine the parameters.

θ represents the joint variable for a revolute joint and d represents joint variable for a prismatic joint. Since this is an all-revolute robot, all joint variables are angles.

The transformations between each two successive joints can be written by simply substituting the parameters from the parameters table into the A-matrix.

Table 2.4 *Parameters for the Robot of Example 2.25.*

#	θ	d	a	α
0–1	θ_1	0	0	90
1–2	θ_2	0	a_2	0
2–3	θ_3	0	a_3	0
3–4	θ_4	0	a_4	−90
4–5	θ_5	0	0	90
5–6	θ_6	0	0	0

We get:

$$
A_1 = \begin{bmatrix} C_1 & 0 & S_1 & 0 \\ S_1 & 0 & -C_1 & 0 \\ 0 & 1 & 0 & 0 \\ 0 & 0 & 0 & 1 \end{bmatrix} \quad
A_2 = \begin{bmatrix} C_2 & -S_2 & 0 & C_2 a_2 \\ S_2 & C_2 & 0 & S_2 a_2 \\ 0 & 0 & 1 & 0 \\ 0 & 0 & 0 & 1 \end{bmatrix} \quad
A_3 = \begin{bmatrix} C_3 & -S_3 & 0 & C_3 a_3 \\ S_3 & C_3 & 0 & S_3 a_3 \\ 0 & 0 & 1 & 0 \\ 0 & 0 & 0 & 1 \end{bmatrix}
$$

$$
A_4 = \begin{bmatrix} C_4 & 0 & -S_4 & C_4 a_4 \\ S_4 & 0 & C_4 & S_4 a_4 \\ 0 & -1 & 0 & 0 \\ 0 & 0 & 0 & 1 \end{bmatrix} \quad
A_5 = \begin{bmatrix} C_5 & 0 & S_5 & 0 \\ S_5 & 0 & -C_5 & 0 \\ 0 & 1 & 0 & 0 \\ 0 & 0 & 0 & 1 \end{bmatrix} \quad
A_6 = \begin{bmatrix} C_6 & -S_6 & 0 & 0 \\ S_6 & C_6 & 0 & 0 \\ 0 & 0 & 1 & 0 \\ 0 & 0 & 0 & 1 \end{bmatrix}
$$

$$(2.57)$$

Once again, to simplify the final solutions, we will use the following trigonometric functions:

$$
\begin{aligned}
S\theta_1 C\theta_2 + C\theta_1 S\theta_2 &= S(\theta_1 + \theta_2) = S_{12} \\
C\theta_1 C\theta_2 - S\theta_1 S\theta_2 &= C(\theta_1 + \theta_2) = C_{12}
\end{aligned}
$$

$$(2.58)$$

The total transformation between the base of the robot (where the 0-frame is) and the hand will be:

$$
{}^R T_H = A_1 A_2 A_3 A_4 A_5 A_6
$$

$$
= \begin{bmatrix}
\begin{array}{l} C_1(C_{234}C_5C_6 - S_{234}S_6) \\ \quad - S_1 S_5 C_6 \end{array} &
\begin{array}{l} C_1(-C_{234}C_5C_6 - S_{234}C_6) \\ \quad + S_1 S_5 S_6 \end{array} &
C_1(C_{234}S_5) + S_1 C_5 &
C_1(C_{234}a_4 + C_{23}a_3 + C_2 a_2) \\[2ex]
\begin{array}{l} S_1(C_{234}C_5C_6 - S_{234}S_6) \\ \quad + C_1 S_5 C_6 \end{array} &
\begin{array}{l} S_1(-C_{234}C_5C_6 - S_{234}C_6) \\ \quad - C_1 S_5 S_6 \end{array} &
S_1(C_{234}S_5) - C_1 C_5 &
S_1(C_{234}a_4 + C_{23}a_3 + C_2 a_2) \\[2ex]
S_{234}C_5C_6 + C_{234}S_6 &
-S_{234}C_5C_6 + C_{234}C_6 &
S_{234}S_5 &
S_{234}a_4 + S_{23}a_3 + S_2 a_2 \\[2ex]
0 & 0 & 0 & 1
\end{bmatrix}
$$

$$(2.59)$$

∎

Note the following important insights:

1. In assigning the x- and z-axes, you may choose either direction along the chosen line of action. Ultimately, the result of the total transformation will be the same. However, your individual matrices and parameters are similarly affected.
2. It is acceptable to use additional frames to make things easier to follow. However, you may not have any fewer or more unknown variables than you have joints.
3. The D-H representation does not use a transformation along the y-axis. Therefore, if you find that you need to move along the y-axis to transform from one frame to another, you either have made a mistake, or you need an additional frame in between.
4. In reality, there may be small angles between parallel z-axes due to manufacturing errors or tolerances. To represent these errors between seemingly parallel z-axes, it

will be necessary to make transformations along the y-axis. Therefore, the D-H methodology cannot represent these errors.

5. Note that frame $x_n - z_n$ represents link n before itself. It is attached to link n and moves with it relative to frame $n-1$. Motions about joint n are relative to frame $n-1$.

6. Obviously, you may use other representations to develop the kinematic equations of a robot. However, in order to be able to use subsequent derivations that will be used for differential motions, dynamic analysis, and so on—which are all based on the D-H representation—you may benefit from following this methodology.

7. So far, in all of our examples in this section, we derived the transformation between the base of the robot and the end effector (^0T_H). It is also possible to desire the transformation between the Universe frame and the end effector (^UT_H). In that case, we will need to pre-multiply ^0T_H by the transformation between the base and the Universe frames, or $^UT_H = {}^UT_0 \times {}^0T_H$. Since the location of the base of the robot is always known, this will not add to the number of unknowns (or complexity of the problem). The transformation $^UT_0$ usually involves simple translations and rotations about the Universe frame to get to the base frame. This process is not based on the D-H representation; it is a simple set of rotations and translations.

8. As you have probably noticed, the D-H representation can be used for any configuration of joints and links, whether or not they follow known coordinates such as rectangular, spherical, Euler, and so on. Additionally, you cannot use those representations if any twist angles or joint offsets are present. In reality, twist angles and joint offsets are very common. The derivation of kinematic equations based on rectangular, cylindrical, spherical, RPY, and Euler was presented only for teaching purposes. Therefore, you should normally use the D-H for analysis.

Example 2.26

The Stanford Arm: Assign coordinate frames to the Stanford Arm (Figure 2.34) and fill out the parameters table. The Stanford Arm is a spherical coordinate arm: the first two joints are revolute, the third is prismatic, and the last three wrist joints are revolute joints.

Solution: To allow you to work on this before you see the solution, the answer to this problem is included at the end of this chapter. It is recommended that before you look at the assignment of the frames and the solution of the Arm, you try to do this on your own.

The final forward kinematic solution of the Arm[5] is the product of the six matrices representing the transformation between successive joints, as follows:

$$
{}^RT_{H\,Stanford} = {}^0T_6 = \begin{bmatrix} n_x & o_x & a_x & p_x \\ n_y & o_y & a_y & p_y \\ n_z & o_z & a_z & p_z \\ 0 & 0 & 0 & 1 \end{bmatrix}
$$

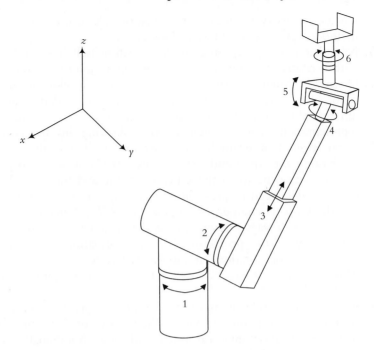

Figure 2.34   Schematic drawing of the Stanford Arm.

where

$$n_x = C_1[C_2(C_4C_5C_6 - S_4S_6) - S_2S_5C_6] - S_1(S_4C_5C_6 + C_4S_6)$$

$$n_y = S_1[C_2(C_4C_5C_6 - S_4S_6) - S_2S_5C_6] + C_1(S_4C_5C_6 + C_4S_6)$$

$$n_z = -S_2(C_4C_5C_6 - S_4S_6) - C_2S_5C_6$$

$$o_x = C_1[-C_2(C_4C_5S_6 + S_4C_6) + S_2S_5S_6] - S_1(-S_4C_5S_6 + C_4C_6)$$

$$o_y = S_1[-C_2(C_4C_5S_6 + S_4C_6) + S_2S_5S_6] + C_1(-S_4C_5S_6 + C_4C_6)$$

$$o_z = S_2(C_4C_5S_6 + S_4C_6) + C_2S_5S_6$$

$$a_x = C_1(C_2C_4S_5 + S_2C_5) - S_1S_4S_5$$

$$a_y = S_1(C_2C_4S_5 + S_2C_5) + C_1S_4S_5$$

$$a_z = -S_2C_4S_5 + C_2C_5$$

$$p_x = C_1S_2d_3 - S_1d_2$$

$$p_y = S_1S_2d_3 + C_1d_2$$

$$p_z = C_2d_3$$

(2.60)

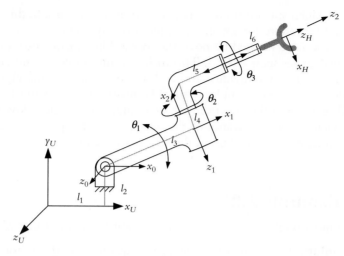

Figure 2.35 4-axis robot of Example 2.27.

Example 2.27

Assign required frames to the 4-axis robot of Figure 2.35 and write an equation describing $^U T_H$.

Solution: This example shows a robot with a twist angle, a joint offset, and a double-action joint represented by the same z-axis. Applying the standard procedure, we assign the frames. The parameters table is shown in Table 2.5.
    The total transformation is:

$$^U T_H = {}^U T_0 \times {}^0 T_H = \begin{bmatrix} 1 & 0 & 0 & l_1 \\ 0 & 1 & 0 & l_2 \\ 0 & 0 & 1 & 0 \\ 0 & 0 & 0 & 1 \end{bmatrix} \times A_1 A_2 A_H$$

Table 2.5 *Parameters for the Robot of Example 2.27.*

#	θ	d	a	α
0–1	θ_1	0	l_3	90
1–2	θ_2	$-l_4$	0	90
2–H	θ_3	$l_5 + l_6$	0	0

∎

2.13 The Inverse Kinematic Solution of Robots

As we mentioned earlier, we are actually interested in the inverse kinematic solutions. With inverse kinematic solutions, we will be able to determine the value of each joint in order to place the robot at a desired position and orientation. We have already seen the inverse kinematic solutions of specific coordinate systems. In this section, we will learn a general procedure for solving the kinematic equations.

As you have noticed by now, the forward kinematic equations have a multitude of coupled angles such as C_{234}. This makes it impossible to find enough elements in the matrix to solve for individual sines and cosines to calculate the angles. To de-couple some of the angles, we may multiply the $^R T_H$ matrix with individual A_n^{-1} matrices. This will yield one side of the equation free of an individual angle, allowing us to find elements that yield sines and cosines of the angle, and subsequently, the angle itself. We will demonstrate the procedure in the following section.

Example 2.28

Find a symbolic expression for the joint variables of the robot of Example 2.23.

Solution: The forward kinematic equation for the robot is shown as Equation (2.56), repeated here. Assume that we desire to place the robot at a position—and consequently, an orientation—given as **n**, **o**, **a**, **p** vectors:

$$^0 T_H = A_1 \times A_2 = \begin{bmatrix} C_{12} & -S_{12} & 0 & a_2 C_{12} + a_1 C_1 \\ S_{12} & C_{12} & 0 & a_2 S_{12} + a_1 S_1 \\ 0 & 0 & 1 & 0 \\ 0 & 0 & 0 & 1 \end{bmatrix} = \begin{bmatrix} n_x & o_x & a_x & p_x \\ n_y & o_y & a_y & p_y \\ n_z & o_z & a_z & p_z \\ 0 & 0 & 0 & 1 \end{bmatrix}$$

$$(2.56)$$

Since this robot has only two degrees of freedom, its solution is relatively simple. We can solve for the angles either algebraically, or by de-coupling the unknowns. We will do both for comparison. Remember that whenever possible, we should look for values of both the sine and cosine of an angle in order to correctly identify the quadrant in which the angle falls.

I. **Algebraic solution:** Equating elements (2,1), (1,1), (1,4), and (2,4) of the two matrices, we get:

$$S_{12} = n_y \text{ and } C_{12} = n_x \rightarrow \theta_{12} = ATAN2(n_y, n_x)$$

$$a_2 C_{12} + a_1 C_1 = p_x \text{ or } a_2 n_x + a_1 C_1 = p_x \rightarrow C_1 = \frac{p_x - a_2 n_x}{a_1}$$

$$a_2 S_{12} + a_1 S_1 = p_y \text{ or } a_2 n_y + a_1 S_1 = p_y \rightarrow S_1 = \frac{p_y - a_2 n_y}{a_1}$$

$$\theta_1 = ATAN2(S_1, C_1) = ATAN2\left(\frac{p_y - a_2 n_y}{a_1}, \frac{p_x - a_2 n_x}{a_1}\right)$$

Since θ_1 and θ_{12} are known, θ_2 can also be calculated.

II. Alternative solution: In this case, we will post-multiply both sides of Equation (2.56) by A_2^{-1} to de-couple θ_1 from θ_2. We get:

$$
A_1 \times A_2 \times A_2^{-1} = \begin{bmatrix} n_x & o_x & a_x & p_x \\ n_y & o_y & a_y & p_y \\ n_z & o_z & a_z & p_z \\ 0 & 0 & 0 & 1 \end{bmatrix} \times A_2^{-1} \quad \text{or} \quad A_1 = \begin{bmatrix} n_x & o_x & a_x & p_x \\ n_y & o_y & a_y & p_y \\ n_z & o_z & a_z & p_z \\ 0 & 0 & 0 & 1 \end{bmatrix} \times A_2^{-1}
$$

$$
\begin{bmatrix} C_1 & -S_1 & 0 & a_1 C_1 \\ S_1 & C_1 & 0 & a_1 S_1 \\ 0 & 0 & 1 & 0 \\ 0 & 0 & 0 & 1 \end{bmatrix} = \begin{bmatrix} n_x & o_x & a_x & p_x \\ n_y & o_y & a_y & p_y \\ n_z & o_z & a_z & p_z \\ 0 & 0 & 0 & 1 \end{bmatrix} \times \begin{bmatrix} C_2 & -S_2 & 0 & -a_2 \\ S_2 & C_2 & 0 & 0 \\ 0 & 0 & 1 & 0 \\ 0 & 0 & 0 & 1 \end{bmatrix}
$$

$$
\begin{bmatrix} C_1 & -S_1 & 0 & a_1 C_1 \\ S_1 & C_1 & 0 & a_1 S_1 \\ 0 & 0 & 1 & 0 \\ 0 & 0 & 0 & 1 \end{bmatrix} = \begin{bmatrix} C_2 n_x - S_2 o_x & S_2 n_x + C_2 o_x & a_x & p_x - a_2 n_x \\ C_2 n_y - S_2 o_y & S_2 n_y + C_2 o_y & a_y & p_y - a_2 n_y \\ C_2 n_z - S_2 o_z & S_2 n_z + C_2 o_z & a_z & p_z - a_2 n_z \\ 0 & 0 & 0 & 1 \end{bmatrix}
$$

From elements 1,4 and 2,4 we get $a_1 C_1 = p_x - a_2 n_x$ and $a_1 S_1 = p_y - a_2 n_y$ which is exactly what we got from the other method. Knowing S_1 and C_1, we can find expressions for S_2 and C_2. ∎

2.13.1 General Solution for Articulated Robot Arms

In this section, a summary of a technique is presented that may generally be used for inverse kinematic analysis of manipulators.[5] The process is applied to the simple manipulator arm of Example 2.25. Although this solution is for this particular robot with the given configuration, it may similarly be repeated for other robots. As we saw in Example 2.25, the final equation representing the robot, repeated here, is:

$^R T_H = A_1 A_2 A_3 A_4 A_5 A_6$

$$
= \begin{bmatrix} C_1(C_{234}C_5 C_6 - S_{234}S_6) & C_1(-C_{234}C_5 C_6 - S_{234}C_6) & C_1(C_{234}S_5) + S_1 C_5 & C_1(C_{234}a_4 + C_{23}a_3 + C_2 a_2) \\ -S_1 S_5 C_6 & +S_1 S_5 S_6 & & \\ S_1(C_{234}C_5 C_6 - S_{234}S_6) & S_1(-C_{234}C_5 C_6 - S_{234}C_6) & S_1(C_{234}S_5) - C_1 C_5 & S_1(C_{234}a_4 + C_{23}a_3 + C_2 a_2) \\ +C_1 S_5 C_6 & -C_1 S_5 S_6 & & \\ S_{234}C_5 C_6 + C_{234}S_6 & -S_{234}C_5 C_6 + C_{234}C_6 & S_{234}S_5 & S_{234}a_4 + S_{23}a_3 + S_2 a_2 \\ 0 & 0 & 0 & 1 \end{bmatrix}
$$

We will denote the above matrix as [RHS] (Right-Hand Side) for simplicity in writing. Let's, once again, express the desired location and orientation of the robot with:

$$
{}^{R}T_{H} = \begin{bmatrix} n_x & o_x & a_x & p_x \\ n_y & o_y & a_y & p_y \\ n_z & o_z & a_z & p_z \\ 0 & 0 & 0 & 1 \end{bmatrix}
$$

(2.61)

To solve for the angles, we will pre-multiply these two matrices with selected A_n^{-1} matrices, first with A_1^{-1}:

$$
A_1^{-1} \times \begin{bmatrix} n_x & o_x & a_x & p_x \\ n_y & o_y & a_y & p_y \\ n_z & o_z & a_z & p_z \\ 0 & 0 & 0 & 1 \end{bmatrix} = A_1^{-1}[\text{RHS}] = A_2 A_3 A_4 A_5 A_6
$$

(2.62)

$$
\begin{bmatrix} C_1 & S_1 & 0 & 0 \\ 0 & 0 & 1 & 0 \\ S_1 & -C_1 & 0 & 0 \\ 0 & 0 & 0 & 1 \end{bmatrix} \times \begin{bmatrix} n_x & o_x & a_x & p_x \\ n_y & o_y & a_y & p_y \\ n_z & o_z & a_z & p_z \\ 0 & 0 & 0 & 1 \end{bmatrix} = A_2 A_3 A_4 A_5 A_6
$$

$$
\begin{bmatrix} n_x C_1 + n_y S_1 & o_x C_1 + o_y S_1 & a_x C_1 + a_y S_1 & p_x C_1 + p_y S_1 \\ n_z & o_z & a_z & p_z \\ n_x S_1 - n_y C_1 & o_x S_1 - o_y C_1 & a_x S_1 - a_y C_1 & p_x S_1 - p_y C_1 \\ 0 & 0 & 0 & 1 \end{bmatrix}
$$

$$
= \begin{bmatrix} C_{234} C_5 C_6 - S_{234} S_6 & -C_{234} C_5 C_6 - S_{234} C_6 & C_{234} S_5 & C_{234} a_4 + C_{23} a_3 + C_2 a_2 \\ S_{234} C_5 C_6 + C_{234} S_6 & -S_{234} C_5 C_6 + C_{234} C_6 & S_{234} S_5 & S_{234} a_4 + S_{23} a_3 + S_2 a_2 \\ -S_5 C_6 & S_5 S_6 & C_5 & 0 \\ 0 & 0 & 0 & 1 \end{bmatrix}
$$

(2.63)

From the 3,4 elements of Equation (2.63):

$$
p_x S_1 - p_y C_1 = 0 \rightarrow \theta_1 = \tan^{-1}\left(\frac{p_y}{p_x}\right) \text{ and } \theta_1 = \theta_1 + 180°
$$

(2.64)

From the 1,4 and 2,4 elements, we will get:

$$
\begin{aligned} p_x C_1 + p_y S_1 &= C_{234} a_4 + C_{23} a_3 + C_2 a_2 \\ p_z &= S_{234} a_4 + S_{23} a_3 + S_2 a_2 \end{aligned}
$$

(2.65)

We will rearrange the two expressions in Equation (2.65) and square and add them to get:

$$\left(p_x C_1 + p_y S_1 - C_{234} a_4\right)^2 = \left(C_{23} a_3 + C_2 a_2\right)^2$$

$$\left(p_z - S_{234} a_4\right)^2 = \left(S_{23} a_3 + S_2 a_2\right)^2$$

$$\left(p_x C_1 + p_y S_1 - C_{234} a_4\right)^2 + \left(p_z - S_{234} a_4\right)^2 = a_2^2 + a_3^2 + 2a_2 a_3 (S_2 S_{23} + C_2 C_{23})$$

Referring to the trigonometric functions of Equation (2.58):

$$S_2 S_{23} + C_2 C_{23} = \text{Cos}[(\theta_2 + \theta_3) - \theta_2] = \text{Cos}\theta_3$$

Therefore:

$$C_3 = \frac{\left(p_x C_1 + p_y S_1 - C_{234} a_4\right)^2 + \left(p_z - S_{234} a_4\right)^2 - a_2^2 - a_3^2}{2a_2 a_3} \tag{2.66}$$

In this equation, everything is known except for S_{234} and C_{234}, which we will find next. Knowing that $S_3 = \pm\sqrt{1 - C_3^2}$, we can then say that:

$$\theta_3 = \tan^{-1}\frac{S_3}{C_3} \tag{2.67}$$

Since joints 2, 3, and 4 are parallel, additional pre-multiplications by A_2^{-1} and A_3^{-1} will not yield useful results. The next step is to pre-multiply by the inverses of A_1 through A_4, which results in:

$$A_4^{-1}A_3^{-1}A_2^{-1}A_1^{-1} \times \begin{bmatrix} n_x & o_x & a_x & p_x \\ n_y & o_y & a_y & p_y \\ n_z & o_z & a_z & p_z \\ 0 & 0 & 0 & 1 \end{bmatrix} = A_4^{-1}A_3^{-1}A_2^{-1}A_1^{-1}[\text{RHS}] = A_5 A_6 \tag{2.68}$$

which yields:

$$\begin{bmatrix} C_{234}(C_1 n_x + S_1 n_y) \\ +S_{234} n_z \\ C_1 n_y - S_1 n_x \\ -S_{234}(C_1 n_x + S_1 n_y) \\ +C_{234} n_z \\ 0 \end{bmatrix}$$

$$\begin{bmatrix} C_{234}(C_1 n_x + S_1 n_y) & C_{234}(C_1 o_x + S_1 o_y) & C_{234}(C_1 a_x + S_1 a_y) & C_{234}\left(C_1 p_x + S_1 p_y\right)+ \\ +S_{234} n_z & +S_{234} o_z & +S_{234} a_z & S_{234} p_z - C_{34} a_2 - C_4 a_3 - a_4 \\ C_1 n_y - S_1 n_x & C_1 o_y - S_1 o_x & C_1 a_y - S_1 a_x & 0 \\ -S_{234}(C_1 n_x + S_1 n_y) & -S_{234}(C_1 o_x + S_1 o_y) & -S_{234}(C_1 a_x + S_1 a_y) & -S_{234}\left(C_1 p_x + S_1 p_y\right)+ \\ +C_{234} n_z & +C_{234} o_z & +C_{234} a_z & C_{234} p_z + S_{34} a_2 + S_4 a_3 \\ 0 & 0 & 0 & 1 \end{bmatrix}$$

$$= \begin{bmatrix} C_5 C_6 & -C_5 S_6 & S_5 & 0 \\ S_5 C_6 & -S_5 S_6 & -C_5 & 0 \\ S_6 & C_6 & 0 & 0 \\ 0 & 0 & 0 & 1 \end{bmatrix}$$

$$\tag{2.69}$$

From the 3,3 elements of the matrices in Equation (2.69):

$$- S_{234}\left(C_1 a_x + S_1 a_y\right) + C_{234} a_z = 0 \rightarrow \theta_{234} = \tan^{-1}\left(\frac{a_z}{C_1 a_x + S_1 a_y}\right) \text{ and } \theta_{234} = \theta_{234} + 180°$$

$$(2.70)$$

and we can calculate S_{234} and C_{234}, which are used to calculate θ_3, as previously discussed.

Now, referring again to Equation (2.65), repeated here, we can calculate the sine and cosine of θ_2 as follows:

$$\begin{cases} p_x C_1 + p_y S_1 = C_{234} a_4 + C_{23} a_3 + C_2 a_2 \\ p_z = S_{234} a_4 + S_{23} a_3 + S_2 a_2 \end{cases}$$

Since $C_{12} = C_1 C_2 - S_1 S_2$ and $S_{12} = S_1 C_2 + C_1 S_2$, we get:

$$\begin{cases} p_x C_1 + p_y S_1 - C_{234} a_4 = (C_2 C_3 - S_2 S_3) a_3 + C_2 a_2 \\ p_z - S_{234} a_4 = (S_2 C_3 + C_2 S_3) a_3 + S_2 a_2 \end{cases}$$

$$(2.71)$$

Treating this as a set of two equations and two unknowns and solving for C_2 and S_2, we get:

$$\begin{cases} S_2 = \dfrac{(C_3 a_3 + a_2)(p_z - S_{234} a_4) - S_3 a_3 (p_x C_1 + p_y S_1 - C_{234} a_4)}{(C_3 a_3 + a_2)^2 + S_3^2 a_3^2} \\[4mm] C_2 = \dfrac{(C_3 a_3 + a_2)(p_x C_1 + p_y S_1 - C_{234} a_4) + S_3 a_3 (p_z - S_{234} a_4)}{(C_3 a_3 + a_2)^2 + S_3^2 a_3^2} \end{cases}$$

$$(2.72)$$

Although this is a large equation, all its elements are known and it can be evaluated. Then:

$$\theta_2 = \tan^{-1}\frac{(C_3 a_3 + a_2)(p_z - S_{234} a_4) - S_3 a_3 (p_x C_1 + p_y S_1 - C_{234} a_4)}{(C_3 a_3 + a_2)(p_x C_1 + p_y S_1 - C_{234} a_4) + S_3 a_3 (p_z - S_{234} a_4)}$$

$$(2.73)$$

Now that θ_2 and θ_3 are known:

$$\theta_4 = \theta_{234} - \theta_2 - \theta_3 \tag{2.74}$$

Remember that since there are two solutions for θ_{234} (Equation (2.70)), there will be two solutions for θ_4 as well. From 1,3 and 2,3 elements of Equation (2.69), we get:

$$\begin{cases} S_5 = C_{234}\left(C_1 a_x + S_1 a_y\right) + S_{234} a_z \\ C_5 = -C_1 a_y + S_1 a_x \end{cases}$$

$$(2.75)$$

$$\text{and} \quad \theta_5 = \tan^{-1}\frac{C_{234}\left(C_1 a_x + S_1 a_y\right) + S_{234} a_z}{S_1 a_x - C_1 a_y}$$

$$(2.76)$$

As you have probably noticed, there is no de-coupled equation for θ_6. As a result, we have to pre-multiply Equation (2.69) by the inverse of A_5 to de-couple it. We get:

$$
\begin{bmatrix}
\begin{array}{l} C_5\left[C_{234}\left(C_1 n_x + S_1 n_y\right) + S_{234} n_z\right] \\ -S_5\left(S_1 n_x - C_1 n_y\right) \\ -S_{234}\left(C_1 n_x + S_1 n_y\right) + C_{234} n_z \end{array} & \begin{array}{l} C_5\left[C_{234}\left(C_1 o_x + S_1 o_y\right) + S_{234} o_z\right] \\ -S_5\left(S_1 o_x - C_1 o_y\right) \\ -S_{234}\left(C_1 o_x + S_1 o_y\right) + C_{234} o_z \end{array} & 0 & 0 \\
0 & 0 & 1 & 0 \\
0 & 0 & 0 & 1
\end{bmatrix}
$$

$$
=
\begin{bmatrix}
C_6 & -S_6 & 0 & 0 \\
S_6 & C_6 & 0 & 0 \\
0 & 0 & 1 & 0 \\
0 & 0 & 0 & 1
\end{bmatrix}
\tag{2.77}
$$

From 2,1 and 2,2 elements of Equation (2.77) we get:

$$
\theta_6 = \tan^{-1} \frac{-S_{234}\left(C_1 n_x + S_1 n_y\right) + C_{234} n_z}{-S_{234}\left(C_1 o_x + S_1 o_y\right) + C_{234} o_z}
\tag{2.78}
$$

Therefore, we have found six equations that collectively yield the values needed to place and orientate the robot at any desired location. Although this solution is only good for the given robot, a similar approach may be taken for any other robot.

It is important to notice that this solution is only possible because the last three joints of the robot are intersecting at a common point. Otherwise, it will not be possible to solve for this kind of solution, and as a result, we would have to solve the matrices directly or by calculating the inverse of the matrix and solving for the unknowns. Most industrial robots have intersecting wrist joints.

2.14 Inverse Kinematic Programming of Robots

The equations we found for solving the inverse kinematic problem of robots can directly be used to drive the robot to a desired position. In fact, no robot would actually use the forward kinematic equations in order to solve for these results. The only equations that are used are the set of six (or less, depending on the number of joints) equations that calculate the joint values. In other words, the robot designer must calculate the inverse solution and derive these equations and, in turn, use them to drive the robot to position. This is necessary for the practical reason that it takes a long time

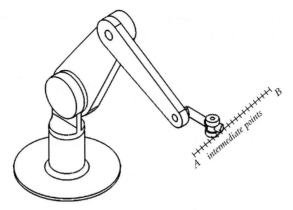

Figure 2.36   Small sections of movement for straight line motions.

for a computer to calculate the inverse of the forward kinematic equations or to substitute values into them and calculate the unknowns (joint variables) by methods such as Gaussian elimination.

For a robot to move in a predictable path, say a straight line, it is necessary to recalculate joint variables many times a second. Imagine that a robot needs to move in a straight line between a starting point A and a destination point B. If no other action is taken and the robot moves from point A to point B, the path is unpredictable. The robot moves all its joints until they are at the final value, which will place the robot at the destination point B. However, depending on the rate of change in each joint, the hand will follow an unknown path in between the two points. To make the robot follow a straight line, it is necessary to break the line into many small sections (Figure 2.36) and make the robot follow those very small sections sequentially between the two points. This means that a new solution must be calculated for each small section. Typically, the location may be recalculated between 50 to 200 times a second. This means that if calculating a solution takes more than 5 to 20 ms, the robot will lose accuracy or will not follow the specified path.[10] The shorter the time it takes to calculate a new solution, the more accurate the robot. As a result, it is vital to eliminate as many unnecessary computations as possible to allow the computer controller to calculate more solutions. This is why the designer must do all mathematical manipulations beforehand and only program the robot controller to calculate the final solutions. This will be discussed in more detail in Chapter 5.

For the 6-axis robot discussed earlier, given the final desired location and orientation as:

$$
{}^{R}T_{H_{Desired}} = \begin{bmatrix} n_x & o_x & a_x & p_x \\ n_y & o_y & a_y & p_y \\ n_z & o_z & a_z & p_z \\ 0 & 0 & 0 & 1 \end{bmatrix}
$$

all the controller needs to use to calculate the unknown angles is the set of inverse solutions as summarized below:

$$\theta_1 = \tan^{-1}\left(\frac{p_y}{p_x}\right) \quad \text{and} \quad \theta_1 = \theta_1 + 180°$$

$$\theta_{234} = \tan^{-1}\left(\frac{a_z}{C_1 a_x + S_1 a_y}\right) \quad \text{and} \quad \theta_{234} = \theta_{234} + 180°$$

$$C_3 = \frac{(p_x C_1 + p_y S_1 - C_{234} a_4)^2 + (p_z - S_{234} a_4)^2 - a_2^2 - a_3^2}{2 a_2 a_3}$$

$$S_3 = \pm\sqrt{1 - C_3^2}$$

$$\theta_3 = \tan^{-1}\frac{S_3}{C_3} \tag{2.79}$$

$$\theta_2 = \tan^{-1}\frac{(C_3 a_3 + a_2)(p_z - S_{234} a_4) - S_3 a_3 (p_x C_1 + p_y S_1 - C_{234} a_4)}{(C_3 a_3 + a_2)(p_x C_1 + p_y S_1 - C_{234} a_4) + S_3 a_3 (p_z - S_{234} a_4)}$$

$$\theta_4 = \theta_{234} - \theta_2 - \theta_3$$

$$\theta_5 = \tan^{-1}\frac{C_{234}(C_1 a_x + S_1 a_y) + S_{234} a_z}{S_1 a_x - C_1 a_y}$$

$$\theta_6 = \tan^{-1}\frac{-S_{234}(C_1 n_x + S_1 n_y) + C_{234} n_z}{-S_{234}(C_1 o_x + S_1 o_y) + C_{234} o_z}$$

Although this is not trivial, it is much quicker to use these equations and calculate the angles than it is to invert the matrices or do Gaussian elimination. Notice that all operations in this computation are simple arithmetic or trigonometric operations.

2.15 Degeneracy and Dexterity

2.15.1 Degeneracy

Degeneracy occurs when the robot loses a degree of freedom, and therefore, cannot perform as desired.[11] This occurs under two conditions: (1) when the robot's joints reach their physical limits and as a result, cannot move any further; (2) a robot may become degenerate in the middle of its workspace if the z-axes of two similar joints become colinear. This means that, at this instant, whichever joint moves, the same motion will result, and consequently, the controller does not know which joint to move. Since in either case the total number of degrees of freedom available is less than six, there is no solution for the robot. In the case of colinear joints, the determinant of the position matrix is zero as well. Figure 2.37 shows a simple robot in a vertical configuration, where joints 1 and 6 are colinear. As you can see, whether joint 1 or joint 6 rotate, the end effector will rotate the same amount. In practice, it is important to direct the controller to

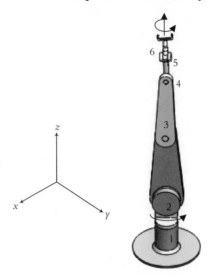

Figure 2.37 An example of a robot in a degenerate position.

take an emergency action; otherwise the robot will stop. Please note that this condition occurs if the two joints are similar. Otherwise, if one joint is prismatic and one is revolute (as in joints 3 and 4 of the Stanford arm), although the z-axes are colinear, the robot will not be in degenerate condition. Paul[11] has shown that if $\sin\alpha_4$, $\sin\alpha_5$ or $\sin\theta_5$ are zero, the robot will be degenerate (this occurs if joints 4 and 5, or 5 and 6 are parallel, and therefore, result in similar motions). Obviously, α_4 and α_5 can be designed to prevent the degeneracy of the robot. However, anytime θ_5 approaches zero or $180°$, the robot will become degenerate.

2.15.2 Dexterity

We should be able to position and orientate a 6-DOF robot at any desired location within its work envelope by specifying the position and the orientation of the hand. However, as the robot gets increasingly closer to the limits of its workspace, it will get to a point where, although it is possible to locate it at a desired point, it will be impossible to orientate it at desired orientations. The volume of points where we can position the robot as desired but not orientate it is called nondexterous volume.

2.16 The Fundamental Problem with the Denavit-Hartenberg Representation

Although Denavit-Hartenberg representation has been extensively used in modeling and analysis of robot motions, and although it has become a standard method for doing so, there is a fundamental flaw with this technique, which many researchers have tried to solve by modifying the process.[12] The fundamental problem is that since all motions are about the x- and z-axes, the method cannot represent any motion about the y-axis. Therefore, if there is any motion about the y-axis, the method will fail. This occurs in a number of circumstances. For example, suppose two joint axes that are supposed to be parallel are assembled with a slight deviation. The small angle between the two axes will require a motion about the y-axis. Since all real industrial robots have some degree of inaccuracy in their manufacture, their inaccuracy cannot be modeled with the D-H representation.

Example 2.26 (Continued)

Reference Frames for the Stanford Arm: Figure 2.38 is the solution for the Stanford Arm in Example 2.26 (Figure 2.34). It is simplified for improved visibility. Table 2.6 shows the corresponding parameters.

For the derivation of the inverse kinematic solution of Stanford Arm, refer to References 5 and 13 at the end of the chapter. The following is a summary of the inverse kinematic solution for the Stanford Arm:

$$\theta_1 = \tan^{-1}\left(\frac{p_y}{p_x}\right) - \tan^{-1}\frac{d_2}{\pm\sqrt{r^2 - d_2^2}} \text{ where } r = \sqrt{p_x^2 + p_y^2} \qquad (2.80)$$

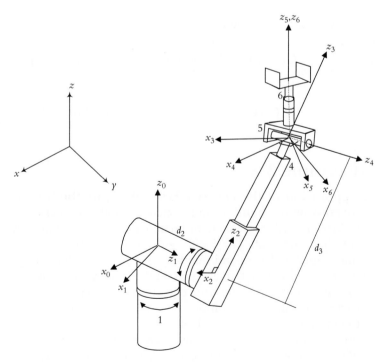

Figure 2.38 The frames of the Stanford Arm.

Table 2.6   *The Parameters Table for the Stanford Arm.*

#	θ	d	a	α
0–1	θ_1	0	0	-90
1–2	θ_2	d_2	0	90
2–3	0	d_3	0	0
3–4	θ_4	0	0	-90
4–5	θ_5	0	0	90
5–6	θ_6	0	0	0

$$\theta_2 = \tan^{-1} \frac{C_1 p_x + S_1 p_y}{p_z} \tag{2.81}$$

$$d_3 = S_2(C_1 p_x + S_1 p_y) + C_2 p_z \tag{2.82}$$

$$\theta_4 = \tan^{-1} \frac{-S_1 a_x + C_1 a_y}{C_2(C_1 a_x + S_1 a_y) - S_2 a_z} \quad \text{and} \quad \theta_4 = \theta_4 + 180° \text{ if } \theta_5 < 0 \tag{2.83}$$

$$\theta_5 = \tan^{-1} \frac{C_4[C_2(C_1 a_x + S_1 a_y) - S_2 a_z] + S_4[-S_1 a_x + C_1 a_y]}{S_2(C_1 a_x + S_1 a_y) + C_2 a_z} \tag{2.84}$$

$$\theta_6 = \tan^{-1} \frac{S_6}{C_6} \quad \text{where}$$

$$S_6 = -C_5\{C_4[C_2(C_1 o_x + S_1 o_y) - S_2 o_z] + S_4[-S_1 o_x + C_1 o_y]\} \tag{2.85}$$
$$\quad + S_5\{S_2(C_1 o_x + S_1 o_y) + C_2 o_z\}$$
$$C_6 = -S_4[C_2(C_1 o_x + S_1 o_y) - S_2 o_z] + C_4[-S_1 o_x + C_1 o_y] \qquad\blacksquare$$

Example 2.29

Application of the Denavit-Hartenberg methodology in the design of a finger-spelling hand: A finger-spelling hand[14] was designed at Cal Poly, San Luis Obispo, in order to enable ordinary users to communicate with individuals who are blind and deaf. The hand, with its 17 degrees of freedom, can form all the finger-spelling letters and numbers (Figure 2.39). Each finger-wrist combination was assigned a set of frames based on the D-H representation in order to derive the forward and inverse kinematic equations of the hand. These equations may be used to drive the fingers to position. This application shows that in addition to modeling the motions of a robot, the D-H technique may be used to represent transformations, rotations, and movements between different kinematic elements, regardless of whether or not a robot is involved. You may also find other applications for this representation. $\blacksquare$

Figure 2.39   Cal Poly finger-spelling hand. (Supported by the Smith-Kettlewell Eye Research Institute, San Francisco.)

2.17 Design Projects

Starting with this chapter and continuing with the rest of the book, we will apply the current information in each chapter to the design of simple robots. This will help you to apply the current material to the design of a robot of your own. Common 6-DOF robots are too complicated to be considered simple; therefore, we will use 3-DOF robots. The intention is to design a simple robot that can possibly be built by you from readily available parts from hobby shops, hardware stores, and surplus dealers.

In this section, you may consider the preliminary design of the robot and its configuration, keeping in mind the possible types of actuators you may want to consider later. Although we will study this subject later, it is a good idea to consider the types of actuators now. You should also consider the types of links and joints you may want to use, possible lengths, types of joints, and material (for example, wood dowels, hollow aluminum or brass tubes available in hardware stores, and so on).

2.17.1 A 3-DOF Robot

For this project, you may want to design your own preferred robot with your own preferred configuration. Creativity is always encouraged. However, we will discuss a simple robot as a guideline for you to design and build. After the configuration of the robot is finalized, you should proceed with the derivation of forward and inverse kinematic equations. The final result of this part of the design project will be a set of inverse kinematic equations for the simple 3-DOF robot that can later be used to drive the robot to desired positions. You must realize that the price we pay for this simplicity is that we may only specify the position of the robot, but not its orientation.

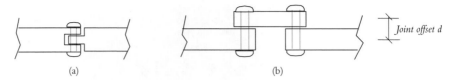

(a)   (b)

Figure 2.40   Two simple designs for a joint.

One of the important considerations in the design of the robot is its joints. Figure 2.40(a) shows a simple design that has no joint offset d. This would apparently simplify the analysis of the robot, since the A matrix related to the joint would be simpler. However, manufacturing such a joint is not as simple as the design in Figure 2.40(b). The latter allows a larger range of motion too. On the other hand, although we apparently have to deal with a joint offset d with the joint design in Figure 2.40(b), you must remember that in most cases, there will be a second joint with the same joint offset in the opposite direction, which cancels the former in the robot's overall equation. As a result, we will assume that the joints of our robot can be built as in Figure 2.40(b) without having to worry about joint offset d.

We will discuss actuators in Chapter 7. However, for this design project, you should probably consider the use of a servomotor or a stepper motor. While you are designing your robot, consider what type of actuators you will use and how you will connect the actuators to the links and joints. Remember that at this point, you are only designing the robot configuration; you can always change your actuators and adapt the new design to your robot.

When the preliminary sketch of the robot is finished, assign coordinate frames to each joint, fill out the parameters table for the frames, develop the matrices for each frame transformation, and calculate the final $^{U}T_{H}$. Then, using the methods learned in this chapter, develop the inverse kinematic equations of the robot. This means that using these equations, if you actually build the robot, you will be able to run it and control its position (since the robot is 3-DOF, you will not be able to control its orientation).

Figure 2.41 shows a simple design for a 3-DOF robot you may use as a guide for your design. In one student design, the lengths were 8, 2, 9, 2, and 9 inches respectively. The links were made of hollow aluminum bars, actuated by three DC gear-motors with encoder feedback and connected to the joint through worm gears.

Figure 2.41 also shows one possible set of frames assigned to the joints. The end of the robot has its own frame. Frame 3 is needed in order to transform from frame 2 to the hand frame. To be able to correctly develop forward and inverse kinematic equations of the robot, it is crucial to define the reset position of the robot, where all joint angles are zero. In this example, the reset position is defined as the robot pointing up and x_0 parallel to x_U. At this point, there is a 90° angle between x_1 and x_2. Therefore, the actual angle for this joint should be $-90 + \theta_2$. The same is true for x_0 and x_1, where a 90° angle exists between the two when θ_1 is zero; therefore, the angle between the two is $90 + \theta_1$. Also notice the permanent angles between other frames.

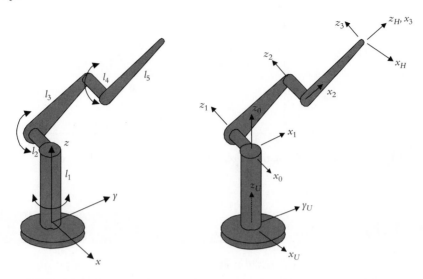

Figure 2.41 A simple 3-DOF robot design that may be used for the design project.

This exercise is left for you to complete. The inverse kinematic equations for the robot relative to frame 0 are:

$$\theta_1 = \tan^{-1}(-p_x/p_y)$$
$$\theta_3 = \cos^{-1}[((p_y/C_1)^2 + (p_z)^2 - 162)/162] \qquad (2.86)$$
$$\theta_2 = \cos^{-1}[(p_z C_1(1 + C_3) + p_y S_3)/(18(1 + C_3)C_1)]$$

Note: If $\cos\theta_1$ is zero, use p_x/S_1 instead of p_y/C_1.

2.17.2 A 3-DOF Mobile Robot

Another project you may consider is a mobile robot. These robots are very common and are used in autonomous navigation and developing artificial intelligence for robots. In general, you may assume the robot is capable of moving in a plane that may be represented by translations along the x- and y-axes or a translation and rotation in a polar form (r, θ). Additionally, the orientation of the robot may be changed by rotating it about the z-axis (α). Therefore, the kinematic equations of the motion of the robot can be developed and used to control its motions. A schematic representation of the robot is shown in Figure 2.42. (See Chapter 7 for a design project involving a single-axis robot that may also be used for this project).

In the next chapters, we will continue with the design of your robots.

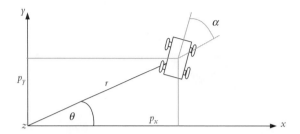

Figure 2.42   Schematic representation of a 3-DOF mobile robot.

Summary

In this chapter, we discussed methods for representation of points, vectors, frames, and transformations by matrices. Using matrices, we discussed forward and inverse kinematic equations for specific types of robots such as Cartesian, cylindrical, and spherical robots as well as Euler and RPY orientation angles. However, the main thrust of this chapter was to learn how to represent the motions of a multi-DOF robot in space and how to derive the forward and inverse kinematic equations of the robot using the Denavit-Hartenberg (D-H) representation technique. This method can be used to represent any type of robot configuration, regardless of the number and type of joints or joint and link offsets or twists.

In the next chapter, we will continue with the differential motions of robots, which in effect is the equivalent of velocity analysis of robots.

References

1. Niku, S., "Scheme for Active Positional Correction of Robot Arms," Proceedings of the 5th International Conference on CAD/CAM, Robotics and Factories of Future, Springer Verlag, pp. 590–593, 1991.

2. Puopolo, Michael G., Saeed B. Niku, "Robot Arm Positional Deflection Control with a Laser Light," Proceedings of the Mechatronics '98 Conference, Skovde, Sweden, Adolfsson and Karlsen, Editors, Pergamon Press, September 1998, pp. 281–286.

3. Ardayfio, D. D., Q. Danwen, "Kinematic Simulation of Novel Robotic Mechanisms Having Closed Kinematic Chains," Paper # 85-DET-81, American Society of Mechanical Engineers, 1985.

4. Denavit, J., R. S. Hartenberg, "A Kinematic Notation for Lower-Pair Mechanisms Based on Matrices," *ASME Journal of Applied Mechanics*, June 1955, 215–221.

5. Paul, Richard P., "Robot Manipulators, Mathematics, Programming, and Control," MIT Press, 1981.

6. Craig, John J., "Introduction to Robotics: Mechanics and Control", 2nd Edition, Addison Wesley, 1989.

7. Shahinpoor, Mohsen, "A Robot Engineering Textbook," Harper & Row, 1987.

8. Koren, Yoram, "Robotics for Engineers," McGraw-Hill, 1985.

9. Fu, K. S., R.C. Gonzalez, C. S. G. Lee, "Robotics: Control, Sensing, Vision, and Intelligence," McGraw-Hill, 1987.

10. Eman, Kornel F., "Trajectories," International Encyclopedia of Robotics: Applications and Automation," Richard C. Dorf, Editor, John Wiley and Sons, NY, 1988, pp. 1796–1810.

11. Paul, Richard P., C. N. Stevenson, "Kinematics of Robot Wrists," The International Journal of Robotics Research, Vol. 2, No. 1, Spring 1983, pp. 31–38.

12. Barker, Keith, "Improved Robot-Joint Calculations," NASA Tech Briefs, September 1988, p. 79.

13. Ardayfio, D. D., R. Kapur, S. B. Yang, W. A. Watson, "Micras, Microcomputer Interactive Codes for Robot Analysis and Simulation," Mechanisms and Machine Theory, Vol. 20, No. 4, 1985, pp. 271–284.

14. Garcia, Mario, S. B. Niku, "Finger-Spelling Hand," masters thesis, Mechanical Engineering, Cal Poly, San Luis Obispo, CA, 2009.

Problems

The isometric grid (Figure 2.43) is provided to you for use with the problems in this chapter. It is meant to be used as a tracing grid for drawing 3-D shapes and objects such as robots, frames, and transformations. Please make copies of the grid for each problem that requires graphical representation of the results. The grid is also available commercially.

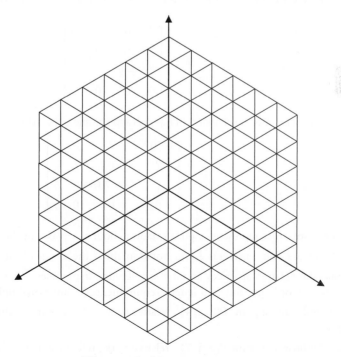

Figure 2.43 Isometric grid.

2.1. Write a unit vector in matrix form that describes the direction of the cross product of $\mathbf{p} = 5\mathbf{i} + 3\mathbf{k}$ and $\mathbf{q} = 3\mathbf{i} + 4\mathbf{j} + 5\mathbf{k}$.

2.2. A vector $\mathbf{p}$ is 8 units long and is perpendicular to vectors $\mathbf{q}$ and $\mathbf{r}$ described below. Express the vector in matrix form.

$$\mathbf{q}_{unit} = \begin{bmatrix} 0.3 \\ q_y \\ 0.4 \\ 0 \end{bmatrix} \quad \mathbf{r}_{unit} = \begin{bmatrix} r_x \\ 0.5 \\ 0.4 \\ 0 \end{bmatrix}$$

2.3. Will the three vectors $\mathbf{p}$, $\mathbf{q}$, and $\mathbf{r}$ in Problem 2.2 form a traditional frame? If not, find the necessary unit vector $\mathbf{s}$ to form a frame between $\mathbf{p}$, $\mathbf{q}$, and $\mathbf{s}$.

2.4. Suppose that instead of a frame, a point $P = (3, 5, 7)^T$ in space was translated a distance of $d = (2, 3, 4)^T$. Find the new location of the point relative to the reference frame.

2.5. The following frame B was moved a distance of $d = (5, 2, 6)^T$. Find the new location of the frame relative to the reference frame.

$$B = \begin{bmatrix} 0 & 1 & 0 & 2 \\ 1 & 0 & 0 & 4 \\ 0 & 0 & -1 & 6 \\ 0 & 0 & 0 & 1 \end{bmatrix}$$

2.6. For frame F, find the values of the missing elements and complete the matrix representation of the frame.

$$F = \begin{bmatrix} ? & 0 & -1 & 5 \\ ? & 0 & 0 & 3 \\ ? & -1 & 0 & 2 \\ 0 & 0 & 0 & 1 \end{bmatrix}$$

2.7. Find the values of the missing elements of frame B and complete the matrix representation of the frame.

$$B = \begin{bmatrix} 0.707 & ? & 0 & 2 \\ ? & 0 & 1 & 4 \\ ? & -0.707 & 0 & 5 \\ 0 & 0 & 0 & 1 \end{bmatrix}$$

2.8. Derive the matrix that represents a pure rotation about the y-axis of the reference frame.

2.9. Derive the matrix that represents a pure rotation about the z-axis of the reference frame.

2.10. Verify that the rotation matrices about the reference frame axes follow the required constraint equations set by orthogonality and length requirements of directional unit vectors.

2.11. Find the coordinates of point $P(2, 3, 4)^T$ relative to the reference frame after a rotation of $45°$ about the x-axis.

2.12. Find the coordinates of point $P(3, 5, 7)^T$ relative to the reference frame after a rotation of $30°$ about the z-axis.

2.13. Find the new location of point $P(1, 2, 3)^T$ relative to the reference frame after a rotation of $30°$ about the z-axis followed by a rotation of $60°$ about the y-axis.

2.14. A point P in space is defined as $^BP = (5, 3, 4)^T$ relative to frame B which is attached to the origin of the reference frame A and is parallel to it. Apply the following transformations to frame B and find ^AP. Using the 3-D grid, plot the transformations and the result and verify it. Also verify graphically that you would not get the same results if you apply the transformations relative to the current frame:

- Rotate $90°$ about x-axis; then
- Translate 3 units about y-axis, 6 units about z-axis, and 5 units about x-axis; then,
- Rotate $90°$ about z-axis.

2.15. A point P in space is defined as $^BP = (2, 3, 5)^T$ relative to frame B which is attached to the origin of the reference frame A and is parallel to it. Apply the following transformations to frame B and find ^AP. Using the 3-D grid, plot the transformations and the result and verify it:

- Rotate $90°$ about x-axis, then
- Rotate $90°$ about local a-axis, then
- Translate 3 units about y-, 6 units about z-, and 5 units about x-axes.

2.16. A frame B is rotated $90°$ about the z-axis, then translated 3 and 5 units relative to the n- and o-axes respectively, then rotated another $90°$ about the n-axis, and finally, $90°$ about the y-axis. Find the new location and orientation of the frame.

$$B = \begin{bmatrix} 0 & 1 & 0 & 1 \\ 1 & 0 & 0 & 1 \\ 0 & 0 & -1 & 1 \\ 0 & 0 & 0 & 1 \end{bmatrix}$$

2.17. The frame B of Problem 2.16 is rotated $90°$ about the a-axis, $90°$ about the y-axis, then translated 2 and 4 units relative to the x- and y-axes respectively, then rotated another $90°$ about the n-axis. Find the new location and orientation of the frame.

$$B = \begin{bmatrix} 0 & 1 & 0 & 1 \\ 1 & 0 & 0 & 1 \\ 0 & 0 & -1 & 1 \\ 0 & 0 & 0 & 1 \end{bmatrix}$$

2.18. Show that rotation matrices about the y- and the z-axes are unitary.

2.19. Calculate the inverse of the following transformation matrices:

$$T_1 = \begin{bmatrix} 0.527 & -0.574 & 0.628 & 2 \\ 0.369 & 0.819 & 0.439 & 5 \\ -0.766 & 0 & 0.643 & 3 \\ 0 & 0 & 0 & 1 \end{bmatrix} \text{ and } T_2 = \begin{bmatrix} 0.92 & 0 & 0.39 & 5 \\ 0 & 1 & 0 & 6 \\ -0.39 & 0 & 0.92 & 2 \\ 0 & 0 & 0 & 1 \end{bmatrix}$$

2.20. Calculate the inverse of the matrix B of Problem 2.17.

2.21. Write the correct sequence of movements that must be made in order to restore the original orientation of the spherical coordinates and make it parallel to the reference frame. About what axes are these rotations supposed to be?

2.22. A spherical coordinate system is used to position the hand of a robot. In a certain situation, the hand orientation of the frame is later restored in order to be parallel to the reference frame, and the matrix representing it is described as:

$$T_{sph} = \begin{bmatrix} 1 & 0 & 0 & 3.1375 \\ 0 & 1 & 0 & 2.195 \\ 0 & 0 & 1 & 3.214 \\ 0 & 0 & 0 & 1 \end{bmatrix}$$

- Find the necessary values of r, β, γ to achieve this location.
- Find the components of the original matrix **n**, **o**, **a** vectors for the hand before the orientation was restored.

2.23. Suppose that a robot is made of a Cartesian and RPY combination of joints. Find the necessary RPY angles to achieve the following:

$$T = \begin{bmatrix} 0.527 & -0.574 & 0.628 & 4 \\ 0.369 & 0.819 & 0.439 & 6 \\ -0.766 & 0 & 0.643 & 9 \\ 0 & 0 & 0 & 1 \end{bmatrix}$$

2.24. Suppose that a robot is made of a Cartesian and Euler combination of joints. Find the necessary Euler angles to achieve the following:

$$T = \begin{bmatrix} 0.527 & -0.574 & 0.628 & 4 \\ 0.369 & 0.819 & 0.439 & 6 \\ -0.766 & 0 & 0.643 & 9 \\ 0 & 0 & 0 & 1 \end{bmatrix}$$

2.25. Assume that the three Euler angles used with a robot are 30°, 40°, 50° respectively. Determine what angles should be used to achieve the same result if RPY is used instead.

2.26. A frame ^UB was moved along its own o-axis a distance of 6 units, then rotated about its n-axis an angle of 60°, then translated about the z-axis for 3 units, followed by a rotation of 60° about the z-axis, and finally rotated about x-axis for 45°.

- Calculate the total transformation performed.
- What angles and movements would we have to make if we were to create the same location and orientation using Cartesian and Euler configurations?

2.27. A frame ^UF was moved along its own n-axis a distance of 5 units, then rotated about its o-axis an angle of 60°, followed by a rotation of 60° about the z-axis, then translated about its a-axis for 3 units, and finally rotated 45° about the x-axis.

- Calculate the total transformation performed.
- What angles and movements would we have to make if we were to create the same location and orientation using Cartesian and RPY configurations?

2.28. Frames describing the base of a robot and an object are given relative to the Universe frame.

- Find a transformation $^R T_H$ of the robot configuration if the hand of the robot is to be placed on the object.
- By inspection, show whether this robot can be a 3-axis spherical robot, and if so, find α, β, r.
- Assuming that the robot is a 6-axis robot with Cartesian and Euler coordinates, find $p_x, p_y, p_z, \phi, \theta, \psi$.

$$
^U T_{obj} = \begin{bmatrix} 1 & 0 & 0 & 1 \\ 0 & 0 & -1 & 4 \\ 0 & 1 & 0 & 0 \\ 0 & 0 & 0 & 1 \end{bmatrix}
\qquad
^U T_R = \begin{bmatrix} 0 & -1 & 0 & 2 \\ 1 & 0 & 0 & -1 \\ 0 & 0 & 1 & 0 \\ 0 & 0 & 0 & 1 \end{bmatrix}
$$

2.29. A 3-DOF robot arm has been designed for applying paint on flat walls, as shown.

- Assign coordinate frames as necessary based on the D-H representation.
- Fill out the parameters table.
- Find the $^U T_H$ matrix.

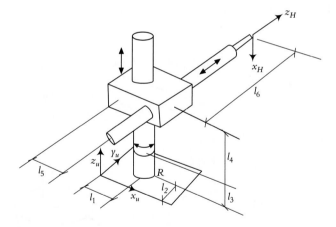

Figure P.2.29

2.30. In the 2-DOF robot shown, the transformation matrix $^0 T_H$ is given in symbolic form, as well as in numerical form for a specific location. The length of each link l_1 and l_2 is 1 ft. Calculate the values of θ_1 and θ_2 for the given location.

$$
^0 T_H = \begin{bmatrix} C_{12} & -S_{12} & 0 & l_2 C_{12} + l_1 C_1 \\ S_{12} & C_{12} & 0 & l_2 S_{12} + l_1 S_1 \\ 0 & 0 & 1 & 0 \\ 0 & 0 & 0 & 1 \end{bmatrix}
= \begin{bmatrix} -0.2924 & -0.9563 & 0 & 0.6978 \\ 0.9563 & -0.2924 & 0 & 0.8172 \\ 0 & 0 & 1 & 0 \\ 0 & 0 & 0 & 1 \end{bmatrix}
$$

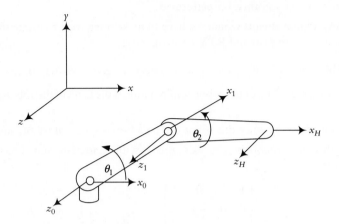

Figure P.2.30

2.31. For the following SCARA-type robot:

- Assign the coordinate frames based on the D-H representation.
- Fill out the parameters table.
- Write all the A matrices.
- Write the $^{U}T_{H}$ matrix in terms of the A matrices.

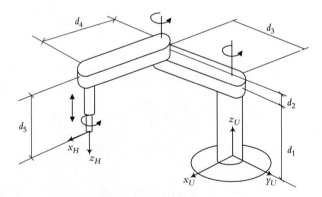

Figure P.2.31

2.32. A special 3-DOF spraying robot has been designed as shown:

- Assign the coordinate frames based on the D-H representation.
- Fill out the parameters table.
- Write all the A matrices.
- Write the $^{U}T_{H}$ matrix in terms of the A matrices.

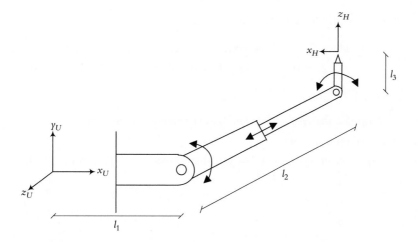

Figure P.2.32

2.33. For the Unimation Puma 562, 6-axis robot shown,

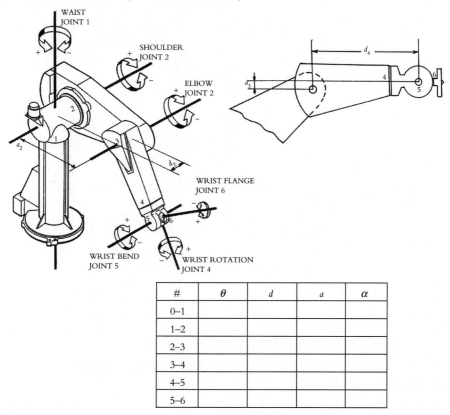

#	θ	d	a	α
0–1				
1–2				
2–3				
3–4				
4–5				
5–6				

Figure P.2.33 Puma 562. (Reprinted with permission from Staubli Robotics.)

- Assign the coordinate frames based on the D-H representation.
- Fill out the parameters table.
- Write all the A matrices.

- Find the ^RT_H matrix for the following values:

$$\text{Base height} = 27 \text{ in.}, \ d_2 = 6 \text{ in.}, \ a_2 = 15 \text{ in.}, \ a_3 = 1 \text{ in}, \ d_4 = 18 \text{ in.}$$

$$\theta_1 = 0°, \ \theta_2 = 45°, \ \theta_3 = 0°, \ \theta_4 = 0°, \ \theta_5 = -45°, \ \theta_6 = 0°$$

2.34. For the given 4-DOF robot:

- Assign appropriate frames for the Denavit-Hartenberg representation.
- Fill out the parameters table.
- Write an equation in terms of A matrices that shows how ^UT_H can be calculated.

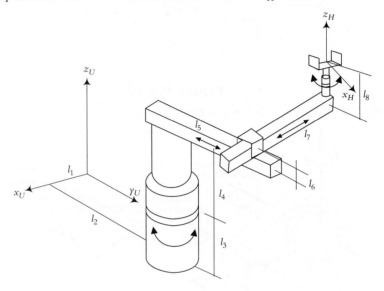

Figure P.2.34

#	θ	d	a	α
0–1				
1–2				
2–3				
3–				

2.35. For the given 4-DOF robot designed for a specific operation:

- Assign appropriate frames for the Denavit-Hartenberg representation.
- Fill out the parameters table.
- Write an equation in terms of A matrices that shows how ^UT_H can be calculated.

#	θ	d	a	α
0–1				
1–2				
2–3				
3–				

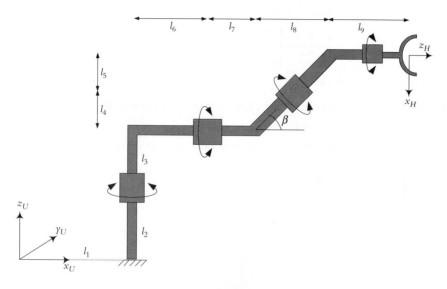

Figure P.2.35

2.36. For the given specialty designed 4–DOF robot:

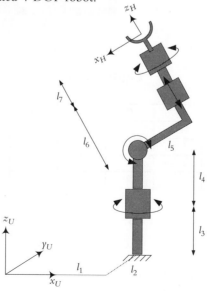

Figure P.2.36

- Assign appropriate frames for the Denavit-Hartenberg representation.
- Fill out the parameters table.
- Write an equation in terms of A matrices that shows how $^U T_H$ can be calculated.

#	θ	d	a	α
0–1				
1–2				
2–3				
3–				

2.37. For the given 3-DOF robot:

- Assign appropriate frames for the Denavit-Hartenberg representation.
- Fill out the parameters table.
- Write an equation in terms of A matrices that shows how $^U T_H$ can be calculated.

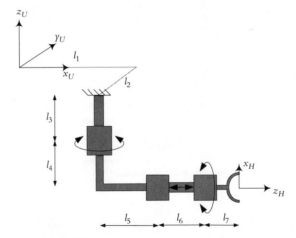

Figure P.2.37

#	θ	d	a	α
0–1				
1–2				
2–				

2.38. For the given 4-DOF robot:

- Assign appropriate frames for the Denavit-Hartenberg representation.
- Fill out the parameters table.
- Write an equation in terms of A matrices that shows how ^UT_H can be calculated.

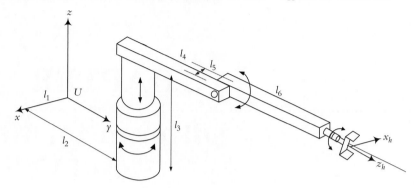

Figure P.2.38

#	θ	d	a	α
0–1				
1–2				
2–3				
3–				

2.39. Derive the inverse kinematic equations for the robot of Problem 2.36.

2.40. Derive the inverse kinematic equations for the robot of Problem 2.37.

CHAPTER 3

Differential Motions and Velocities

3.1 Introduction

Differential motions are small movements of mechanisms (e.g., robots) that can be used to derive velocity relationships between different parts of the mechanism. A differential motion is, by definition, a small movement. Therefore, if it is measured in—or calculated for—a small period of time (a differential or small time), a velocity relationship can be found.

In this chapter, we will learn about differential motions of frames relative to a fixed frame, differential motions of robot joints relative to a fixed frame, Jacobians, and robot velocity relationships. This chapter contains a fair amount of velocity terms that you have seen in your dynamics course. If you do not remember the material well, a review of it is recommended before you continue.

3.2 Differential Relationships

First, let's see what the differential relationships are. To do this, we will consider a simple 2-DOF mechanism, as shown in Figure 3.1, where each link can independently rotate. The rotation of the first link θ_1 is measured relative to the reference frame, whereas the rotation of the second link θ_2 is measured relative to the first link. This would be similar to a robot, where each link's movement is measured relative to a current frame attached to the previous link.

The velocity of point B can be calculated as follows:

$$
\begin{aligned}
\mathbf{v}_B &= \mathbf{v}_A + \mathbf{v}_{B/A} \\
&= l_1\dot{\theta}_1[\perp \text{ to } l_1] + l_2(\dot{\theta}_1 + \dot{\theta}_2)[\perp \text{ to } l_2] \\
&= -l_1\dot{\theta}_1\sin\theta_1\mathbf{i} + l_1\dot{\theta}_1\cos\theta_1\mathbf{j} - l_2(\dot{\theta}_1 + \dot{\theta}_2)\sin(\theta_1 + \theta_2)\mathbf{i} + l_2(\dot{\theta}_1 + \dot{\theta}_2)\cos(\theta_1 + \theta_2)\mathbf{j}
\end{aligned}
\tag{3.1}
$$

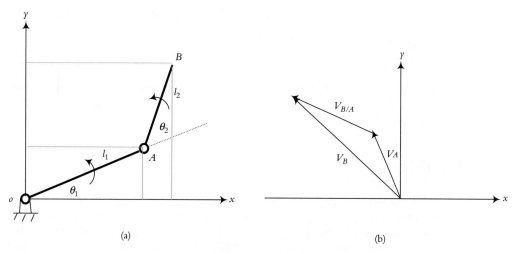

Figure 3.1   (a) 2-DOF planar mechanism, (b) Velocity diagram.

Writing the above velocity equation in matrix form will yield the following:

$$\begin{bmatrix} v_{B_x} \\ v_{B_y} \end{bmatrix} = \begin{bmatrix} -l_1 \sin \theta_1 - l_2 \sin (\theta_1 + \theta_2) & -l_2 \sin (\theta_1 + \theta_2) \\ l_1 \cos \theta_1 + l_2 \cos (\theta_1 + \theta_2) & l_2 \cos (\theta_1 + \theta_2) \end{bmatrix} \begin{bmatrix} \dot{\theta}_1 \\ \dot{\theta}_2 \end{bmatrix} \tag{3.2}$$

The left side of the equation represents the x and y components of the velocity of point B. As you can see, if the elements of the right-hand side of the equation are multiplied by the corresponding angular velocities of the two links, the velocity of point B can be found.

Next, instead of deriving the velocity equation directly from the velocity relationship, we will try to find the same velocity relationship by differentiating the equations that describe the position of point B, as follows:

$$\begin{cases} x_B \doteq l_1 \cos \theta_1 + l_2 \cos (\theta_1 + \theta_2) \\ y_B = l_1 \sin \theta_1 + l_2 \sin (\theta_1 + \theta_2) \end{cases} \tag{3.3}$$

Differentiating the above equation with respect to the two variables θ_1 and θ_2 will yield:

$$\begin{cases} dx_B = -l_1 \sin \theta_1 \, d\theta_1 - l_2 \sin (\theta_1 + \theta_2)(d\theta_1 + d\theta_2) \\ dy_B = l_1 \cos \theta_1 \, d\theta_1 + l_2 \cos (\theta_1 + \theta_2)(d\theta_1 + d\theta_2) \end{cases} \tag{3.4}$$

and in matrix form:

$$\begin{bmatrix} dx_B \\ dy_B \end{bmatrix} = \begin{bmatrix} -l_1 \sin \theta_1 - l_2 \sin (\theta_1 + \theta_2) & -l_2 \sin (\theta_1 + \theta_2) \\ l_1 \cos \theta_1 + l_2 \cos (\theta_1 + \theta_2) & l_2 \cos (\theta_1 + \theta_2) \end{bmatrix} \begin{bmatrix} d\theta_1 \\ d\theta_2 \end{bmatrix} \tag{3.5}$$

| Differential motion of B | Jacobian | Differential motion of joints |

Notice the similarities between Equation (3.2) and Equation (3.5). Although the two equations are similar in content and form, the difference is that Equation (3.2) is the velocity relationship, whereas Equation (3.5) is the differential motion relationship. If both sides of Equation (3.5) are divided by dt, since dx_B/dt is v_{B_x} and $d\theta_1/dt$ is $\dot{\theta}_1$, etc.,

Equation (3.6) will be exactly the same as Equation (3.2), as follows:

$$\begin{bmatrix} dx_B \\ dy_B \end{bmatrix} \bigg/ dt = \begin{bmatrix} -l_1 \sin \theta_1 - l_2 \sin (\theta_1 + \theta_2) & -l_2 \sin (\theta_1 + \theta_2) \\ l_1 \cos \theta_1 + l_2 \cos (\theta_1 + \theta_2) & l_2 \cos (\theta_1 + \theta_2) \end{bmatrix} \begin{bmatrix} d\theta_1 \\ d\theta_2 \end{bmatrix} \bigg/ dt \quad (3.6)$$

Similarly, in a robot with many degrees of freedom, the joint differential motions, or velocities, can be related to the differential motion, or velocity, of the hand by the same technique.

3.3 Jacobian

The Jacobian is a representation of the geometry of the elements of a mechanism in time. It allows the conversion of differential motions or velocities of individual joints to differential motions or velocities of points of interest (e.g., the end effector). It also relates the individual joint motions to overall mechanism motions. Jacobian is time-related; since the values of joint angles vary in time, the magnitude of the elements of the Jacobian vary in time as well.

To better understand Jacobians, let's consider a simple 2-DOF robot in three positions, as shown in Figure 3.2. Obviously, if joint 1 of the robot moves an angle of θ, the magnitude and direction of the resulting motion of the end of the robot will be very different in each case. This dependence of the resulting motion on the geometry of the mechanism is expressed by the Jacobian. Therefore, the Jacobian is a representation of the geometry and the interrelationship between different parts of the mechanism and where they are at any given time. Clearly, as time goes on and the relative positions of the different parts of the mechanism change, the Jacobian will also change.

As you noted in Section 3.2, the Jacobian was formed from the position equations that were differentiated with respect to θ_1 and θ_2. Therefore, the Jacobian can be calculated by taking the derivatives of each position equation with respect to all variables.

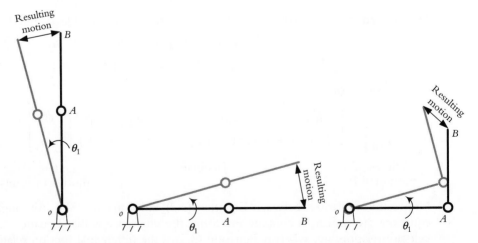

Figure 3.2 Resulting motions of the robot are dependent on the geometry of the robot.

Suppose we have a set of equations y_i in terms of a set of variables x_j as:

$$y_i = f_i(x_1, x_2, x_3, \ldots, x_j) \tag{3.7}$$

The differential change in y_i as a result of a differential change in x_j will be:

$$\begin{cases} \delta y_1 = \dfrac{\partial f_1}{\partial x_1}\delta x_1 + \dfrac{\partial f_1}{\partial x_2}\delta x_2 + \cdots + \dfrac{\partial f_1}{\partial x_j}\delta x_j \\[2ex] \delta y_2 = \dfrac{\partial f_2}{\partial x_1}\delta x_1 + \dfrac{\partial f_2}{\partial x_2}\delta x_2 + \cdots + \dfrac{\partial f_2}{\partial x_j}\delta x_j \\[2ex] \vdots \\[1ex] \delta y_i = \dfrac{\partial f_i}{\partial x_1}\delta x_1 + \dfrac{\partial f_i}{\partial x_2}\delta x_2 + \cdots + \dfrac{\partial f_i}{\partial x_j}\delta x_j \end{cases} \tag{3.8}$$

Equation (3.8) can be written in matrix form, representing the differential relationship between individual variables and the functions. The matrix encompassing this relationship is the Jacobian, as shown in Equation (3.9). Therefore, the Jacobian can be calculated by taking the derivative of each equation with respect to all variables. We will apply the same principle for the calculation of the Jacobian of a robot.

$$\begin{bmatrix} \delta y_1 \\ \delta y_2 \\ \vdots \\ \delta y_i \end{bmatrix} = \begin{bmatrix} \dfrac{\partial f_1}{\partial x_1} & \dfrac{\partial f_1}{\partial x_2} & & \dfrac{\partial f_1}{\partial x_j} \\[2ex] \dfrac{\partial f_2}{\partial x_1} & \cdots & \cdots & \cdots \\[1ex] \vdots & & & \vdots \\[1ex] \dfrac{\partial f_i}{\partial x_1} & & \dfrac{\partial f_i}{\partial x_j} & \end{bmatrix} \begin{bmatrix} \delta x_1 \\ \delta x_2 \\ \vdots \\ \delta x_j \end{bmatrix} \quad \text{or} \quad [\delta y_i] = \left[\dfrac{\partial f_i}{\partial x_j}\right][\delta x_j] \tag{3.9}$$

Using the same relationship as before and differentiating the position equations of a robot, we can write the following equation that relates joint differential motions of a robot to the differential motion of its hand frame:

$$\begin{bmatrix} dx \\ dy \\ dz \\ \delta x \\ \delta y \\ \delta z \end{bmatrix} = \begin{bmatrix} Robot \\ Jacobian \end{bmatrix} \begin{bmatrix} d\theta_1 \\ d\theta_2 \\ d\theta_3 \\ d\theta_4 \\ d\theta_5 \\ d\theta_6 \end{bmatrix} \quad \text{or} \quad [D] = [J][D_\theta] \tag{3.10}$$

where dx, dy, and dz in $[D]$ represent the differential motions of the hand along the x-, y-, and z-axes, δx, δy, and δz in $[D]$ represent the differential rotations of the hand around the x-, y-, and z-axes, and $[D_\theta]$ represents the differential motions of the joints. As was mentioned earlier, if these two matrices are divided by dt, they will represent velocities instead of differential motions. In this chapter, we will work with the differential motions rather than velocities, knowing that in all relationships, by simply dividing the differential motions by dt we can get the velocities.

Example 3.1

The Jacobian of a robot at a particular time is given. Calculate the linear and angular differential motions of the robot's hand frame for the given joint differential motions.

$$J = \begin{bmatrix} 2 & 0 & 0 & 0 & 1 & 0 \\ -1 & 0 & 1 & 0 & 0 & 0 \\ 0 & 1 & 0 & 0 & 0 & 0 \\ 0 & 0 & 0 & 2 & 0 & 0 \\ 0 & 0 & 1 & 0 & 0 & 0 \\ 0 & 0 & 0 & 0 & 0 & 1 \end{bmatrix} \quad D_\theta = \begin{bmatrix} 0 \\ 0.1 \\ -0.1 \\ 0 \\ 0 \\ 0.2 \end{bmatrix}$$

Solution: Substituting the above matrices into Equation (3.10), we get:

$$[D] = [J][D_\theta] = \begin{bmatrix} 2 & 0 & 0 & 0 & 1 & 0 \\ -1 & 0 & 1 & 0 & 0 & 0 \\ 0 & 1 & 0 & 0 & 0 & 0 \\ 0 & 0 & 0 & 2 & 0 & 0 \\ 0 & 0 & 1 & 0 & 0 & 0 \\ 0 & 0 & 0 & 0 & 0 & 1 \end{bmatrix} \begin{bmatrix} 0 \\ 0.1 \\ -0.1 \\ 0 \\ 0 \\ 0.2 \end{bmatrix} = \begin{bmatrix} 0 \\ -0.1 \\ 0.1 \\ 0 \\ -0.1 \\ 0.2 \end{bmatrix} = \begin{bmatrix} dx \\ dy \\ dz \\ \delta x \\ \delta y \\ \delta z \end{bmatrix} \blacksquare$$

3.4 Differential versus Large-Scale Motions

In our discussion of transformations so far, we only considered rotations and translations that occurred at one particular time, but not continually. For example, we would consider a rotation of θ about an axis, then a translation of a certain amount along another axis, etc. All these transformations were singular and happened once. But what if transformations happen continually? To understand the difference between the two, let's consider the following scenario.

Imagine that a wheelchair rider starts at the origin, first moves a distance of l along a straight line, then rotates θ (Figure 3.3(a)) and ends up at point A. Now imagine that the rider first rotates θ and then moves in a straight line similar to the previous case

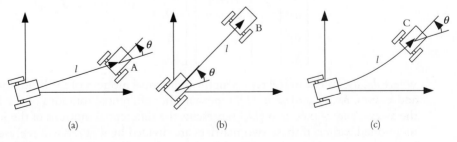

(a)   (b)   (c)

Figure 3.3 Differential motions versus non-differential motions.

(Figure 3.3(b)), ending up at point B. Obviously, although the final orientations are the same, the rider follows a different path and ends up in a different position. In both cases, each transformation was performed once. Now imagine that the rider would simultaneously rotate and translate the same total amounts (Figure 3.3(c)). Depending on whether or not he or she would start and stop both motions simultaneously, and depending on how fast these motions were, the rider follows a different path and ends up in a different place. The difference relates to the small motions that the wheelchair makes relative to time, and therefore, is the basis for differential motion analysis. The path and the final state of the wheelchair are functions of differential motions (or velocities) and their sequence with respect to time. The same analysis applies to robots depending on whether motions occur at one instant or are continuous, regardless of whether the robot is a mobile robot or a manipulator.

3.5 Differential Motions of a Frame versus a Robot

Suppose that a frame moves a differential amount relative to the reference frame. We can either look at the differential motions of the frame without regard to what causes the motions, or we can include the mechanism that causes the motion. In the first case, we will study the motions of the frame and the changes in the representation of the frame (Figure 3.4(a)). In the second case, we analyze the differential motions of the mechanism that causes the motions and how it relates to the motions of the frame (Figure 3.4(b)). As you can see in Figure 3.4(c), the differential motions of the hand frame of the robot are caused by the differential motions in each of the joints of the robot. Therefore, as the joints of the robot move a differential amount, the hand moves a differential amount, consequently moving the frame representing it a differential amount. In this way, we relate the motions of the robot to the motions of the frame.

In reality, this means the following: suppose you have a robot welding two pieces together. For best results, the robot should move at a constant speed. This means that the differential motions of the hand frame must be defined to represent a constant speed in a particular direction. This relates to the differential motion of the frame. However, the motion is caused by the robot (it could actually be caused by something else; we are using a robot, so we must relate it to the robot's motions). Therefore, we have to calculate the

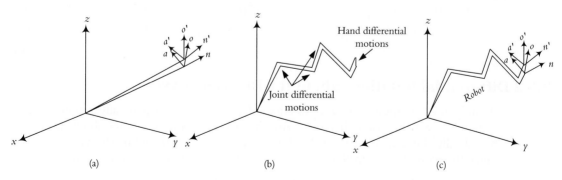

Figure 3.4 (a) Differential motions of a frame, (b) differential motions of the robot joints and the end-plate, (c) differential motions of a frame caused by the differential motions of a robot.

speeds of each and every joint at any instant, such that the total motion caused by the robot will be equal to the desired speed of the frame. In this section, we will first study the differential motions of a frame. Then, we will study the differential motions of a robot mechanism. Finally, we will relate the two together.

3.6 Differential Motions of a Frame

Differential motions of a frame can be divided into the following:

- Differential translations
- Differential rotations
- Differential transformations (combinations of translations and rotations)

3.6.1 Differential Translations

A differential translation is the translation of a frame at differential values. Therefore, it can be represented by *Trans(dx, dy, dz)*. This means the frame has moved a differential amount along the *x-*, *y-*, and *z*-axes.

Example 3.2

A frame *B* has translated a differential amount of *Trans*(0.01, 0.05, 0.03) units. Find its new location and orientation.

$$B = \begin{bmatrix} 0.707 & 0 & -0.707 & 5 \\ 0 & 1 & 0 & 4 \\ 0.707 & 0 & 0.707 & 9 \\ 0 & 0 & 0 & 1 \end{bmatrix}$$

Solution: Since the differential motion is only a translation, the orientation of the frame should not be affected. The new location of the frame is:

$$B = \begin{bmatrix} 1 & 0 & 0 & 0.01 \\ 0 & 1 & 0 & 0.05 \\ 0 & 0 & 1 & 0.03 \\ 0 & 0 & 0 & 1 \end{bmatrix} \times \begin{bmatrix} 0.707 & 0 & -0.707 & 5 \\ 0 & 1 & 0 & 4 \\ 0.707 & 0 & 0.707 & 9 \\ 0 & 0 & 0 & 1 \end{bmatrix} = \begin{bmatrix} 0.707 & 0 & -0.707 & 5.01 \\ 0 & 1 & 0 & 4.05 \\ 0.707 & 0 & 0.707 & 9.03 \\ 0 & 0 & 0 & 1 \end{bmatrix}$$

∎

3.6.2 Differential Rotations about the Reference Axes

A differential rotation is a small rotation of the frame. It is generally represented by *Rot(q, dθ)*, which means that the frame has rotated an angle of *dθ* about an axis *q*. Specifically, differential rotations about the *x-*, *y-*, and *z*-axes are defined by *δx*, *δy*, and *δz*. Since the rotations are small amounts, we can use the following approximations:

$$\sin \delta x = \delta x \text{ (in radians)}$$
$$\cos \delta x = 1$$

Then, the rotation matrices representing differential rotations about the x-, y-, and z-axes will be:

$$Rot(x, \delta x) = \begin{bmatrix} 1 & 0 & 0 & 0 \\ 0 & 1 & -\delta x & 0 \\ 0 & \delta x & 1 & 0 \\ 0 & 0 & 0 & 1 \end{bmatrix}, \ Rot(y, \delta y) = \begin{bmatrix} 1 & 0 & \delta y & 0 \\ 0 & 1 & 0 & 0 \\ -\delta y & 0 & 1 & 0 \\ 0 & 0 & 0 & 1 \end{bmatrix}, \ Rot(z, \delta z) = \begin{bmatrix} 1 & -\delta z & 0 & 0 \\ \delta z & 1 & 0 & 0 \\ 0 & 0 & 1 & 0 \\ 0 & 0 & 0 & 1 \end{bmatrix}$$

$$(3.11)$$

Similarly, we can also define differential rotations about the current axes as:

$$Rot(n, \delta n) = \begin{bmatrix} 1 & 0 & 0 & 0 \\ 0 & 1 & -\delta n & 0 \\ 0 & \delta n & 1 & 0 \\ 0 & 0 & 0 & 1 \end{bmatrix}, \ Rot(o, \delta o) = \begin{bmatrix} 1 & 0 & \delta o & 0 \\ 0 & 1 & 0 & 0 \\ -\delta o & 0 & 1 & 0 \\ 0 & 0 & 0 & 1 \end{bmatrix}, \ Rot(a, \delta a) = \begin{bmatrix} 1 & -\delta a & 0 & 0 \\ \delta a & 1 & 0 & 0 \\ 0 & 0 & 1 & 0 \\ 0 & 0 & 0 & 1 \end{bmatrix}$$

$$(3.12)$$

Notice that these matrices defy the rule we had established previously about the magnitude of unit vectors. For example, $\sqrt{1^2 + (\delta x)^2} > 1$. However, as you may remember, a differential value is assumed to be very small. In mathematics, higher-order differentials are considered negligible and are usually neglected. If we do neglect the higher-order differentials such as $(\delta x)^2$, the magnitude of the vectors remain acceptable.

As we have already seen, if the order of multiplication of matrices changes, the result will change as well. Therefore, in matrix multiplication, maintaining the order of matrices is very important. If we multiply two differential motions in different orders, we expectedly get two different results, as shown below:

$$Rot(x, \delta x)Rot(y, \delta y) = \begin{bmatrix} 1 & 0 & 0 & 0 \\ 0 & 1 & -\delta x & 0 \\ 0 & \delta x & 1 & 0 \\ 0 & 0 & 0 & 1 \end{bmatrix}\begin{bmatrix} 1 & 0 & \delta y & 0 \\ 0 & 1 & 0 & 0 \\ -\delta y & 0 & 1 & 0 \\ 0 & 0 & 0 & 1 \end{bmatrix} = \begin{bmatrix} 1 & 0 & \delta y & 0 \\ \delta x \delta y & 1 & -\delta x & 0 \\ -\delta y & \delta x & 1 & 0 \\ 0 & 0 & 0 & 1 \end{bmatrix}$$

$$Rot(y, \delta y)Rot(x, \delta x) = \begin{bmatrix} 1 & 0 & \delta y & 0 \\ 0 & 1 & 0 & 0 \\ -\delta y & 0 & 1 & 0 \\ 0 & 0 & 0 & 1 \end{bmatrix}\begin{bmatrix} 1 & 0 & 0 & 0 \\ 0 & 1 & -\delta x & 0 \\ 0 & \delta x & 1 & 0 \\ 0 & 0 & 0 & 1 \end{bmatrix} = \begin{bmatrix} 1 & \delta x \delta y & \delta y & 0 \\ 0 & 1 & -\delta x & 0 \\ -\delta y & \delta x & 1 & 0 \\ 0 & 0 & 0 & 1 \end{bmatrix}$$

However, if as before, we set higher-order differentials such as $\delta x \delta y$ to zero, the results are exactly the same. Consequently, for differential motions, the order of multiplication is no longer important and $Rot(x, \delta x)Rot(y, \delta y) = Rot(y, \delta y)Rot(x, \delta x)$. The same is true for other combinations of rotations, including differential rotations about the z-axis.

You may remember from your dynamics course that large-angle rotations about different axes are not commutative, and therefore, cannot be added in different orders. For example, as we have already seen, if you rotate an object 90° about the x-axis, followed by a 90° rotation about the z-axis, the result will be different if you reverse the order. However, velocities are commutative and can be added as vectors; therefore, $\Omega = \omega_x \mathbf{i} + \omega_y \mathbf{j} + \omega_z \mathbf{k}$, regardless of the order. This is true because, as the above discussion shows, if we neglect the higher-order differentials, the order of multiplication

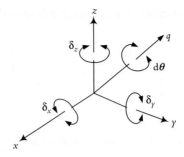

Figure 3.5   Differential rotations about a general axis q.

will be unimportant. Since velocities are in fact differential motions divided by time, the same is true for velocities.

3.6.3 Differential Rotation about a General Axis q

Based on the above, since the order of multiplication for differential rotations is not important, we can multiply differential rotations in any order. As a result, we can assume that a differential rotation about a general axis q is composed of three differential rotations about the three axes, in any order, or $(d\theta)\mathbf{q} = (\delta x)\mathbf{i} + (\delta y)\mathbf{j} + (\delta z)\mathbf{k}$ (Figure 3.5).

Consequently, a differential motion about any general axis q can be expressed as:

$$Rot(q, d\theta) = Rot(x, \delta x)Rot(y, \delta y)Rot(z, \delta z)$$

$$= \begin{bmatrix} 1 & 0 & 0 & 0 \\ 0 & 1 & -\delta x & 0 \\ 0 & \delta x & 1 & 0 \\ 0 & 0 & 0 & 1 \end{bmatrix} \begin{bmatrix} 1 & 0 & \delta y & 0 \\ 0 & 1 & 0 & 0 \\ -\delta y & 0 & 1 & 0 \\ 0 & 0 & 0 & 1 \end{bmatrix} \begin{bmatrix} 1 & -\delta z & 0 & 0 \\ \delta z & 1 & 0 & 0 \\ 0 & 0 & 1 & 0 \\ 0 & 0 & 0 & 1 \end{bmatrix} \tag{3.13}$$

$$= \begin{bmatrix} 1 & -\delta z & \delta y & 0 \\ \delta x \delta y + \delta z & -\delta x \delta y \delta z + 1 & -\delta x & 0 \\ -\delta y + \delta x \delta z & \delta x + \delta y \delta z & 1 & 0 \\ 0 & 0 & 0 & 1 \end{bmatrix}$$

If we neglect all higher-order differentials, we get:

$$Rot(q, d\theta) = Rot(x, \delta x)Rot(y, \delta y)Rot(z, \delta z) = \begin{bmatrix} 1 & -\delta z & \delta y & 0 \\ \delta z & 1 & -\delta x & 0 \\ -\delta y & \delta x & 1 & 0 \\ 0 & 0 & 0 & 1 \end{bmatrix} \tag{3.14}$$

Example 3.3

Find the total differential transformation caused by small rotations about the three axes of $\delta x = 0.1$, $\delta y = 0.05$, $\delta z = 0.02$ radians.

Solution: Substituting the given rotations in Equation (3.14), we get:

$$Rot(q, d\theta) = \begin{bmatrix} 1 & -\delta z & \delta y & 0 \\ \delta z & 1 & -\delta x & 0 \\ -\delta y & \delta x & 1 & 0 \\ 0 & 0 & 0 & 1 \end{bmatrix} = \begin{bmatrix} 1 & -0.02 & 0.05 & 0 \\ 0.02 & 1 & -0.1 & 0 \\ -0.05 & 0.1 & 1 & 0 \\ 0 & 0 & 0 & 1 \end{bmatrix}$$

Note that the lengths of the three directional unit vectors are 1.001, 1.005, and 1.006, respectively. If we assume that 0.1 radians (about 5.7°) is small (differential), these values are acceptably close to 1. Otherwise, we should use smaller values for differential angles. ∎

3.6.4 Differential Transformations of a Frame

The differential transformation of a frame is a combination of differential translations and rotations in any order. If we denote the original frame as T and assume that dT is the *change* in the frame T as a result of a differential transformation, then:

$$[T + dT] = [Trans(dx, dy, dz)\, Rot(q, d\theta)][T]$$

or

$$[dT] = [Trans(dx, dy, dz)\, Rot(q, d\theta) - I][T] \qquad (3.15)$$

where I is a unit matrix. Equation (3.15) can be written as:

$$[dT] = [\Delta][T]$$

where

$$[\Delta] = [Trans(dx, dy, dz) \times Rot(q, d\theta) - I] \qquad (3.16)$$

$[\Delta]$ (or simply Δ) is called *differential operator*. It is the product of differential translations and rotations, minus a unit matrix. Multiplying a frame by the differential operator $[\Delta]$ will yield the change in the frame. The differential operator can be found by multiplying the matrices and subtracting the unit matrix, as follows:

$$\Delta = Trans(dx, dy, dz) \times Rot(q, d\theta) - I$$

$$= \begin{bmatrix} 1 & 0 & 0 & dx \\ 0 & 1 & 0 & dy \\ 0 & 0 & 1 & dz \\ 0 & 0 & 0 & 1 \end{bmatrix} \begin{bmatrix} 1 & -\delta z & \delta y & 0 \\ \delta z & 1 & -\delta x & 0 \\ -\delta y & \delta x & 1 & 0 \\ 0 & 0 & 0 & 1 \end{bmatrix} - \begin{bmatrix} 1 & 0 & 0 & 0 \\ 0 & 1 & 0 & 0 \\ 0 & 0 & 1 & 0 \\ 0 & 0 & 0 & 1 \end{bmatrix} \qquad (3.17)$$

$$\Delta = \begin{bmatrix} 0 & -\delta z & \delta y & dx \\ \delta z & 0 & -\delta x & dy \\ -\delta y & \delta x & 0 & dz \\ 0 & 0 & 0 & 0 \end{bmatrix}$$

As you can see, the differential operator is *not* a transformation matrix, or a frame. It does not follow the required format either; it is only an operator, and it yields the changes in a frame.

Example 3.4

Write the differential operator matrix for the following differential transformations: $dx = 0.05$, $dy = 0.03$, $dz = 0.01$ units and $\delta x = 0.02$, $\delta y = 0.04$, $\delta z = 0.06$ radians.

Solution: Substituting the given values into Equation (3.17), we get:

$$\Delta = \begin{bmatrix} 0 & -0.06 & 0.04 & 0.05 \\ 0.06 & 0 & -0.02 & 0.03 \\ -0.04 & 0.02 & 0 & 0.01 \\ 0 & 0 & 0 & 0 \end{bmatrix}$$

∎

Example 3.5

Find the effect of a differential rotation of 0.1 rad about the y-axis followed by a differential translation of $[0.1, 0, 0.2]$ on the given frame B.

$$B = \begin{bmatrix} 0 & 0 & 1 & 10 \\ 1 & 0 & 0 & 5 \\ 0 & 1 & 0 & 3 \\ 0 & 0 & 0 & 1 \end{bmatrix}$$

Solution: As we saw before, the change in the frame can be found by pre-multiplying the frame with the differential operator. Substituting the given information and multiplying the matrices, we will get:

For $dx = 0.1$, $dy = 0$, $dz = 0.2$, $\delta x = 0$, $\delta y = 0.1$, $\delta z = 0$

$$[dB] = [\Delta][B] = \begin{bmatrix} 0 & 0 & 0.1 & 0.1 \\ 0 & 0 & 0 & 0 \\ -0.1 & 0 & 0 & 0.2 \\ 0 & 0 & 0 & 0 \end{bmatrix} \begin{bmatrix} 0 & 0 & 1 & 10 \\ 1 & 0 & 0 & 5 \\ 0 & 1 & 0 & 3 \\ 0 & 0 & 0 & 1 \end{bmatrix} = \begin{bmatrix} 0 & 0.1 & 0 & 0.4 \\ 0 & 0 & 0 & 0 \\ 0 & 0 & -0.1 & -0.8 \\ 0 & 0 & 0 & 0 \end{bmatrix}$$

∎

3.7 Interpretation of the Differential Change

The matrix dT in Equations (3.15) and (3.16) represents the changes in a frame as a result of differential motions. The elements of this matrix are:

$$dT = \begin{bmatrix} dn_x & do_x & da_x & dp_x \\ dn_y & do_y & da_y & dp_y \\ dn_z & do_z & da_z & dp_z \\ 0 & 0 & 0 & 0 \end{bmatrix} \tag{3.18}$$

The dB matrix in Example 3.5 represents the change in the frame B, as shown in Equation (3.18). Therefore, each element of the matrix represents the change in the corresponding element of the frame. As an example, this means that the frame moved a differential amount of 0.4 units along the x-axis, zero along the y-axis, and a differential amount of -0.8 along the z-axis. It also rotated such that there was no change in its **n**-vector, there was a change of 0.1 in the o_x component of the **o**-vector, and a change of -0.1 in the a_z component of the **a**-vector.

The new location and orientation of the frame after the differential motions can be found by adding the change to the frame:

$$T_{new} = dT + T_{old} \tag{3.19}$$

Example 3.6

Find the location and the orientation of frame B of Example 3.5 after the move.

Solution: The new location and orientation of the frame can be found by adding the changes to the original values. The result is:

$B_{new} = B_{original} + dB$

$$
= \begin{bmatrix} 0 & 0 & 1 & 10 \\ 1 & 0 & 0 & 5 \\ 0 & 1 & 0 & 3 \\ 0 & 0 & 0 & 1 \end{bmatrix} + \begin{bmatrix} 0 & 0.1 & 0 & 0.4 \\ 0 & 0 & 0 & 0 \\ 0 & 0 & -0.1 & -0.8 \\ 0 & 0 & 0 & 0 \end{bmatrix} = \begin{bmatrix} 0 & 0.1 & 1 & 10.4 \\ 1 & 0 & 0 & 5 \\ 0 & 1 & -0.1 & 2.2 \\ 0 & 0 & 0 & 1 \end{bmatrix}
$$

■

3.8 Differential Changes between Frames

The differential operator Δ in Equation (3.16) represents a differential operator relative to the fixed reference frame and is technically $^U\Delta$. However, it is possible to define another differential operator, this time relative to the current frame itself, that will enable us to calculate the same changes in the frame. Since the differential operator relative to the frame $(^T\Delta)$ is relative to a current frame, to find the changes in the frame we must post-multiply the frame by $^T\Delta$ (as we did in Chapter 2). The result will be the same, since both operations represent the same changes in the frame. Then:

$$
\begin{aligned}
[dT] &= [\Delta][T] = [T]\left[{}^T\Delta\right] \\
\rightarrow \quad \left[T^{-1}\right][\Delta][T] &= \left[T^{-1}\right][T]\left[{}^T\Delta\right] \\
\rightarrow \quad \left[{}^T\Delta\right] &= \left[T^{-1}\right][\Delta][T]
\end{aligned}
\tag{3.20}
$$

Therefore, Equation (3.20) can be used to calculate the differential operator relative to the frame, $^T\Delta$. We can multiply the matrices in Equation (3.20) and simplify the result as follows. Assuming that the frame T is represented by an **n**, **o**, **a**, **p** matrix, we get:

$$T^{-1} = \begin{bmatrix} n_x & n_y & n_z & -\mathbf{p} \cdot \mathbf{n} \\ o_x & o_y & o_z & -\mathbf{p} \cdot \mathbf{o} \\ a_x & a_y & a_z & -\mathbf{p} \cdot \mathbf{a} \\ 0 & 0 & 0 & 1 \end{bmatrix} \quad \text{and} \quad \Delta = \begin{bmatrix} 0 & -\delta z & \delta y & dx \\ \delta z & 0 & -\delta x & dy \\ -\delta y & \delta x & 0 & dz \\ 0 & 0 & 0 & 0 \end{bmatrix}$$

$$[T^{-1}][\Delta][T] = {}^T\Delta = \begin{bmatrix} 0 & -{}^T\delta z & {}^T\delta y & {}^Tdx \\ {}^T\delta z & 0 & -{}^T\delta x & {}^Tdy \\ -{}^T\delta y & {}^T\delta x & 0 & {}^Tdz \\ 0 & 0 & 0 & 0 \end{bmatrix}$$

(3.21)

As you can see, the ${}^T\Delta$ is *made* to look exactly like the Δ matrix, but all elements are relative to the current frame, where these elements are found from the above multiplication of matrices, and are summarized as follows:

$$\begin{aligned} {}^T\delta x &= \boldsymbol{\delta} \cdot \mathbf{n} \\ {}^T\delta y &= \boldsymbol{\delta} \cdot \mathbf{o} \\ {}^T\delta z &= \boldsymbol{\delta} \cdot \mathbf{a} \\ {}^Tdx &= \mathbf{n} \cdot [\boldsymbol{\delta} \times \mathbf{p} + \mathbf{d}] \\ {}^Tdy &= \mathbf{o} \cdot [\boldsymbol{\delta} \times \mathbf{p} + \mathbf{d}] \\ {}^Tdz &= \mathbf{a} \cdot [\boldsymbol{\delta} \times \mathbf{p} + \mathbf{d}] \end{aligned}$$

(3.22)

See Paul[1] for the derivation of the above equations.

Example 3.7

Find ${}^B\Delta$ for Example 3.5.

Solution: We have the following vectors from the given information. We will substitute these values into Equation (3.22) and will calculate vectors ${}^Bd$ and ${}^B\delta$.

$$\mathbf{n} = [0, 1, 0] \quad \mathbf{o} = [0, 0, 1] \quad \mathbf{a} = [1, 0, 0] \quad \mathbf{p} = [10, 5, 3]$$

$$\boldsymbol{\delta} = [0, 0.1, 0] \quad \mathbf{d} = [0.1, 0, 0.2]$$

$$\boldsymbol{\delta} \times \mathbf{p} = \begin{vmatrix} \mathbf{i} & \mathbf{j} & \mathbf{k} \\ 0 & 0.1 & 0 \\ 10 & 5 & 3 \end{vmatrix} = [0.3, 0, -1]$$

$$\boldsymbol{\delta} \times \mathbf{p} + \mathbf{d} = [0.3, 0, -1] + [0.1, 0, 0.2] = [0.4, 0, -0.8]$$

thus

$$\begin{aligned} {}^Bdx &= \mathbf{n} \cdot [\boldsymbol{\delta} \times \mathbf{p} + \mathbf{d}] = 0(0.4) + 1(0) + 0(-0.8) = 0 \\ {}^Bdy &= \mathbf{o} \cdot [\boldsymbol{\delta} \times \mathbf{p} + \mathbf{d}] = 0(0.4) + 0(0) + 1(-0.8) = -0.8 \\ {}^Bdz &= \mathbf{a} \cdot [\boldsymbol{\delta} \times \mathbf{p} + \mathbf{d}] = 1(0.4) + 0(0) + 0(-0.8) = 0.4 \\ {}^B\delta x &= \boldsymbol{\delta} \cdot \mathbf{n} = 0(0) + 0.1(1) + 0(0) = 0.1 \\ {}^B\delta y &= \boldsymbol{\delta} \cdot \mathbf{o} = 0(0) + 0.1(0) + 0(1) = 0 \\ {}^B\delta z &= \boldsymbol{\delta} \cdot \mathbf{a} = 0(1) + 0.1(0) + 0(0) = 0 \end{aligned}$$

Substituting into Equation (3.21) yields:

$$^Bd = [0, -0.8, 0.4] \quad \text{and} \quad ^B\delta = [0.1, 0, 0]$$

$$^B\Delta = \begin{bmatrix} 0 & 0 & 0 & 0 \\ 0 & 0 & -0.1 & -0.8 \\ 0 & 0.1 & 0 & 0.4 \\ 0 & 0 & 0 & 0 \end{bmatrix}$$

As you see, these values for $^B\Delta$ are not the same as Δ. However, post-multiplying the B matrix by $^B\Delta$ will yield the same result dB as before. ∎

Example 3.8

Calculate $^B\Delta$ of Example 3.7 directly from the differential operator.

Solution: Using Equation (3.20), we can calculate the $^B\Delta$ directly as:

$$[^B\Delta] = [B^{-1}][\Delta][B] = \begin{bmatrix} 0 & 1 & 0 & -5 \\ 0 & 0 & 1 & -3 \\ 1 & 0 & 0 & -10 \\ 0 & 0 & 0 & 1 \end{bmatrix} \begin{bmatrix} 0 & 0 & 0.1 & 0.1 \\ 0 & 0 & 0 & 0 \\ -0.1 & 0 & 0 & 0.2 \\ 0 & 0 & 0 & 0 \end{bmatrix} \begin{bmatrix} 0 & 0 & 1 & 10 \\ 1 & 0 & 0 & 5 \\ 0 & 1 & 0 & 3 \\ 0 & 0 & 0 & 1 \end{bmatrix}$$

$$= \begin{bmatrix} 0 & 0 & 0 & 0 \\ 0 & 0 & -0.1 & -0.8 \\ 0 & 0.1 & 0 & 0.4 \\ 0 & 0 & 0 & 0 \end{bmatrix}$$

which, of course, is the same result as in Example 3.7. ∎

3.9 Differential Motions of a Robot and its Hand Frame

In the previous section, we saw the changes made to a frame as a result of differential motions. This only relates to the frame changes, but not how they were accomplished. In this section, we will relate the changes to the mechanism, in this case the robot, that accomplishes the differential motions. We will learn how the robot's movements are translated into the frame changes at the hand.

The frame we discussed previously may be any frame, including the hand frame of a robot. dT describes the changes in the components of the **n**, **o**, **a**, **p** vectors. If the frame were the hand frame of the robot, we would need to find out how the differential motions of the joints of the robot relate to differential motions of the hand frame, and specifically, to dT. Of course, this relationship is a function of the robot's configuration and design, but also a function of its instantaneous location and orientation. For example, the simple revolute robot and the Stanford Arm of Chapter 2 would require very different joint velocities for similar hand velocities since their configurations are different.

However, for either robot, whether the arm is completely extended or not, and whether it is pointed upward or downward, we would need very different joint velocities in order to achieve the same hand velocity. Of course, as we discussed before, the Jacobian of the robot will create this link between the joint movements and the hand movement as:

$$
\begin{bmatrix} dx \\ dy \\ dz \\ \delta x \\ \delta y \\ \delta z \end{bmatrix} = \begin{bmatrix} & & \\ & \text{Robot} & \\ & \text{Jacobian} & \\ & & \end{bmatrix} \begin{bmatrix} d\theta_1 \\ d\theta_2 \\ d\theta_3 \\ d\theta_4 \\ d\theta_5 \\ d\theta_6 \end{bmatrix} \quad \text{or} \quad [D] = [J][D_\theta] \tag{3.10}
$$

Since the elements of matrix $[D]$ are the same information as in $[\Delta]$, the differential motions of the frame and the robot are related to each other.

3.10 Calculation of the Jacobian

Each element in the Jacobian is the derivative of a corresponding kinematic equation with respect to one of the variables. Referring to Equation (3.10), the first element in $[D]$ is dx. This means that the first kinematic equation must represent movements along the x-axis, which, of course, would be p_x. In other words, p_x expresses the motion of the hand frame along the x-axis; consequently, its derivative will be dx. The same will be true for dy and dz. Considering the **n, o, a, p** matrix, we may pick the corresponding elements of p_x, p_y, and p_z and differentiate them to get dx, dy, and dz.

As an example, consider the simple revolute arm of Example 2.25. The last column of the forward kinematic equation of the robot is:

$$
\begin{bmatrix} p_x \\ p_y \\ p_z \\ 1 \end{bmatrix} = \begin{bmatrix} C_1(C_{234}a_4 + C_{23}a_3 + C_2a_2) \\ S_1(C_{234}a_4 + C_{23}a_3 + C_2a_2) \\ S_{234}a_4 + S_{23}a_3 + S_2a_2 \\ 1 \end{bmatrix} \tag{3.23}
$$

Taking the derivative of p_x will yield:

$$
p_x = C_1(C_{234}a_4 + C_{23}a_3 + C_2a_2)
$$

$$
dp_x = \frac{\partial p_x}{\partial \theta_1} d\theta_1 + \frac{\partial p_x}{\partial \theta_2} d\theta_2 + \cdots + \frac{\partial p_x}{\partial \theta_6} d\theta_6
$$

$$
dp_x = -S_1[C_{234}a_4 + C_{23}a_3 + C_2a_2]d\theta_1 + C_1[-S_{234}a_4 - S_{23}a_3 - S_2a_2]d\theta_2
$$
$$
+ C_1[-S_{234}a_4 - S_{23}a_3]d\theta_3 + C_1[-S_{234}a_4]d\theta_4
$$

From this, we can write the first row of the Jacobian as:

$$\frac{\partial p_x}{\partial \theta_1} = J_{11} = -S_1[C_{234}a_4 + C_{23}a_3 + C_2 a_2]$$

$$\frac{\partial p_x}{\partial \theta_2} = J_{12} = C_1[-S_{234}a_4 - S_{23}a_3 - S_2 a_2]$$

$$\frac{\partial p_x}{\partial \theta_3} = J_{13} = C_1[-S_{234}a_4 - S_{23}a_3]$$

$$\frac{\partial p_x}{\partial \theta_4} = J_{14} = C_1[-S_{234}a_4] \qquad (3.24)$$

$$\frac{\partial p_x}{\partial \theta_5} = J_{15} = 0$$

$$\frac{\partial p_x}{\partial \theta_6} = J_{16} = 0$$

The same can be done for the next two rows. However, since there is no unique equation that describes the rotations about the axes (we only have the components of the orientation vectors about the three axes), there is no single equation available for differential rotations about the three axes, namely δx, δy, and δz. As a result, we will have to calculate these differently.

In reality, it is actually a lot simpler to calculate the Jacobian relative to T_6, the last frame, than it is to calculate it relative to the first frame. Therefore, we will instead use the following approach. Paul[1] has shown that we can write the velocity equation relative to the last frame as:

$$[^{T_6}D] = [^{T_6}J][D_\theta] \qquad (3.25)$$

This means that for the same joint differential motions, pre-multiplied with the Jacobian relative to the last frame, we get the hand differential motions relative to the last frame. Paul[1] has also shown that we can calculate the Jacobian with respect to the last frame using the following formulae in Equations (3.26), (3.27), and (3.28):

- The differential motion relationship of Equation (3.25) can be written as:

$$\begin{bmatrix} ^{T_6}d_x \\ ^{T_6}d_y \\ ^{T_6}d_z \\ ^{T_6}\delta_x \\ ^{T_6}\delta_y \\ ^{T_6}\delta_z \end{bmatrix} = \begin{bmatrix} ^{T_6}J_{11} & ^{T_6}J_{12} & . & . & . & ^{T_6}J_{16} \\ ^{T_6}J_{21} & ^{T_6}J_{22} & . & . & . & ^{T_6}J_{26} \\ ^{T_6}J_{31} & . & . & . & . & ^{T_6}J_{36} \\ ^{T_6}J_{41} & . & . & . & . & ^{T_6}J_{46} \\ ^{T_6}J_{51} & . & . & . & . & ^{T_6}J_{56} \\ ^{T_6}J_{61} & . & . & . & . & ^{T_6}J_{66} \end{bmatrix} \begin{bmatrix} d\theta_1 \\ d\theta_2 \\ . \\ . \\ . \\ d\theta_6 \end{bmatrix}$$

- Assuming that any combination of $A_1 A_2 \ldots A_n$ can be expressed with a corresponding $\mathbf{n}$, $\mathbf{o}$, $\mathbf{a}$, $\mathbf{p}$ matrix, the corresponding elements of the matrix will be used to calculate the Jacobian.

- If joint i under consideration is a revolute joint, then:

$$^{T_6}J_{1i} = (-n_x p_y + n_y p_x) \quad ^{T_6}J_{2i} = (-o_x p_y + o_y p_x) \quad ^{T_6}J_{3i} = (-a_x p_y + a_y p_x)$$
$$^{T_6}J_{4i} = n_z \qquad\qquad\qquad ^{T_6}J_{5i} = o_z \qquad\qquad\qquad ^{T_6}J_{6i} = a_z \tag{3.26}$$

- If joint i under consideration is a prismatic joint, then:

$$^{T_6}J_{1i} = n_z \qquad ^{T_6}J_{2i} = o_z \qquad ^{T_6}J_{3i} = a_z$$
$$^{T_6}J_{4i} = 0 \qquad ^{T_6}J_{5i} = 0 \qquad ^{T_6}J_{6i} = 0 \tag{3.27}$$

- For Equations (3.26) and (3.27), for column i use $^{i-1}T_6$, meaning:

$$
\begin{aligned}
&\text{For column 1, use } {}^{0}T_6 = A_1 A_2 A_3 A_4 A_5 A_6 \\
&\text{For column 2, use } {}^{1}T_6 = A_2 A_3 A_4 A_5 A_6 \\
&\text{For column 3, use } {}^{2}T_6 = A_3 A_4 A_5 A_6 \\
&\text{For column 4, use } {}^{3}T_6 = A_4 A_5 A_6 \\
&\text{For column 5, use } {}^{4}T_6 = A_5 A_6 \\
&\text{For column 6, use } {}^{5}T_6 = A_6
\end{aligned}
\tag{3.28}
$$

Example 3.9

Using Equation (3.23), find the elements of the second row of the Jacobian for the simple revolute robot.

Solution: For the second row of the Jacobian, we will have to differentiate the p_y expression of Equation (3.23) as follows:

$$p_y = S_1(C_{234} a_4 + C_{23} a_3 + C_2 a_2)$$

$$dp_y = \frac{\partial p_y}{\partial \theta_1} d\theta_1 + \frac{\partial p_y}{\partial \theta_2} d\theta_2 + \cdots + \frac{\partial p_y}{\partial \theta_6} d\theta_6$$

$$dp_y = C_1(C_{234} a_4 + C_{23} a_3 + C_2 a_2) d\theta_1$$
$$\quad + S_1[-S_{234} a_4 (d\theta_2 + d\theta_3 + d\theta_4) - S_{23} a_3 (d\theta_2 + d\theta_3) - S_2 a_2 (d\theta_2)]$$

Rearranging the terms will yield:

$$\frac{\partial p_y}{\partial \theta_1} d\theta_1 = J_{21} d\theta_1 = C_1(C_{234} a_4 + C_{23} a_3 + C_2 a_2) d\theta_1$$

$$\frac{\partial p_y}{\partial \theta_2} d\theta_2 = J_{22} d\theta_2 = S_1(-S_{234} a_4 - S_{23} a_3 - S_2 a_2) d\theta_2$$

$$\frac{\partial p_y}{\partial \theta_3} d\theta_3 = J_{23} d\theta_3 = S_1(-S_{234} a_4 - S_{23} a_3) d\theta_3$$

$$\frac{\partial p_y}{\partial \theta_4} d\theta_4 = J_{24} d\theta_4 = S_1(-S_{234} a_4) d\theta_4$$

$$\frac{\partial p_y}{\partial \theta_5} d\theta_5 = J_{25} d\theta_5 = 0 \quad \text{and} \quad \frac{\partial p_y}{\partial \theta_6} d\theta_6 = J_{26} d\theta_6 = 0$$

■

Example 3.10

Find the $^{T_6}J_{11}$ and $^{T_6}J_{41}$ elements of the Jacobian for the simple revolute robot.

Solution: To calculate two elements of the first column of the Jacobian, we need to use $A_1 A_2 \ldots A_6$ matrix. From Example 2.25, we get:

$$^{R}T_H = A_1 A_2 A_3 A_4 A_5 A_6 = \begin{bmatrix} n_x & o_x & a_x & 0 \\ n_y & o_y & a_y & 0 \\ n_z & o_z & a_z & 0 \\ 0 & 0 & 0 & 1 \end{bmatrix}$$

$$= \begin{bmatrix} C_1(C_{234}C_5C_6 - S_{234}S_6) \\ -S_1S_5C_6 & C_1\begin{pmatrix} -C_{234}C_5C_6 \\ -S_{234}C_6 \\ +S_1S_5S_6 \end{pmatrix} & C_1(C_{234}S_5) \\ +S_1C_5 & C_1\begin{pmatrix} C_{234}a_4 + C_{23}a_3 \\ +C_2a_2 \end{pmatrix} \\ S_1(C_{234}C_5C_6 - S_{234}S_6) \\ +C_1S_5C_6 & S_1\begin{pmatrix} -C_{234}C_5C_6 \\ -S_{234}C_6 \\ -C_1S_5S_6 \end{pmatrix} & S_1(C_{234}S_5) \\ -C_1C_5 & S_1\begin{pmatrix} C_{234}a_4 + C_{23}a_3 \\ +C_2a_2 \end{pmatrix} \\ S_{234}C_5C_6 + C_{234}S_6 & -S_{234}C_5C_6 + C_{234}C_6 & S_{234}S_5 & S_{234}a_4 + S_{23}a_3 \\ +S_2a_2 \\ 0 & 0 & 0 & 1 \end{bmatrix}$$

Using the corresponding values of **n**, **o**, **a**, **p** and Equation (3.26) for revolute joints, we get:

$$
\begin{aligned}
^{T_6}J_{11} &= (-n_x p_y + n_y p_x) \\
&= -[C_1(C_{234}C_5C_6 - S_{234}S_6) - S_1S_5C_6] \times [S_1(C_{234}a_4 + C_{23}a_3 + C_2a_2)] \\
&\quad + [S_1(C_{234}C_5C_6 - S_{234}S_6) + C_1S_5C_6] \times [C_1(C_{234}a_4 + C_{23}a_3 + C_2a_2)] \quad (3.29) \\
&= S_5C_6(C_{234}a_4 + C_{23}a_3 + C_2a_2) \\
^{T_6}J_{41} &= n_z = S_{234}C_5C_6 + C_{234}S_6
\end{aligned}
$$

As you can see, the results in Equations (3.24) and (3.29) are different for the J_{11} elements. This is because one is relative to the reference frame, and the other is relative to the current or T_6 frame. ∎

3.11 How to Relate the Jacobian and the Differential Operator

Now that we have seen the Jacobians and the differential operators separately, we need to relate the two together.

Suppose a robot's joints are moved a differential amount. Using Equation (3.10), and knowing the Jacobian, we can calculate the [D] matrix, which contains values of

dx, dy, dz, δx, δy, δz (differential motions of the hand). These can be substituted in Equation (3.17) to form the differential operator. Next, Equation (3.16) can be used to calculate dT. This can be used to locate the new position and orientation of the robot's hand. Therefore, the differential motions of the robot's joints are ultimately related to the hand frame of the robot. Alternately, Equation (3.25) and the Jacobian can be used to calculate the ^{T_6}D matrix, which contains values of ^{T_6}dx, ^{T_6}dy, ^{T_6}dz, $^{T_6}\delta x$, $^{T_6}\delta y$, $^{T_6}\delta z$ (differential motions of the hand relative to the current frame). These can be substituted in Equation (3.21) to form the differential operator $^{T_6}\Delta$. Next, Equation (3.20) can be used to calculate dT, the same as before.

Example 3.11

The hand frame of a 5-DOF robot, its numerical Jacobian for this instance, and a set of differential motions are given. The robot has a 2RP2R configuration. Find the new location of the hand after the differential motion.

$$T_6 = \begin{bmatrix} 1 & 0 & 0 & 5 \\ 0 & 0 & -1 & 3 \\ 0 & 1 & 0 & 2 \\ 0 & 0 & 0 & 1 \end{bmatrix} \quad J = \begin{bmatrix} 3 & 0 & 0 & 0 & 0 \\ -2 & 0 & 1 & 0 & 0 \\ 0 & 4 & 0 & 0 & 0 \\ 0 & 1 & 0 & 1 & 0 \\ -1 & 0 & 0 & 0 & 1 \end{bmatrix} \quad \begin{bmatrix} d\theta_1 \\ d\theta_2 \\ ds_1 \\ d\theta_4 \\ d\theta_5 \end{bmatrix} = \begin{bmatrix} 0.1 \\ -0.1 \\ 0.05 \\ 0.1 \\ 0 \end{bmatrix}$$

Solution: It is assumed that the robot can only rotate about the x- and y-axes, since it has only 5 DOF. Using Equation (3.10), we will calculate the $[D]$ matrix, which is then substituted into Equation (3.17) as follows:

$$[D] = \begin{bmatrix} dx \\ dy \\ dz \\ \delta x \\ \delta y \end{bmatrix} = [J][D_\theta] = \begin{bmatrix} 3 & 0 & 0 & 0 & 0 \\ -2 & 0 & 1 & 0 & 0 \\ 0 & 4 & 0 & 0 & 0 \\ 0 & 1 & 0 & 1 & 0 \\ -1 & 0 & 0 & 0 & 1 \end{bmatrix} \begin{bmatrix} 0.1 \\ -0.1 \\ 0.05 \\ 0.1 \\ 0 \end{bmatrix} = \begin{bmatrix} 0.3 \\ -0.15 \\ -0.4 \\ 0 \\ -0.1 \end{bmatrix}$$

$$\rightarrow \Delta = \begin{bmatrix} 0 & 0 & -0.1 & 0.3 \\ 0 & 0 & 0 & -0.15 \\ 0.1 & 0 & 0 & -0.4 \\ 0 & 0 & 0 & 0 \end{bmatrix}$$

From Equation (3.16), we get:

$$[dT_6] = [\Delta][T_6] = \begin{bmatrix} 0 & 0 & -0.1 & 0.3 \\ 0 & 0 & 0 & -0.15 \\ 0.1 & 0 & 0 & -0.4 \\ 0 & 0 & 0 & 0 \end{bmatrix} \begin{bmatrix} 1 & 0 & 0 & 5 \\ 0 & 0 & -1 & 3 \\ 0 & 1 & 0 & 2 \\ 0 & 0 & 0 & 1 \end{bmatrix}$$

$$= \begin{bmatrix} 0 & -0.1 & 0 & 0.1 \\ 0 & 0 & 0 & -0.15 \\ 0.1 & 0 & 0 & 0.1 \\ 0 & 0 & 0 & 0 \end{bmatrix}$$

The new location of the frame after the differential motion is:

$$T_6 = dT_6 + T_{6 Original} = \begin{bmatrix} 0 & -0.1 & 0 & 0.1 \\ 0 & 0 & 0 & -0.15 \\ 0.1 & 0 & 0 & 0.1 \\ 0 & 0 & 0 & 0 \end{bmatrix} + \begin{bmatrix} 1 & 0 & 0 & 5 \\ 0 & 0 & -1 & 3 \\ 0 & 1 & 0 & 2 \\ 0 & 0 & 0 & 1 \end{bmatrix}$$

$$= \begin{bmatrix} 1 & -0.1 & 0 & 5.1 \\ 0 & 0 & -1 & 2.85 \\ 0.1 & 1 & 0 & 2.1 \\ 0 & 0 & 0 & 1 \end{bmatrix}$$

∎

Example 3.12

The differential motions applied to a frame (T_1), described as $D = [dx, \delta y, \delta z]^T$ and resulting (T_2) positions and orientations of the end of a 3-DOF robot are given. The corresponding Jacobian is also given.

(a) Find the original frame T_1 before the differential motions were applied to it.
(b) Find $^T\Delta$. Is it possible to achieve the same resulting change in T_1 by performing the differential motions relative to the frame?

$$D = \begin{bmatrix} 0.01 \\ 0.02 \\ 0.03 \end{bmatrix} \quad T_2 = \begin{bmatrix} -0.03 & 1 & -0.02 & 4.97 \\ 1 & 0.03 & 0 & 8.15 \\ 0 & -0.02 & -1 & 9.9 \\ 0 & 0 & 0 & 1 \end{bmatrix} \quad J = \begin{bmatrix} 5 & 10 & 0 \\ 3 & 0 & 0 \\ 0 & 1 & 1 \end{bmatrix}$$

Solution: Using Equations (3.16) and (3.21), we get:

$$dT = T_2 - T_1 = \Delta \cdot T_1 \rightarrow T_2 = (\Delta + I) \cdot T_1 \quad \text{and} \quad T_1 = (\Delta + I)^{-1} \cdot T_2$$

Substituting the values from the D matrix into Δ, adding I to it, and then inverting it, we get:

$$\Delta = \begin{bmatrix} 0 & -0.03 & 0.02 & 0.01 \\ 0.03 & 0 & 0 & 0 \\ -0.02 & 0 & 0 & 0 \\ 0 & 0 & 0 & 0 \end{bmatrix} \text{ and}$$

$$(\Delta + I)^{-1} = \begin{bmatrix} 0.999 & 0.03 & -0.02 & -0.01 \\ -0.03 & 0.999 & 0.001 & 0.0003 \\ 0.02 & 0.001 & 1 & -0.002 \\ 0 & 0 & 0 & 1 \end{bmatrix}$$

$$\text{and } T_1 = \begin{bmatrix} 0 & 1 & 0 & 5 \\ 1 & 0 & 0 & 8 \\ 0 & 0 & -1 & 10 \\ 0 & 0 & 0 & 1 \end{bmatrix} \text{ (approximately)}.$$

$$\text{Then: } {}^{T}\Delta = T_1^{-1} \cdot \Delta \cdot T_1 = \begin{bmatrix} 0 & 0.03 & 0 & 0.15 \\ -0.03 & 0 & -0.02 & -0.03 \\ 0 & 0.02 & 0 & 0.1 \\ 0 & 0 & 0 & 0 \end{bmatrix}$$

Since these differential motions relate to δx, δz, dx, dy, and dz, requiring 5 DOF, it is impossible to achieve the same results by performing the differential motions relative to the frame. ■

3.12 Inverse Jacobian

In order to calculate the differential motions (or velocities) needed at the joints of the robot for a desired hand differential motion (or velocity), we need to calculate the inverse of the Jacobian and use it in the following equation:

$$[D] = [J][D_\theta]$$

$$[J^{-1}][D] = [J^{-1}][J][D_\theta] \;\rightarrow\; [D_\theta] = [J^{-1}][D] \tag{3.30}$$

and similarly:

$$\left[{}^{T_6}J^{-1}\right]\left[{}^{T_6}D\right] = \left[{}^{T_6}J^{-1}\right]\left[{}^{T_6}J\right][D_\theta] \;\rightarrow\; D_\theta = \left[{}^{T_6}J^{-1}\right]\left[{}^{T_6}D\right] \tag{3.31}$$

This means that knowing the inverse of the Jacobian, we can calculate how fast each joint must move, such that the robot's hand will yield a desired differential motion or velocity. In reality, this analysis, and not the forward differential motion calculations, is the main purpose of the differential motion analysis. Imagine that there is a robot laying glue on a plate. It is necessary that the robot not only follows a particular path on a flat plane, but also that it goes at a constant speed. Otherwise, the glue will not be uniform, and the operation will be useless. In this case, similar to the case with inverse kinematic equations—where we have to divide the path into very small sections and calculate joint values at all times to make sure the robot follows a desired path—we have to calculate joint velocities continuously in order to ensure that the robot's hand maintains a desired velocity.

As we discussed earlier, with the robot moving and its configuration changing, the actual magnitudes of all elements of the Jacobian of a robot change continuously. As a result, although the symbolic equations describing the Jacobian remain the same, their numerical values change. Consequently, it is necessary to calculate the Jacobian's numerical values continuously. This means that, to calculate enough joint velocities per second to have accurate velocities, the process must be very efficient and quick; otherwise, the results will be inaccurate and useless.

Inverting the Jacobian may be done in two ways; both are very difficult, computationally intensive, and time consuming. The first way is to find the symbolic inverse of the Jacobian and then substitute the values into it to compute the velocities. The second technique is to substitute the numbers in the Jacobian and then invert the numerical matrix through techniques such as Gaussian elimination or other similar approaches. Although these are both possible, they are usually not done. Remember that we are dealing with a matrix as large as 6×6.

Instead, we may use the inverse kinematic equations to calculate the joint velocities. Consider Equation (2.64), repeated here, which yields the value of θ_1 for the simple revolute robot:

$$p_x S_1 - p_y C_1 = 0 \ \rightarrow \ \theta_1 = \tan^{-1}\left(\frac{p_y}{p_x}\right) \quad \text{and} \quad \theta_1 = \theta_1 + 180° \tag{2.64}$$

We can differentiate the relationship to find $d\theta_1$, which is the differential value of θ_1, as:

$$p_x S_1 = p_y C_1$$
$$dp_x S_1 + p_x C_1 d\theta_1 = dp_y C_1 - p_y S_1 d\theta_1$$
$$d\theta_1 (p_x C_1 + p_y S_1) = -dp_x S_1 + dp_y C_1$$

$$d\theta_1 = \frac{-dp_x S_1 + dp_y C_1}{(p_x C_1 + p_y S_1)} \tag{3.32}$$

Similarly, from Equation (2.70), repeated here, we get:

$$S_{234}(C_1 a_x + S_1 a_y) = C_{234} a_z$$
$$C_{234}(d\theta_2 + d\theta_3 + d\theta_4)(C_1 a_x + S_1 a_y) + S_{234}\left[-a_x S_1 d\theta_1 + C_1 da_x + a_y C_1 d\theta_1 + S_1 da_y\right]$$
$$= -S_{234}(d\theta_2 + d\theta_3 + d\theta_4)a_z + C_{234} da_z$$

$$(d\theta_2 + d\theta_3 + d\theta_4) = \frac{S_{234}\left[a_x S_1 d\theta_1 - C_1 da_x - a_y C_1 d\theta_1 - S_1 da_y\right] + C_{234} da_z}{C_{234}(C_1 a_x + S_1 a_y) + S_{234} a_z} \tag{3.33}$$

Equation (3.33) gives the combination of three differential motions in terms of known values. Remember that da_x, da_y, and so on are all known from the dT matrix, since dT is the differential change of the **n**, **o**, **a**, **p** matrix:

$$dT = \begin{bmatrix} dn_x & do_x & da_x & dp_x \\ dn_y & do_y & da_y & dp_y \\ dn_z & do_z & da_z & dp_z \\ 0 & 0 & 0 & 0 \end{bmatrix} \tag{3.34}$$

Next, we can differentiate Equation (2.64) to find a relationship for $d\theta_3$ as follows:

$$2a_2a_3C_3 = (p_xC_1 + p_yS_1 - C_{234}a_4)^2 + (p_z - S_{234}a_4)^2 - a_2^2 - a_3^2$$

$$-2a_2a_3S_3d\theta_3 = 2(p_xC_1 + p_yS_1 - C_{234}a_4)$$
$$\times [C_1dp_x - p_xS_1d\theta_1 + S_1dp_y + p_yC_1d\theta_1 \tag{3.35}$$
$$+ a_4S_{234}(d\theta_2 + d\theta_3 + d\theta_4)]$$
$$+2(p_z - S_{234}a_4)[dp_z - a_4C_{234}(d\theta_2 + d\theta_3 + d\theta_4)]$$

Although Equation (3.35) is long, all elements in it are already known, and $d\theta_3$ can be calculated. Next, differentiating Equation (2.72) we get:

$$S_2[(C_3a_3 + a_2)^2 + S_3^2a_3^2] = (C_3a_3 + a_2)(p_z - S_{234}a_4) - S_3a_3(p_xC_1 + p_yS_1 - C_{234}a_4)$$

$$C_2d\theta_2[(C_3a_3 + a_2)^2 + S_3^2a_3^2] + S_2[2(C_3a_3 + a_2)(-a_3S_3d\theta_3) + 2a_3^2S_3C_3d\theta_3]$$
$$= -a_3S_3d\theta_3(p_z - S_{234}a_4) + (C_3a_3 + a_2)[dp_z - a_4C_{234}(d\theta_2 + d\theta_3 + d\theta_4)] \tag{3.36}$$
$$- a_3C_3d\theta_3(p_xC_1 + p_yS_1 - C_{234}a_4)$$
$$- S_3a_3[dp_xC_1 - p_xS_1d\theta_1 + dp_yS_1 + p_yC_1d\theta_1 + S_{234}a_4(d\theta_2 + d\theta_3 + d\theta_4)]$$

which will yield $d\theta_2$, since all other elements are known. This will also enable us to calculate $d\theta_4$ from Equation (3.33). Next, we will differentiate C_5 from Equation (2.75) to get:

$$C_5 = -C_1a_y + S_1a_x$$
$$-S_5d\theta_5 = S_1a_yd\theta_1 - C_1da_y + C_1a_xd\theta_1 + S_1da_x \tag{3.37}$$

which results in $d\theta_5$. Lastly, we will differentiate the 2,1 elements of Equation (2.77) to calculate $d\theta_6$ as:

$$S_6 = -S_{234}(C_1n_x + S_1n_y) + C_{234}n_z$$

$$C_6d\theta_6 = -C_{234}(C_1n_x + S_1n_y)(d\theta_2 + d\theta_3 + d\theta_4)$$
$$- S_{234}(-S_1n_xd\theta_1 + C_1dn_x + C_1n_yd\theta_1 + S_1dn_y) \tag{3.38}$$
$$- S_{234}n_z(d\theta_2 + d\theta_3 + d\theta_4) + C_{234}dn_z$$

As you can see, there are six differential equations that result in six differential joint values from which velocities can be calculated. The robot controller may be programmed by these six equations, which enable the controller to quickly calculate velocities and run the robot joints accordingly.

Example 3.13

For the robot of Example 3.12, find the values of joint differential motions for the three joints (we will call them ds_1, $d\theta_2$, $d\theta_3$) of the robot that caused the given frame change.

Solution: From Equation (3.30), we get:

$$D_\theta = J^{-1} \cdot D = \begin{bmatrix} 0 & 0.333 & 0 \\ 0.1 & -0.167 & 0 \\ -0.1 & 0.167 & 1 \end{bmatrix} \times \begin{bmatrix} 0.01 \\ 0.02 \\ 0.03 \end{bmatrix} = \begin{bmatrix} 0.0067 \\ -0.0023 \\ 0.0323 \end{bmatrix}$$ ∎

Example 3.14

A camera is attached to the hand frame T_H of a robot as given. The corresponding inverse Jacobian of the robot at this location is also shown. The robot makes a differential motion described as $D = \begin{bmatrix} 0.05 & 0 & -0.1 & 0 & 0.1 & 0.03 \end{bmatrix}^T$.

(a) Find which joints must make a differential motion, and by how much, in order to create the indicated differential motions.
(b) Find the change in the Hand frame.
(c) Find the new location of the camera after the differential motion.
(d) Find how much the differential motions should have been instead, if measured relative to Frame T_H, to move the robot to the same new location as in part (c).

$$T_H = \begin{bmatrix} 0 & 1 & 0 & 3 \\ 1 & 0 & 0 & 2 \\ 0 & 0 & -1 & 8 \\ 0 & 0 & 0 & 1 \end{bmatrix} \qquad J^{-1} = \begin{bmatrix} 1 & 0 & 0 & 0 & 0 & 0 \\ 2 & 0 & -1 & 0 & 0 & 0 \\ 0 & -0.2 & 0 & 0 & 0 & 0 \\ 0 & -1 & 0 & 0 & 1 & 0 \\ 0 & 0 & 0 & 1 & 0 & 0 \\ 1 & 0 & 0 & 0 & 0 & 1 \end{bmatrix}$$

Solution: Substituting the values into the corresponding equations, we get:

(a) $$D_\theta = J^{-1} \cdot D = \begin{bmatrix} 1 & 0 & 0 & 0 & 0 & 0 \\ 2 & 0 & -1 & 0 & 0 & 0 \\ 0 & -0.2 & 0 & 0 & 0 & 0 \\ 0 & -1 & 0 & 0 & 1 & 0 \\ 0 & 0 & 0 & 1 & 0 & 0 \\ 1 & 0 & 0 & 0 & 0 & 1 \end{bmatrix} \times \begin{bmatrix} 0.05 \\ 0 \\ -0.1 \\ 0 \\ 0.1 \\ 0.03 \end{bmatrix} = \begin{bmatrix} 0.05 \\ 0.2 \\ 0 \\ 0.1 \\ 0 \\ 0.08 \end{bmatrix}$$

From this, we can tell that joints 1, 2, 4, and 6 need to move as shown.

(b) The change in the hand frame is:

$$
dT = \Delta \cdot T =
\begin{bmatrix}
0 & -0.03 & 0.1 & 0.05 \\
0.03 & 0 & 0 & 0 \\
-0.1 & 0 & 0 & -0.1 \\
0 & 0 & 0 & 0
\end{bmatrix}
\times
\begin{bmatrix}
0 & 1 & 0 & 3 \\
1 & 0 & 0 & 2 \\
0 & 0 & -1 & 8 \\
0 & 0 & 0 & 1
\end{bmatrix}
$$

$$
=
\begin{bmatrix}
-0.03 & 0 & -0.1 & 0.79 \\
0 & 0.03 & 0 & 0.09 \\
0 & -0.1 & 0 & -0.4 \\
0 & 0 & 0 & 0
\end{bmatrix}
$$

(c) The new location of the camera is:

$$
T_{new} = T_{old} + dT =
\begin{bmatrix}
0 & 1 & 0 & 3 \\
1 & 0 & 0 & 2 \\
0 & 0 & -1 & 8 \\
0 & 0 & 0 & 1
\end{bmatrix}
+
\begin{bmatrix}
-0.03 & 0 & -0.1 & 0.79 \\
0 & 0.03 & 0 & 0.09 \\
0 & -0.1 & 0 & -0.4 \\
0 & 0 & 0 & 0
\end{bmatrix}
$$

$$
=
\begin{bmatrix}
-0.03 & 1 & -0.1 & 3.79 \\
1 & 0.03 & 0 & 2.09 \\
0 & -0.1 & -1 & 7.6 \\
0 & 0 & 0 & 1
\end{bmatrix}
$$

(d) $ {}^{T}\Delta = T^{-1} \cdot \Delta \cdot T = T^{-1} \cdot dT $

$$
{}^{T}\Delta =
\begin{bmatrix}
0 & 1 & 0 & -2 \\
1 & 0 & 0 & -3 \\
0 & 0 & -1 & 8 \\
0 & 0 & 0 & 1
\end{bmatrix}
\times
\begin{bmatrix}
-0.03 & 0 & -0.1 & 0.79 \\
0 & 0.03 & 0 & 0.09 \\
0 & -0.1 & 0 & -0.4 \\
0 & 0 & 0 & 0
\end{bmatrix}
$$

$$
=
\begin{bmatrix}
0 & 0.03 & 0 & 0.09 \\
-0.03 & 0 & -0.1 & 0.79 \\
0 & 0.1 & 0 & 0.4 \\
0 & 0 & 0 & 0
\end{bmatrix}
$$

and therefore, the differential motions relative to the frame will be:

$$^TD = [0.09 \quad 0.79 \quad 0.4 \quad 0.1 \quad 0 \quad -0.03]^T$$ ∎

Example 3.15

A spherical robot, at joint values of $\beta = 0°$, $\gamma = 90°$ and $r = 5$ units, has moved a differential amount $D = [0.1, 0, 0.1, 0.05, 0, 0.1]^T$. Differentiating different elements of the T_{sph} matrix has yielded the following equations for the joint differential motions:

$$d\beta = \frac{-d(a_z)}{\sin\beta}, \quad d\gamma = \frac{d(o_y)}{-\sin\gamma}, \quad dr = \frac{d(p_z) + r\sin\beta(d\beta)}{\cos\beta},$$

Where $d(a_z)$ represents a change in a_z, etc.

(a) Write the differential operator matrix representing the differential motions.
(b) Find the initial position and orientation of the robot before the differential motion.
(c) Find the values of $d\beta$, $d\gamma$, dr.
(d) Identify the elements of T_{sph} that could have been used to derive the equations above.

Solution:

(a) We substitute the differential motion values given into Δ to get:

$$\Delta = \begin{bmatrix} 0 & -0.1 & 0 & 0.1 \\ 0.1 & 0 & -0.05 & 0 \\ 0 & 0.05 & 0 & 0.1 \\ 0 & 0 & 0 & 0 \end{bmatrix}$$

(b) Substituting the given joint values into Equation 2.36, we get the initial position and orientation of the robot:

$$T_{sph} = \begin{bmatrix} C\beta\, C\gamma & -S\gamma & S\beta\, C\gamma & rS\beta\, C\gamma \\ C\beta\, S\gamma & C\gamma & S\beta\, S\gamma & rS\beta\, S\gamma \\ -S\beta & 0 & C\beta & rC\beta \\ 0 & 0 & 0 & 1 \end{bmatrix} = \begin{bmatrix} 0 & -1 & 0 & 0 \\ 1 & 0 & 0 & 0 \\ 0 & 0 & 1 & 5 \\ 0 & 0 & 0 & 1 \end{bmatrix}$$

(c) We find the changes in the **n, o, a, p** values from dT:

$$dT = \Delta \cdot T = \begin{bmatrix} -0.1 & 0 & 0 & 0.1 \\ 0 & -0.1 & -0.05 & -0.25 \\ 0.05 & 0 & 0 & 0.1 \\ 0 & 0 & 0 & 0 \end{bmatrix}$$

and therefore, $d(a_z) = 0$, $d(o_y) = -0.1$, $d(p_z) = 0.1$. From preceding equations given, we get: $d\beta = 0/0$ (undefined, but basically zero), $d\gamma = -0.1/-1 = 0.1$, and $dr = 0.1 + (5)(0)(0)/1 = 0.1$.

(d) Elements 3,1 and 1,2 and 2,2 and 3,3 and 3,4 can be used for this purpose. ∎

Example 3.16

The revolute robot of Example 2.25 is in the following configuration. Calculate the angular velocity of the first joint for the given values such that the hand frame will have the following linear and angular velocities:

$$dx/dt = 1 \text{ in/sec} \quad dy/dt = -2 \text{ in/sec} \quad \delta x/dt = 0.1 \text{ rad/sec}$$

$$\theta_1 = 0°, \quad \theta_2 = 90°, \quad \theta_3 = 0°, \quad \theta_4 = 90°, \quad \theta_5 = 0°, \quad \theta_6 = 45°$$
$$a_2 = 15'', \quad a_3 = 15'', \quad a_4 = 5''$$

The parameters of the robot are shown in Table 3.1. The robot is shown in Figure 3.6.

Solution: First, we will substitute these values into Equation (2.59), repeated here, to obtain the final position and orientation of the robot. Please notice that the actual position and orientation of the robot depends on what is considered the reset (rest) position of the robot, or from where an angle is measured. Assuming the reset

Table 3.1 *Parameters for the Robot of Example 2.25.*

#	θ	d	a	α
1	θ_1	0	0	90
2	θ_2	0	a_2	0
3	θ_3	0	a_3	0
4	θ_4	0	a_4	−90
5	θ_5	0	0	90
6	θ_6	0	0	0

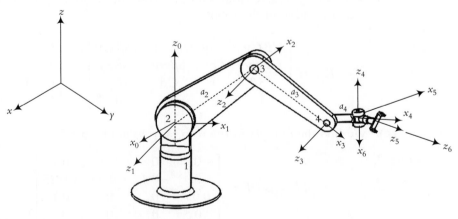

Figure 3.6 Reference frames for the simple 6–DOF articulate robot.

position of this robot is along the *x*-axis, then:

$^R T_H = A_1 A_2 A_3 A_4 A_5 A_6$

$$= \begin{bmatrix} \begin{matrix} C_1(C_{234}C_5C_6 - S_{234}S_6) \\ -S_1S_5C_6 \end{matrix} & \begin{matrix} C_1(-C_{234}C_5C_6 - S_{234}C_6) \\ +S_1S_5S_6 \end{matrix} & \begin{matrix} C_1(C_{234}S_5) \\ +S_1C_5 \end{matrix} & C_1(C_{234}a_4 + C_{23}a_3 + C_2a_2) \\ \begin{matrix} S_1(C_{234}C_5C_6 - S_{234}S_6) \\ +C_1S_5C_6 \end{matrix} & \begin{matrix} S_1(-C_{234}C_5C_6 - S_{234}C_6) \\ -C_1S_5S_6 \end{matrix} & \begin{matrix} S_1(C_{234}S_5) \\ -C_1C_5 \end{matrix} & S_1(C_{234}a_4 + C_{23}a_3 + C_2a_2) \\ S_{234}C_5C_6 + C_{234}S_6 & -S_{234}C_5C_6 + C_{234}C_6 & S_{234}S_5 & S_{234}a_4 + S_{23}a_3 + S_2a_2 \\ 0 & 0 & 0 & 1 \end{bmatrix}$$

$$^R T_H = \begin{bmatrix} n_x & o_x & a_x & p_x \\ n_y & o_y & a_y & p_y \\ n_z & o_z & a_z & p_z \\ 0 & 0 & 0 & 1 \end{bmatrix} = \begin{bmatrix} -0.707 & 0.707 & 0 & -5 \\ 0 & 0 & -1 & 0 \\ -0.707 & -0.707 & 0 & 30 \\ 0 & 0 & 0 & 1 \end{bmatrix}$$

Substituting the desired differential motion values into Equations (3.16) and (3.17), we get:

$$\Delta = \begin{bmatrix} 0 & -\delta z & \delta y & dx \\ \delta z & 0 & -\delta x & dy \\ -\delta y & \delta x & 0 & dz \\ 0 & 0 & 0 & 0 \end{bmatrix} = \begin{bmatrix} 0 & 0 & 0 & 1 \\ 0 & 0 & -0.1 & -2 \\ 0 & 0.1 & 0 & 0 \\ 0 & 0 & 0 & 0 \end{bmatrix}$$

$$[dT] = [\Delta][T] = \begin{bmatrix} 0 & 0 & 0 & 1 \\ 0.0707 & 0.0707 & 0 & -5 \\ 0 & 0 & -0.1 & 0 \\ 0 & 0 & 0 & 0 \end{bmatrix}$$

Substituting the values from the *dT* and *T* matrices into Equation (3.32), we get:

$$\frac{d\theta_1}{dt} = \frac{-dp_x S_1 + dp_y C_1}{(p_x C_1 + p_y S_1)} = \frac{-1(0) - 5(1)}{-5(1) + 0(0)} = 1 \text{ rad/sec}$$

Note that the above given number for θ_5 causes a degenerate condition for the robot and, as a result, other angular velocities cannot be calculated for this configuration. ∎

3.13 Design Projects

3.13.1 The 3-DOF Robot

This is a continuation of the design project we started in Chapter 2. If you designed a 3-DOF robot, and if you developed its forward and inverse kinematic equations, you can now continue with the project.

In this part of the project, you may continue with the differential motion calculations of the robot. Using the forward and inverse kinematic equations, calculate the forward and inverse differential motions of your robot. Since this is a 3-DOF robot, the calculations are relatively easy. Remember that with three degrees of freedom, you can only position the hand of the robot, but you may not pick a desired orientation. Similarly, you can only calculate three differential motion equations that are relative to the three axes, or $d(p_x)$, $d(p_y)$, and $d(p_z)$. From Equation (2.86), we can derive an equation for $d\theta_1$ as follows:

$$\tan(\theta_1) = -\frac{p_x}{p_y} \;\rightarrow\; p_x C_1 = -p_y S_1$$

$$d(p_x) C_1 - p_x S_1 \cdot d(\theta_1) = -d(p_y) S_1 - p_y C_1 \cdot d(\theta_1)$$

$$d(\theta_1)[-p_x S_1 + p_y C_1] = -d(p_y) S_1 - d(p_x) C_1$$

$$d(\theta_1) = \frac{d(p_y) S_1 + d(p_x) C_1}{p_x S_1 - p_y C_1}$$

You may continue deriving equations for the other two joints. With these equations, if you eventually build your robot, and if you use actuators that respond to velocity control commands (such as a servomotor or a stepper motor), you will be able to control the velocity of the robot relative to the three axes. Since in this process you may have calculated the Jacobian of your robot as well, you may use it to find whether there are any degenerate points in its workspace. Do you expect to have any degenerate points?

We will continue with this project in the next chapters.

3.13.2 The 3-DOF Mobile Robot

Similarly, if you designed a 3-DOF mobile robot, you may either differentiate its position and orientation equations and use them for controlling the robot's velocity, or you may differentiate the inverse equations directly to derive the same. In either case, these equations allow you to relate the joint differential motions with the differential motions of the robot.

Summary

In this chapter, we first discussed the differential motions of a frame and the effects of these motions on the frame and its location and orientation. Later, we discussed the differential motions of a robot and how the differential motions of the robot's joints are related to the differential motions of the robot's hand. We then related the two together. Through this, we can calculate how fast a robot's hand moves in space if the joint velocities are known. We also discussed inverse differential motion equations of a robot. Using these equations, we can determine how fast each joint of a robot must move in order to generate a desired hand velocity. Together with the inverse kinematic equations of motion, we can control both the motions and the velocity of a multi-DOF robot in space. We can also follow the location of the hand frame as it moves in space.

In the next chapter, we will continue with the derivation of dynamic equations of motion. Through this, we are able to design and choose appropriate actuators capable of running the robot joints at desired velocities and accelerations.

References

1. Paul, Richard P., "Robot Manipulators, Mathematics, Programming, and Control," The MIT Press, 1981.

Suggestions for Further Reading

I. Craig, John J., "Introduction to Robotics: Mechanics and Control," 2nd Edition, Addison Wesley, 1989.

II. Shahinpoor, Mohsen, "A Robot Engineering Textbook," Harper & Row, 1987.

III. Koren, Yoram, "Robotics for Engineers," McGraw-Hill, 1985.

IV. Fu, K. S., R. C. Gonzalez, C. S. G. Lee, "Robotics: Control, Sensing, Vision, and Intelligence," McGraw-Hill, 1987.

V. Asada, Haruhiko, J. J. E. Slotine, "Robot Analysis and Control," John Wiley and Sons, NY, 1986.

VI. Sciavicco, Lorenzo, B. Siciliano, "Modeling and Control of Robot Manipulators," McGraw-Hill, NY, 1996.

Problems

3.1. Suppose the location and orientation of a hand frame is expressed by the following matrix. What is the effect of a differential rotation of 0.15 radians about the z-axis, followed by a differential translation of $[0.1, 0.1, 0.3]$? Find the new location of the hand.

$$^{R}T_{H} = \begin{bmatrix} 0 & 0 & 1 & 2 \\ 1 & 0 & 0 & 7 \\ 0 & 1 & 0 & 5 \\ 0 & 0 & 0 & 1 \end{bmatrix}$$

3.2. As a result of applying a set of differential motions to frame T shown, it has changed an amount dT as shown. Find the magnitude of the differential changes made $(dx, dy, dz, \delta x, \delta y, \delta z)$ and the differential operator with respect to frame T.

$$T = \begin{bmatrix} 1 & 0 & 0 & 5 \\ 0 & 0 & 1 & 3 \\ 0 & -1 & 0 & 8 \\ 0 & 0 & 0 & 1 \end{bmatrix} \quad dT = \begin{bmatrix} 0 & -0.1 & -0.1 & 0.6 \\ 0.1 & 0 & 0 & 0.5 \\ -0.1 & 0 & 0 & -0.5 \\ 0 & 0 & 0 & 0 \end{bmatrix}$$

3.3. Suppose the following frame was subjected to a differential translation of $d = [1 \quad 0 \quad 0.5]$ units and a differential rotation of $\delta = [0 \quad 0.1 \quad 0]$.

(a) What is the differential operator relative to the reference frame?

(b) What is the differential operator relative to the frame A?

$$A = \begin{bmatrix} 0 & 0 & 1 & 10 \\ 1 & 0 & 0 & 5 \\ 0 & 1 & 0 & 0 \\ 0 & 0 & 0 & 1 \end{bmatrix}$$

3.4. The initial location and orientation of a robot's hand are given by T_1, and its new location and orientation after a change are given by T_2.

(a) Find a transformation matrix Q that will accomplish this transform (in the Universe frame).

(b) Assuming the change is small, find a differential operator Δ that will do the same.

(c) By inspection, find a differential translation and a differential rotation that constitute this operator.

$$T_1 = \begin{bmatrix} 1 & 0 & 0 & 5 \\ 0 & 0 & -1 & 3 \\ 0 & 1 & 0 & 6 \\ 0 & 0 & 0 & 1 \end{bmatrix} \quad T_2 = \begin{bmatrix} 1 & 0 & 0.1 & 4.8 \\ 0.1 & 0 & -1 & 3.5 \\ 0 & 1 & 0 & 6.2 \\ 0 & 0 & 0 & 1 \end{bmatrix}$$

3.5. The hand frame of a robot and the corresponding Jacobian are given. For the given differential changes of the joints, compute the change in the hand frame, its new location, and corresponding Δ.

$$T_6 = \begin{bmatrix} 0 & 1 & 0 & 10 \\ 1 & 0 & 0 & 5 \\ 0 & 0 & -1 & 0 \\ 0 & 0 & 0 & 1 \end{bmatrix} \quad {}^{T_6}J = \begin{bmatrix} 8 & 0 & 0 & 0 & 0 & 0 \\ -3 & 0 & 1 & 0 & 0 & 0 \\ 0 & 10 & 0 & 0 & 0 & 0 \\ 0 & 1 & 0 & 0 & 1 & 0 \\ 0 & 0 & 0 & 1 & 0 & 0 \\ -1 & 0 & 0 & 0 & 0 & 1 \end{bmatrix} \quad D_\theta = \begin{bmatrix} 0 \\ 0.1 \\ -0.1 \\ 0.2 \\ 0.2 \\ 0 \end{bmatrix}$$

3.6. Two consecutive frames describe the old (T_1) and new (T_2) positions and orientations of the end of a 3-DOF robot. The corresponding Jacobian relative to T_1, relating to ${}^{T_1}dz$, ${}^{T_1}\delta x$, ${}^{T_1}\delta z$, is also given. Find values of joint differential motions ds_1, $d\theta_2$, $d\theta_3$ of the robot that caused the given frame change.

$$T_1 = \begin{bmatrix} 0 & 0 & 1 & 8 \\ 1 & 0 & 0 & 5 \\ 0 & 1 & 0 & 2 \\ 0 & 0 & 0 & 1 \end{bmatrix} \quad T_2 = \begin{bmatrix} 0 & 0.01 & 1 & 8.1 \\ 1 & -0.05 & 0 & 5 \\ 0.05 & 1 & -0.01 & 2 \\ 0 & 0 & 0 & 1 \end{bmatrix} \quad {}^{T_1}J = \begin{bmatrix} 5 & 10 & 0 \\ 3 & 0 & 0 \\ 0 & 1 & 1 \end{bmatrix}$$

3.7. Two consecutive frames describe the old (T_1) and new (T_2) positions and orientations of the end of a 3-DOF robot. The corresponding Jacobian, relating to dz, δx, δz, is also given. Find values of joint differential motions ds_1, $d\theta_2$, $d\theta_3$ of the robot that caused the given frame change.

$$T_1 = \begin{bmatrix} 0 & 0 & 1 & 10 \\ 1 & 0 & 0 & 5 \\ 0 & 1 & 0 & 3 \\ 0 & 0 & 0 & 1 \end{bmatrix} \quad T_2 = \begin{bmatrix} -0.05 & 0 & 1 & 9.75 \\ 1 & -0.1 & 0.05 & 5.2 \\ 0.1 & 1 & 0 & 3.7 \\ 0 & 0 & 0 & 1 \end{bmatrix} \quad J = \begin{bmatrix} 5 & 10 & 0 \\ 3 & 0 & 0 \\ 0 & 1 & 1 \end{bmatrix}$$

3.8. A camera is attached to the hand frame T of a robot as given. The corresponding inverse Jacobian of the robot at this location is also given. The robot makes a differential motion, as a result of which, the change in the frame dT is recorded as given.

(a) Find the new location of the camera after the differential motion.

(b) Find the differential operator.

(c) Find the joint differential motion values associated with this move.

(d) Find how much the differential motions of the hand frame ($^T D$) should have been instead, if measured relative to frame T, to move the robot to the same new location as in part (a).

$$T = \begin{bmatrix} 0 & 1 & 0 & 3 \\ 1 & 0 & 0 & 2 \\ 0 & 0 & -1 & 8 \\ 0 & 0 & 0 & 1 \end{bmatrix} \quad J^{-1} = \begin{bmatrix} 1 & 0 & 0 & 0 & 0 & 0 \\ 2 & 0 & -1 & 0 & 0 & 0 \\ 0 & -0.2 & 0 & 0 & 0 & 0 \\ 0 & -1 & 0 & 0 & 1 & 0 \\ 0 & 0 & 0 & 1 & 0 & 0 \\ 1 & 0 & 0 & 0 & 0 & 1 \end{bmatrix} \quad dT = \begin{bmatrix} -0.03 & 0 & -0.1 & 0.79 \\ 0 & 0.03 & 0 & 0.09 \\ 0 & -0.1 & 0 & -0.4 \\ 0 & 0 & 0 & 0 \end{bmatrix}$$

3.9. A camera is attached to the hand frame T of a robot as given. The corresponding inverse Jacobian of the robot relative to the frame at this location is also given. The robot makes a differential motion, as a result of which, the change dT in the frame is recorded as given.

(a) Find the new location of the camera after the differential motion.

(b) Find the differential operator.

(c) Find the joint differential motion values D_θ associated with this move.

$$T = \begin{bmatrix} 0 & 1 & 0 & 3 \\ 1 & 0 & 0 & 2 \\ 0 & 0 & -1 & 8 \\ 0 & 0 & 0 & 1 \end{bmatrix} \quad {}^T J^{-1} = \begin{bmatrix} 1 & 0 & 0 & 0 & 0 & 0 \\ 2 & 0 & -1 & 0 & 0 & 0 \\ 0 & -0.1 & 0 & 0 & 0 & 0 \\ 0 & -1 & 0 & 0 & 1 & 0 \\ 0 & 0 & 0 & 1 & 0 & 0 \\ 1 & 0 & 0 & 0 & 0 & 1 \end{bmatrix} \quad dT = \begin{bmatrix} -0.02 & 0 & -0.1 & 0.7 \\ 0 & 0.02 & 0 & 0.08 \\ 0 & -0.1 & 0 & -0.3 \\ 0 & 0 & 0 & 0 \end{bmatrix}$$

3.10. The hand frame T_H of a robot is given. The corresponding inverse Jacobian of the robot at this location relative to this frame is also shown. The robot makes a differential motion relative to this frame described as:

$$^{T_H} D = \begin{bmatrix} 0.05 & 0 & -0.1 & 0 & 0.1 & 0.1 \end{bmatrix}^T.$$

(a) Find which joints must make a differential motion, and by how much, in order to create the indicated differential motions.

(b) Find the change in the frame.

(c) Find the new location of the frame after the differential motion.

(d) Find how much the differential motions (given above) should have been, if measured relative to the Universe, to move the robot to the same new location as in part (c).

$$T_H = \begin{bmatrix} 0 & 1 & 0 & 3 \\ 1 & 0 & 0 & 3 \\ 0 & 0 & -1 & 8 \\ 0 & 0 & 0 & 1 \end{bmatrix} \quad {}^{T_H} J^{-1} = \begin{bmatrix} 5 & 0 & 0 & 0 & 0 & 0 \\ 2 & 0 & -1 & 0 & 0 & 0 \\ 0 & -0.2 & 0 & 0 & 0 & 0 \\ 0 & -1 & 0 & 0 & 1 & 0 \\ 0 & 0 & 0 & 1 & 0 & 0 \\ 1 & 0 & 0 & 0 & 0 & 1 \end{bmatrix}$$

3.11. The hand frame T of a robot is given. The corresponding inverse Jacobian of the robot at this location is also shown. The robot makes a differential motion described as $D = \begin{bmatrix} 0.05 & 0 & -0.1 & 0 & 0.1 & 0.1 \end{bmatrix}^T$.

(a) Find which joints must make a differential motion, and by how much, in order to create the indicated differential motions.

(b) Find the change in the frame.

(c) Find the new location of the frame after the differential motion.

(d) Find how much the differential motions (given above) should have been, if measured relative to Frame T, to move the robot to the same new location as in part (c).

$$T = \begin{bmatrix} 0 & 1 & 0 & 3 \\ 1 & 0 & 0 & 3 \\ 0 & 0 & -1 & 8 \\ 0 & 0 & 0 & 1 \end{bmatrix} \qquad J^{-1} = \begin{bmatrix} 5 & 0 & 0 & 0 & 0 & 0 \\ 2 & 0 & -1 & 0 & 0 & 0 \\ 0 & -0.2 & 0 & 0 & 0 & 0 \\ 0 & -1 & 0 & 0 & 1 & 0 \\ 0 & 0 & 0 & 1 & 0 & 0 \\ 1 & 0 & 0 & 0 & 0 & 1 \end{bmatrix}$$

3.12. Calculate the $^{T_6}J_{21}$ element of the Jacobian for the revolute robot of Example 2.25.

3.13. Calculate the $^{T_6}J_{16}$ element of the Jacobian for the revolute robot of Example 2.25.

3.14. Using Equation (2.34), differentiate proper elements of the matrix to develop a set of symbolic equations for joint differential motions of a cylindrical robot and write the corresponding Jacobian.

3.15. Using Equation (2.36), differentiate proper elements of the matrix to develop a set of symbolic equations for joint differential motions of a spherical robot and write the corresponding Jacobian.

3.16. For a cylindrical robot, the three joint velocities are given for a corresponding location. Find the three components of the velocity of the hand frame.

$$\dot{r} = 0.1 \text{ in/sec}, \ \dot{\alpha} = 0.05 \text{ rad/sec}, \ \dot{l} = 0.2 \text{ in/sec}, \ r = 15 \text{ in}, \ \alpha = 30°, \ l = 10 \text{ in}.$$

3.17. For a spherical robot, the three joint velocities are given for a corresponding location. Find the three components of the velocity of the hand frame.

$$\dot{r} = 2 \text{ in/sec}, \ \dot{\beta} = 0.05 \text{ rad/sec}, \ \dot{\gamma} = 0.1 \text{ rad/sec}, \ r = 20 \text{ in}, \ \beta = 60°, \ \gamma = 30°.$$

3.18. For a spherical robot, the three joint velocities are given for a corresponding location. Find the three components of the velocity of the hand frame.

$$\dot{r} = 1 \text{ unit/sec}, \ \dot{\beta} = 1 \text{ rad/sec}, \ \dot{\gamma} = 1 \text{ rad/sec}, \ r = 5 \text{ units}, \ \beta = 45°, \ \gamma = 45°.$$

3.19. For a cylindrical robot, the three components of the velocity of the hand frame are given for a corresponding location. Find the required three joint velocities that will generate the given hand frame velocity.

$$\dot{x} = 1 \text{ in/sec}, \ \dot{y} = 3 \text{ in/sec}, \ \dot{z} = 5 \text{ in/sec}, \ \alpha = 45°, \ r = 20 \text{ in}, \ l = 25 \text{ in}.$$

3.20. For a spherical robot, the three components of the velocity of the hand frame are given for a corresponding location. Find the required three joint velocities that will generate the given hand frame velocity.

$$\dot{x} = 5 \text{ in/sec}, \ \dot{y} = 9 \text{ in/sec}, \ \dot{z} = 6 \text{ in/sec}, \ \beta = 60°, \ r = 20 \text{ in}, \ \gamma = 30°.$$

CHAPTER 4

Dynamic Analysis and Forces

4.1 Introduction

In previous chapters, we studied the kinematic position and differential motions of robots. In this chapter, we will look at the dynamics of robots as it relates to accelerations, loads, and masses and inertias. We will also study the static force relationships of robots.

As you may remember from your dynamics course, in order to be able to accelerate a mass, we need to exert a force on it. Similarly, to cause an angular acceleration in a rotating body, a torque must be exerted on it (Figure 4.1), as:

$$\sum \mathbf{F} = m \cdot \mathbf{a} \quad \text{and} \quad \sum \mathbf{T} = I \cdot \boldsymbol{\alpha} \tag{4.1}$$

To accelerate a robot's links, it is necessary to have actuators capable of exerting large enough forces and torques on the links and joints to move them at a desired acceleration and velocity. Otherwise, the links may not be moving as fast as necessary and, consequently, the robot may not maintain its desired positional accuracy. To calculate how strong each actuator must be, it is necessary to determine the dynamic relationships that govern the motions of the robot. These relationships are the force-mass-acceleration and the torque-inertia-angular-acceleration equations. Based on these equations, and considering the external loads on the robot, the designer can calculate the largest loads to which the actuators may be subjected, thereby designing the actuators to be able to deliver the necessary forces and torques.

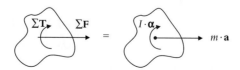

Figure 4.1 Force-mass-acceleration and torque-inertia-angular-acceleration relationships for a rigid body.

147

In general, the dynamic equations may be used to find the equations of motion of mechanisms. This means that, by knowing the forces and torques, we can predict how a mechanism will move. However, in our case, we have already found the equations of motion; besides, in all but the simplest cases, solving the dynamic equations of multi-axis robots is very complicated and involved. Instead, we will use these equations to find what forces and torques may be needed to induce desired accelerations in the robot's joints and links. These equations are also used to see the effects of different inertial loads on the robot, and depending on the desired accelerations, whether certain loads are important or not. As an example, consider a robot in space. Although objects are weightless in space, they do have inertia. As a result, the weight of objects that a robot in space may handle may be trivial, but their inertia is not. So long as the movements are very slow, a light robot may be able to move very large loads in space with little effort. This is why the very slender robots used in the Space Shuttle program are able to handle very large satellites. The dynamic equations allow the designer to investigate the relationship between different elements of the robot and design its components appropriately.

In general, techniques such as Newtonian mechanics can be used to find the dynamic equations for robots. However, due to the fact that robots are 3-D and multi-DOF mechanisms with distributed masses, it is very difficult to use Newtonian mechanics. Instead, we may opt to use other techniques such as Lagrangian mechanics, which is based on energy terms only, and therefore, in many cases, easier to use. Although Newtonian mechanics and other techniques can be used for this derivation, most references are based on Lagrangian mechanics. In this chapter, we will briefly study Lagrangian mechanics with some examples, and then we will see how it can be used to solve for robot equations. Since this is an introductory book, these equations will not be completely derived, but only the results will be demonstrated and discussed. Interested students are encouraged to refer to other references for more detail.[1–7]

4.2 Lagrangian Mechanics: A Short Overview

Lagrangian mechanics is based on the differentiation of the energy terms with respect to the system's variables and time, as shown below. For simple cases, it may take longer to use this technique than Newtonian mechanics. However, as the complexity of the system increases, the Lagrangian method becomes relatively simpler to use. Lagrangian mechanics is based on the following two generalized equations: one for linear motions and one for rotational motions. First, we define a Lagrangian as:

$$L = K - P \tag{4.2}$$

where L is the Lagrangian, K is the kinetic energy of the system, and P is the potential energy of the system. Then:

$$F_i = \frac{\partial}{\partial t}\left(\frac{\partial L}{\partial \dot{x}_i}\right) - \frac{\partial L}{\partial x_i} \tag{4.3}$$

$$T_i = \frac{\partial}{\partial t}\left(\frac{\partial L}{\partial \dot{\theta}_i}\right) - \frac{\partial L}{\partial \theta_i} \tag{4.4}$$

where F_i is the summation of all external forces for a linear motion, T_i is the summation of all external torques for a rotational motion, and θ_i and x_i are system variables. As a result, in order to get the equations of motion, we need to derive energy equations for the system and then differentiate the Lagrangian according to Equations (4.3) and (4.4). The following five examples demonstrate the application of Lagrangian mechanics in deriving equations of motion. Notice how the complexity of the terms increases as the number of degrees of freedom (and variables) increases.

Example 4.1

Derive the force–acceleration relationship for the 1-DOF system shown in Figure 4.2, using both the Lagrangian mechanics as well as the Newtonian mechanics. Assume the wheels have negligible inertia.

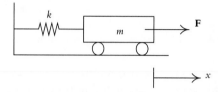

Figure 4.2 Schematic of a simple cart–spring system.

Solution: The x-axis denotes the motion of the cart and is used as the only variable in this system. Since this is a 1-DOF system, there will be only one equation describing the motion. Because the motion is linear, we will only use Equation (4.3), as follows:

$$K = \frac{1}{2}mv^2 = \frac{1}{2}m\dot{x}^2 \quad \text{and} \quad P = \frac{1}{2}kx^2 \rightarrow L = K - P = \frac{1}{2}m\dot{x}^2 - \frac{1}{2}kx^2$$

The derivatives of the Lagrangian are:

$$\frac{\partial L}{\partial \dot{x}} = m\dot{x} \quad \text{and} \quad \frac{d}{dt}(m\dot{x}) = m\ddot{x} \quad \text{and} \quad \frac{\partial L}{\partial x} = -kx$$

Therefore, the equation of motion for the cart will be:

$$F = m\ddot{x} + kx$$

To solve the problem with Newtonian mechanics, we will draw the free-body-diagram of the cart (Figure 4.3) and solve for forces as follows:

$$\sum \mathbf{F} = m\mathbf{a}$$

$$F_x - kx = ma_x \quad \rightarrow \quad F_x = ma_x + kx$$

Figure 4.3 Free body diagram for the cart–spring system.

which is exactly the same result. For this simple system, it appears that Newtonian mechanics is simpler. ■

Example 4.2

Derive the equations of motion for the 2-DOF system shown in Figure 4.4.

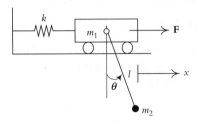

Figure 4.4 Schematic of a cart-pendulum system.

Solution: In this problem, there are two degrees of freedom, two coordinates x and θ, and there will be two equations of motion: one for the linear motion of the system and one for the rotation of the pendulum.

The kinetic energy of the system is comprised of the kinetic energies of the cart and the pendulum. Notice that the velocity of the pendulum is the summation of the velocity of the cart and of the pendulum relative to the cart, or:

$$\mathbf{v}_p = \mathbf{v}_c + \mathbf{v}_{p/c} = (\dot{x})\mathbf{i} + \left(l\dot{\theta}\cos\theta\right)\mathbf{i} + \left(l\dot{\theta}\sin\theta\right)\mathbf{j} = \left(\dot{x} + l\dot{\theta}\cos\theta\right)\mathbf{i} + \left(l\dot{\theta}\sin\theta\right)\mathbf{j}$$

and $v_p^2 = \left(\dot{x} + l\dot{\theta}\cos\theta\right)^2 + \left(l\dot{\theta}\sin\theta\right)^2$

Therefore:

$$K = K_{cart} + K_{pendulum}$$

$$K_{cart} = \frac{1}{2}m_1\dot{x}^2$$

$$K_{pendulum} = \frac{1}{2}m_2\left(\left(\dot{x} + l\dot{\theta}\cos\theta\right)^2 + \left(l\dot{\theta}\sin\theta\right)^2\right)$$

$$K = \frac{1}{2}(m_1 + m_2)\dot{x}^2 + \frac{1}{2}m_2(l^2\dot{\theta}^2 + 2l\dot{\theta}\dot{x}\cos\theta)$$

Likewise, the potential energy is the summation of the potential energies in the spring and in the pendulum, or:

$$P = \frac{1}{2}kx^2 + m_2gl(1 - \cos\theta)$$

Notice that the zero potential energy line (datum) is chosen at $\theta = 0°$. The Lagrangian will be:

$$L = K - P = \frac{1}{2}(m_1 + m_2)\dot{x}^2 + \frac{1}{2}m_2(l^2\dot{\theta}^2 + 2l\dot{\theta}\dot{x}\cos\theta) - \frac{1}{2}kx^2 - m_2gl(1 - \cos\theta)$$

The derivatives and the equation of motion related to the linear motion will be:

$$\frac{\partial L}{\partial \dot{x}} = (m_1 + m_2)\dot{x} + m_2 l\dot{\theta}\cos\theta$$

$$\frac{d}{dt}\left(\frac{\partial L}{\partial \dot{x}}\right) = (m_1 + m_2)\ddot{x} + m_2 l\ddot{\theta}\cos\theta - m_2 l\dot{\theta}^2\sin\theta$$

$$\frac{\partial L}{\partial x} = -kx$$

$$F = (m_1 + m_2)\ddot{x} + m_2 l\ddot{\theta}\cos\theta - m_2 l\dot{\theta}^2\sin\theta + kx$$

and for the rotational motion, it will be:

$$\frac{\partial L}{\partial \dot{\theta}} = m_2 l^2\dot{\theta} + m_2 l\dot{x}\cos\theta$$

$$\frac{d}{dt}\left(\frac{\partial L}{\partial \dot{\theta}}\right) = m_2 l^2\ddot{\theta} + m_2 l\ddot{x}\cos\theta - m_2 l\dot{x}\dot{\theta}\sin\theta$$

$$\frac{\partial L}{\partial \theta} = -m_2 gl\sin\theta - m_2 l\dot{\theta}\dot{x}\sin\theta$$

$$T = m_2 l^2\ddot{\theta} + m_2 l\ddot{x}\cos\theta + m_2 gl\sin\theta$$

If we write the two equations of motion in matrix form, we will get:

$$F = (m_1 + m_2)\ddot{x} + m_2 l\ddot{\theta}\cos\theta - m_2 l\dot{\theta}^2\sin\theta + kx$$
$$T = m_2 l^2\ddot{\theta} + m_2 l\ddot{x}\cos\theta + m_2 gl\sin\theta$$

$$\begin{bmatrix} F \\ T \end{bmatrix} = \begin{bmatrix} m_1 + m_2 & m_2 l\cos\theta \\ m_2 l\cos\theta & m_2 l^2 \end{bmatrix}\begin{bmatrix} \ddot{x} \\ \ddot{\theta} \end{bmatrix} + \begin{bmatrix} 0 & -m_2 l\sin\theta \\ 0 & 0 \end{bmatrix}\begin{bmatrix} \dot{x}^2 \\ \dot{\theta}^2 \end{bmatrix} + \begin{bmatrix} kx \\ m_2 gl\sin\theta \end{bmatrix}$$

$$(4.5)$$

∎

Example 4.3

Derive the equations of motion for the 2-DOF system shown in Figure 4.5.

Solution: Notice that this example is somewhat more similar to a robot, except that the mass of each link is assumed to be concentrated at the end of each link and that there are only two degrees of freedom. However, in this example, we will see many more acceleration terms, as we would expect to see with robots, including centripetal and Coriolis accelerations.

We follow the same format as before. First, we calculate the kinetic and potential energies of the system, as follows:

$$K = K_1 + K_2$$

where $K_1 = \dfrac{1}{2}m_1 l_1^2\dot{\theta}_1^2$

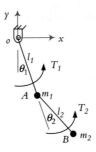

Figure 4.5   A two-link mechanism with concentrated masses.

To calculate K_2, first we write the position equation for m_2 at B and, subsequently, we differentiate it for the velocity:

$$\begin{cases} x_B = l_1 \sin\theta_1 + l_2 \sin(\theta_1 + \theta_2) = l_1 S_1 + l_2 S_{12} \\ y_B = -l_1 C_1 - l_2 C_{12} \end{cases}$$

$$\begin{cases} \dot{x}_B = l_1 C_1 \dot{\theta}_1 + l_2 C_{12}(\dot{\theta}_1 + \dot{\theta}_2) \\ \dot{y}_B = l_1 S_1 \dot{\theta}_1 + l_2 S_{12}(\dot{\theta}_1 + \dot{\theta}_2) \end{cases}$$

Since $v^2 = \dot{x}^2 + \dot{y}^2$, we get:

$$\begin{aligned} v_B^2 &= l_1^2 \dot{\theta}_1^{\,2}\left(S_1^2 + C_1^2\right) + l_2^2(\dot{\theta}_1^{\,2} + \dot{\theta}_2^{\,2} + 2\dot{\theta}_1\dot{\theta}_2)\left(S_{12}^2 + C_{12}^2\right) \\ &\quad + 2l_1 l_2 (C_1 C_{12} + S_1 S_{12})(\dot{\theta}_1^{\,2} + \dot{\theta}_1\dot{\theta}_2) \\ &= l_1^2 \dot{\theta}_1^{\,2} + l_2^2(\dot{\theta}_1^{\,2} + \dot{\theta}_2^{\,2} + 2\dot{\theta}_1\dot{\theta}_2) + 2l_1 l_2 C_2(\dot{\theta}_1^{\,2} + \dot{\theta}_1\dot{\theta}_2) \end{aligned}$$

Then the kinetic energy for the second mass will be:

$$K_2 = \frac{1}{2}m_2 l_1^2 \dot{\theta}_1^{\,2} + \frac{1}{2}m_2 l_2^2(\dot{\theta}_1^{\,2} + \dot{\theta}_2^{\,2} + 2\dot{\theta}_1\dot{\theta}_2) + m_2 l_1 l_2 C_2(\dot{\theta}_1^{\,2} + \dot{\theta}_1\dot{\theta}_2)$$

and the total kinetic energy will be:

$$K = \frac{1}{2}(m_1 + m_2) l_1^2 \dot{\theta}_1^{\,2} + \frac{1}{2}m_2 l_2^2(\dot{\theta}_1^{\,2} + \dot{\theta}_2^{\,2} + 2\dot{\theta}_1\dot{\theta}_2) + m_2 l_1 l_2 C_2(\dot{\theta}_1^{\,2} + \dot{\theta}_1\dot{\theta}_2)$$

With the datum (zero potential energy) at the axis of rotation "o", the potential energy of the system can be written as:

$$P_1 = -m_1 g l_1 C_1$$
$$P_2 = -m_2 g l_1 C_1 - m_2 g l_2 C_{12}$$
$$P = P_1 + P_2 = -(m_1 + m_2)g l_1 C_1 - m_2 g l_2 C_{12}$$

The Lagrangian for the system will be:

$$L = K - P = \frac{1}{2}(m_1 + m_2)l_1^2\dot{\theta}_1^2 + \frac{1}{2}m_2l_2^2(\dot{\theta}_1^2 + \dot{\theta}_2^2 + 2\dot{\theta}_1\dot{\theta}_2)$$

$$+ m_2l_1l_2C_2(\dot{\theta}_1^2 + \dot{\theta}_1\dot{\theta}_2) + (m_1 + m_2)gl_1C_1 + m_2gl_2C_{12}$$

The derivatives of the Lagrangian are:

$$\frac{\partial L}{\partial \dot{\theta}_1} = (m_1 + m_2)l_1^2\dot{\theta}_1 + m_2l_2^2(\dot{\theta}_1 + \dot{\theta}_2) + 2m_2l_1l_2C_2\dot{\theta}_1 + m_2l_1l_2C_2\dot{\theta}_2$$

$$\frac{d}{dt}\frac{\partial L}{\partial \dot{\theta}_1} = \left[(m_1 + m_2)l_1^2 + m_2l_2^2 + 2m_2l_1l_2C_2\right]\ddot{\theta}_1 + \left[m_2l_2^2 + m_2l_1l_2C_2\right]\ddot{\theta}_2$$

$$- 2m_2l_1l_2S_2\dot{\theta}_1\dot{\theta}_2 - m_2l_1l_2S_2\dot{\theta}_2^2$$

$$\frac{\partial L}{\partial \theta_1} = -(m_1 + m_2)gl_1S_1 - m_2gl_2S_{12}$$

From Equation (4.4), the first equation of motion will be:

$$T_1 = \left[(m_1 + m_2)l_1^2 + m_2l_2^2 + 2m_2l_1l_2C_2\right]\ddot{\theta}_1 + \left[m_2l_2^2 + m_2l_1l_2C_2\right]\ddot{\theta}_2$$

$$- 2m_2l_1l_2S_2\dot{\theta}_1\dot{\theta}_2 - m_2l_1l_2S_2\dot{\theta}_2^2 + (m_1 + m_2)gl_1S_1 + m_2gl_2S_{12}$$

Similarly:

$$\frac{\partial L}{\partial \dot{\theta}_2} = m_2l_2^2(\dot{\theta}_1 + \dot{\theta}_2) + m_2l_1l_2C_2\dot{\theta}_1$$

$$\frac{d}{dt}\frac{\partial L}{\partial \dot{\theta}_2} = m_2l_2^2(\ddot{\theta}_1 + \ddot{\theta}_2) + m_2l_1l_2C_2\ddot{\theta}_1 - m_2l_1l_2S_2\dot{\theta}_1\dot{\theta}_2$$

$$\frac{\partial L}{\partial \theta_2} = -m_2l_1l_2S_2(\dot{\theta}_1^2 + \dot{\theta}_1\dot{\theta}_2) - m_2gl_2S_{12}$$

$$T_2 = \left(m_2l_2^2 + m_2l_1l_2C_2\right)\ddot{\theta}_1 + m_2l_2^2\ddot{\theta}_2 + m_2l_1l_2S_2\dot{\theta}_1^2 + m_2gl_2S_{12}$$

Writing these two equations in matrix form, we get:

$$\begin{bmatrix} T_1 \\ T_2 \end{bmatrix} = \begin{bmatrix} (m_1 + m_2)l_1^2 + m_2l_2^2 + 2m_2l_1l_2C_2 & m_2l_2^2 + m_2l_1l_2C_2 \\ \left(m_2l_2^2 + m_2l_1l_2C_2\right) & m_2l_2^2 \end{bmatrix} \begin{bmatrix} \ddot{\theta}_1 \\ \ddot{\theta}_2 \end{bmatrix}$$

$$+ \begin{bmatrix} 0 & -m_2l_1l_2S_2 \\ m_2l_1l_2S_2 & 0 \end{bmatrix} \begin{bmatrix} \dot{\theta}_1^2 \\ \dot{\theta}_2^2 \end{bmatrix} + \begin{bmatrix} -m_2l_1l_2S_2 & -m_2l_1l_2S_2 \\ 0 & 0 \end{bmatrix} \begin{bmatrix} \dot{\theta}_1\dot{\theta}_2 \\ \dot{\theta}_2\dot{\theta}_1 \end{bmatrix}$$

$$+ \begin{bmatrix} (m_1 + m_2)gl_1S_1 + m_2gl_2S_{12} \\ m_2gl_2S_{12} \end{bmatrix}$$

$$(4.6)$$

Note that in Equation (4.6), the $\ddot{\theta}$ terms are related to the angular accelerations of the links, the $\dot{\theta}^2$ terms are centripetal accelerations, and the $\dot{\theta}_1\dot{\theta}_2$ terms are Coriolis accelerations. In this example, the first link acts as a rotating frame for link 2; therefore, Coriolis acceleration is present, whereas in Example 4.2, the cart is not rotating, therefore, there is no Coriolis acceleration. Based on this, we should expect to have multiple Coriolis acceleration terms for a multi-axis, 3-D manipulator arm because each link acts as a rotating frame for the links succeeding it. ■

Example 4.4

Using the Lagrangian method, derive the equations of motion for the 2-DOF robot arm, as shown in Figure 4.6. The center of mass for each link is at the center of the link. The moments of inertia are I_1 and I_2.

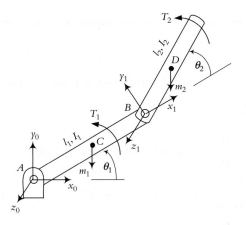

Figure 4.6   A 2-DOF robot arm.

Solution: The solution of this example robot arm is in fact similar to the solution of Example 4.3. However, in addition to a change in the coordinate frames, the two links have distributed masses, requiring the use of moments of inertia in the calculation of the kinetic energy. We will follow the same steps as before. First, we calculate the velocity of the center of mass of link 2 by differentiating its position:

$$x_D = l_1 C_1 + 0.5 l_2 C_{12} \quad \rightarrow \quad \dot{x}_D = -l_1 S_1 \dot{\theta}_1 - 0.5 l_2 S_{12}\left(\dot{\theta}_1 + \dot{\theta}_2\right)$$
$$y_D = l_1 S_1 + 0.5 l_2 S_{12} \quad \rightarrow \quad \dot{y}_D = l_1 C_1 \dot{\theta}_1 + 0.5 l_2 C_{12}\left(\dot{\theta}_1 + \dot{\theta}_2\right)$$

Therefore, the total velocity of the center of mass of link 2 is:

$$v_D^2 = \dot{x}_D^2 + \dot{y}_D^2 = \dot{\theta}_1^{\,2}\left(l_1^2 + 0.25 l_2^2 + l_1 l_2 C_2\right) + \dot{\theta}_2^{\,2}\left(0.25 l_2^2\right) + \dot{\theta}_1\dot{\theta}_2\left(0.5 l_2^2 + l_1 l_2 C_2\right)$$

$$(4.7)$$

The total kinetic energy of the system is the sum of the kinetic energies of links 1 and 2. Remembering that the kinetic energy for a link rotating about a fixed point (for link 1) and about the center of mass (for link 2) is given below, we will have:

$$
\begin{aligned}
K = K_1 + K_2 &= \left[\frac{1}{2} I_A \dot{\theta}_1^2\right] + \left[\frac{1}{2} I_D (\dot{\theta}_1 + \dot{\theta}_2)^2 + \frac{1}{2} m_2 v_D^2\right] \\
&= \left[\frac{1}{2}\left(\frac{1}{3} m_1 l_1^2\right)\dot{\theta}_1^2\right] + \left[\frac{1}{2}\left(\frac{1}{12} m_2 l_2^2\right)(\dot{\theta}_1 + \dot{\theta}_2)^2 + \frac{1}{2} m_2 v_D^2\right]
\end{aligned}
$$

(4.8)

Substituting Equation (4.7) into Equation (4.8) and regrouping, we get:

$$
\begin{aligned}
K = \dot{\theta}_1^2 \left(\frac{1}{6} m_1 l_1^2 + \frac{1}{6} m_2 l_2^2 + \frac{1}{2} m_2 l_1^2 + \frac{1}{2} m_2 l_1 l_2 C_2\right) \\
+ \dot{\theta}_2^2 \left(\frac{1}{6} m_2 l_2^2\right) + \dot{\theta}_1 \dot{\theta}_2 \left(\frac{1}{3} m_2 l_2^2 + \frac{1}{2} m_2 l_1 l_2 C_2\right)
\end{aligned}
$$

(4.9)

The potential energy of the system is the sum of the potential energies of the two links:

$$
P = m_1 g \frac{l_1}{2} S_1 + m_2 g \left(l_1 S_1 + \frac{l_2}{2} S_{12}\right)
$$

(4.10)

The Lagrangian for the two-link robot arm will be:

$$
\begin{aligned}
L = K - P = \dot{\theta}_1^2 \left(\frac{1}{6} m_1 l_1^2 + \frac{1}{6} m_2 l_2^2 + \frac{1}{2} m_2 l_1^2 + \frac{1}{2} m_2 l_1 l_2 C_2\right) + \dot{\theta}_2^2 \left(\frac{1}{6} m_2 l_2^2\right) \\
+ \dot{\theta}_1 \dot{\theta}_2 \left(\frac{1}{3} m_2 l_2^2 + \frac{1}{2} m_2 l_1 l_2 C_2\right) - m_1 g \frac{l_1}{2} S_1 - m_2 g \left(l_1 S_1 + \frac{l_2}{2} S_{12}\right)
\end{aligned}
$$

Taking the derivatives of the Lagrangian and substituting the terms into Equation (4.4) will yield the following two equations of motion:

$$
\begin{aligned}
T_1 = \left(\frac{1}{3} m_1 l_1^2 + m_2 l_1^2 + \frac{1}{3} m_2 l_2^2 + m_2 l_1 l_2 C_2\right)\ddot{\theta}_1 + \left(\frac{1}{3} m_2 l_2^2 + \frac{1}{2} m_2 l_1 l_2 C_2\right)\ddot{\theta}_2 \\
- (m_2 l_1 l_2 S_2)\dot{\theta}_1 \dot{\theta}_2 - \left(\frac{1}{2} m_2 l_1 l_2 S_2\right)\dot{\theta}_2^2 + \left(\frac{1}{2} m_1 + m_2\right) g l_1 C_1 + \frac{1}{2} m_2 g l_2 C_{12}
\end{aligned}
$$

(4.11)

$$
T_2 = \left(\frac{1}{3} m_2 l_2^2 + \frac{1}{2} m_2 l_1 l_2 C_2\right)\ddot{\theta}_1 + \left(\frac{1}{3} m_2 l_2^2\right)\ddot{\theta}_2 + \left(\frac{1}{2} m_2 l_1 l_2 S_2\right)\dot{\theta}_1^2 + \frac{1}{2} m_2 g l_2 C_{12}
$$

(4.12)

Equations (4.11) and (4.12) can be written in matrix form as:

$$
\begin{bmatrix} T_1 \\ T_2 \end{bmatrix} = \begin{bmatrix} \left(\dfrac{1}{3}m_1 l_1^2 + m_2 l_1^2 + \dfrac{1}{3}m_2 l_2^2 + m_2 l_1 l_2 C_2\right) & \left(\dfrac{1}{3}m_2 l_2^2 + \dfrac{1}{2}m_2 l_1 l_2 C_2\right) \\ \left(\dfrac{1}{3}m_2 l_2^2 + \dfrac{1}{2}m_2 l_1 l_2 C_2\right) & \left(\dfrac{1}{3}m_2 l_2^2\right) \end{bmatrix} \begin{bmatrix} \ddot\theta_1 \\ \ddot\theta_2 \end{bmatrix}
$$

$$
+ \begin{bmatrix} 0 & -\left(\dfrac{1}{2}m_2 l_1 l_2 S_2\right) \\ \left(\dfrac{1}{2}m_2 l_1 l_2 S_2\right) & 0 \end{bmatrix} \begin{bmatrix} \dot\theta_1^{\,2} \\ \dot\theta_2^{\,2} \end{bmatrix} + \begin{bmatrix} -(m_2 l_1 l_2 S_2) & 0 \\ 0 & 0 \end{bmatrix} \begin{bmatrix} \dot\theta_1 \dot\theta_2 \\ \dot\theta_2 \dot\theta_1 \end{bmatrix}
$$

$$
+ \begin{bmatrix} \left(\dfrac{1}{2}m_1 + m_2\right)g l_1 C_1 + \dfrac{1}{2}m_2 g l_2 C_{12} \\ \dfrac{1}{2}m_2 g l_2 C_{12} \end{bmatrix}
$$

$$(4.13)$$

∎

Example 4.5

Using the Lagrangian method, derive the equations of motion for the 2-DOF polar robot arm, as shown in Figure 4.7. The center of mass for each link is at the center of the link. The moments of inertia are I_1 and I_2.

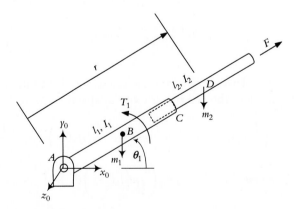

Figure 4.7 A 2-DOF polar robot arm.

Solution: Note that in this case, the arm extends/retracts linearly as well. We denote the length of the robot to the center of the outer arm as r, which will serve as one of our variables. The total length of the arm is $r + (l_2/2)$. As before, we derive the Lagrangian and take the proper derivatives as follows:

$$K = K_1 + K_2$$

$$K_1 = \frac{1}{2}I_{1,A}\dot{\theta}^2 = \frac{1}{2}\frac{1}{3}m_1l_1^2\dot{\theta}^2 = \frac{1}{6}m_1l_1^2\dot{\theta}^2$$

$$x_D = rC\theta \quad \rightarrow \quad \dot{x}_D = \dot{r}C\theta - r\dot{\theta}S\theta$$

$$\text{and} \quad v_D^2 = \dot{r}^2 + r^2\dot{\theta}^2$$

$$y_D = rS\theta \quad \rightarrow \quad \dot{y}_D = \dot{r}S\theta + r\dot{\theta}C\theta$$

$$K_2 = \frac{1}{2}I_{2,D}\dot{\theta}^2 + \frac{1}{2}m_2v_D^2 = \frac{1}{2}\frac{1}{12}m_2l_2^2\dot{\theta}^2 + \frac{1}{2}m_2(\dot{r}^2 + r^2\dot{\theta}^2)$$

$$K = \left(\frac{1}{6}m_1l_1^2 + \frac{1}{24}m_2l_2^2 + \frac{1}{2}m_2r^2\right)\dot{\theta}^2 + \frac{1}{2}m_2\dot{r}^2$$

$$P = m_1g\frac{l_1}{2}S\theta + m_2grS\theta$$

$$L = \left(\frac{1}{6}m_1l_1^2 + \frac{1}{24}m_2l_2^2 + \frac{1}{2}m_2r^2\right)\dot{\theta}^2 + \frac{1}{2}m_2\dot{r}^2 - \left(m_1g\frac{l_1}{2} + m_2gr\right)S\theta$$

$$\frac{d}{dt}\frac{\partial L}{\partial \dot{\theta}} = \left(\frac{1}{3}m_1l_1^2 + \frac{1}{12}m_2l_2^2 + m_2r^2\right)\ddot{\theta} + 2m_2\dot{r}\dot{\theta}$$

$$\frac{\partial L}{\partial \theta} = -\left(m_1g\frac{l_1}{2} + m_2gr\right)C\theta$$

$$T = \left(\frac{1}{3}m_1l_1^2 + \frac{1}{12}m_2l_2^2 + m_2r^2\right)\ddot{\theta} + 2m_2r\dot{r}\dot{\theta} + \left(m_1g\frac{l_1}{2} + m_2gr\right)C\theta \quad (4.14)$$

$$\frac{d}{dt}\frac{\partial L}{\partial \dot{r}} = m_2\ddot{r}$$

$$\frac{\partial L}{\partial r} = m_2r\dot{\theta}^2 - m_2gS\theta$$

$$F = m_2\ddot{r} - m_2r\dot{\theta}^2 + m_2gS\theta \quad (4.15)$$

Writing these two equations in matrix form, we get:

$$\begin{bmatrix} T \\ F \end{bmatrix} = \begin{bmatrix} \frac{1}{3}m_1l_1^2 + \frac{1}{12}m_2l_2^2 + m_2r^2 & 0 \\ 0 & m_2 \end{bmatrix}\begin{bmatrix} \ddot{\theta} \\ \ddot{r} \end{bmatrix} + \begin{bmatrix} 0 & 0 \\ -m_2r & 0 \end{bmatrix}\begin{bmatrix} \dot{\theta}^2 \\ \dot{r}^2 \end{bmatrix}$$

$$+ \begin{bmatrix} m_2r & m_2r \\ 0 & 0 \end{bmatrix}\begin{bmatrix} \dot{r}\dot{\theta} \\ \dot{\theta}\dot{r} \end{bmatrix} + \begin{bmatrix} \left(m_1g\frac{l_1}{2} + m_2gr\right)C\theta \\ m_2gS\theta \end{bmatrix} \quad (4.16)$$

■

4.3 Effective Moments of Inertia

To simplify the writing of the equations of motion, Equations (4.5), (4.6), (4.13), or (4.16) can be rewritten in symbolic form as follows:

$$
\begin{bmatrix} T_i \\ T_j \end{bmatrix} = \begin{bmatrix} D_{ii} & D_{ij} \\ D_{ji} & D_{jj} \end{bmatrix} \begin{bmatrix} \ddot{\theta}_i \\ \ddot{\theta}_j \end{bmatrix} + \begin{bmatrix} D_{iii} & D_{ijj} \\ D_{jii} & D_{jjj} \end{bmatrix} \begin{bmatrix} \dot{\theta}_i^2 \\ \dot{\theta}_j^2 \end{bmatrix} + \begin{bmatrix} D_{iij} & D_{iji} \\ D_{jij} & D_{jji} \end{bmatrix} \begin{bmatrix} \dot{\theta}_i \dot{\theta}_j \\ \dot{\theta}_j \dot{\theta}_i \end{bmatrix} + \begin{bmatrix} D_i \\ D_j \end{bmatrix}
$$

$$(4.17)$$

In this equation, which is written for a 2-DOF system, a coefficient in the form of D_{ii} is known as effective inertia at joint i, such that an acceleration at joint i causes a torque at joint i equal to $D_{ii}\ddot{\theta}_i$. A coefficient in the form D_{ij} is known as coupling inertia between joints i and j as an acceleration at joint i or j causes a torque at joint j or i equal to $D_{ij}\ddot{\theta}_j$ or $D_{ji}\ddot{\theta}_i$. $D_{iij}\dot{\theta}_j^2$ terms represent centripetal forces acting at joint i due to a velocity at joint j. All terms with $\dot{\theta}_1\dot{\theta}_2$ represent Coriolis accelerations and, when multiplied by corresponding inertias, represent Coriolis forces. The remaining terms in the form D_i represent gravity forces at joint i.

4.4 Dynamic Equations for Multiple-DOF Robots

As you can see, the dynamic equations for a 2-DOF system are much more complicated than a 1-DOF system. Similarly, these equations for a multiple-DOF robot are very long and complicated but can be found by calculating the kinetic and potential energies of the links and the joints, defining the Lagrangian, and differentiating the Lagrangian equation with respect to the joint variables. The following is a summary of this procedure. For more information, refer to References 1 through 7 at the end of the chapter.

4.4.1 Kinetic Energy

As you may remember from your dynamics course,[8] the kinetic energy of a rigid body in 3-D motion (Figure 4.8(a)) is:

$$
K = \frac{1}{2}mv_G^2 + \frac{1}{2}\boldsymbol{\omega} \cdot \mathbf{h}_G
$$

$$(4.18)$$

where $\mathbf{h}_G$ is the angular momentum of the body about G.

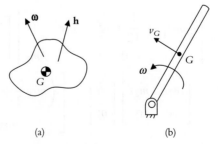

(a)   (b)

Figure 4.8 A rigid body in 3-D motion and in plane motion.

The kinetic energy of a rigid body in plane motion (Figure 4.8(b)) will simplify to:

$$K = \frac{1}{2}mv_G^2 + \frac{1}{2}\bar{I}\omega^2 \qquad (4.19)$$

Therefore, we will need to derive expressions for the velocity of a point on a rigid body (e.g., the center of mass G) as well as the moments of inertia.

The velocity of a point on a robot's link can be defined by differentiating the position equation of the point, which in our notation is expressed by a frame relative to the robot's base, $^R T_P$. Here, we will use the D-H transformation matrices A_i, to find the velocity terms for points along the robot's links. In Chapter 2, we defined the transformation between the hand frame and the base frame of the robot in terms of the A matrices as:

$$^R T_H = {}^R T_1 \,^1 T_2 \,^2 T_3 \ldots \,^{n-1} T_n = A_1 A_2 A_3 \ldots A_n \qquad (2.55)$$

For a 6-axis robot, this equation can be written as:

$$^0 T_6 = {}^0 T_1 \,^1 T_2 \,^2 T_3 \ldots \,^5 T_6 = A_1 A_2 A_3 \ldots A_6 \qquad (4.20)$$

Referring to Equation (2.53), the derivative of an A_i matrix for a revolute joint with respect to its joint variable θ_i is:

$$\frac{\partial A_i}{\partial \theta_i} = \frac{\partial}{\partial \theta_i}
\begin{bmatrix}
C\theta_i & -S\theta_i C\alpha_i & S\theta_i S\alpha_i & a_i C\theta_i \\
S\theta_i & C\theta_i C\alpha_i & -C\theta_i S\alpha_i & a_i S\theta_i \\
0 & S\alpha_i & C\alpha_i & d_i \\
0 & 0 & 0 & 1
\end{bmatrix}
=
\begin{bmatrix}
-S\theta_i & -C\theta_i C\alpha_i & C\theta_i S\alpha_i & -a_i S\theta_i \\
C\theta_i & -S\theta_i C\alpha_i & S\theta_i S\alpha_i & a_i C\theta_i \\
0 & 0 & 0 & 0 \\
0 & 0 & 0 & 0
\end{bmatrix}$$

$$(4.21)$$

However, this matrix can be broken into a constant matrix Q_i and the A_i matrix such that:

$$\begin{bmatrix}
-S\theta_i & -C\theta_i C\alpha_i & C\theta_i S\alpha_i & -a_i S\theta_i \\
C\theta_i & -S\theta_i C\alpha_i & S\theta_i S\alpha_i & a_i C\theta_i \\
0 & 0 & 0 & 0 \\
0 & 0 & 0 & 0
\end{bmatrix}$$

$$(4.22)$$

$$=
\begin{bmatrix}
0 & -1 & 0 & 0 \\
1 & 0 & 0 & 0 \\
0 & 0 & 0 & 0 \\
0 & 0 & 0 & 0
\end{bmatrix}
\times
\begin{bmatrix}
C\theta_i & -S\theta_i C\alpha_i & S\theta_i S\alpha_i & a_i C\theta_i \\
S\theta_i & C\theta_i C\alpha_i & -C\theta_i S\alpha_i & a_i S\theta_i \\
0 & S\alpha_i & C\alpha_i & d_i \\
0 & 0 & 0 & 1
\end{bmatrix}$$

or:

$$\frac{\partial A_i}{\partial \theta_i} = Q_i A_i \qquad (4.23)$$

Similarly, the derivative of an A_i matrix for a prismatic joint with respect to its joint variable d_i is:

$$\frac{\partial A_i}{\partial d_i} = \frac{\partial}{\partial d_i} \begin{bmatrix} C\theta_i & -S\theta_i C\alpha_i & S\theta_i S\alpha_i & a_i C\theta_i \\ S\theta_i & C\theta_i C\alpha_i & -C\theta_i S\alpha_i & a_i S\theta_i \\ 0 & S\alpha_i & C\alpha_i & d_i \\ 0 & 0 & 0 & 1 \end{bmatrix} = \begin{bmatrix} 0 & 0 & 0 & 0 \\ 0 & 0 & 0 & 0 \\ 0 & 0 & 0 & 1 \\ 0 & 0 & 0 & 0 \end{bmatrix} \qquad (4.24)$$

which, as before, can be broken into a constant matrix Q_i and the A_i matrix such that:

$$\begin{bmatrix} 0 & 0 & 0 & 0 \\ 0 & 0 & 0 & 0 \\ 0 & 0 & 0 & 1 \\ 0 & 0 & 0 & 0 \end{bmatrix} = \begin{bmatrix} 0 & 0 & 0 & 0 \\ 0 & 0 & 0 & 0 \\ 0 & 0 & 0 & 1 \\ 0 & 0 & 0 & 0 \end{bmatrix} \times \begin{bmatrix} C\theta_i & -S\theta_i C\alpha_i & S\theta_i S\alpha_i & a_i C\theta_i \\ S\theta_i & C\theta_i C\alpha_i & -C\theta_i S\alpha_i & a_i S\theta_i \\ 0 & S\alpha_i & C\alpha_i & d_i \\ 0 & 0 & 0 & 1 \end{bmatrix}$$

$$(4.25)$$

or:

$$\frac{\partial A_i}{\partial \theta_i} = Q_i A_i \qquad (4.26)$$

In both Equations (4.23) and (4.26), the Q_i matrices are always constant, as shown, and can be summarized as:

$$Q_i(revolute) = \begin{bmatrix} 0 & -1 & 0 & 0 \\ 1 & 0 & 0 & 0 \\ 0 & 0 & 0 & 0 \\ 0 & 0 & 0 & 0 \end{bmatrix} \qquad Q_i(prismatic) = \begin{bmatrix} 0 & 0 & 0 & 0 \\ 0 & 0 & 0 & 0 \\ 0 & 0 & 0 & 1 \\ 0 & 0 & 0 & 0 \end{bmatrix} \qquad (4.27)$$

Using q_i to represent the joint variables (θ_1, θ_2 . . . for revolute joints and d_1, d_2 . . . for prismatic joints), and extending the same differentiation principle to the ^0T_i matrix of Equation (4.20) with multiple joint variables (θ's and d's), differentiated with respect to only one variable q_j will result in:

$$U_{ij} = \frac{\partial^0 T_i}{\partial q_j} = \frac{\partial (A_1 A_2..A_j..A_i)}{\partial q_j} = A_1 A_2..Q_j A_j..A_i \quad j \leq i \qquad (4.28)$$

Note that since ^0T_i is differentiated only with respect to one variable q_j, there is only one Q_j. Higher-order derivatives can similarly be formulated from:

$$U_{ijk} = \partial U_{ij}/\partial q_k \qquad (4.29)$$

Let's see how this method works before we continue with this subject.

Example 4.6

Find the expression for the derivative of the transformation of the fifth link of the Stanford Arm relative to the base frame, with respect to the second and third joint variables.

Solution: The Stanford Arm is a spherical robot, where the second joint is revolute and the third joint is prismatic. Therefore:

$$^0T_5 = A_1 A_2 A_3 A_4 A_5$$

$$U_{52} = \frac{\partial\, ^0T_5}{\partial\theta_2} = A_1 Q_2 A_2 A_3 A_4 A_5$$

$$U_{53} = \frac{\partial\, ^0T_5}{\partial d_3} = A_1 A_2 Q_3 A_3 A_4 A_5$$

where Q_2 and Q_3 are as defined in Equation (4.27). ■

Example 4.7

Find an expression for U_{635} of the Stanford Arm.

Solution:

$$^0T_6 = A_1 A_2 A_3 A_4 A_5 A_6$$

$$U_{63} = \frac{\partial\, ^0T_6}{\partial d_3} = A_1 A_2 Q_3 A_3 A_4 A_5 A_6$$

$$U_{635} = \frac{\partial U_{63}}{\partial q_5} = \frac{\partial(A_1 A_2 Q_3 A_3 A_4 A_5 A_6)}{\partial q_5} = A_1 A_2 Q_3 A_3 A_4 Q_5 A_5 A_6$$

■

4.4.1 Continued

We now continue with the derivation of the velocity term for a point on a link of a robot. Denoting r_i to represent a point on any link i of the robot relative to frame i, the position of the point can be expressed by pre-multiplying the vector with the transformation matrix representing its frame, or:

$$p_i = {}^R T_i r_i = {}^0 T_i r_i \tag{4.30}$$

The velocity of the point is a function of the velocities of all the joints $\dot{q}_1, \dot{q}_2, \ldots, \dot{q}_6$. Therefore, differentiating Equation (4.30) with respect to all the joint variables q_j will yield the velocity of the point as:

$$v_i = \frac{d}{dt}\left({}^0 T_i r_i\right) = \sum_{j=1}^{i}\left(\frac{\partial({}^0 T_i)}{\partial q_j}\frac{dq_j}{dt}\right) r_i = \sum_{j=1}^{i}\left(U_{ij}\frac{dq_i}{dt}\right)\cdot r_i \tag{4.31}$$

The kinetic energy of an element of mass m_i on a link will be:

$$dK_i = \frac{1}{2}\left(\dot{x}_i^2 + \dot{y}_i^2 + \dot{z}_i^2\right) dm \tag{4.32}$$

Since v_i has three components of $\dot{x}_i$, $\dot{y}_i$, $\dot{z}_i$, it can be written as a 3×1 matrix:

$$v_i v_i^T = \begin{bmatrix} \dot{x}_i \\ \dot{y}_i \\ \dot{z}_i \end{bmatrix} \begin{bmatrix} \dot{x}_i & \dot{y}_i & \dot{z}_i \end{bmatrix} = \begin{bmatrix} \dot{x}_i^2 & \dot{x}_i\dot{y}_i & \dot{x}_i\dot{z}_i \\ \dot{y}_i\dot{x}_i & \dot{y}_i^2 & \dot{y}_i\dot{z}_i \\ \dot{z}_i\dot{x}_i & \dot{z}_i\dot{y}_i & \dot{z}_i^2 \end{bmatrix}$$

and

$$Trace\left(v_i\,v_i^T\right) = Trace \begin{bmatrix} \dot{x}_i^2 & \dot{x}_i\dot{y}_i & \dot{x}_i\dot{z}_i \\ \dot{y}_i\dot{x}_i & \dot{y}_i^2 & \dot{y}_i\dot{z}_i \\ \dot{z}_i\dot{x}_i & \dot{z}_i\dot{y}_i & \dot{z}_i^2 \end{bmatrix} = \dot{x}_i^2 + \dot{y}_i^2 + \dot{z}_i^2 \qquad (4.33)$$

Combining Equations (4.31), (4.32), and (4.33) will yield the following equation for the kinetic energy of the element:

$$dK_i = \frac{1}{2} Trace \left[\left(\sum_{p=1}^{i} \left(U_{ip} \frac{dq_p}{dt} \right) \cdot r_i \right) \left(\sum_{r=1}^{i} \left(U_{ir} \frac{dq_r}{dt} \right) \cdot r_i \right)^T \right] dm_i \qquad (4.34)$$

where p and r represent the different joint numbers. This allows us to add the contributions made to the final velocity of a point on any link i from other joints' movements. Integrating the above equation and rearranging it will yield the total kinetic energy as:

$$K_i = \int dK_i = \frac{1}{2} Trace \left[\sum_{p=1}^{i} \sum_{r=1}^{i} U_{ip} \left(\int r_i r_i^T dm_i \right) U_{ir}^T \dot{q}_p \dot{q}_r \right] \qquad (4.35)$$

Writing r_i in terms of its coordinates relative to its frame, we can derive the following inertia terms:

$$r_i = \begin{bmatrix} x_i \\ y_i \\ z_i \\ 1 \end{bmatrix} \text{ and } r_i^T = \begin{bmatrix} x_i & y_i & z_i & 1 \end{bmatrix} \text{ and } r_i r_i^T = \begin{bmatrix} x_i^2 & x_i y_i & x_i z_i & x_i \\ x_i y_i & y_i^2 & y_i z_i & y_i \\ x_i z_i & y_i z_i & z_i^2 & z_i \\ x_i & y_i & z_i & 1 \end{bmatrix}$$

Therefore:

$$\int r_i r_i^T dm_i = \begin{bmatrix} x_i \\ y_i \\ z_i \\ 1 \end{bmatrix} \begin{bmatrix} x_i & y_i & z_i & 1 \end{bmatrix} \int dm_i = \begin{bmatrix} \int x_i^2 dm_i & \int x_i y_i dm_i & \int x_i z_i dm_i & \int x_i dm_i \\ \int x_i y_i dm_i & \int y_i^2 dm_i & \int y_i z_i dm_i & \int y_i dm_i \\ \int x_i z_i dm_i & \int y_i z_i dm_i & \int z_i^2 dm_i & \int z_i dm_i \\ \int x_i dm_i & \int y_i dm_i & \int z_i dm_i & \int dm_i \end{bmatrix}$$
$$(4.36)$$

Through the following manipulations of Equation (4.36), it is possible to derive the *Pseudo Inertia Matrix* as shown:

$$2x^2 = x^2 + x^2 + y^2 - y^2 + z^2 - z^2 \rightarrow x^2 = \frac{1}{2}\left[-\left(y^2 + z^2\right) + \left(x^2 + z^2\right) + \left(x^2 + y^2\right)\right]$$

and: $I_{xx} = \int (y^2 + z^2)dm \quad I_{yy} = \int (x^2 + z^2)dm \quad I_{zz} = \int (x^2 + y^2)dm$

$\quad\quad I_{xy} = \int xydm \quad I_{xz} = \int xzdm \quad I_{yz} = \int yzdm$

$\quad\quad m\bar{x} = \int xdm \quad\quad m\bar{y} = \int ydm \quad\quad m\bar{z} = \int zdm$

then: $\int x^2 dm = -\frac{1}{2}\int (y^2 + z^2)dm + \frac{1}{2}\int (x^2 + z^2)dm + \frac{1}{2}\int (x^2 + y^2)dm = \frac{1}{2}\left(-I_{xx} + I_{yy} + I_{zz}\right)$

$\quad\quad \int y^2 dm = \frac{1}{2}\int (y^2 + z^2)dm - \frac{1}{2}\int (x^2 + z^2)dm + \frac{1}{2}\int (x^2 + y^2)dm = \frac{1}{2}\left(I_{xx} - I_{yy} + I_{zz}\right)$

$\quad\quad \int z^2 dm = \frac{1}{2}\int (y^2 + z^2)dm + \frac{1}{2}\int (x^2 + z^2)dm - \frac{1}{2}\int (x^2 + y^2)dm = \frac{1}{2}\left(I_{xx} + I_{yy} - I_{zz}\right)$

Therefore, Equation (4.36) can be written as:

$$
J_i = \begin{bmatrix}
\frac{1}{2}\left(-I_{xx} + I_{yy} + I_{zz}\right)_i & I_{ixy} & I_{ixz} & m_i\bar{x}_i \\[2ex]
I_{ixy} & \frac{1}{2}\left(I_{xx} - I_{yy} + I_{zz}\right)_i & I_{iyz} & m_i\bar{y}_i \\[2ex]
I_{ixz} & I_{iyz} & \frac{1}{2}\left(I_{xx} + I_{yy} - I_{zz}\right)_i & m_i\bar{z}_i \\[2ex]
m_i\bar{x}_i & m_i\bar{y}_i & m_i\bar{z}_i & m_i
\end{bmatrix} \quad (4.37)
$$

Since this matrix is independent of joint angles and velocities, it must be evaluated only once. Substituting Equation (4.36) into Equation (4.35) will result in the final form for kinetic energy of the robot manipulator as:

$$
K = \frac{1}{2}\sum_{i=1}^{n}\sum_{p=1}^{i}\sum_{r=1}^{i} Trace\left(U_{ip}J_iU_{ir}^T\right)\dot{q}_p\dot{q}_r \quad (4.38)
$$

The kinetic energy of the actuators can also be added to this equation. Assuming that each actuator has an inertia of $I_{i(act)}$, the kinetic energy of the actuator will be $\frac{1}{2}I_{i(act)}\dot{q}_i^2$, and the total kinetic energy of the robot will be:

$$
K = \frac{1}{2}\sum_{i=1}^{n}\sum_{p=1}^{i}\sum_{r=1}^{i} Trace\left(U_{ip}J_iU_{ir}^T\right)\dot{q}_p\dot{q}_r + \frac{1}{2}\sum_{i=1}^{n} I_{i(act)}\dot{q}_i^2 \quad (4.39)
$$

4.4.2 Potential Energy

The potential energy of the system is the sum of the potential energies of each link, and can be written as:

$$
P = \sum_{i=1}^{n} P_i = \sum_{i=1}^{n}\left[-m_i g^T \cdot \left({}^{0}T_i\bar{r}_i\right)\right] \quad (4.40)
$$

where $g^T = \begin{bmatrix} g_x & g_y & g_z & 0 \end{bmatrix}$ is the gravity matrix and $\bar{r}_i$ is the location of the center of mass of a link relative to the frame representing the link. Obviously, the potential energy must be a scalar quantity, and therefore g^T, which is a (1×4) matrix, when multiplied by the position vector $\left({}^{0}T_i\,\bar{r}_i\right)$, which is a (4×1) matrix, will yield a single scalar quantity. Notice that the values in the gravity matrix are dependent on the orientation of the reference frame.

4.4.3 The Lagrangian

The Lagrangian will then be:

$$L = K - P = \frac{1}{2}\sum_{i=1}^{n}\sum_{p=1}^{i}\sum_{r=1}^{i} Trace\left(U_{ip}J_iU_{ir}^T\right)\dot{q}_p\dot{q}_r + 1/2\sum_{i=1}^{n} I_{i(act)}\dot{q}_i^2 - \sum_{i=1}^{n}\left[-m_i g^T \cdot \left({}^0T_i\, \bar{r}_i\right)\right]$$

$$(4.41)$$

4.4.4 Robot's Equations of Motion

The Lagrangian can now be differentiated in order to form the dynamic equations of motion. Although this process is not shown, the final equations of motion for a general multi-axis robot can be summarized as follows:

$$T_i = \sum_{j=1}^{n} D_{ij}\,\ddot{q}_j + I_{i(act)}\,\ddot{q}_i + \sum_{j=1}^{n}\sum_{k=1}^{n} D_{ijk}\dot{q}_j\dot{q}_k + D_i \qquad (4.42)$$

where:

$$D_{ij} = \sum_{p=\max(i,j)}^{n} Trace\left(U_{pj}J_pU_{pi}^T\right) \qquad (4.43)$$

and:

$$D_{ijk} = \sum_{p=\max(i,j,k)}^{n} Trace\left(U_{pjk}J_pU_{pi}^T\right) \qquad (4.44)$$

and:

$$D_i = \sum_{p=i}^{n} -m_pg^T U_{pi}\bar{r}_p \qquad (4.45)$$

In Equation (4.42), the first part is the angular acceleration-inertia terms, the second part is the actuator inertia term, the third part is the Coriolis and centrifugal terms, and the last part is the gravity term. This equation can be expanded for a 6-axis revolute robot as follows:

$$\begin{aligned}
T_i \;=\;& D_{i1}\ddot{\theta}_1 + D_{i2}\ddot{\theta}_2 + D_{i3}\ddot{\theta}_3 + D_{i4}\ddot{\theta}_4 + D_{i5}\ddot{\theta}_5 + D_{i6}\ddot{\theta}_6 + I_{i(act)}\ddot{\theta}_i \\
&+ D_{i11}\dot{\theta}_1^2 + D_{i22}\dot{\theta}_2^2 + D_{i33}\dot{\theta}_3^2 + D_{i44}\dot{\theta}_4^2 + D_{i55}\dot{\theta}_5^2 + D_{i66}\dot{\theta}_6^2 \\
&+ D_{i12}\dot{\theta}_1\dot{\theta}_2 + D_{i13}\dot{\theta}_1\dot{\theta}_3 + D_{i14}\dot{\theta}_1\dot{\theta}_4 + D_{i15}\dot{\theta}_1\dot{\theta}_5 + D_{i16}\dot{\theta}_1\dot{\theta}_6 \\
&+ D_{i21}\dot{\theta}_2\dot{\theta}_1 + D_{i23}\dot{\theta}_2\dot{\theta}_3 + D_{i24}\dot{\theta}_2\dot{\theta}_4 + D_{i25}\dot{\theta}_2\dot{\theta}_5 + D_{i26}\dot{\theta}_2\dot{\theta}_6 \\
&+ D_{i31}\dot{\theta}_3\dot{\theta}_1 + D_{i32}\dot{\theta}_3\dot{\theta}_2 + D_{i34}\dot{\theta}_3\dot{\theta}_4 + D_{i35}\dot{\theta}_3\dot{\theta}_5 + D_{i36}\dot{\theta}_3\dot{\theta}_6 \\
&+ D_{i41}\dot{\theta}_4\dot{\theta}_1 + D_{i42}\dot{\theta}_4\dot{\theta}_2 + D_{i43}\dot{\theta}_4\dot{\theta}_3 + D_{i45}\dot{\theta}_4\dot{\theta}_5 + D_{i46}\dot{\theta}_4\dot{\theta}_6 \\
&+ D_{i51}\dot{\theta}_5\dot{\theta}_1 + D_{i52}\dot{\theta}_5\dot{\theta}_2 + D_{i53}\dot{\theta}_5\dot{\theta}_3 + D_{i54}\dot{\theta}_5\dot{\theta}_4 + D_{i56}\dot{\theta}_5\dot{\theta}_6 \\
&+ D_{i61}\dot{\theta}_6\dot{\theta}_1 + D_{i62}\dot{\theta}_6\dot{\theta}_2 + D_{i63}\dot{\theta}_6\dot{\theta}_3 + D_{i64}\dot{\theta}_6\dot{\theta}_4 + D_{i65}\dot{\theta}_6\dot{\theta}_5 + D_i
\end{aligned}$$

$$(4.46)$$

Notice that in Equation (4.46) there are two terms with $\dot{\theta}_1\dot{\theta}_2$. The two coefficients are D_{i21} and D_{i12}. To see what these terms look like, let's calculate them for $i = 5$. From

Equation (4.44), for D_{512} we have $i = 5, j = 1, k = 2, n = 6, p = 5$, and for D_{521} we have $i = 5, j = 2, k = 1, n = 6, p = 5$, resulting in:

$$D_{512} = Trace\left(U_{512}J_5U_{55}^T\right) + Trace\left(U_{612}J_6U_{65}^T\right)$$
$$D_{521} = Trace\left(U_{521}J_5U_{55}^T\right) + Trace\left(U_{621}J_6U_{65}^T\right)$$

(4.47)

and from Equation (4.28), we have:

$$U_{51} = \frac{\partial A_1 A_2 A_3 A_4 A_5}{\partial \theta_1} = Q_1 A_1 A_2 A_3 A_4 A_5 \rightarrow U_{512} = U_{(51)2} = \frac{\partial(Q_1 A_1 A_2 A_3 A_4 A_5)}{\partial \theta_2}$$

$$= Q_1 A_1 Q_2 A_2 A_3 A_4 A_5$$

$$U_{52} = \frac{\partial A_1 A_2 A_3 A_4 A_5}{\partial \theta_2} = A_1 Q_2 A_2 A_3 A_4 A_5 \rightarrow U_{521} = U_{(52)1} = \frac{\partial(A_1 Q_2 A_2 A_3 A_4 A_5)}{\partial \theta_1}$$

$$= Q_1 A_1 Q_2 A_2 A_3 A_4 A_5$$

(4.48)

$$U_{61} = \frac{\partial A_1 A_2 A_3 A_4 A_5 A_6}{\partial \theta_1} = Q_1 A_1 A_2 A_3 A_4 A_5 A_6 \rightarrow U_{612} = U_{(61)2} = \frac{\partial(Q_1 A_1 A_2 A_3 A_4 A_5 A_6)}{\partial \theta_2}$$

$$= Q_1 A_1 Q_2 A_2 A_3 A_4 A_5 A_6$$

$$U_{62} = \frac{\partial A_1 A_2 A_3 A_4 A_5 A_6}{\partial \theta_2} = A_1 Q_2 A_2 A_3 A_4 A_5 A_6 \rightarrow U_{621} = U_{(62)1} = \frac{\partial(A_1 Q_2 A_2 A_3 A_4 A_5 A_6)}{\partial \theta_1}$$

$$= Q_1 A_1 Q_2 A_2 A_3 A_4 A_5 A_6$$

Note that in these equations, Q_1 and Q_2 are the same. The indices are only used to clarify the relationship with the derivatives. Substituting the result from Equation (4.48) into (4.47) shows that $D_{512} = D_{521}$. Clearly, the summation of the two similar terms yields the corresponding Coriolis acceleration term for $\dot\theta_1\dot\theta_2$. This is true for all similar coefficients in Equation (4.46). Therefore, we can simplify this equation for all joints as follows:

$$T_1 = D_{11}\ddot\theta_1 + D_{12}\ddot\theta_2 + D_{13}\ddot\theta_3 + D_{14}\ddot\theta_4 + D_{15}\ddot\theta_5 + D_{16}\ddot\theta_6 + I_{1(act)}\ddot\theta_1$$
$$+ D_{111}\dot\theta_1^2 + D_{122}\dot\theta_2^2 + D_{133}\dot\theta_3^2 + D_{144}\dot\theta_4^2 + D_{155}\dot\theta_5^2 + D_{166}\dot\theta_6^2$$
$$+ 2D_{112}\dot\theta_1\dot\theta_2 + 2D_{113}\dot\theta_1\dot\theta_3 + 2D_{114}\dot\theta_1\dot\theta_4 + 2D_{115}\dot\theta_1\dot\theta_5 + 2D_{116}\dot\theta_1\dot\theta_6$$
$$+ 2D_{123}\dot\theta_2\dot\theta_3 + 2D_{124}\dot\theta_2\dot\theta_4 + 2D_{125}\dot\theta_2\dot\theta_5 + 2D_{126}\dot\theta_2\dot\theta_6 + 2D_{134}\dot\theta_3\dot\theta_4$$
$$+ 2D_{135}\dot\theta_3\dot\theta_5 + 2D_{136}\dot\theta_3\dot\theta_6 + 2D_{145}\dot\theta_4\dot\theta_5 + 2D_{146}\dot\theta_4\dot\theta_6 + 2D_{156}\dot\theta_5\dot\theta_6 + D_1$$

(4.49)

$$T_2 = D_{21}\ddot\theta_1 + D_{22}\ddot\theta_2 + D_{23}\ddot\theta_3 + D_{24}\ddot\theta_4 + D_{25}\ddot\theta_5 + D_{26}\ddot\theta_6 + I_{2(act)}\ddot\theta_2$$
$$+ D_{211}\dot\theta_1^2 + D_{222}\dot\theta_2^2 + D_{233}\dot\theta_3^2 + D_{244}\dot\theta_4^2 + D_{255}\dot\theta_5^2 + D_{266}\dot\theta_6^2$$
$$+ 2D_{212}\dot\theta_1\dot\theta_2 + 2D_{213}\dot\theta_1\dot\theta_3 + 2D_{214}\dot\theta_1\dot\theta_4 + 2D_{215}\dot\theta_1\dot\theta_5 + 2D_{216}\dot\theta_1\dot\theta_6$$
$$+ 2D_{223}\dot\theta_2\dot\theta_3 + 2D_{224}\dot\theta_2\dot\theta_4 + 2D_{225}\dot\theta_2\dot\theta_5 + 2D_{226}\dot\theta_2\dot\theta_6 + 2D_{234}\dot\theta_3\dot\theta_4$$
$$+ 2D_{235}\dot\theta_3\dot\theta_5 + 2D_{236}\dot\theta_3\dot\theta_6 + 2D_{245}\dot\theta_4\dot\theta_5 + 2D_{246}\dot\theta_4\dot\theta_6 + 2D_{256}\dot\theta_5\dot\theta_6 + D_2$$

(4.50)

$$T_3 = D_{31}\ddot{\theta}_1 + D_{32}\ddot{\theta}_2 + D_{33}\ddot{\theta}_3 + D_{34}\ddot{\theta}_4 + D_{35}\ddot{\theta}_5 + D_{36}\ddot{\theta}_6 + I_{3(act)}\ddot{\theta}_3$$

$$+D_{311}\dot{\theta}_1^2 + D_{322}\dot{\theta}_2^2 + D_{333}\dot{\theta}_3^2 + D_{344}\dot{\theta}_4^2 + D_{355}\dot{\theta}_5^2 + D_{366}\dot{\theta}_6^2$$

$$+2D_{312}\dot{\theta}_1\dot{\theta}_2 + 2D_{313}\dot{\theta}_1\dot{\theta}_3 + 2D_{314}\dot{\theta}_1\dot{\theta}_4 + 2D_{315}\dot{\theta}_1\dot{\theta}_5 + 2D_{316}\dot{\theta}_1\dot{\theta}_6 \quad (4.51)$$

$$+2D_{323}\dot{\theta}_2\dot{\theta}_3 + 2D_{324}\dot{\theta}_2\dot{\theta}_4 + 2D_{325}\dot{\theta}_2\dot{\theta}_5 + 2D_{326}\dot{\theta}_2\dot{\theta}_6 + 2D_{334}\dot{\theta}_3\dot{\theta}_4$$

$$+2D_{335}\dot{\theta}_3\dot{\theta}_5 + 2D_{336}\dot{\theta}_3\dot{\theta}_6 + 2D_{345}\dot{\theta}_4\dot{\theta}_5 + 2D_{346}\dot{\theta}_4\dot{\theta}_6 + 2D_{356}\dot{\theta}_5\dot{\theta}_6 + D_3$$

$$T_4 = D_{41}\ddot{\theta}_1 + D_{42}\ddot{\theta}_2 + D_{43}\ddot{\theta}_3 + D_{44}\ddot{\theta}_4 + D_{45}\ddot{\theta}_5 + D_{46}\ddot{\theta}_6 + I_{4(act)}\ddot{\theta}_4$$

$$+D_{411}\dot{\theta}_1^2 + D_{422}\dot{\theta}_2^2 + D_{433}\dot{\theta}_3^2 + D_{444}\dot{\theta}_4^2 + D_{455}\dot{\theta}_5^2 + D_{466}\dot{\theta}_6^2$$

$$+2D_{412}\dot{\theta}_1\dot{\theta}_2 + 2D_{413}\dot{\theta}_1\dot{\theta}_3 + 2D_{414}\dot{\theta}_1\dot{\theta}_4 + 2D_{415}\dot{\theta}_1\dot{\theta}_5 + 2D_{416}\dot{\theta}_1\dot{\theta}_6 \quad (4.52)$$

$$+2D_{423}\dot{\theta}_2\dot{\theta}_3 + 2D_{424}\dot{\theta}_2\dot{\theta}_4 + 2D_{425}\dot{\theta}_2\dot{\theta}_5 + 2D_{426}\dot{\theta}_2\dot{\theta}_6 + 2D_{434}\dot{\theta}_3\dot{\theta}_4$$

$$+2D_{435}\dot{\theta}_3\dot{\theta}_5 + 2D_{436}\dot{\theta}_3\dot{\theta}_6 + 2D_{445}\dot{\theta}_4\dot{\theta}_5 + 2D_{446}\dot{\theta}_4\dot{\theta}_6 + 2D_{456}\dot{\theta}_5\dot{\theta}_6 + D_4$$

$$T_5 = D_{51}\ddot{\theta}_1 + D_{52}\ddot{\theta}_2 + D_{53}\ddot{\theta}_3 + D_{54}\ddot{\theta}_4 + D_{55}\ddot{\theta}_5 + D_{56}\ddot{\theta}_6 + I_{5(act)}\ddot{\theta}_5$$

$$+D_{511}\dot{\theta}_1^2 + D_{522}\dot{\theta}_2^2 + D_{533}\dot{\theta}_3^2 + D_{544}\dot{\theta}_4^2 + D_{555}\dot{\theta}_5^2 + D_{566}\dot{\theta}_6^2$$

$$+2D_{512}\dot{\theta}_1\dot{\theta}_2 + 2D_{513}\dot{\theta}_1\dot{\theta}_3 + 2D_{514}\dot{\theta}_1\dot{\theta}_4 + 2D_{515}\dot{\theta}_1\dot{\theta}_5 + 2D_{516}\dot{\theta}_1\dot{\theta}_6 \quad (4.53)$$

$$+2D_{523}\dot{\theta}_2\dot{\theta}_3 + 2D_{524}\dot{\theta}_2\dot{\theta}_4 + 2D_{525}\dot{\theta}_2\dot{\theta}_5 + 2D_{526}\dot{\theta}_2\dot{\theta}_6 + 2D_{534}\dot{\theta}_3\dot{\theta}_4$$

$$+2D_{535}\dot{\theta}_3\dot{\theta}_5 + 2D_{536}\dot{\theta}_3\dot{\theta}_6 + 2D_{545}\dot{\theta}_4\dot{\theta}_5 + 2D_{546}\dot{\theta}_4\dot{\theta}_6 + 2D_{556}\dot{\theta}_5\dot{\theta}_6 + D_5$$

$$T_6 = D_{61}\ddot{\theta}_1 + D_{62}\ddot{\theta}_2 + D_{63}\ddot{\theta}_3 + D_{64}\ddot{\theta}_4 + D_{65}\ddot{\theta}_5 + D_{66}\ddot{\theta}_6 + I_{6(act)}\ddot{\theta}_6$$

$$+D_{611}\dot{\theta}_1^2 + D_{622}\dot{\theta}_2^2 + D_{633}\dot{\theta}_3^2 + D_{644}\dot{\theta}_4^2 + D_{655}\dot{\theta}_5^2 + D_{666}\dot{\theta}_6^2$$

$$+2D_{612}\dot{\theta}_1\dot{\theta}_2 + 2D_{613}\dot{\theta}_1\dot{\theta}_3 + 2D_{614}\dot{\theta}_1\dot{\theta}_4 + 2D_{615}\dot{\theta}_1\dot{\theta}_5 + 2D_{616}\dot{\theta}_1\dot{\theta}_6 \quad (4.54)$$

$$+2D_{623}\dot{\theta}_2\dot{\theta}_3 + 2D_{624}\dot{\theta}_2\dot{\theta}_4 + 2D_{625}\dot{\theta}_2\dot{\theta}_5 + 2D_{626}\dot{\theta}_2\dot{\theta}_6 + 2D_{634}\dot{\theta}_3\dot{\theta}_4$$

$$+2D_{635}\dot{\theta}_3\dot{\theta}_5 + 2D_{636}\dot{\theta}_3\dot{\theta}_6 + 2D_{645}\dot{\theta}_4\dot{\theta}_5 + 2D_{646}\dot{\theta}_4\dot{\theta}_6 + 2D_{656}\dot{\theta}_5\dot{\theta}_6 + D_6$$

Substituting the numerical values related to the robot in these equations yields the equations of motion for the robot. These equations can also show how each term can affect the dynamics of the robot or whether a particular term is important or not. For example, in the absence of gravity, such as in space, the gravity terms may be neglected. However, inertia terms will be important. On the other hand, if a robot moves slowly, many terms in these equations that relate to centrifugal and Coriolis accelerations may become negligible. In general, using these equations, the robot can be properly designed and controlled.

Example 4.8

Using the above equations, derive the equations of motion for the 2-DOF robot arm of Example 4.4, shown in Figure 4.9. The two links are assumed to be of equal length.

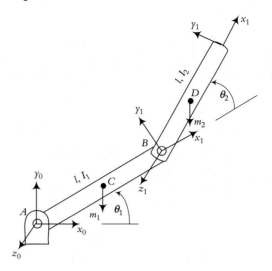

Figure 4.9 The 2-DOF robot arm of Example 4.8.

Solution: To use the above equations of motion for a 2-DOF robot, we first write the A matrices for the two links, then we develop the D_{ij}, D_{ijk}, and D_i terms for the robot. Finally, we substitute the results into Equations (4.49) and (4.50) to get the final equations of motion. The joint and link parameters of the robot are $d_1 = 0, d_2 = 0, a_1 = l$, $a_2 = l, \alpha_1 = 0, \quad \alpha_2 = 0$.

$$
A_1 = \begin{bmatrix} C_1 & -S_1 & 0 & lC_1 \\ S_1 & C_1 & 0 & lS_1 \\ 0 & 0 & 1 & 0 \\ 0 & 0 & 0 & 1 \end{bmatrix} \quad A_2 = \begin{bmatrix} C_2 & -S_2 & 0 & lC_2 \\ S_2 & C_2 & 0 & lS_2 \\ 0 & 0 & 1 & 0 \\ 0 & 0 & 0 & 1 \end{bmatrix}
$$

$$
{}^0T_2 = A_1 A_2 = \begin{bmatrix} C_{12} & -S_{12} & 0 & l(C_{12} + C_1) \\ S_{12} & C_{12} & 0 & l(S_{12} + S_1) \\ 0 & 0 & 1 & 0 \\ 0 & 0 & 0 & 1 \end{bmatrix}
$$

From Equation (4.27), $Q(revolute) = \begin{bmatrix} 0 & -1 & 0 & 0 \\ 1 & 0 & 0 & 0 \\ 0 & 0 & 0 & 0 \\ 0 & 0 & 0 & 0 \end{bmatrix}$

From Equation (4.28) we have $U_{ij} = \dfrac{\partial {}^0T_i}{\partial q_j} = \dfrac{\partial (A_1 A_2 .. A_j .. A_i)}{\partial q_j} = A_1 A_2 .. Q_j A_j .. A_i.$

Therefore:

$$U_{11} = QA_1 = \begin{bmatrix} -S_1 & -C_1 & 0 & -lS_1 \\ C_1 & -S_1 & 0 & lC_1 \\ 0 & 0 & 0 & 0 \\ 0 & 0 & 0 & 0 \end{bmatrix} \rightarrow U_{111} = \frac{\partial(QA_1)}{\partial\theta_1} = QQA_1 \text{ and}$$

$$U_{112} = \frac{\partial(QA_1)}{\partial\theta_2} = 0$$

$$U_{21} = QA_1A_2 = \begin{bmatrix} -S_{12} & -C_{12} & 0 & -l(S_{12} + S_1) \\ C_{12} & -S_{12} & 0 & l(C_{12} + C_1) \\ 0 & 0 & 0 & 0 \\ 0 & 0 & 0 & 0 \end{bmatrix} \rightarrow U_{211} = \frac{\partial(QA_1A_2)}{\partial\theta_1} = QQA_1A_2 \text{ and}$$

$$U_{212} = \frac{\partial(QA_1A_2)}{\partial\theta_2} = QA_1QA_2$$

$$U_{22} = A_1QA_2 = \begin{bmatrix} -S_{12} & -C_{12} & 0 & -lS_{12} \\ C_{12} & -S_{12} & 0 & lC_{12} \\ 0 & 0 & 0 & 0 \\ 0 & 0 & 0 & 0 \end{bmatrix} \rightarrow U_{221} = \frac{\partial(A_1QA_2)}{\partial\theta_1} = QA_1QA_2 \text{ and}$$

$$U_{222} = \frac{\partial(A_1QA_2)}{\partial\theta_2} = A_1QQA_2$$

$$U_{12} = \frac{\partial A_1}{\partial\theta_2} = 0$$

From Equation (4.36), assuming that all products of inertia are zero, we get:

$$J_1 = \begin{bmatrix} \frac{1}{3}m_1l^2 & 0 & 0 & \frac{1}{2}m_1l \\ 0 & 0 & 0 & 0 \\ 0 & 0 & 0 & 0 \\ \frac{1}{2}m_1l & 0 & 0 & m_1 \end{bmatrix} \quad J_2 = \begin{bmatrix} \frac{1}{3}m_2l^2 & 0 & 0 & \frac{1}{2}m_2l \\ 0 & 0 & 0 & 0 \\ 0 & 0 & 0 & 0 \\ \frac{1}{2}m_2l & 0 & 0 & m_2 \end{bmatrix}$$

From Equation (4.49) and (4.50), for a 2-DOF robot, we get:

$$T_1 = D_{11}\ddot{\theta}_1 + D_{12}\ddot{\theta}_2 + D_{111}\dot{\theta}_1^2 + D_{122}\dot{\theta}_2^2 + 2D_{112}\dot{\theta}_1\dot{\theta}_2 + D_1 + I_{1(act)}\ddot{\theta}_1 \quad (4.55)$$

$$T_2 = D_{21}\ddot{\theta}_1 + D_{22}\ddot{\theta}_2 + D_{211}\dot{\theta}_1^2 + D_{222}\dot{\theta}_2^2 + 2D_{212}\dot{\theta}_1\dot{\theta}_2 + D_2 + I_{2(act)}\ddot{\theta}_2 \quad (4.56)$$

From Equations (4.43), (4.44), and (4.45), we have:

$$\begin{aligned} D_{11} &= Trace\left(U_{11}J_1U_{11}^T\right) + Trace\left(U_{21}J_2U_{21}^T\right) && \text{for } i = 1, j = 1, p = 1, 2 \\ D_{12} &= Trace\left(U_{22}J_2U_{21}^T\right) && \text{for } i = 1, j = 2, p = 2 \\ D_{21} &= Trace\left(U_{21}J_2U_{22}^T\right) && \text{for } i = 2, j = 1, p = 2 \\ D_{22} &= Trace\left(U_{22}J_2U_{22}^T\right) && \text{for } i = 2, j = 2, p = 2 \end{aligned}$$

$$D_{111} = Trace\left(U_{111}J_1U_{11}^T\right) + Trace\left(U_{211}J_2U_{21}^T\right) \quad \text{for } i = 1, j = 1, k = 1, p = 1, 2$$

$$D_{122} = Trace\left(U_{222}J_2U_{21}^T\right) \quad \text{for } i = 1, j = 2, k = 2, p = 2$$

$$D_{112} = Trace\left(U_{212}J_2U_{21}^T\right) \quad \text{for } i = 1, j = 1, k = 2, p = 2$$

$$D_{211} = Trace\left(U_{211}J_2U_{22}^T\right) \quad \text{for } i = 2, j = 1, k = 1, p = 2$$

$$D_{222} = Trace\left(U_{222}J_2U_{22}^T\right) \quad \text{for } i = 2, j = 2, k = 2, p = 2$$

$$D_{212} = Trace\left(U_{212}J_2U_{22}^T\right) \quad \text{for } i = 2, j = 1, k = 2, p = 2$$

$$D_1 = -m_1g^T U_{11}\bar{r}_1 - m_2g^T U_{21}\bar{r}_2 \quad \text{for } i = 1, p = 1, 2$$

$$D_2 = -m_1g^T U_{12}\bar{r}_1 - m_2 g^T U_{22}\bar{r}_2 \quad \text{for } i = 2, p = 1, 2$$

Although forbiddingly long, even for a 2-DOF robot, substituting all given matrices into these equations yields:

$$D_{11} = \frac{1}{3}m_1l^2 + \frac{4}{3}m_2l^2 + m_2l^2C_2$$

$$D_{12} = D_{21} = \frac{1}{3}m_2l^2 + \frac{1}{2}m_2l^2C_2$$

$$D_{22} = \frac{1}{3}m_2l^2$$

$$D_{111} = 0 \qquad\qquad D_{112} = D_{121} = -\frac{1}{2}m_2l^2S_2$$

$$D_{122} = -\frac{1}{2}m_2l^2S_2 \qquad D_{211} = \frac{1}{2}m_2l^2S_2$$

$$D_{212} = 0 \qquad\qquad D_{221} = 0 \quad D_{222} = 0$$

and from Equation (4.45) for $g^T = \begin{bmatrix} 0 & -g & 0 & 0 \end{bmatrix}$ (because acceleration of gravity is in the minus direction of y-axis) and $\bar{r}_1^T = \bar{r}_2^T = \begin{bmatrix} -l/2 & 0 & 0 & 1 \end{bmatrix}$ (because the center of mass of the bar is at $-\frac{l}{2}$), we get:

$$D_1 = -m_1g^T U_{11}\bar{r}_1 - m_2g^T U_{21}\bar{r}_2$$

$$= -m_1\begin{bmatrix} 0 & -g & 0 & 0 \end{bmatrix}\begin{bmatrix} -S_1 & -C_1 & 0 & -lS_1 \\ C_1 & -S_1 & 0 & lC_1 \\ 0 & 0 & 0 & 0 \\ 0 & 0 & 0 & 0 \end{bmatrix}\begin{bmatrix} -\frac{l}{2} \\ 0 \\ 0 \\ 1 \end{bmatrix}$$

$$-m_2\begin{bmatrix} 0 & -g & 0 & 0 \end{bmatrix}\begin{bmatrix} -S_{12} & -C_{12} & 0 & -l(S_{12} + S_1) \\ C_{12} & -S_{12} & 0 & l(C_{12} + C_1) \\ 0 & 0 & 0 & 0 \\ 0 & 0 & 0 & 0 \end{bmatrix}\begin{bmatrix} -\frac{l}{2} \\ 0 \\ 0 \\ 1 \end{bmatrix}$$

and similarly for D_2, we get:

$$D_1 = \frac{1}{2}m_1glC_1 + \frac{1}{2}m_2glC_{12} + m_2glC_1$$

$$D_2 = \frac{1}{2}m_2glC_{12}$$

Substituting the results into Equations (4.55) and (4.56) will result in the final equations of motion as:

$$T_1 = \left(\frac{1}{3}m_1l^2 + \frac{4}{3}m_2l^2 + m_2l^2C_2\right)\ddot{\theta}_1 + \left(\frac{1}{3}m_2l^2 + \frac{1}{2}m_2l^2C_2\right)\ddot{\theta}_2 + \left(\frac{1}{2}m_2l^2S_2\right)\dot{\theta}_2^2$$

$$+ \left(m_2l^2S_2\right)\dot{\theta}_1\dot{\theta}_2 + \frac{1}{2}m_1glC_1 + \frac{1}{2}m_2glC_{12} + m_2glC_1 + I_{1(act)}\ddot{\theta}_1$$

$$T_2 = \left(\frac{1}{3}m_2l^2 + \frac{1}{2}m_2l^2C_2\right)\ddot{\theta}_1 + \left(\frac{1}{3}m_2l^2\right)\ddot{\theta}_2 + \left(\frac{1}{2}m_2l^2S_2\right)\dot{\theta}_1^2 + \frac{1}{2}m_2glC_{12} + I_{2(act)}\ddot{\theta}_1$$

which, except for the actuator inertia terms, are the same as Equations (4.11) and (4.12).   ■

4.5 Static Force Analysis of Robots

Robots may be under either position control or force control. Imagine a robot that is following a line, say, on the flat surface of a panel and is cutting a groove in the surface. If the robot follows a prescribed path, it is under position control. So long as the surface is flat, and the robot is following the line on the flat surface, the groove will be uniform. However, if the surface is not flat, since the robot is following a given path, it will either cut deeper into the surface, or it will not cut deep enough. Alternately, suppose the robot were to measure the force it is exerting on the surface while cutting the groove. If the force becomes too large or too small, indicating that the tool is cutting too deep or not deep enough, the robot could adjust the depth until it cuts uniformly. In this case, the robot is under force control.

Similarly, suppose it is required that a robot tap a hole in a machine part. The robot would need to exert a known axial force along the axis of the hole as well as rotate the tap by exerting a moment on it. To be able to do this, the controller would need to move the joints and rotate them at particular rates to create the desired forces and moments at the hand frame. To relate the joint forces and torques to forces and moments generated at the hand frame of the robot,[1,9,10] we will define the following:

$$\left[^HF\right] = \begin{bmatrix} f_x & f_y & f_z & m_x & m_y & m_z \end{bmatrix}^T \tag{4.57}$$

where f_x, f_y, f_z are the forces along the x-, y-, and z-axes of the hand frame, and m_x, m_y, m_z are the moments about the x-, y-, and z-axes of the hand frame. Similarly, we define the following:

$$\left[^HD\right] = \begin{bmatrix} dx & dy & dz & \delta x & \delta y & \delta z \end{bmatrix}^T \tag{4.58}$$

which express displacements and rotations about the x-, y-, and z-axes of the hand frame. We can also define similar entities for the joints as:

$$[T] = [\,T_1 \quad T_2 \quad T_3 \quad T_4 \quad T_5 \quad T_6\,]^T \tag{4.59}$$

which are the torques (for revolute joints) and forces (for prismatic joints) at each joint, and:

$$[D_\theta] = [\,d\theta_1 \quad d\theta_2 \quad d\theta_3 \quad d\theta_4 \quad d\theta_5 \quad d\theta_6\,]^T \tag{4.60}$$

which describes the differential movements at the joints, either an angle for a revolute joint, or a linear displacement for a prismatic joint.

Using the method of virtual work,[11] which indicates that the total virtual work at the joints must be the same as the total virtual work at the hand frame, we get:

$$\delta W = [^H F]^T [^H D] = [T]^T [D_\theta] \tag{4.61}$$

or that the forces and moments times the displacements at the hand frame are equal to the torques or forces times the displacements at the joints. Can you tell why transpose of the force and torque matrices are used? Substituting the values, we will get the following for the left-hand side of Equation (4.61):

$$\begin{bmatrix} f_x & f_y & f_z & m_x & m_y & m_z \end{bmatrix} \begin{bmatrix} dx \\ dy \\ dz \\ \delta x \\ \delta y \\ \delta z \end{bmatrix} = f_x dx + f_y dy + \cdots + m_z \delta z \tag{4.62}$$

However, from Equation (3.26), we have:

$$[^{T_6} D] = [^{T_6} J][D_\theta] \quad \text{or} \quad [^H D] = [^H J][D_\theta] \tag{3.24}$$

Substituting this into Equation (4.61) results in:

$$[^H F]^T [^H J][D_\theta] = [T]^T [D_\theta] \rightarrow [^H F]^T [^H J] = [T]^T \tag{4.63}$$

Referring to Appendix A, this equation can be written as:

$$[T] = [^H J]^T [^H F] \tag{4.64}$$

which indicates that the joint forces and moments can be determined from the desired set of forces and moments at the hand frame. Since the Jacobian is already known from previous analysis for differential motions, the controller can calculate the forces and moments at the joints and control the robot based on the desired values.

Force control of robots may also be accomplished through the use of sensors such as force and torque sensors. This includes robots that can "feel" the object they are handling and that can relay this information back to the controller or the "master" operator.[12] We will discuss this later in Chapter 8.

Example 4.9

The numerical value of the Jacobian of a spherical-RPY robot (like the Stanford Arm) is given below. It is desired to apply a force of 1 lb along the z-axis of the hand frame as well as a moment of 20 lb.in about the z-axis of the hand frame to drill a hole in a block. Find the necessary joint forces and torques.

$$
{}^{H}J = \begin{bmatrix}
20 & 0 & 0 & 0 & 0 & 0 \\
-5 & 0 & 1 & 0 & 0 & 0 \\
0 & 20 & 0 & 0 & 0 & 0 \\
0 & 1 & 0 & 0 & 1 & 0 \\
0 & 0 & 0 & 1 & 0 & 0 \\
-1 & 0 & 0 & 0 & 0 & 1
\end{bmatrix}
$$

Solution: Substituting the values given into Equation (4.64), we get:

$$[T] = [{}^{H}J]^{T}[{}^{H}F]$$

$$
[T] = \begin{bmatrix}
T_1 \\
T_2 \\
F_3 \\
T_4 \\
T_5 \\
T_6
\end{bmatrix}
=
\begin{bmatrix}
20 & -5 & 0 & 0 & 0 & -1 \\
0 & 0 & 20 & 1 & 0 & 0 \\
0 & 1 & 0 & 0 & 0 & 0 \\
0 & 0 & 0 & 0 & 1 & 0 \\
0 & 0 & 0 & 1 & 0 & 0 \\
0 & 0 & 0 & 0 & 0 & 1
\end{bmatrix}
\begin{bmatrix}
0 \\
0 \\
1 \\
0 \\
0 \\
20
\end{bmatrix}
=
\begin{bmatrix}
-20 \\
20 \\
0 \\
0 \\
0 \\
20
\end{bmatrix}
$$

As you can see, for this particular configuration of the robot and with the robot's specific dimensions, it is necessary to exert the indicated torques at the first, second, and sixth joints in order to create the desired force and torque at the hand frame. There is no need for the third joint—the prismatic joint—to exert any force, even though we want a force at the hand frame. Can you visualize why?

Obviously, as the configuration of the robot changes, the Jacobian changes as well. Therefore, for continued exertion of the same force and moment at the hand frame as the robot moves, the joint torques will have to change as well, requiring continuous calculation of joint torques by the controller. ∎

4.6 Transformation of Forces and Moments between Coordinate Frames

Suppose two coordinate frames are attached to an object, a force and moment are acting on the object and they are described with respect to one of the coordinate frames. The principle of virtual work can also be used here to find an equivalent force and moment with respect to the other coordinate frame such that they will have the same effect on the object. To do this, we define F as forces and moments acting on the object, and D as the displacements caused by these forces and moments, also relative to the same reference frame, as:

$$[F]^T = \left[f_x, f_y, f_z, m_x, m_y, m_z \right] \tag{4.65}$$

$$[D]^T = \left[d_x, d_y, d_z, \delta_x, \delta_y, \delta_z \right] \tag{4.66}$$

We also define ^BF to be the forces and moments acting on the object relative to a coordinate frame B, and ^BD to be the displacements caused by these forces and moments, also relative to the coordinate frame B, as:

$$\left[^BF \right]^T = \left[^Bf_x, \ ^Bf_y, \ ^Bf_z, \ ^Bm_x, \ ^Bm_y, \ ^Bm_z \right] \tag{4.67}$$

$$\left[^BD \right]^T = \left[^Bd_x, \ ^Bd_y, \ ^Bd_z, \ ^B\delta_x, \ ^B\delta_y, \ ^B\delta_z \right] \tag{4.68}$$

Since the total virtual work performed on the object in either frame must be the same, then:

$$\delta W = [F]^T [D] = \left[^BF \right]^T \left[^BD \right] \tag{4.69}$$

Paul[1] has shown that displacements relative to the two frames are related to each other by the following relationship:

$$\left[^BD \right] = \left[^BJ \right] [D] \tag{4.70}$$

Substituting Equation (4.70) into Equation (4.69) will result in:

$$[F]^T [D] = \left[^BF \right]^T \left[^BJ \right] [D] \quad \text{or} \quad [F]^T = \left[^BF \right]^T \left[^BJ \right] \tag{4.71}$$

which can be rearranged to:

$$[F] = \left[^BJ \right]^T \left[^BF \right] \tag{4.72}$$

Paul[1] has also shown that instead of calculating the Jacobian with respect to Frame B, the forces and moments with respect to frame B can be directly found from the following equations:

$$
\begin{aligned}
^Bf_x &= \mathbf{n} \cdot \mathbf{f} \\
^Bf_y &= \mathbf{o} \cdot \mathbf{f} \\
^Bf_z &= \mathbf{a} \cdot \mathbf{f} \\
^Bm_x &= \mathbf{n} \cdot [(\mathbf{f} \times \mathbf{p}) + \mathbf{m}] \\
^Bm_y &= \mathbf{o} \cdot [(\mathbf{f} \times \mathbf{p}) + \mathbf{m}] \\
^Bm_z &= \mathbf{a} \cdot [(\mathbf{f} \times \mathbf{p}) + \mathbf{m}]
\end{aligned}
\tag{4.73}
$$

Using Equations (4.73), we may find equivalent forces and moments in different frames that will have the same effect on an object.

Example 4.10

An object, attached to a frame B, is subjected to the forces and moments given relative to the reference frame. Find the equivalent forces and moments

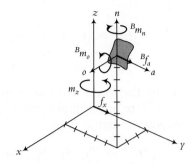

Figure 4.10   Equivalent force-moment systems in two different frames.

in frame B.

$$F^T = [0, 10(\text{lb}), 0, 0, 0, 20(\text{lb·in})]$$

$$B = \begin{bmatrix} 0 & 1 & 0 & 3 \\ 0 & 0 & 1 & 5 \\ 1 & 0 & 0 & 8 \\ 0 & 0 & 0 & 1 \end{bmatrix}$$

Solution: From the information above, we have:

$$\mathbf{f}^T = [0, 10, 0] \quad \mathbf{m}^T = [0, 0, 20] \quad \mathbf{p}^T = [3, 5, 8]$$
$$\mathbf{n}^T = [0, 0, 1] \quad \mathbf{o}^T = [1, 0, 0] \quad \mathbf{a}^T = [0, 1, 0]$$

$$\mathbf{f} \times \mathbf{p} = \begin{vmatrix} \mathbf{i} & \mathbf{j} & \mathbf{k} \\ 0 & 10 & 0 \\ 3 & 5 & 8 \end{vmatrix} = \mathbf{i}(80) - \mathbf{j}(0) + \mathbf{k}(-30)$$

$$(\mathbf{f} \times \mathbf{p}) + \mathbf{m} = 80\mathbf{i} - 10\mathbf{k}$$

From Equations (4.73), we get:

$$^B f_x = \mathbf{n} \cdot \mathbf{f} = 0$$
$$^B f_y = \mathbf{o} \cdot \mathbf{f} = 0$$
$$^B f_z = \mathbf{a} \cdot \mathbf{f} = 10 \qquad\qquad \rightarrow {}^B f = [0, 0, 10]$$
$$^B m_x = \mathbf{n} \cdot [(\mathbf{f} \times \mathbf{p}) + \mathbf{m}] = -10$$
$$^B m_y = \mathbf{o} \cdot [(\mathbf{f} \times \mathbf{p}) + \mathbf{m}] = 80$$
$$^B m_z = \mathbf{a} \cdot [(\mathbf{f} \times \mathbf{p}) + \mathbf{m}] = 0 \qquad \rightarrow {}^B m = [-10, 80, 0]$$

This means that a 10-lb force along the a-axis of the frame B, together with two moments of -10 lb.in, along the n-axis and 80 lb.in along the o-axis will have the same effect on the object as the original force and moment relative to the reference frame. Does this match what you learned in your statics course? Figure 4.10 shows the two equivalent force-moment systems. ∎

4.7 Design Project

You may want to continue with the analysis of your robot. You will have to develop the dynamic equations of motion, which will enable you to calculate the power needed at each joint to move the robot at desired accelerations. This information will be used for choosing the appropriate actuators as well as in controlling the robot's motions.

Alternately, since your robot will not be moving too fast, you can calculate the torque needed at each joint by trying to find the worst case situations for each joint. For example, try to model the robot with both links extended outwardly. In this case, the first actuator acting on the first joint will experience the largest load. This estimate is not very accurate, since we are eliminating all coupling inertia terms, Coriolis accelerations, and so on. But as we discussed earlier, under low load conditions and with low velocities, you can get a relatively acceptable estimate of the torques needed.

Summary

In this chapter, we discussed the derivation of the dynamic equations of motion of robots. These equations can be used to estimate the necessary power needed at each joint to drive the robot with desired velocity and accelerations. They can also be used to choose appropriate actuators for a robot.

Dynamic equations of multi-DOF, 3-D mechanisms such as robots are complicated and, at times, very difficult to use. As a result, they are mostly used in simplified forms with simplifying assumptions. As an example, we may determine the importance of a particular term and its contribution to the total torque or power needed by considering how large it is relative to other terms. For instance, we may determine the importance of Coriolis terms in these equations by knowing how large the velocity terms are. Conversely, the importance of gravity terms in space robots may be determined and, if appropriate, dropped from the equations of motion.

In the next chapter, we will discuss how a robot's motions are controlled and planned in order to yield a desired trajectory.

References

1. Paul, Richard P., "Robot Manipulators, Mathematics, Programming, and Control," The MIT Press, 1981.

2. Shahinpoor, Mohsen, "A Robot Engineering Textbook," Harper & Row, NY, 1987.

3. Asada, Haruhiko, J. J. E., Slotine, "Robot Analysis and Control," John Wiley and Sons, NY, 1986.

4. Sciavicco, Lorenzo, B. Siciliano, "Modeling and Control of Robot Manipulators," McGraw-Hill, NY, 1996.

5. Fu, K. S., R. C. Gonzalez, C. S. G. Lee, "Robotics: Control, Sensing, Vision, and Intelligence," McGraw-Hill, 1987.

6. Featherstone, R., "The Calculation of Robot Dynamics Using Articulated-Body Inertias," The International Journal of Robotics Research, Vol. 2, No. 1, Spring 1983, pp. 13–30.

7. Shahinpoor, M., "Dynamics," International Encyclopedia of Robotics: Applications and Automation, Richard C. Dorf, Editor, John Wiley and Sons, NY, 1988, pp. 329–347.

8. Pytel, Andrew, J. Kiusalaas, "Engineering Mechanics, Dynamics," 2nd Edition, Brooks/ Cole Publishing, Pacific Grove, 1999.

9. Paul, Richard, C. N. Stevenson, "Kinematics of Robot Wrists," The International Journal of Robotics Research, Vol. 2, No. 1, Spring 1983, pp. 31–38.

10. Whitney, D. E., "The Mathematics of Coordinated Control of Prosthetic Arms and Manipulators," Transactions of ASME, *Journal of Dynamic Systems, Measurement, and Control*, 94G(4), 1972, pp. 303–309.

11. Pytel, Andrew, J. Kiusalaas, "Engineering Mechanics, Statics," 2nd Edition, Brooks/Cole Publishing, Pacific Grove, 1999.

12. Chicurel, Marina, "Once More, With Feeling," *Stanford Magazine*, March/April 2000, pp. 70–73.

Problems

4.1. Using Lagrangian mechanics, derive the equations of motion of a cart with two tires under the cart as shown:

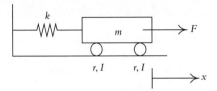

Figure P.4.1

4.2. Calculate the total kinetic energy of the link AB, attached to a roller with negligible mass, as shown:

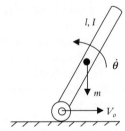

Figure P.4.2

4.3. Derive the equations of motion for the 2-link mechanism with distributed mass, as shown:

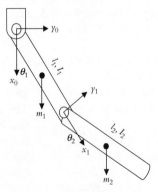

Figure P.4.3

4.4. Write the equations that express U_{62}, U_{35}, U_{53}, U_{623}, and U_{534} for a 6-axis cylindrical-RPY robot in terms of the A and Q matrices.

4.5. Using Equations (4.49) to (4.54), write the equations of motion for a 3-DOF revolute robot and describe each term.

4.6. Expand the D_{134} and D_{15} terms of Equation (4.49) in terms of their constituent matrices.

4.7. An object is subjected to the following forces and moments relative to the reference frame. Attached to the object is a frame, which describes the orientation and the location of the object. Find the equivalent forces and torques acting on the object relative to the current frame.

$$B = \begin{bmatrix} 0.707 & 0.707 & 0 & 2 \\ 0 & 0 & 1 & 5 \\ 0.707 & -0.707 & 0 & 3 \\ 0 & 0 & 0 & 1 \end{bmatrix} \quad F^T = [10, 0, 5, 12, 20, 0]\, \text{N}, \text{N.m}$$

4.8. In order to assemble two parts together, one part must be pushed into the other with a force of 10 lb in the x-axis direction, 5 lb in the y-direction, and be turned with a torque of 5 lb·in along the x-axis direction. The object's location relative to the base frame of a robot is described by ${}^R T_0$. Assuming that the two parts must be aligned together for this purpose, find the necessary forces and moments that the robot must apply to the part relative to its hand frame.

$$^R T_0 = \begin{bmatrix} 0 & -0.707 & 0.707 & 4 \\ 1 & 0 & 0 & 6 \\ 0 & 0.707 & 0.707 & 3 \\ 0 & 0 & 0 & 1 \end{bmatrix}$$

Chapter 5

Trajectory Planning

5.1 Introduction

In the previous chapters, we studied the kinematics and dynamics of robots. This means that, using the equations of motion of the robot, we can determine where the robot will be if we have the joint variables or we can determine what the joint variables must be in order to place the robot at a desired position and orientation with a desired velocity. Path and trajectory planning relates to the way a robot is moved from one location to another in a controlled manner. In this chapter, we will study the sequence of movements that must be made to create a controlled movement between motion segments, whether in straight-line motions or sequential motions. Path and trajectory planning requires the use of both kinematics and dynamics of robots. In practice, precise motion requirements are so intensive that approximations are always necessary.

5.2 Path versus Trajectory

A path is defined as the collection of a sequence of configurations a robot makes to go from one place to another without regard to the timing of these configurations. So, in Figure 5.1, if a robot goes from point (configuration) A to point B to point C, the sequence of the configurations between A and B and C constitutes a path.[1] A trajectory is related to the timing at which each part of the path must be attained. As a result, regardless of when points B and C in Figure 5.1 are reached, the path is the same, whereas depending on how fast each portion of the path is traversed, the trajectory may differ. Therefore, the points at which the robot may be on a path and on a trajectory at a given time may be different, even if the robot traverses the same points. On a trajectory, depending on the velocities and accelerations, points B and C may be reached at different times, creating different trajectories. In this chapter, we are not only concerned about the path a robot takes, but also its velocities and accelerations.

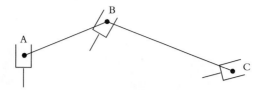

Figure 5.1   Sequential robot movements in a path.

5.3 Joint-Space versus Cartesian-Space Descriptions

Consider a 6-axis robot at a point A in space, which is directed to move to another point B. Using the inverse kinematic equations of the robot, derived in Chapter 2, we may calculate the total joint displacements the robot needs to make to get to the new location. The joint values thus calculated can be used by the controller to drive the robot joints to their new values and, consequently, move the robot arm to its new position. The description of the motion to be made by the robot by its joint values is called *joint-space* description. In this case, although the robot will eventually reach the desired position, as we will see later, the motion between the two points is unpredictable.

Now assume that a straight line is drawn between points A and B, and it is desirable to have the robot move from point A to point B in a straight line. To do this, it will be necessary to divide the line into small portions, as shown in Figure 5.2, and to move the robot through all intermediate points. To accomplish this task, at each intermediate location, the robot's inverse kinematic equations are solved, a set of joint variables is calculated, and the controller is directed to drive the robot to those values. When all segments are completed, the robot will be at point B, as desired. However, in this case, unlike the joint-space case discussed above, the motion is known at all times. The sequence of movements the robot makes is described in *Cartesian-space* and is converted to joint-space at each segment. As is clear from this simple example, the Cartesian-space description is much more computationally intensive than the joint-space description, but yields a controlled and known path. Both joint-space and Cartesian-space descriptions are very useful and are used in industry. However, each one has its own advantages and disadvantages.

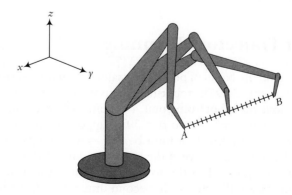

Figure 5.2   Sequential motions of a robot to follow a straight line.

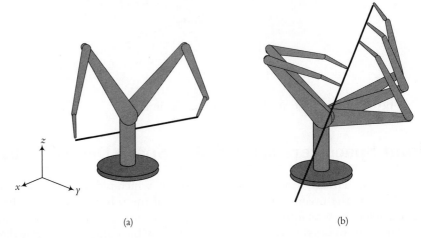

(a)   (b)

Figure 5.3   Cartesian-space trajectory problems. In (a), the trajectory specified in Cartesian coordinates may force the robot to run into itself; in (b), the trajectory may require a sudden change in the joint angles.

Cartesian-space trajectories are very easy to visualize. Since the trajectories are in the common Cartesian space in which we all operate, it is easy to visualize what the end effector's trajectory must be. However, Cartesian-space trajectories are computationally expensive and require a faster processing time for similar resolution than joint-space trajectories. Additionally, although it is easy to visualize the trajectory, it is difficult to visually ensure that singularities will not occur. For example, consider the situation in Figure 5.3(a). If not careful, we may specify a trajectory that requires the robot to run into itself or to reach a point outside of the work envelope—which, of course, is impossible—and yields an unsatisfactory solution.[2] This is true because it may be impossible to know whether the robot can actually make a particular location and orientation before the motion is made. Also, as shown in Figure 5.3(b), the motion between two points may require an instantaneous change in the joint values (we discussed why this may happen in Chapter 2), which is impossible to predict. Some of these problems may be solved by specifying via points the robot must pass through in order to avoid obstacles and other similar singularities.

5.4 Basics of Trajectory Planning

To understand the basics of planning a trajectory in joint-space and Cartesian-space, let's consider a simple 2-DOF robot (mechanism). In this case, as shown in Figure 5.4, we desire to move the robot from point A to point B. The configuration of the robot at point A is shown, with $\alpha = 20°$ and $\beta = 30°$. Suppose it has been calculated that in order for the robot to be at point B, it must be at $\alpha = 40°$ and $\beta = 80°$. Also suppose that both joints of the robot can move at the maximum rate of 10 degrees/sec. One way to move the robot from point A to B is to run both joints at their maximum angular velocities. This means that at the end of the second time interval, the lower link of the robot will have finished its motion, while the upper link continues for another three seconds, as

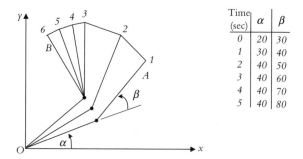

Time (sec)	α	β
0	20	30
1	30	40
2	40	50
3	40	60
4	40	70
5	40	80

Figure 5.4 Joint-space, non-normalized movements of a 2-DOF robot.

shown in Figure 5.4. The trajectory of the end of the robot is also shown. As indicated, the path is irregular, and the distances traveled by the robot's end are not uniform.

Now suppose the motions of both joints of the robot are normalized such that the joint with smaller motion will move proportionally slower so that both joints will start and stop their motion simultaneously. In this case, both joints move at different speeds, but move continuously together. α changes 4 degrees/second while β changes 10 degrees/second. The resulting trajectory will be different, as shown in Figure 5.5. Notice that the segments of the movement are much more similar to each other than before, but the path is still irregular (and different from the previous case). Both of these cases were planned in joint-space as we were only concerned with the values of the joints, not the location of the end of the mechanism. The only calculation needed was the joint values for the destination and, in the second case, normalization of the joint velocities.

Now suppose we want the robot's hand to follow a known path between points A and B, say, in a straight line. The simplest solution would be to draw a line between points A and B, divide the line into, say, 5 segments, and solve for necessary angles α and β at each point, as shown in Figure 5.6. This is called interpolation between points A and B.[3–5] Notice that in this case, the path is a straight line, but the joint angles are not uniformly changing. Although the resulting motion is a straight (and consequently, known) trajectory, it is necessary to solve for the joint values at each point. Obviously, many more points must be calculated for better accuracy; with so few segments the robot will not exactly follow the lines at each segment. This trajectory is in Cartesian-space since all

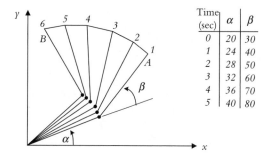

Time (sec)	α	β
0	20	30
1	24	40
2	28	50
3	32	60
4	36	70
5	40	80

Figure 5.5 Joint-space, normalized movements of a robot with 2 DOF.

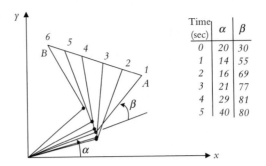

Time (sec)	α	β
0	20	30
1	14	55
2	16	69
3	21	77
4	29	81
5	40	80

Figure 5.6 Cartesian-space movements of a 2-DOF robot.

segments of the motion must be calculated based on the information expressed in a Cartesian frame.

In this case, it is assumed that the robot's actuators are strong enough to provide the large forces necessary to accelerate and decelerate the joints as needed. For example, notice that we are assuming the arm will be instantaneously accelerated to have the desired velocity right at the beginning of the motion in segment 1. If this is not true, the robot will follow a trajectory different from our assumption; it will be slightly behind as it accelerates to the desired speed. Additionally, note how the difference between two consecutive values is larger than the maximum specified joint velocity of 10 degrees/second (e.g., between times 0 and 1, the joint must move 25 degrees). Obviously, this is not attainable. Also note how, in this case, joint 1 moves downward first before moving up.

To improve the situation, we may actually divide the segments differently by starting the arm with smaller segments and, as we speed up the arm, going at a constant cruising rate and finally decelerating with smaller segments as we approach point B. This is schematically shown in Figure 5.7. Of course, we still need to solve the inverse kinematic equations of the robot at each point, which is similar to the previous case. However, in this case, instead of dividing the straight line AB into equal segments, we may divide it based on $x = \frac{1}{2}at^2$ until such time t when we attain the cruising velocity of $v = at$. Similarly, the end portion of the motion can be divided based on a decelerating regiment.

Another variation to this trajectory planning is to plan a path that is not straight, but one that follows some desired path, for example a quadratic equation. To do this, the

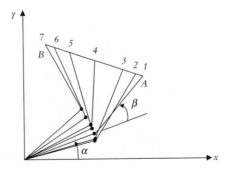

Figure 5.7 Trajectory planning with an acceleration/deceleration regiment.

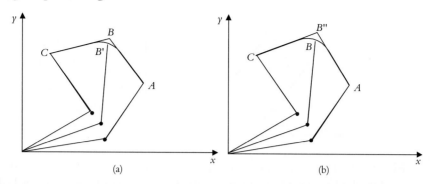

Figure 5.8 Blending of different motion segments in a path.

coordinates of each segment are calculated based on the desired path and are used to calculate joint variables at each segment; therefore, the trajectory of the robot can be planned for any desired path.

So far, we have only considered the movement of the robot between two points A and B. However, in other cases, the robot may be going through many consecutive points, including intermediate or via points. The next level of trajectory planning is between multiple points and, eventually, for continuous movements.

Assume the robot is to go from point A to B and then to C, as shown in Figure 5.8. One way to run the robot is to accelerate from point A toward point B, cruise, decelerate and stop at point B, start again at point B and accelerate toward C, cruise, and decelerate in order to stop at C. This stop-and-go motion will create jerky motions with unnecessary stops. An alternative way is to blend the two portions of the motion at point B such that the robot will approach point B, decelerate if necessary, follow the blended path, accelerate once again toward point C, and eventually stop at C. This creates a much more graceful motion, reduces the stress on the robot, and requires less energy. If the motion consists of more segments, all intermediate segments can be blended together. Notice how, due to this blending of the segments, the robot will go through a different point B' (Figure 5.8(a)) and not B. If it is necessary that the robot goes through point B exactly, then a different point B'' must be specified such that after blending, the robot still goes through point B (Figure 5.8(b)). Another alternative[2] is to specify two via points C and D before and after point B, as shown in Figure 5.9 such that point B falls on the straight-line portion of the motion between the via points and thus ensure that the robot goes through point B.

In the next sections, we will discuss different methods of trajectory planning in more detail. Generally, higher-order polynomials are used to match positions, velocities, and accelerations at each point between two segments. When the path is planned, the controller uses the path information (coordinates) in calculating joint variables from the inverse kinematic equations and runs the robot accordingly. Ultimately, if the path of the robot is very complicated and involved, such that it cannot be easily expressed by an equation, it may become necessary to physically move the robot by hand and record the motions at each joint. The recorded joint values can be used later to run the robot. This is commonly done for teaching robots tasks such as spray painting automobiles, seam welding complicated shapes, and other similar tasks.

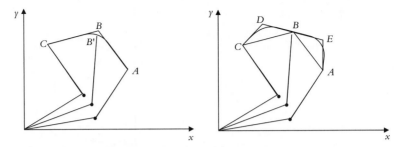

Figure 5.9   An alternative scheme for ensuring that the robot will go through a specified point during blending of motion segments. Two via points C and D are picked such that point B will fall on the straight-line section of the segment ensuring that the robot will pass through point B.

5.5 Joint-Space Trajectory Planning

In this section, we will see how the motions of a robot can be planned in joint-space with controlled characteristics. A number of different schemes such as polynomials of different orders and linear functions with parabolic blends can be used for this purpose. The following is a discussion of some of these schemes used in joint-space trajectory planning. Notice that these schemes specify joint values, not Cartesian values. We will discuss Cartesian-space later.

5.5.1 Third-Order Polynomial Trajectory Planning

In this application, the initial location and orientation of the robot are known and, using the inverse kinematic equations, the final joint angles for the desired position and orientation are found. However, the motions of *each joint* of the robot must be planned individually. Therefore, consider one of the joints, which at the beginning of the motion segment at time t_i is at θ_i, and which we want to move to a new value of θ_f at time t_f. One way to do this is to use a polynomial to plan the trajectory such that the initial and final boundary conditions match what we already know, namely that θ_i and θ_f are known and the velocities at the beginning and the end of the motion segment are zero (or other known values). These four pieces of information allow us to solve for four unknowns (or a third-order polynomial) in the form of:

$$\theta(t) = c_0 + c_1 t + c_2 t^2 + c_3 t^3 \tag{5.1}$$

where the initial and final conditions are:

$$\begin{aligned}
\theta(t_i) &= \theta_i \\
\theta(t_f) &= \theta_f \\
\dot{\theta}(t_i) &= 0 \\
\dot{\theta}(t_f) &= 0
\end{aligned} \tag{5.2}$$

Taking the first derivative of the polynomial of Equation (5.1), we get:

$$\dot{\theta}(t) = c_1 + 2c_2 t + 3c_3 t^2 \tag{5.3}$$

Substituting the initial and final conditions into Equations (5.1) and (5.3) yields:

$$\begin{aligned}
\theta(t_i) &= c_0 = \theta_i \\
\theta(t_f) &= c_0 + c_1 t_f + c_2 t_f^2 + c_3 t_f^3 \\
\dot{\theta}(t_i) &= c_1 = 0 \\
\dot{\theta}(t_f) &= c_1 + 2c_2 t_f + 3c_3 t_f^2 = 0
\end{aligned} \tag{5.4}$$

which can also be written in matrix form as:

$$\begin{bmatrix} \theta_i \\ \dot{\theta}_i \\ \theta_f \\ \dot{\theta}_f \end{bmatrix} = \begin{bmatrix} 1 & 0 & 0 & 0 \\ 0 & 1 & 0 & 0 \\ 1 & t_f & t_f^2 & t_f^3 \\ 0 & 1 & 2t_f & 3t_f^2 \end{bmatrix} \begin{bmatrix} c_0 \\ c_1 \\ c_2 \\ c_3 \end{bmatrix} \tag{5.5}$$

By solving these four equations simultaneously, we get the necessary values for the constants. This allows us to calculate the joint position at any interval of time, which can be used by the controller to drive the joint to position. The same process must be used for each joint individually, but they are all driven together from start to finish. Obviously, if the initial and final velocities are not zero, the given values can be used in these equations. Therefore, applying this third-order polynomial to each joint motion creates a motion profile that can be used to drive each joint.

If more than two points are specified, such that the robot will go through the points successively, the final velocities and positions at the conclusion of each segment can be used as the initial values for the next segments. Similar third-order polynomials can be used to plan each section. However, although positions and velocities are continuous, accelerations are not, which may cause problems.

Example 5.1

It is desired to have the first joint of a 6-axis robot go from initial angle of 30° to a final angle of 75° in 5 seconds. Using a third-order polynomial, calculate the joint angle at 1, 2, 3, and 4 seconds.

Solution: Substituting the boundary conditions into Equation (5.4), we get:

$$\begin{cases} \theta(t_i) = c_0 = 30 \\ \theta(t_f) = c_0 + c_1(5) + c_2(5^2) + c_3(5^3) = 75 \\ \dot{\theta}(t_i) = c_1 = 0 \\ \dot{\theta}(t_f) = c_1 + 2c_2(5) + 3c_3(5^2) = 0 \end{cases} \rightarrow \begin{cases} c_0 = 30 \\ c_1 = 0 \\ c_2 = 5.4 \\ c_3 = -0.72 \end{cases}$$

This results in the following cubic polynomial equation for position as well as the velocity and acceleration equations for joint 1:

$$\begin{aligned}
\theta(t) &= 30 + 5.4t^2 - 0.72t^3 \\
\dot{\theta}(t) &= 10.8t - 2.16t^2 \\
\ddot{\theta}(t) &= 10.8 - 4.32t
\end{aligned}$$

Substituting the desired time intervals into the motion equation will result in:

$$\theta(1) = 34.68°, \quad \theta(2) = 45.84°, \quad \theta(3) = 59.16°, \quad \theta(4) = 70.32°$$

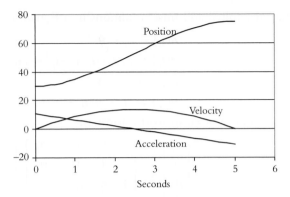

Figure 5.10 Joint positions, velocities, and accelerations for Example 5.1.

The joint angles, velocities, and accelerations are shown in Figure 5.10. Notice that in this case, the acceleration needed at the beginning of the motion is $10.8°/sec^2$ (as well as $-10.8°/sec^2$ deceleration at the conclusion of the motion). ■

Example 5.2

Suppose the robot arm of Example 5.1 is to continue to the next point, where the joint is to reach $105°$ in another 3 seconds. Draw the position, velocity, and acceleration curves for the motion.

Solution: At the conclusion of the first segment, we know the position and velocity of the joint. Using these values as initial conditions for the next segment, we get:

$$\theta(t) = c_0 + c_1 t + c_2 t^2 + c_3 t^3$$
$$\dot{\theta}(t) = c_1 + 2c_2 t + 3c_3 t^2$$
$$\ddot{\theta}(t) = 2c_2 + 6c_3 t$$

where:

$$\text{at } t_i = 0 \quad \theta_i = 75 \quad \dot{\theta}_i = 0$$
$$\text{at } t_f = 3 \quad \theta_f = 105 \quad \dot{\theta}_f = 0$$

which will yield:

$$c_0 = 75 \quad c_1 = 0 \quad c_2 = 10 \quad c_3 = -2.222$$
$$\theta(t) = 75 + 10t^2 - 2.222t^3$$
$$\dot{\theta}(t) = 20t - 6.666t^2$$
$$\ddot{\theta}(t) = 20 - 13.332t$$

Figure 5.11 shows the positions, velocities, and accelerations for the entire motion. As you can see, the boundary conditions are as expected. However, notice that although the velocity curve is continuous, the slope of the velocity curve changes from negative to positive at the intermediate point, creating an instantaneous change in acceleration. Whether the robot is capable of creating such accelerations or

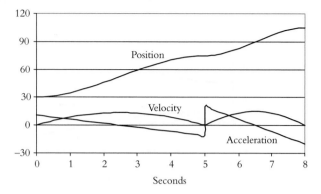

Figure 5.11 Joint positions, velocities, and accelerations for Example 5.2.

not is a question that must be answered depending on the robot's capabilities. To ensure that the robot's accelerations will not exceed its capabilities, acceleration limits may be used to calculate the necessary time to reach the target. In that case, for $\dot{\theta}_i = 0$ and $\dot{\theta}_f = 0$, the maximum acceleration will be:[4]

$$\left|\ddot{\theta}\right|_{max} = \left|\frac{6\left(\theta_f - \theta_i\right)}{\left(t_f - t_i\right)^2}\right|$$

from which the time-to-target can be calculated. You should also notice that the velocity at the intermediate point does *not* have to be zero. In that case, the concluding velocity at the intermediate point will be the same as the initial velocity of the next segment. These values must be used in calculating the coefficients of the third-order polynomial. ■

5.5.2 Fifth-Order Polynomial Trajectory Planning

In the previous section, we encountered accelerations that may be impossible to achieve during motion trajectories; therefore, we may need to specify maximum accelerations as well. Specifying the initial and ending positions, velocities, and accelerations of a segment yields six pieces of information, enabling us to use a fifth-order polynomial to plan a trajectory, as follows:

$$\theta(t) = c_0 + c_1 t + c_2 t^2 + c_3 t^3 + c_4 t^4 + c_5 t^5 \tag{5.6}$$

$$\dot{\theta}(t) = c_1 + 2c_2 t + 3c_3 t^2 + 4c_4 t^3 + 5c_5 t^4 \tag{5.7}$$

$$\ddot{\theta}(t) = 2c_2 + 6c_3 t + 12c_4 t^2 + 20c_5 t^3 \tag{5.8}$$

These equations allow us to calculate the coefficients of a fifth-order polynomial with position, velocity, and acceleration boundary conditions.

Example 5.3

Repeat Example 5.1, but assume the initial acceleration and final deceleration will be $5°/\sec^2$.

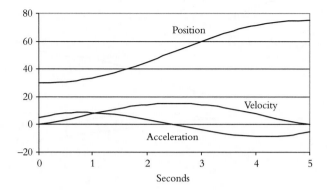

Figure 5.12 Joint positions, velocities, and accelerations for Example 5.3.

Solution: From Example 5.1 and the given accelerations, we have:

$$\theta_i = 30° \quad \dot{\theta}_i = 0°/\text{sec} \quad \ddot{\theta}_i = 5°/\text{sec}^2$$
$$\theta_f = 75° \quad \dot{\theta}_f = 0°/\text{sec} \quad \ddot{\theta}_f = -5°/\text{sec}^2$$

Using Equations (5.6) through (5.8), with the given initial and final boundary conditions, we get:

$$c_0 = 30 \quad c_1 = 0 \quad c_2 = 2.5$$
$$c_3 = 1.6 \quad c_4 = -0.58 \quad c_5 = 0.0464$$

This results in the following motion equations:

$$\theta(t) = 30 + 2.5t^2 + 1.6t^3 - 0.58t^4 + 0.0464t^5$$
$$\dot{\theta}(t) = 5t + 4.8t^2 - 2.32t^3 + 0.232t^4$$
$$\ddot{\theta}(t) = 5 + 9.6t - 6.96t^2 + 0.928t^3$$

Figure 5.12 shows the position, velocity, and acceleration graphs for the joint. The maximum acceleration is $8.7°/\text{sec}^2$. ∎

5.5.3 Linear Segments with Parabolic Blends

Another alternative for joint-space trajectory planning is to run the joints at constant speed between the initial and final locations, as we discussed in Section 5.4. This is equivalent to a first-order polynomial, where the velocity is constant and acceleration is zero. However, this also means that at the beginning and at the end of the motion segment, accelerations must be infinite in order to create instantaneous velocities at the boundaries. To prevent this, the linear segment can be blended with parabolic sections at the beginning and at the end of the motion segment, creating a continuous position and velocity, as shown in Figure 5.13. Assuming the initial and final positions are θ_i and θ_f at times $t_i = 0$ and t_f and the parabolic segments are symmetrically blended with the linear

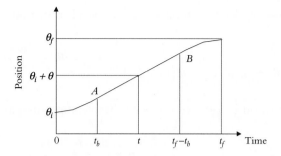

Figure 5.13 Scheme for linear segments with parabolic blends.

section at blending times t_b and $t_f - t_b$, we can write:

$$\theta(t) = c_0 + c_1 t + \frac{1}{2}c_2 t^2$$
$$\dot{\theta}(t) = c_1 + c_2 t \tag{5.9}$$
$$\ddot{\theta}(t) = c_2$$

Obviously, in this scenario, acceleration is constant for the parabolic sections, yielding a continuous velocity at the common points (called knot points) A and B. Substituting the boundary conditions into the parabolic equation segment yields:

$$\begin{cases} \theta(t=0) = \theta_i = c_0 \\ \dot{\theta}(t=0) = 0 = c_1 \\ \ddot{\theta}(t) = c_2 \end{cases} \rightarrow \begin{cases} c_0 = \theta_i \\ c_1 = 0 \\ c_2 = \ddot{\theta} \end{cases}$$

This will result in parabolic segments in the form:

$$\theta(t) = \theta_i + \frac{1}{2}c_2 t^2 \tag{5.10}$$

$$\dot{\theta}(t) = c_2 t \tag{5.11}$$

$$\ddot{\theta}(t) = c_2 \tag{5.12}$$

Clearly, for the linear segment, the velocity will be constant and can be chosen based on the physical capabilities of the actuators. Substituting zero initial velocity, a constant known joint velocity ω in the linear portion and zero final velocity in Equation (5.11), we find the joint positions and velocities for points A, B and the final point as follows:

$$\begin{aligned} \theta_A &= \theta_i + \frac{1}{2}c_2 t_b^2 \\ \dot{\theta}_A &= c_2 t_b = \omega \\ \theta_B &= \theta_A + \omega((t_f - t_b) - t_b) = \theta_A + \omega(t_f - 2t_b) \\ \dot{\theta}_B &= \dot{\theta}_A = \omega \\ \theta_f &= \theta_B + (\theta_A - \theta_i) \\ \dot{\theta}_f &= 0 \end{aligned} \tag{5.13}$$

The necessary blending time t_b can be found from Equation (5.13):

$$\begin{cases} c_2 = \dfrac{\omega}{t_b} \\ \theta_f = \theta_i + c_2 t_b^2 + \omega(t_f - 2t_b) \end{cases} \rightarrow \theta_f = \theta_i + \left(\dfrac{\omega}{t_b}\right) t_b^2 + \omega(t_f - 2t_b) \qquad (5.14)$$

From Equation (5.14), we calculate the blending time as:

$$t_b = \frac{\theta_i - \theta_f + \omega t_f}{\omega} \qquad (5.15)$$

Obviously, t_b cannot be bigger than half of the total time t_f, which results in a parabolic speed-up and a parabolic slowdown, with no linear segment. A corresponding maximum velocity of $\omega_{\max} = 2(\theta_f - \theta_i)/t_f$ can be found from Equation (5.15). It should be mentioned here that if, for any segment, the initial time is not zero but t_a, to simplify the mathematics, we can always shift the time axis by t_a to make the initial time zero.

The final parabolic segment is symmetrical with the initial parabola, but with a negative acceleration; therefore, it can be expressed as follows:

$$\theta(t) = \theta_f - \frac{1}{2}c_2(t_f - t)^2 \text{ where } c_2 = \frac{\omega}{t_b} \rightarrow \begin{cases} \theta(t) = \theta_f - \dfrac{\omega}{2t_b}(t_f - t)^2 \\ \dot{\theta}(t) = \dfrac{\omega}{t_b}(t_f - t) \\ \ddot{\theta}(t) = -\dfrac{\omega}{t_b} \end{cases} \qquad (5.16)$$

Example 5.4

Joint 1 of the 6-axis robot of Example 5.1 is to go from initial angle of $\theta_i = 30°$ to the final angle of $\theta_f = 70°$ in 5 seconds with a cruising velocity of $\omega_1 = 10°/\text{sec}$. Find the necessary time for blending and plot the joint positions, velocities, and accelerations.

Solution: From Equations (5.10) through (5.12), (5.15), and (5.16), we get:

$$t_c = \frac{\theta_i - \theta_f + \omega_1 t_f}{\omega_1} = \frac{30 - 70 + 10(5)}{10} = 1 \text{ sec}$$

$$\text{For } \theta = \theta_i \text{ to } \theta_A \qquad \text{For } \theta = \theta_A \text{ to } \theta_B \qquad \text{For } \theta = \theta_B \text{ to } \theta_f$$

$$\begin{cases} \theta = 30 + 5t^2 \\ \dot{\theta} = 10t \\ \ddot{\theta} = 10 \end{cases} \qquad \begin{cases} \theta = \theta_A + 10(t - 1) \\ \dot{\theta} = 10 \\ \ddot{\theta} = 0 \end{cases} \qquad \begin{cases} \theta = 70 - 5(5 - t)^2 \\ \dot{\theta} = 10(5 - t) \\ \ddot{\theta} = -10 \end{cases}$$

Figure 5.14 shows the position, velocity, and acceleration graphs for this joint.

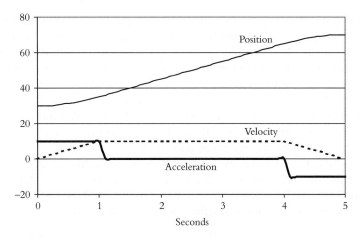

Figure 5.14 Position, velocity, and acceleration graphs for joint 1 in Example 5.4.

5.5.4 Linear Segments with Parabolic Blends and Via Points

Suppose there is more than one motion segment such that at the conclusion of the first segment, the robot will continue to move on to the next point, either another via point, or the destination. As we discussed earlier, we would like to blend the motion segments together to prevent stop-and-go motions. In this case too, we know where the robot is at time t_0, and using the inverse kinematic equations of the robot, we can calculate the joint angles at via points and at the end of the motion. To blend the motion segments together, we use the boundary conditions of each point to calculate the coefficients of the parabolic segments. As an example, at the beginning of the motion, we know the velocity and position of the joint. At the conclusion of the first segment, the position and velocity must be continuous. This will be the boundary condition for the via point and, consequently, a new segment can be calculated. The process continues until all segments are calculated and the destination is reached. Obviously, for each motion segment, a new t_b must be calculated based on the given joint velocity. We should also check to make sure that maximum allowable accelerations are not exceeded.

5.5.5 Higher-Order Trajectories

When, in addition to the initial and final destination points, other via points (including a lift-off and set-down point) are specified, we may match the positions, velocities, and accelerations of the two segments at each point to plan one continuous trajectory. Incorporating the initial and final boundary conditions together with this information enables us to use higher-order polynomials in the form:

$$\theta(t) = c_0 + c_1 t + c_2 t^2 + \cdots + c_{n-1}t^{n-1} + c_n t^n \tag{5.17}$$

such that the trajectory will pass through all specified points. However, solving a high-order polynomial for each joint will require extensive calculations. Instead, it is possible to use combinations of lower-order polynomials for different segments of the trajectory and blend them together to satisfy all required boundary conditions,[6] including a 4-3-4 trajectory, a 3-5-3 trajectory, and a 5-cubic trajectory to replace a seventh-order polynomial. For example, for a 4-3-4 trajectory, a fourth-order polynomial is used to plan a trajectory between the initial point and the first via point (e.g., lift-off), a third-order polynomial is used to plan a trajectory between two via points (e.g., lift off and set-down), and another fourth-order polynomial is used to plan the trajectory between the last via point (e.g., set-down) and the final destination. Similarly, a 3-5-3 trajectory may be planned between the initial and the first via point, between the successive via points, and between the last via point and the final destination.

Note that we must solve for 4 coefficients for a third-order polynomial, 5 coefficients for a fourth-order polynomial, and 6 coefficients for a fifth-order polynomial. Both 4-3-4 and 3-5-3 trajectories require solving for a total of 14 coefficients. For the 4-3-4 trajectory, the unknown coefficients are in the form:

$$\begin{aligned}
\theta(t)_1 &= a_0 + a_1 t + a_2 t^2 + a_3 t^3 + a_4 t^4 \\
\theta(t)_2 &= b_0 + b_1 t + b_2 t^2 + b_3 t^3 \\
\theta(t)_3 &= c_0 + c_1 t + c_2 t^2 + c_3 t^3 + c_4 t^4
\end{aligned} \tag{5.18}$$

However, there are also 14 boundary and blending conditions available that can be used to solve for all unknown coefficients and to plan the trajectory:

1. Initial position of θ_1 is known.
2. Initial velocity may be specified.
3. Initial acceleration may be specified.
4. Position of the first via point θ_2 is known and is the same as the final position of the first fourth-order segment.
5. The first via point's position is the same as the initial position of the third-order segment for continuity.
6. Continuous velocity must be maintained at the via point.
7. Continuous acceleration must be maintained at the via point.
8. Position of a second (and other) via point θ_n is specified and is the same as the final position of the third-order segment.
9. The position of the second (and other) via point is the same as the initial position of the next segment for continuity.
10. Continuous velocity must be maintained at the next via point.
11. Continuous acceleration must be maintained at the next via point.
12. Position of destination θ_f is specified.
13. Velocity of the destination is specified.
14. Acceleration of the destination is specified.

A similar set of requirements may be specified for the 3-5-3 trajectory. We denote time t as the global, normalized variable for the whole motion and τ_j as specific local times for each segment j. We also assume the local initial starting time τ_{ji} for each segment is zero and the local final ending time τ_{jf} for each segment is specified. This means that all segments start at a local time zero and end at a specified, given, local time, where the next segment starts at its local time $\tau_{ji} = 0$. Based on the above, the 4-3-4 segments and their derivatives can be written as follows:

1. The first fourth-order segment at local time $\tau_1 = 0$ yields the initial known position of θ_1. Therefore:

$$\theta_1 = a_0 \qquad (5.19)$$

2. The starting velocity at $\tau_1 = 0$ for the first segment is known. Therefore:

$$\dot\theta_1 = a_1 \qquad (5.20)$$

3. The starting acceleration at $\tau_1 = 0$ for the first segment is known. Thus:

$$\ddot\theta_1 = 2a_2 \qquad (5.21)$$

4. The position of the first via point θ_2 at the conclusion of the first segment at local time τ_{1f} is known. Consequently:

$$\theta_2 = a_0 + a_1(\tau_{1f}) + a_2(\tau_{1f})^2 + a_3(\tau_{1f})^3 + a_4(\tau_{1f})^4 \qquad (5.22)$$

5. The position of the first via point must be the same as the initial position of the third-order polynomial at time $\tau_2 = 0$. Thus:

$$\theta_2 = b_0 \qquad (5.23)$$

6. Continuous velocity must be maintained at the via point. Therefore:

$$a_1 + 2a_2(\tau_{1f}) + 3a_3(\tau_{1f})^2 + 4a_4(\tau_{1f})^3 = b_1 \qquad (5.24)$$

7. Continuous acceleration must be maintained at the via point. Thus:

$$2a_2 + 6a_3(\tau_{1f}) + 12a_4(\tau_{1f})^2 = 2b_2 \qquad (5.25)$$

8. The position of a second via point θ_3 at the conclusion of the third-order segment at time τ_{2f} is specified. Consequently:

$$\theta_3 = b_0 + b_1(\tau_{2f}) + b_2(\tau_{2f})^2 + b_3(\tau_{2f})^3 \qquad (5.26)$$

9. The position of the via point θ_3 must be the same as the initial position of the next segment at $\tau_3 = 0$ for continuity. Thus:

$$\theta_3 = c_0 \qquad (5.27)$$

10. Continuous velocity must be maintained at the via point. Thus:

$$b_1 + 2b_2(\tau_{2f}) + 3b_3(\tau_{2f})^2 = c_1 \qquad (5.28)$$

11. Continuous acceleration must be maintained at the via point. Therefore:

$$2b_2 + 6b_3(\tau_{2f}) = 2c_2 \qquad (5.29)$$

12. The position of destination at the conclusion of the last segment τ_{3f} is specified as θ_f. Thus:

$$\theta_4 = c_0 + c_1(\tau_{3f}) + c_2(\tau_{3f})^2 + c_3(\tau_{3f})^3 + c_4(\tau_{3f})^4 \qquad (5.30)$$

13. Velocity of the destination at the conclusion of the last segment at time τ_{3f} is specified. Consequently:

$$\dot\theta_4 = c_1 + 2c_2(\tau_{3f}) + 3c_3(\tau_{3f})^2 + 4c_4(\tau_{3f})^3 \qquad (5.31)$$

14. Acceleration of the destination at the conclusion of the last segment at time τ_{3f} is specified. Therefore:

$$\ddot\theta_4 = 2c_2 + 6c_3(\tau_{3f}) + 12c_4(\tau_{3f})^2 \qquad (5.32)$$

Equations (5.19) through (5.32) can be rewritten in matrix form as:

$$
\begin{bmatrix} \theta_1 \\ \dot{\theta}_1 \\ \ddot{\theta}_1 \\ \theta_2 \\ \theta_2 \\ 0 \\ 0 \\ \theta_3 \\ \theta_3 \\ 0 \\ 0 \\ \theta_4 \\ \dot{\theta}_4 \\ \ddot{\theta}_4 \end{bmatrix}
=
\begin{bmatrix}
1 & 0 & 0 & 0 & 0 & 0 & 0 & 0 & 0 & 0 & 0 & 0 & 0 & 0 \\
0 & 1 & 0 & 0 & 0 & 0 & 0 & 0 & 0 & 0 & 0 & 0 & 0 & 0 \\
0 & 0 & 2 & 0 & 0 & 0 & 0 & 0 & 0 & 0 & 0 & 0 & 0 & 0 \\
1 & \tau_{1f} & \tau_{1f}^2 & \tau_{1f}^3 & \tau_{1f}^4 & 0 & 0 & 0 & 0 & 0 & 0 & 0 & 0 & 0 \\
0 & 0 & 0 & 0 & 0 & 1 & 0 & 0 & 0 & 0 & 0 & 0 & 0 & 0 \\
0 & 1 & 2\tau_{1f} & 3\tau_{1f}^2 & 4\tau_{1f}^3 & 0 & -1 & 0 & 0 & 0 & 0 & 0 & 0 & 0 \\
0 & 0 & 2 & 6\tau_{1f} & 12\tau_{1f}^2 & 0 & 0 & -2 & 0 & 0 & 0 & 0 & 0 & 0 \\
0 & 0 & 0 & 0 & 0 & 1 & \tau_{2f} & \tau_{2f}^2 & \tau_{2f}^3 & 0 & 0 & 0 & 0 & 0 \\
0 & 0 & 0 & 0 & 0 & 0 & 0 & 0 & 0 & 1 & 0 & 0 & 0 & 0 \\
0 & 0 & 0 & 0 & 0 & 0 & 1 & 2\tau_{2f} & 3\tau_{2f}^2 & 0 & -1 & 0 & 0 & 0 \\
0 & 0 & 0 & 0 & 0 & 0 & 0 & 2 & 6\tau_{2f} & 0 & 0 & -2 & 0 & 0 \\
0 & 0 & 0 & 0 & 0 & 0 & 0 & 0 & 0 & 1 & \tau_{3f} & \tau_{3f}^2 & \tau_{3f}^3 & \tau_{3f}^4 \\
0 & 0 & 0 & 0 & 0 & 0 & 0 & 0 & 0 & 0 & 1 & 2\tau_{3f} & 3\tau_{3f}^2 & 4\tau_{3f}^3 \\
0 & 0 & 0 & 0 & 0 & 0 & 0 & 0 & 0 & 0 & 0 & 2 & 6\tau_{3f} & 12\tau_{3f}^2
\end{bmatrix}
\times
\begin{bmatrix} a_0 \\ a_1 \\ a_2 \\ a_3 \\ a_4 \\ b_0 \\ b_1 \\ b_2 \\ b_3 \\ c_0 \\ c_1 \\ c_2 \\ c_3 \\ c_4 \end{bmatrix}
$$

or

$$[\theta] = [M][C] \tag{5.33}$$

and

$$[C] = [M]^{-1}[\theta] \tag{5.34}$$

The unknown coefficients can be found from Equation (5.34) by calculating $[M]^{-1}$. Therefore, the equations of motion for the three segments are known and the robot can be driven through the specified positions. The same must be done for all other joints.

A similar approach may be taken to calculate the coefficients for the other combinations such as the 3-5-3 trajectory or the 5-cubic trajectory.[6]

Example 5.5

A robot is to be driven from an initial position through two via points before it reaches its final destination using a 4-3-4 trajectory. The positions, velocities, and time duration for the three segments for one of the joints are given below. Determine the trajectory equations and plot the position, velocity, and acceleration graphs for the joint.

$$
\begin{aligned}
\theta_1 &= 30° & \dot{\theta}_1 &= 0 & \ddot{\theta}_1 &= 0 & \tau_{1i} &= 0 & \tau_{1f} &= 2 \\
\theta_2 &= 50° & \tau_{2i} &= 0 & \tau_{2f} &= 4 \\
\theta_3 &= 90° & \tau_{3i} &= 0 & \tau_{3f} &= 2 \\
\theta_4 &= 70° & \dot{\theta}_4 &= 0 & \ddot{\theta}_4 &= 0
\end{aligned}
$$

Solution: We can calculate the unknown coefficients of the three segments by substituting the given values directly into the matrices of Equation (5.33) or into Equations (5.19) through (5.32) and solving the resulting set of equations. The result

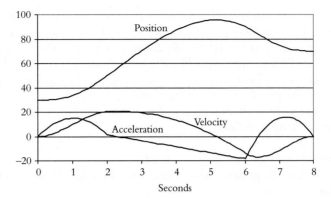

Figure 5.15 The position, velocity, and acceleration graphs for the motion of the joint in Example 5.5, based on a 4-3-4 trajectory.

will be:

$$a_0 = 30 \qquad b_0 = 50 \qquad c_0 = 90$$
$$a_1 = 0 \qquad b_1 = 20.477 \qquad c_1 = -13.81$$
$$a_2 = 0 \qquad b_2 = 0.714 \qquad c_2 = -9.286$$
$$a_3 = 4.881 \qquad b_3 = -0.833 \qquad c_3 = 9.643$$
$$a_4 = -1.191 \qquad\qquad\qquad c_4 = -2.024$$

The three segments will be:

$$\theta(t)_1 = 30 + 4.881t^3 - 1.191t^4 \qquad\qquad 0 < t \le 2$$
$$\theta(t)_2 = 50 + 20.477t + 0.714t^2 - 0.833t^3 \qquad 0 < t \le 4$$
$$\theta(t)_3 = 90 - 13.81t - 9.286t^2 + 9.643t^3 - 2.024t^4 \quad 0 < t \le 2$$

Figure 5.15 shows the position, velocity, and acceleration graphs of this joint for the given motion based on the 4-3-4 trajectory. ∎

5.5.6 Other Trajectories

Many other schemes may also be used to plan trajectories. This includes bang-bang trajectories, square and trapezoidal acceleration profile trajectories, and sine function trajectories. Additionally, it is possible to use other polynomials and other functions to plan a trajectory. For more information about these and other possibilities, refer to the references at the end of this chapter.

5.6 Cartesian-Space Trajectories

As discussed through simple examples in Section 5.4, Cartesian-space trajectories relate to the motions of a robot relative to the Cartesian reference frame, as followed by the position and orientation of the robot's hand. In addition to simple straight-line trajectories, many other schemes may be deployed to drive the robot in its path between different points. In fact, all of the schemes used for joint-space trajectory planning can also be used for Cartesian-space trajectories. The basic difference is that for Cartesian-space, the joint values must be repeatedly calculated through the inverse kinematic equations of the robot. This means that unlike the joint-space schemes in which the generated values relate directly to joint values, in Cartesian-space planning, the calculated values from the functions are positions (and orientations) of the hand, and they must still be converted to joint values through the inverse kinematic equations. This can be simplified into a computer loop as follows:

1. Increment the time by $t = t + \Delta t$.
2. Calculate the position and orientation of the hand based on the selected function for the trajectory.
3. Calculate the joint values for the position and orientation through the inverse kinematic equations of the robot.
4. Send the joint information to the controller.
5. Go to the beginning of the loop.

Straight-line motions between points are the most practical trajectories for industrial applications. However, blending the motions for multiple destinations (such as via points) is also very common.

To accomplish a straight-line trajectory, the transformation between the initial and final positions and orientations must be calculated and divided into small segments. The total transformation R between the initial configuration T_i and final configuration T_f can be calculated as follows:

$$T_f = T_i R$$
$$T_i^{-1} T_f = T_i^{-1} T_i R$$
$$R = T_i^{-1} T_f \tag{5.35}$$

At least three different alternatives may be used to convert this transformation into small segments:

I. Since we want to have a smooth straight-line transformation between the initial and final locations, we want a large number of very small segments. This, in reality, creates a large number of differential motions.[3] Using the equations developed for differential motions in Chapter 3, we can relate the position and orientation of the hand frame at each new segment to the differential motions, the Jacobian, and joint velocities by:

$$D = J D_\theta \qquad \text{and} \qquad D_\theta = J^{-1} D$$
$$dT = \Delta \cdot T$$
$$T_{new} = T_{old} + dT$$

This technique requires extensive calculations and only works if the inverse Jacobian exists.

II. The transformation between the initial and final locations R can be decomposed into a translation and two rotations. The translation involves moving the origin of the frame from the initial position to the final position. The first rotation is to align

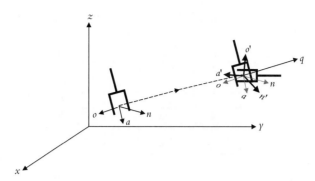

Figure 5.16 Transformation between an initial and final locations in Cartesian-space trajectory planning. The motion can be decomposed into a translation and a rotation about an axis q.

the hand frame to the desired orientation, and the second rotation is to rotate the hand frame about its own axis to the final orientation.[2,3,6] All three transformations are executed simultaneously.

III. The transformation between the initial and final locations R can be decomposed into a translation and one rotation about an axis q. The translation involves moving the origin of the frame from the initial position to the final position. The rotation is to align the hand frame to the final desired orientation.[2,3,6] Both transformations are executed simultaneously (Figure 5.16).

For more information about Cartesian-space trajectory planning, refer to References 7 through 12 at the end of this chapter.

The robot programming language used by Adept Technology Inc. is called V^+. Two common commands for moving the robot from one point (*P1*) to another point (*P2*) are *MOVE P2* and *MOVES P2*. The *MOVE* command instructs the controller to simultaneously move all robot joints to go to point *P2*. This motion is in joint-space, and the path of the robot is the result of the corresponding joint movements. The *MOVES* command stands for "move in straight line" and instructs the controller to move in a straight line in Cartesian-space by dividing the motion into many small segments. Additionally, the robot may be moved with a teach pendant in Joint coordinate, where one joint moves at a time, in World coordinate, where the end effector moves along the reference frame axes, or in Tool coordinate, where the end effector moves along the axes of a frame attached to the end effector. In the last two modes, all joints move simultaneously to accomplish the requested motion (basically in Cartesian-space).

Similarly, the commands *CP-OFF* and *CP-ON* allow the user to turn off and on a "Continuous Path" feature, where the consecutive motions are either blended together into a continuous path or each segment is executed individually by stopping at the end of the segment and starting again with the next segment.

Table 5.1   *The Coordinates and the Joint Angles for Example 5.6.*

#	x	y	θ_1	θ_2
1	3	10	18.8	109
2	3.5	10.4	19	104.0
3	4	10.8	19.5	100.4
4	4.5	11.2	20.2	95.8
5	5	11.6	21.3	90.9
6	5.5	12	22.5	85.7
7	6	12.4	24.1	80.1
8	6.5	12.8	26	74.2
9	7	13.2	28.2	67.8
10	7.5	13.6	30.8	60.7
11	8	14	33.9	52.8

Example 5.6

A 2-DOF planar robot is to follow a straight line between the start (3,10) and the end (8,14) points of the motion segment. Find the joint variables for the robot if the path is divided into 10 sections. Each link is 9 inches long.

Solution: The straight line between the start and the end points in Cartesian-space can be described by:

$$m = \frac{y - 14}{x - 8} = \frac{14 - 10}{8 - 3} = 0.8$$

or: $y = 0.8x + 7.6$

The coordinates of the intermediate points can be found by simply dividing the differences between the x and the y values of the start and the end points. The angles for the two joints are then found for each intermediate point. The results are shown in Table 5.1 and Figure 5.17.

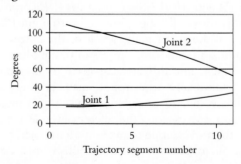

Figure 5.17   The joint positions for the robot in Example 5.6.   ∎

Example 5.7

A 3-DOF robot designed for lab experimentation at Cal Poly has two links, each 9 inches long. As shown in Figure 5.18, the coordinate frames of the joints are such

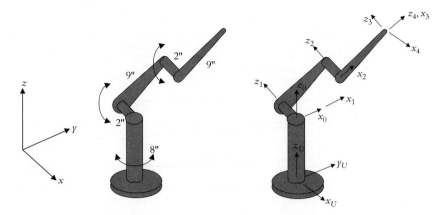

Figure 5.18   Robot of Example 5.7 and its coordinate frames.

that when all angles are zero, the arm is pointed upward. The inverse kinematic equations of the robot are also given below. We want to move the robot from point (9,6,10) to point (3,5,8) along a straight line. Find the angles of the three joints for each intermediate point and plot the results.

$$\theta_1 = \tan^{-1}\left(P_x/P_y\right)$$

$$\theta_3 = \cos^{-1}\left[\left((P_y/C_1)^2 + (P_z - 8)^2 - 162\right)/162\right]$$

$$\theta_2 = \cos^{-1}\left[\left(C_1(P_z - 8)(1 + C_3) + P_yS_3\right)/(18(1 + C_3)C_1)\right]$$

Solution: In this solution, we divide the distance between the start and the end points into 10 segments, although in reality, it is divided into many more sections. The coordinates of each intermediate point are found by dividing the distance between the initial and the end points into 10 equal parts. The inverse kinematic equations are used to calculate the joint angles for each intermediate point, as shown in Table 5.2. The joint angles are shown in Figure 5.19.

Table 5.2   *The Hand Frame Coordinates and Joint Angles for the Robot of Example 5.7.*

P_x	P_y	P_z	θ_1	θ_2	θ_3
9	6	10	56.3	27.2	104.7
8.4	5.9	9.8	54.9	25.4	109.2
7.8	5.8	9.6	53.4	23.8	113.6
7.2	5.7	9.4	51.6	22.4	117.9
6.6	5.6	9.2	49.7	21.2	121.9
6	5.5	9	47.5	20.1	125.8
5.4	5.4	8.8	45	19.3	129.5
4.8	5.3	8.6	42.2	18.7	133
4.2	5.2	8.4	38.9	18.4	136.3
3.6	5.1	8.2	35.2	18.5	139.4
3	5	8	31	18.9	142.2

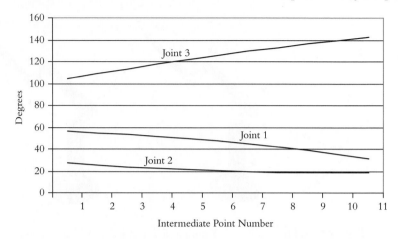

Figure 5.19 Joint angles for Example 5.7.

5.7 Continuous Trajectory Recording

In some operations such as spray painting and deburring, the motions required to accomplish a task may either be too complicated or too intricate to be generated by straight lines or other higher-order polynomials. Instead, it is possible to teach the robot how to move, record the motions, and later replay the motions and execute them. To do this, imagine that a robot can be moved by an operator in the same fashion required to accomplish a task in real time. This can be done by either releasing the joint-brakes and physically moving the robot, or by moving the joints of a robot model that is similar to the real one but is much lighter and can be moved easily. In either case, the joint values are continuously sampled in time and are recorded throughout the motion. Later, by playing back the sampled data and driving the robot joints accordingly, the robot will be forced to follow the same trajectory that was recorded; therefore, the robot will perform the task as planned. Can you tell whether the system records the path or the trajectory?

Obviously, this technique is simple and requires little programming or calculations. However, all motions must be accurately performed, sampled, and recorded for accurate playback. Additionally, every time a part of the motion needs to be changed, the robot must be programmed again. This is particularly difficult for large, heavy robots, especially if they are larger than a human operator.

5.8 Design Project

You may continue with either design project you started in the previous chapters. Here, you may run your robot based on any or all of the methods we discussed in this chapter. Of course, this is only possible if you make your robot and run it. In the next chapter, we will discuss actuators, which enable you to select appropriate actuators for your robot and run it. In that case, you may start with simpler trajectories and continue with more sophisticated ones. As an example, you may first run your robot in a simple point-to-point mode. Next, divide the path between the two (or more) destination points into a small number of segments, then continue with increasingly more segments until an acceptable straight-line motion is achieved. You may also try joint-space trajectory planning methods such as linear segments with parabolic blends or 4-3-4 polynomials. As you continue with this, you will realize that trajectory planning is a very interesting part of creating a robot. It is in trajectory planning that you may create a robot that is more than just a mechanism that moves in space.

Summary

In this chapter, we learned how a robot is actually moved in a predictable manner. Without an appropriately planned trajectory, the robot's motions are not predictable—it may collide with other objects, go through undesirable points, or be inaccurate.

Trajectories may be planned in joint-space or in Cartesian-space. Trajectories in each space may be planned through a number of different methods. Many of these methods may actually be used for both the Cartesian-space and the joint-space. However, although Cartesian-space trajectories are more realistic and can be visualized more easily, they are more difficult to calculate and plan. Obviously, a specific path such as a straight-line motion must be planned in Cartesian-space to be straight. But if the robot is not to follow a specific path, joint-space trajectories are easier to calculate and generate.

In the next chapter, we will discuss how a robot may be controlled.

References

1. Brady, M., J. M. Hollerbach, T. L. Johnson, T. Lozano-Perez, M. T. Mason, editors, "Robot Motion: Planning and Control," MIT Press, Cambridge, Mass., 1982.

2. Craig, John J., "Introduction to Robotics: Mechanics and Control," 2nd Edition, Addison Wesley, 1989.

3. Eman, K. F., Soo-Hun Lee, J. C. Cesarone, "Trajectories," International Encyclopedia of Robotics: Applications and Automation, Richard C. Dorf, Editor, John Wiley and Sons, New York, 1988, pp. 1796–1810.

4. Patel, R. V., Z. Lin, "Trajectory Planning," International Encyclopedia of Robotics: Applications and Automation, Richard C. Dorf, Editor, John Wiley and Sons, New York, 1988, pp. 1810–1820.

5. Selig, J. M., "Introductory Robotics," Prentice Hall, 1992.

6. Fu, K. S., R. C. Gonzales, C. S. G. Lee, "Robotics: Control, Sensing, Vision, and Intelligence," McGraw-Hill, New York, 1987.

7. Paul, Richard P., "Robot Manipulators, Mathematics, Programming, and Control," MIT Press, 1981.

8. Shahinpoor, Mohsen, "A Robot Engineering Textbook," Harper & Row, NY, 1987.

9. Snyder, Wesley, "Industrial Robots: Computer Interfacing and Control," Prentice Hall, 1985.

10. Kudo, Makoto, Y. Nasu, K. Mitobe, B. Borovac, "Multi-arm Robot Control System for Manipulation of Flexible Materials in Sewing Operations," Mechatronics, Vol. 10, No. 3, pp. 371–402.

11. "Path-Planning Program for a Redundant Robotic Manipulator," NASA Tech Briefs, July 2000, pp. 61–62.

12. Derby, Stephen, "Simulating Motion Elements of General-Purpose Robot Arms," The International Journal of Robotics Research, Vol. 2, No. 1, Spring 1983, pp. 3–12.

Problems

5.1. It is desired to have the first joint of a 6-axis robot go from an initial angle of 50° to a final angle of 80° in 3 seconds. Calculate the coefficients for a third-order polynomial joint-space trajectory. Determine the joint angles, velocities, and accelerations at 1, 2, and 3 seconds. It is assumed that the robot starts from rest and stops at its destination.

5.2. It is desired to have the third joint of a 6-axis robot go from an initial angle of 20° to a final angle of 80° in 4 seconds. Calculate the coefficients for a third-order polynomial joint-space trajectory and plot the joint angles, velocities, and accelerations. The robot starts from rest but should have a final velocity of 5°/sec.

5.3. The second joint of a 6-axis robot is to go from an initial angle of 20° to an intermediate angle of 80° in 5 seconds and continue to its destination of 25° in another 5 seconds. Calculate the coefficients for third-order polynomials in joint-space. Plot the joint angles, velocities, and accelerations. Assume the joint stops at intermediate points.

5.4. A fifth-order polynomial is to be used to control the motions of the joints of a robot in joint-space. Find the coefficients of a fifth-order polynomial that will allow a joint to go from an initial angle of 0° to a final joint angle of 75° in 3 seconds, while the initial and final velocities are zero and initial acceleration and final decelerations are $10°/sec^2$.

5.5. Joint 1 of a 6-axis robot is to go from an initial angle of $\theta_i = 30°$ to the final angle of $\theta_f = 120°$ in 4 seconds with a cruising velocity of $\omega_1 = 30°/sec$. Find the necessary blending time for a trajectory with linear segments and parabolic blends and plot the joint positions, velocities, and accelerations.

5.6. A robot is to be driven from an initial position through two via points before it reaches its final destination using a 4-3-4 trajectory. The positions, velocities, and time duration for the three segments for one of the joints are given below. Determine the trajectory equations and plot the position, velocity, and acceleration curves for the joint.

$$
\begin{array}{lllll}
\theta_1 = 20° & \dot{\theta}_1 = 0 & \ddot{\theta}_1 = 0 & \tau_{1i} = 0 & \tau_{1f} = 1 \\
\theta_2 = 60° & \tau_{2i} = 0 & \tau_{2f} = 2 & & \\
\theta_3 = 100° & \tau_{3i} = 0 & \tau_{3f} = 1 & & \\
\theta_4 = 40° & \dot{\theta}_4 = 0 & \ddot{\theta}_4 = 0 & &
\end{array}
$$

5.7. A 2-DOF planar robot is to follow a straight line in Cartesian-space between the start (2, 6) and the end (12, 3) points of the motion segment. Find the joint variables for the robot if the path is divided into 10 sections. Each link is 9 inches long.

5.8. The 3-DOF robot of Example 5.7, as shown in Figure 5.18, is to move from point (3, 5, 5) to point (3, −5, 5) along a straight line, divided into 10 sections. Find the angles of the three joints for each intermediate point and plot the results.

CHAPTER 6

Motion Control Systems

6.1 Introduction

Imagine that the controller of a robot sends a signal to one of the actuators causing it to accelerate toward the next location. Even if a feedback signal is used to measure the movement and stop the motion as soon as the joint reaches the desired destination, the joint may overshoot and go beyond the desired value, requiring that a negative signal be sent to the actuator to return it and perhaps continue this back-and-forth motion until the position is achieved accurately. At worst, with an unstable system, the oscillations may become larger, not smaller, and eventually destroy the system. This happens because the linkage and the actuator have inertia, and they may not stop immediately when the signal is turned off. Obviously, it should be possible to decrease the signal (current, voltage and so on) to the actuator and slow it down as it approaches the destination in order to avoid overshoot. But how early, and at what rate, should we do this? How do we make certain the system does not become unstable? Can we force the actuator to reach the destination as fast as we desire without overshoot and if so, at what rate? All these are basic questions that are answered by designing the control system of the robot to behave as desired. In this chapter, we will learn about fundamental definitions, building blocks, and the theory of motion control systems and how they may apply to robots. We will continue to refer to this chapter as we discuss actuators and sensors later.

What we cover in this chapter is not, and cannot be, complete. The assumption is that you either have had a course on control theory or that you will eventually learn it elsewhere. Instead, we present an introduction so that a student who has not learned the material yet will be able to understand how the motion control theory is applied to robots. Therefore, refer to other sources for complete treatment of the subject.[1–7]

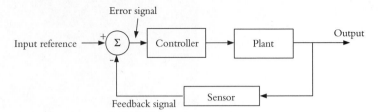

Figure 6.1 Basic components of a control system.

6.2 Basic Components and Terminology

Figure 6.1 shows the basic components of a control system. A control system is used to change (control) the behavior of a device, machine, or process (called a *plant*). The plant may be an air conditioning system, a chemical process, an iron, a robot, and so on. In each case, the plant creates an effect (*output*) such as the changing temperature of the room or the iron, the flow rate in the chemical process, or the motion of the robot arm. To perform its function, the control system uses sensors such as a thermostat, a flow-meter, or a potentiometer and encoder to measure the output of the plant. The controller receives the output signal and, based on its design, controls the plant and its output in relation to the desired input—the desired temperature in a room or on an iron, the rate of flow, or the final destination of the robot linkage and its speed. Note that there is a difference between the nature of different controllers. In the air conditioning system or the iron, the controller regulates the output. In the robot, it tracks the motions and controls their specification. This is called a *servo-controller* system.

An *open-loop* controller lacks a feedback signal; therefore, it is not aware of the output. For example, as we discussed in Chapter 1, a robot's processor calculates the destination's joint variables and sends the information to the controller. If the robot controller were open-loop, it might send a signal to a joint motor proportional to how far it needs to go, but not be aware of whether or not the joint moves at the desired rate. However, with a closed-loop controller, which includes feedback, the controller will receive a signal from the joint indicating its response to the control signal. If the motion is not as desired, the controller increases or decreases the control signal to force the arm to behave as desired.

Figure 6.1 also shows a summing operation between the feedback signal and the input reference signal. As you can see, the feedback signal is subtracted from the reference input signal, resulting in an *error signal*. The error signal is the driving signal for the controller. For a stable system, the feedback signal must be subtracted from the input reference signal. Otherwise, if they are added, the resulting signal becomes larger as the output increases, further increasing the output until the system "blows up."

To better understand the relationship between the different elements of a control system, let's first consider the behavior of a plant or the system's dynamics.

6.3 Block Diagrams

The pictorial representation in Figure 6.2 is called a *block diagram*. A block diagram assists us in finding the relationship between different elements of the system such as the plant,

Figure 6.2 A simple block diagram.

the signals, the controller, the feedback loop, and others. Figure 6.2 shows the simplest block, representing a system with its input and output. Although the actual system in each block is not shown, the relationship between the input and output of the block is represented by an equation. As long as this relationship is known, the details of the block are not needed for analysis. The block diagram also shows how the signals flow between different elements and how they are used. Later, we use block diagrams to represent systems and derive mathematical relationships that govern them.

6.4 System Dynamics

A plant's behavior is a function of its physical characteristics and external influences and how they are related to each other. For example, when an iron is connected to a power source, it starts to heat up. The rate at which the iron heats up is a function of factors such as voltage, the resistance of the heating element, and how the element is attached to the body as well as a function of the heat capacity of the iron and the materials used. Would you expect that, without a control element, the iron would eventually melt due to increased temperature? Perhaps. But as the iron's temperature increases, the heat it dissipates also increases until eventually a balance may be achieved. Therefore, the behavior of the iron is a function of its heat capacity, the input power, rate of heat dissipation, and materials used.

Similarly, if a voltage is applied to a motor, the angular acceleration of the rotor (and therefore, the rate at which its angular speed increases) is a function of the voltage and the inertia of the rotor. However, if the same motor is attached to an arm (such as a robot arm or a fan blade), the moment of inertia of the arm also affects the angular acceleration of the rotor. Therefore, in this case too, the system's behavior is a function of the input voltage, the moments of inertia of the rotor and the arm, and other physical factors (depending on the remaining elements of the system). The relationship between these elements is called *system dynamics*, which is represented by the plant in Figure 6.1. The system's dynamics is generally represented by differential equations and must be known before a control scheme can be designed for the system.

Although superficially it may appear that mechanical, electrical, hydraulic, chemical, or pneumatic systems are very different, they can be represented with differential equations that are very similar in nature. Systems generally include inertia, stiffness, damping, and external forcing functions that can be represented by similar equations, and therefore, in most cases are equivalent. Consequently, in most cases, the control of a system, whether mechanical, electrical, chemical process, hydraulic, or any combination thereof involves the same procedures and basic theory.

Referring to Example 4.1 and Figures 4.2 and 4.3, repeated here, when a force is applied to the mechanical mass-spring system, the mass moves in relation to Equation (6.1). This equation represents both the system's dynamic response to the input force and the plant in Figure 6.1. A controller may be used to control this response based on the elements of Figure 6.1. Therefore, as the controller applies a force and the mass moves,

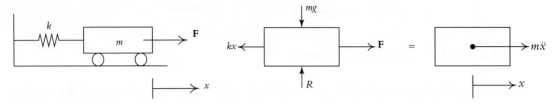

Figure 4.2 Repeated.   **Figure 4.3** Repeated.

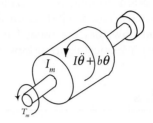

Figure 6.3 Representation of the dynamic behavior of a motor.

a sensor measures the movements and feeds it back to the controller which, in turn, adjusts the force to achieve the desired motion.

$$\sum \mathbf{F} = m\mathbf{a}$$
$$F_x - kx = m\ddot{x} \;\rightarrow\; F_x = m\ddot{x} + kx \tag{6.1}$$

Similarly, consider the rotor of a motor, with its inertia and damping (Figure 6.3). The response of the rotor to the torque generated by the magnetic field and the magnets can be represented as:

$$T = I\ddot{\theta} + b\dot{\theta} \tag{6.2}$$

We will discuss this in more detail in Chapter 7.

Example 6.1

Derive the equations that describe the behavior of the mechanical and electrical systems in Figure 6.4 and compare the results.

Solution: The equation representing the mechanical system can be derived by drawing the free body diagram as shown in Figure 6.5:

$$m\frac{d^2x}{dt^2} + b\frac{dx}{dt} + kx = F \tag{6.3}$$

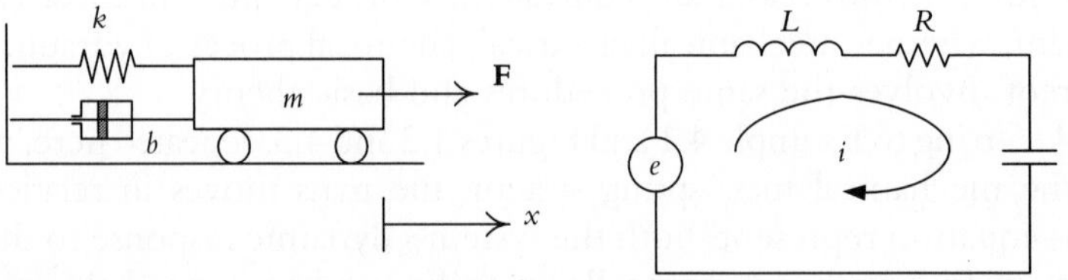

Figure 6.4 A mechanical and an electrical system.

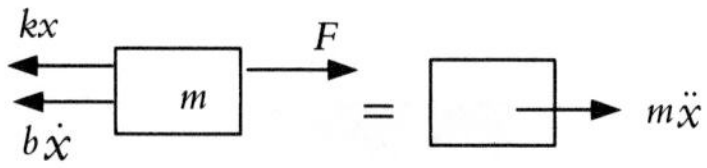

Figure 6.5 Free body diagram for Example 6.1.

Using the Kirchhoff's laws, we derive the equation representing the electrical circuit as:

$$L\frac{di}{dt} + Ri + \frac{1}{C}\int idt = e \tag{6.4}$$

Substituting $i = \dfrac{dq}{dt}$ in Equation (6.4), we get:

$$L\frac{d^2q}{dt^2} + R\frac{dq}{dt} + \frac{1}{C}q = e \tag{6.5}$$

∎

Note how the terms in Equations (6.3) and (6.5) are similar. This indicates that these two systems behave similarly and that the elements in each system are equivalent. The similarity between these elements is summarized in Table 6.1. Similarly, Table 6.2 shows the force–current equivalents between mechanical and electrical systems.

Table 6.1 *Force-Voltage Analogy between Mechanical and Electrical Systems.*

Mechanical Systems	Electrical Systems
Force F or torque T	Voltage e
Mass m or moment of inertia J	Inductance L
Viscous coefficient of friction b	Resistance R
Spring constant k	Reciprocal of capacitance $1/C$
Displacement x or angular displacement θ	Charge q
Velocity $\dot{x}$ or angular velocity $\dot{\theta}$	Current i

Table 6.2 *Force-Current Analogy between Mechanical and Electrical Systems.*

Mechanical Systems	Electrical Systems
Force F or torque T	Current i
Mass m or moment of inertia J	Capacitance C
Viscous coefficient of friction b	Reciprocal of resistance $1/R$
Spring constant k	Reciprocal of inductance $1/L$
Displacement x or angular displacement θ	Magnetic flux linkage
Velocity $\dot{x}$ or angular velocity $\dot{\theta}$	Voltage e

Similar differential equations can also be derived for hydraulic, thermal, and pneumatic systems.

Example 6.2

Figure 6.6 shows a simple hydraulic lift. As we will see later, when a feedback system is added to the hydraulic valve, the same basic system can be used in a hydraulic robot

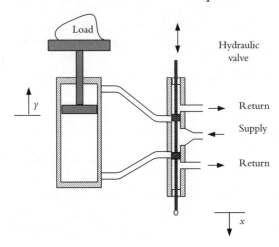

Figure 6.6   Representation of a hydraulic lift.

to move the joints. As the valve stem is pushed down, the pressurized hydraulic fluid is pushed into the lower chamber of the hydraulic ram, lifting the load and pushing out the oil above the piston through the return line, and vice versa. x and y represent the motions of the valve stem and the piston.

The rate of flow q of the fluid through the valve and into the cylinder, which is proportional to x, equals the rate of change of volume of the cylinder under the piston, which is equal to the piston velocity times the area. Therefore, the behavior of the system can be represented by the following first-order differential equation:

$$q = Cx = A\frac{d}{dt}y \;\rightarrow\; \dot{y} = \frac{C}{A}x$$

■

6.5 Laplace Transform

The differential equations describing the behavior of a system may not always be easy to analyze. Transforming a differential equation $f(t)$ in the time domain into $F(s)$ in the Laplace domain allows us to analyze the equation algebraically. Later, through an inverse Laplace transform, we get solutions in the time domain. This will become clear soon. However, the following is an introduction to the Laplace transform.

The Laplace transform $F(s)$ is defined as:

$$\mathcal{L}[f(t)] = F(s) = \int_0^\infty e^{-st}dt[f(t)] = \int_0^\infty f(t)e^{-st}dt \qquad (6.6)$$

where $s \equiv \sigma + j\omega$. To see how this is done, we will derive the Laplace transform for the following. A step function can be defined as:

$$f(t) = \begin{cases} 0 & \text{for } t < 0 \\ A & \text{for } t > 0 \end{cases} \qquad (6.7)$$

Substituting Equation (6.7) into Equation (6.6), we get:

$$\mathcal{L}[A] = F(s) = \int_0^\infty Ae^{-st}dt = \frac{-A}{s}e^{-st}\Big|_0^\infty = \frac{A}{s} \tag{6.8}$$

For a unit step function $A = 1$, this reduces to $F(s) = \frac{1}{s}$. Similarly, the Laplace transform for a ramp function, defined below, is:

$$f(t) = \begin{cases} 0 & \text{for } t < 0 \\ At & \text{for } t \geq 0 \end{cases}$$

$$\mathcal{L}[At] = \int_0^\infty (At)e^{-st}dt = \frac{At}{-s}e^{-st}\Big|_0^\infty - \int_0^\infty \frac{Ae^{-st}}{-s}dt \tag{6.9}$$

$$= \frac{A}{s}\int_0^\infty e^{-st}dt = \frac{A}{s}\frac{1}{s} = \frac{A}{s^2}$$

A sinusoidal function, defined below, can be transformed into the Laplace domain as follows:

$$f(t) = \begin{cases} 0 & \text{for } t < 0 \\ A\sin \omega t & \text{for } t \geq 0 \end{cases} \quad \text{where:} \quad \begin{aligned} e^{j\omega t} &= \cos \omega t + j\sin \omega t \\ e^{-j\omega t} &= \cos \omega t - j\sin \omega t \end{aligned}$$

Therefore, $\sin \omega t = \frac{1}{2j}\left(e^{j\omega t} - e^{-j\omega t}\right)$ and:

$$\mathcal{L}[A\sin \omega t] = \int_0^\infty \frac{A}{2j}\left(e^{j\omega t} - e^{-j\omega t}\right)e^{-st}dt = \frac{A}{2j}\left(\frac{1}{s - j\omega} - \frac{1}{s + j\omega}\right)$$

$$= \frac{A\omega}{s^2 + \omega^2} \tag{6.10}$$

In general, the Laplace transform equations are tabulated and may be used directly from a table.[7] Table 6.3 lists a few select Laplace transforms. For more functions, see References 1, 2, 3, and 7 at the end of the chapter.

The last three equations in Table 6.3, representing the derivatives of functions, are very useful. As shown, when the initial conditions for a function are zero, the Laplace transform of the first and second derivatives of a function are $sF(s)$ and $s^2F(s)$. Therefore, for a variable $x = F(s)$, $\dot{x} = sF(s)$ and $\ddot{x} = s^2F(s)$. Consequently, as we discussed earlier, the Laplace transform simply reduces a differential equation into an algebraic equation.

Example 6.3

Figure 6.4 and Equation (6.3) of Example 6.1 show a mass–spring–damper system and the differential equation representing it. Find the Laplace transform of this equation,

Table 6.3 *Laplace Transform Pairs.*

$f(t)$	$F(s)$
Unit impulse	1
Step $Au(t)$	$\dfrac{A}{s}$
Ramp At	$\dfrac{A}{s^2}$
$\dfrac{t^n}{n!}$	$\dfrac{1}{s^{n+1}}$
$e^{\mp \alpha t}$	$\dfrac{1}{s \pm \alpha}$
$\sin \omega t$	$\dfrac{\omega}{s^2 + \omega^2}$
$\cos \omega t$	$\dfrac{s}{s^2 + \omega^2}$
$e^{-\alpha t} \sin \omega t$	$\dfrac{\omega}{(s+\alpha)^2 + \omega^2}$
$e^{-\alpha t} \cos \omega t$	$\dfrac{s+\alpha}{(s+\alpha)^2 + \omega^2}$
$kf(t)$	$kF(s)$
$f_1(t) \pm f_2(t)$	$F_1(s) \pm F_2(s)$
$f'(t)$	$sF(s) - f(0)$
$f''(t)$	$s^2 F(s) - sf(0) - f'(0)$
$f^n(t)$	$s^n F(s) - s^{n-1}f(0) - \ldots - f^{n-1}(0)$

assuming all initial conditions are zero. Assume that the external forcing function is a step function $A(t)$.

Solution: From Table 6.3, we get:

$$\mathcal{L}[m\ddot{x} + b\dot{x} + kx] = \mathcal{L}[A(t)]$$
$$ms^2 F(s) + bsF(s) + kF(s) = A\frac{1}{s}$$
$$F(s) = \frac{A}{s(ms^2 + bs + k)}$$

■

Example 6.4

Using Table 6.3, derive the Laplace transform of a sine function from a cosine function.

Solution: The Laplace transform of a sine can be derived using $F(s)$ for a cosine and the function for the derivative of a Laplace transform as follows:

$$f(t) = \cos \omega t \rightarrow f'(t) = \frac{d}{dt}(\cos \omega t) = -\omega \sin \omega t \rightarrow \sin \omega t = -\frac{1}{\omega}\frac{d}{dt}\cos \omega t$$

$$\mathcal{L}(A \sin \omega t) = -\frac{A}{\omega}\mathcal{L}\left[\frac{d}{dt}\cos \omega t\right] = -\frac{A}{\omega}[sF(s) - f(0)] = -\frac{A}{\omega}\left[\frac{s^2}{s^2 + \omega^2} - 1\right]$$

$$\mathcal{L}(A \sin \omega t) = \left[\frac{A\omega}{s^2 + \omega^2}\right]$$

∎

Final Value Theorem   The final value theorem allows us to calculate the final value of a time-domain function at $t = \infty$. For example, if the input to a system is a step, the final value theorem enables us to calculate the (eventual) final value of the response of the system or its steady-state value. The final value of a system can be calculated from the following:

$$\lim_{t \to \infty} f(t) = [sF(s)]_{s=0} \tag{6.11}$$

Example 6.5

Find the steady-state value of $F(s) = \dfrac{k}{s + a}$ for a step input P.

Solution: The Laplace transform for a step input with magnitude P is $\dfrac{P}{s}$. Therefore:

$$\lim_{t \to \infty} f(t) = [sF(s)]_{s=0} = s\frac{k}{(s + a)}\frac{P}{s}\bigg|_{s=0} = \frac{kP}{a}$$

∎

6.6 Inverse Laplace Transform

The Laplace transform process was used to convert a differential equation in the time domain into an algebraic equation in the s domain. The inverse Laplace transform refers to the process of inverting an equation from the Laplace domain to the time domain. Two common methods used are the application of Table 6.3 and the application of the partial fraction expansion method. In this process, an equation in the Laplace domain is broken into simple terms, where each term can easily be transformed into the time domain using Table 6.3. Therefore, if:

$$F(s) = F_1(s) + F_2(s) + \cdots + F_n(s)$$

then $$\mathcal{L}^{-1}F(s) = \mathcal{L}^{-1}F_1(s) + \mathcal{L}^{-1}F_2(s) + \cdots + \mathcal{L}^{-1}F_n(s)$$

$$= f_1(t) + f_2(t) + \cdots + f_n(t)$$

Assuming that $F(s) = N(s)_m/D(s)_n$ where $N(s)$ and $D(s)$ are the numerator and denominator and assuming that the order of the denominator n is larger than the order of the numerator m, we can break the equation into the following form where z and p

values are zeros and poles:

$$F(s) = \frac{N(s)_m}{D(s)_n} = \frac{K(s+z_1)(s+z_2)\cdots(s+z_m)}{(s+p_1)(s+p_2)\cdots(s+p_n)} \tag{6.12}$$

Therefore, we should be able to break the equation into simple terms where the inverse Laplace transform can be found. Note that in order to be able to do so, the roots of the denominator $D(s)$ must be known.

6.6.1 Partial Fraction Expansion when *F(s)* Involves Only Distinct Poles

If the roots p of the denominator are all distinct, $F(s)$ can be broken into the following form, where coefficients a (called residues) are constants:

$$F(s) = \frac{N(s)_m}{D(s)_n} = \frac{a_1}{(s+p_1)} + \frac{a_2}{(s+p_2)} + \cdots + \frac{a_n}{(s+p_n)} \tag{6.13}$$

Since multiplying both sides of Equation (6.13) by any $(s+p_k)$ and setting $s = -p_k$ will eliminate all terms except a_k, we can find any of the residues using the following equation:

$$a_k = \left[(s+p_k)\frac{N(s)}{D(s)}\right]_{s=-p_k} \tag{6.14}$$

and since each term can be inverted to the time domain:

$$f(t) = \mathcal{L}^{-1}[F(s)] = a_1 e^{-p_1 t} + a_2 e^{-p_2 t} + \cdots + a_n e^{-p_n t} \tag{6.15}$$

Example 6.6

Derive the inverse Laplace transform of the following equation:

$$F(s) = \frac{(s+5)}{(s^2+4s+3)}$$

Solution: The given equation can be broken into the following:

$$F(s) = \frac{(s+5)}{(s^2+4s+3)} = \frac{(s+5)}{(s+1)(s+3)} = \frac{a_1}{(s+1)} + \frac{a_2}{(s+3)}$$

From Equation (6.14), we get:

$$a_1 = \left[(s+1)\frac{(s+5)}{(s+1)(s+3)}\right]_{s=-1} = \left[\frac{s+5}{s+3}\right]_{s=-1} = \frac{4}{2} = 2$$

$$a_2 = \left[(s+3)\frac{(s+5)}{(s+1)(s+3)}\right]_{s=-3} = \left[\frac{s+5}{s+1}\right]_{s=-3} = \frac{2}{-2} = -1$$

From Equation (6.15), the inverse Laplace transform is:

$$f(t) = \mathcal{L}^{-1}[F(s)] = \mathcal{L}^{-1}\left[\frac{2}{s+1}\right] + \mathcal{L}^{-1}\left[\frac{-1}{s+3}\right] = 2e^{-t} - e^{-3t}$$

■

6.6.2 Partial Fraction Expansion when *F(s)* Involves Repeated Poles

If the roots of Equation (6.13) are repeated, Equation (6.14) will not result in the desired residues. To solve for the residues, assuming that there are q repeated roots $(s + p)^q$, $F(s)$ can be written as:

$$F(s) = \frac{N(s)_m}{D(s)_n} = \frac{b_q}{(s+p)^q} + \frac{b_{q-1}}{(s+p)^{q-1}} + \cdots + \frac{b_1}{(s+p)} + \frac{a_1}{(s+p_1)} + \frac{a_2}{(s+p_2)} + \cdots + \frac{a_n}{(s+p_n)}$$

$$(6.16)$$

where b_q values are constants and can be found from:

$$b_q = [(s+p)^q F(s)]_{s=-p}$$

$$b_{q-1} = \left\{ \frac{d}{ds}[(s+p)^q F(s)] \right\}_{s=-p}$$

$$(6.17)$$

$$b_{q-k} = \left\{ \frac{1}{k!}\frac{d^k}{ds^k}[(s+p)^q F(s)] \right\}_{s=-p}$$

The inverted time domain equation is:

$$f(t) = \left[\frac{b_q t^{q-1}}{(q-1)!} + \frac{b_{q-1} t^{q-2}}{(q-2)!} + \cdots + \frac{b_2 t}{1!} + b_1 \right] e^{-pt} + a_1 e^{-p_1 t} + a_2 e^{-p_2 t} + \cdots + a_n e^{-p_n t}$$

$$(6.18)$$

Example 6.7

Derive the inverse Laplace transform of the following equation:

$$F(s) = \frac{(s+5)}{(s+2)^2(s+3)}$$

Solution: Since there are two repeated roots, we use Equations (6.16), (6.17), and (6.14) to get:

$$F(s) = \frac{b_2}{(s+2)^2} + \frac{b_1}{(s+2)^1} + \frac{a_1}{(s+3)}$$

$$b_2 = \left[(s+2)^2 \left(\frac{(s+5)}{(s+2)^2(s+3)} \right) \right]_{s=-2} = 3$$

$$b_1 = \left\{ \frac{d}{ds} \left[(s+2)^2 \left(\frac{(s+5)}{(s+2)^2(s+3)} \right) \right] \right\} \bigg|_{s=-2} = \frac{d}{ds} \left[\frac{(s+5)}{(s+3)} \right]_{s=-2}$$

$$= \frac{(s+3) - (s+5)}{(s+3)^2} \bigg|_{s=-2} = -2$$

$$a_1 = \left[(s+3) \frac{(s+5)}{(s+2)^2(s+3)} \right]_{s=-3} = 2$$

$$F(s) = \frac{3}{(s+2)^2} + \frac{-2}{(s+2)} + \frac{2}{(s+3)}$$

The time domain equation is:

$$f(t) = \left[\frac{3t}{1} - 2 \right] e^{-2t} + 2e^{-3t} = 3te^{-2t} - 2e^{-2t} + 2e^{-3t}$$

∎

6.6.3 Partial Fraction Expansion when *F(s)* Involves Complex Conjugate Poles

The previous methods apply to complex conjugate poles too, but we will study this in more detail for its unique characteristics. Complex conjugate poles always appear in pairs and have the form $a \pm jb$. This is because, as we know, for a second-order polynomial $f(s)$, the roots are:

$$f(s) = as^2 + bs + c = 0 \quad \rightarrow \quad p_1, p_2 = \frac{-b \pm \sqrt{b^2 - 4ac}}{2a}$$

If $b^2 - 4ac < 0$, there will be complex conjugate poles in the form $f(s) = (s - p_1)(s - p_2)$. For example, for $f(s) = s^2 + 2s + 5$, we get: $f(s) = (s + 1 + 2j)(s + 1 - 2j)$. Therefore, Equation (6.13) can be written as:

$$F(s) = \frac{N(s)_m}{D(s)_n} = \frac{c_1}{(s + \sigma + j\omega)} + \frac{c_2}{(s + \sigma - j\omega)} + \frac{a_1}{(s + p_1)} + \cdots + \frac{a_n}{(s + p_n)} \quad (6.19)$$

The same techniques used for distinct poles and repeated poles may be used to calculate the residues, including the complex conjugate residues. However, the inverse Laplace equation in the time domain is:

$$f(t) = c_1 e^{-(\sigma + j\omega)t} + c_2 e^{-(\sigma - j\omega)t} + a_1 e^{-p_1 t} + \cdots \quad (6.20)$$

Since Equation (6.10) can be rewritten as:

$$\sin \theta = \frac{e^{j\theta} - e^{-j\theta}}{2j} \quad \text{and} \quad \cos \theta = \frac{e^{j\theta} + e^{-j\theta}}{2j} \quad (6.21)$$

the complex portion of Equation (6.20) results in a decaying sinusoidal response. Therefore, as we will see later, when complex conjugate poles are present, the system is underdamped; consequently, it oscillates.

Alternately, we may expand an equation with complex conjugate roots as shown in Example 6.8.

Example 6.8

Derive the inverse Laplace transform of the following equation:

$$F(s) = \frac{1}{s(s^2 + 2s + 2)}$$

Solution: The denominator of the equation contains complex conjugate roots $(s + 1 + j1)$ and $(s + 1 - j1)$. We expand the equation using the following form:

$$F(s) = \frac{a_1}{s} + \frac{a_2 s + a_3}{s^2 + 2s + 2} = \frac{a_1(s^2 + 2s + 2) + s(a_2 s + a_3)}{s(s^2 + 2s + 2)}$$

Setting the numerator equal to the numerator of the original equation, we get:

$$a_1(s^2 + 2s + 2) + s(a_2 s + a_3) = 1$$

$$\begin{cases} a_1 + a_2 = 0 \\ 2a_1 + a_3 = 0 \\ 2a_1 = 1 \end{cases} \rightarrow \begin{cases} a_1 = 1/2 \\ a_2 = -1/2 \\ a_3 = -1 \end{cases}$$

and
$$F(s) = \frac{1}{2s} - \frac{s + 2}{2(s^2 + 2s + 2)} = \frac{1}{2s} - \frac{1}{2[(s + 1)^2 + 1]} - \frac{s + 1}{2[(s + 1)^2 + 1]}$$

From Table 6.3, the inverse Laplace transform of the given equation is:

$$f(t) = \frac{1}{2} - \frac{1}{2}e^{-t}\sin t - \frac{1}{2}e^{-t}\cos t$$

∎

Example 6.9

For the simple system shown in Figure 6.7, derive the equation of motion when a step-force of F_0 is applied at $t = 0$. Assume $b = 2$ and $k = 4$ and all initial conditions are zero.

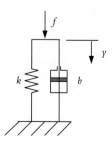

Figure 6.7 The system for Example 6.9.

Solution: As in Example 6.1, the equation of motion can be written as:

$$b\dot{y} + ky = f \quad \text{or} \quad 2\dot{y} + 4 = f$$

$$2sF(s) + 4F(s) = \frac{F_0}{s} \quad \rightarrow \quad F(s) = \frac{F_0}{2s(s+2)}$$

$$F(s) = \frac{a_1}{2s} + \frac{a_2}{(s+2)}$$

where $\quad a_1 = 2sF(s)|_{s=0} = \dfrac{F_0}{2} \quad$ and $\quad a_2 = (s+2)F(s)|_{s=-2} = -\dfrac{F_0}{4}$

Therefore: $\quad F(s) = \dfrac{F_0}{4s} - \dfrac{F_0}{4(s+2)} \quad$ and $\quad f(t) = \dfrac{F_0}{4}\left(1 - e^{-2t}\right)$

The response is exponential, eventually reaching the value of $F_0/4$. Note that the same result may be found from the application of the final value theorem:

$$f(t)_{ss} = sF(s)|_{s=0} = \frac{sF_0}{2s(s+2)}\bigg|_{s=0} = \frac{F_0}{2(s+2)}\bigg|_{s=0} = \frac{F_0}{4}$$

■

6.7 Transfer Function

A transfer function is the equation that represents the ratio of output to input in a system. It may be written across a block or across a complete system. Figure 6.8 shows the block diagram for a simple system, where the output signal is directly fed back to the summing junction. In this system, $R(s)$, $Y(s)$, $G(s)$, and $H(s)$ represent the input, output, the system dynamics plus any controller, and feedback multiplier respectively.

The transfer function for each block is simply the ratio of its output to its input. We define the following transfer functions:

- **Open-loop transfer function:** The open-loop transfer function is defined as the ratio of the feedback signal to the actuating error signal while the feedback loop is open, although the sensor still reads the output. Here, the sensor is used to read the output and report it as the feedback signal. Therefore:

$$Y(s) = E(s)G(s) \quad \text{and} \quad B(s) = Y(s)H(s) = E(s)G(s)H(s)$$

$$OLTF = \frac{B(s)}{E(s)} = \frac{E(s)G(s)H(s)}{E(s)} = G(s)H(s) \tag{6.22}$$

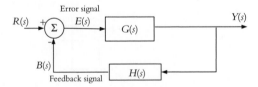

Figure 6.8 The block diagram for a simple control system.

As you can see, if the feedback loop is disconnected from the summing junction, the feedback signal is, in fact, a function of $G(s)H(s)$.

- **Feed-forward transfer function:** The feed-forward transfer function is defined as the ratio of the output to the actuating error signal, or:

$$FFTF = \frac{Y(s)}{E(s)} = G(s) \qquad (6.23)$$

As you can see, if the feedback function is unity, the open-loop and feed-forward transfer functions are the same.

- **Closed-loop transfer function:** Closed-loop transfer function is the ratio of output to input for the system. For the system of Figure 6.8, the closed-loop transfer function is:

$$Y(s) = G(s)E(s)$$
$$E(s) = R(s) - B(s) = R(s) - Y(s)H(s)$$

Eliminating $E(s)$, we get:

$$Y(s) = G(s)[R(s) - Y(s)H(s)]$$
$$Y(s)[1 + G(s)H(s)] = G(s)R(s)$$

and therefore, the closed-loop transfer function is:

$$CLTF = \frac{Y(s)}{R(s)} = \frac{G(s)}{1 + G(s)H(s)} \qquad (6.24)$$

Assuming that both $G(s)$ and $H(s)$ can be represented in ratios of polynomials as $G(s) = \frac{N_G(s)}{D_G(s)}$ and $H(s) = \frac{N_H(s)}{D_H(s)}$, Equation (6.24) can be written as:

$$CLTF = \frac{G(s)}{1 + G(s)H(s)} = \frac{N_G D_H}{N_G N_H + D_G D_H} \qquad (6.25)$$

This form of the closed-loop transfer function can assist in quickly composing the equation if the numerators and denominators of the $G(s)$ and $H(s)$ are known.

Together, the block diagram and the transfer function represent the behavior of a system with feedback mathematically and graphically. We will use the same throughout this book.

Example 6.10

Write the open-loop, feed-forward, and feedback transfer functions for the system in Figure 6.9.

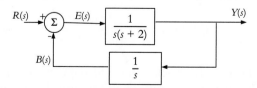

Figure 6.9 The system of Example 6.10.

Solution: We substitute appropriate values into Equations (6.22), (6.23), and (6.24) to get:

$$OLTF = G(s)H(s) = \frac{1}{s(s+2)}\frac{1}{s} = \frac{1}{s^2(s+2)}$$

$$FFTF = G(s) = \frac{1}{s(s+2)}$$

$$CLTF = \frac{G(s)}{1 + G(s)H(s)} = \frac{\frac{1}{s(s+2)}}{1 + \frac{1}{s(s+2)}\frac{1}{s}} = \frac{s}{s^2(s+2)+1}$$

We may also calculate the closed-loop transfer function directly from Equation (6.25) as:

$$CLTF = \frac{N_G D_H}{N_G N_H + D_G D_H} = \frac{1 \times s}{1 \times 1 + s(s+2)s} = \frac{s}{1 + s^2(s+2)} \qquad \blacksquare$$

Example 6.11

Assume that a sensor reads the position of the mass in a mass–spring–damper system and feeds it back to the input-force system, as shown in Figure 6.10. Derive the closed-loop transfer function for this system.

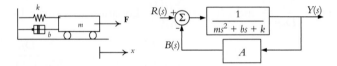

Figure 6.10   A mass-spring-damper system with position feedback.

Solution: Equation (6.3), repeated here, represents the differential motion of the system and its Laplace transform, as:

$$m\frac{d^2x}{dt^2} + b\frac{dx}{dt} + kx = F \quad \text{and} \quad G(s) = \frac{1}{ms^2 + bs + k}$$

Assuming that the feedback gain is A, the block diagram of Figure 6.10 represents the system. From Equation (6.25), the closed-loop transfer function for the system is:

$$CLTF = \frac{1}{A + (ms^2 + bs + k)} = \frac{1}{ms^2 + bs + (k+A)}$$

Notice how the feedback gain is added directly to the spring constant, and therefore, can easily augment the stiffness of the spring. The following application shows how this can be used in a control system.

The jumping robot: Imagine a robot, capable of jumping up and down for locomotion. There are many examples of robots that have this mode of locomotion, including robots that mimic animals such as BigDog, (a robot with four legs that gallops or trots like a horse), a pogo-stick robot with one leg, and one that lowers its body and then jumps by quickly extending its legs. Most robots of this nature are designed for learning how animals and humans perform their tasks, but have utilitarian applications too. However, in all these, the legs contain a spring for energy storage and absorption and controlled landing. Figure 6.11 shows a generic depiction of such a system.

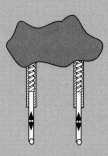

Figure 6.11 A generic leg design for jumping robots.

As a result of different loading conditions, jump characteristics, and control requirements, it may be necessary to change the stiffness of the springs inside the legs. However, readily changing the stiffness of a mechanical spring in-situ is no easy task, if not impossible. As shown in Example 6.11, a simple feedback gain A can be used to control the effective stiffness of the spring through the feedback loop.

The same system can also be used in a car to change the effective stiffness of the springs in the suspension system, therefore forcing it into different modes of operation such a smooth ride, sporty ride, and so on.

■

6.8 Block Diagram Algebra

In reality, block diagrams are not always as simple as the one shown in Figure 6.8. They may contain multiple parallel and series loops, summing junctions, and inputs. However, in general, a block diagram can be reduced to the simple form of Figure 6.8, as long as the results remain the same. In this respect, remember that:

- The products of the feed-forward transfer functions must remain the same, and
- The products of the transfer functions around a loop must remain the same.

Table 6.4 shows some equivalent block diagrams that may be used in reducing larger block diagrams.

Table 6.4   *Equivalent Block Diagrams*

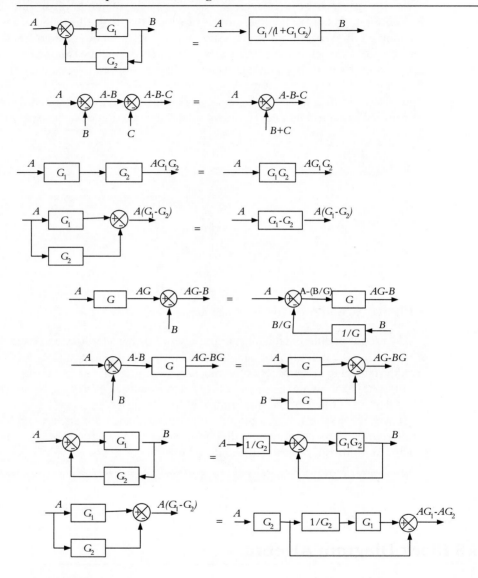

Example 6.12

To reduce the block diagram of Figure 6.12(a) to a simple standard form, do the following:

- Referring to Table 6.4, the first and second loops can be simplified as $G_1/\left(1 + G_1H_1\right)$ and $G_3/\left(1 + G_3H_2\right)$, as shown in Figure 6.12(b).

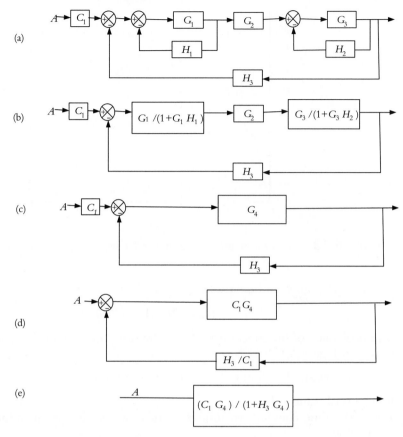

Figure 6.12 The block diagram for Example 6.12.

- Combine the three gains in the forward loop and replace with $G_4 = \dfrac{G_1 G_2 G_3}{(1 + G_1 H_1)(1 + G_3 H_2)}$, as shown in Figure 6.12(c).
- Transfer C_1 into the loop and replace the gains, as shown in Figure 6.12(d).
- Simplify the loop, as shown in Figure 6.12(e). ■

6.9 Characteristics of First-Order Transfer Functions

First-order systems are represented in the following standard forms:

$$G(s) = \frac{K_{ss}}{\tau s + 1} = \frac{K_{ss}/\tau}{s + (1/\tau)} = \frac{K_{ss}a}{s + a} \tag{6.26}$$

where K_{ss} is the steady-state gain and τ is the *time constant*. As you can see, the denominator of this equation is a first-order polynomial with its root (called a *pole*) at $s = -a$. The response of such a system to a step function $Pu(t)$ is:

$$F(s) = \frac{K_{ss}a}{s + a} \times \frac{P}{s} = \frac{PK_{ss}}{s} - \frac{PK_{ss}}{s + a}$$

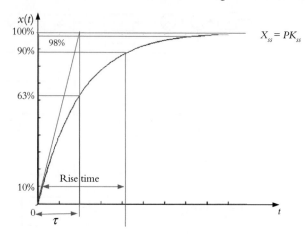

Figure 6.13   The time response of a first-order system to a step function.

The time response can be written as:

$$f(t) = PK_{ss}[1 - e^{-at}]u(t) \tag{6.27}$$

The final value of the function is PK_{ss}. The time response of the first-order system is shown in Figure 6.13.

The following definitions characterize the response:

- The final value is PK_{ss}.
- τ is the time constant, an indication of how fast the system responds to the step function.
- $a = 1/\tau$ is a pole. The location of the pole in a real–imaginary plane relative to the imaginary axis (y-axis) specifies whether or not the system is stable and how fast it responds. If the pole is to the left of the y-axis (negative), the response $1 - e^{-at}$ is bounded. If it is to the right of the imaginary plane, the response is $1 - e^{at}$, which is not bounded. To find the time constant, let $t = \tau$ to get:

$$x(t = \tau) = PK_{ss}(1 - e^{-\tau/\tau}) = PK_{ss}(0.63) = 63\%(PK_{ss}) \tag{6.28}$$

Hence, $x(t)$ reaches 63% of the final value in $t = \tau$ seconds as shown in Figure 6.13. This indicates how fast the response is; a longer τ indicates a longer time to reach the final value.

- Rise time is the time required between 10% and 90% of the final value and can be found by substituting 10% and 90% into Equation (6.27) as $T_r = 2.2\tau$.
- Settling time is the time from 0% to 98% rise and is $T_s = 4\tau$
- The slope at $t=0$ can be found by differentiating Equation (6.27) as:

$$\frac{dx}{dt} = PK_{ss}(0 - (-a)e^{-at})$$

$$\left.\frac{dx}{dt}\right|_{t=0} = PK_{ss}a = \frac{PK_{ss}}{\tau} \tag{6.29}$$

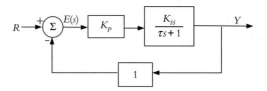

Figure 6.14 A closed-loop first-order system.

This slope is shown in Figure 6.13, which obviously indicates how fast the rise time is. As τ increases, the slope decreases. Since rise time cannot be zero, the slope cannot be infinite.

- In first-order transfer functions, the system responds as soon as $u(t)$ is applied.

Now let's look at a closed-loop first-order transfer function as shown in Figure 6.14. The feedback is unity, but there is a proportional gain K_P added to the feed-forward path.

Using Equation (6.25), the transfer function can be written as:

$$TF(s) = \frac{K_P K_{ss}}{K_P K_{ss} + \tau s + 1} = \frac{K_{sys}}{\tau_{sys} s + 1}$$

where $\tau_{sys} = \dfrac{\tau}{K_P K_{ss} + 1}$ and $K_{sys} = \dfrac{K_P K_{ss}}{K_P K_{ss} + 1}$. As the proportional gain K_P varies, it affects the behavior of the system, although the plant remains the same. For example, when K_P increases, τ_{sys} decreases, causing the system to respond faster. Similarly, K_{sys} approaches 1, making the system more accurate.

6.10 Characteristics of Second-Order Transfer Functions

Second-order transfer functions are represented in the following standard form:

$$G(s) = \frac{\omega_n^2}{s^2 + 2\zeta\omega_n s + \omega_n^2} \tag{6.30}$$

where ζ is the damping ratio and ω_n is the natural frequency. The response of such a system to a step function $u(t)$ is:

$$F(s) = \frac{\omega_n^2}{(s^2 + 2\zeta\omega_n s + \omega_n^2)} \frac{1}{s}$$

After partial fraction expansion, the time response of the system may be written in either of the following forms:

$$f(t) = 1 - e^{-\zeta\omega_n t}\left(\cos\omega_d t + \frac{\zeta}{\sqrt{1 - \zeta^2}}\sin\omega_d t\right) \tag{6.31}$$

or $f(t) = 1 - \dfrac{1}{\sqrt{1 - \zeta^2}} e^{-\zeta\omega_n t}\sin[\omega_d t + \alpha]$ $-1 < \zeta < 1$ $\tag{6.32}$

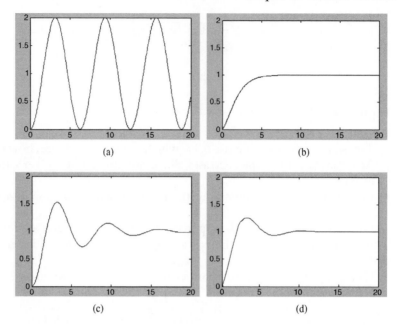

Figure 6.15 The response of a second-order transfer function to a step function at different damping ratios: (a) $\zeta = 0$, (b) $\zeta = 1$, (c) $\zeta = 0.2$, (d) $\zeta = 0.4$.

where $\omega_d = \omega_n \sqrt{1 - \zeta^2}$ and $\alpha = \tan^{-1}\left(\sqrt{1 - \zeta^2}/\zeta\right)$. Equations (6.31) and (6.32) have an exponential portion and sinusoidal portions. Therefore, the response is an oscillatory function influenced by whether the exponential portion is decaying or growing as follows:

- If $\zeta = 0$, indicating no damping, the exponential portion becomes a constant. Consequently, the response is a sinusoidal function that oscillates indefinitely, as shown in Figure 6.15(a).
- If $\zeta = 1$, indicating critical damping, the response is an exponential function that eventually achieves the steady-state value (Figure 6.15(b)).
- If the exponential portion grows, the response will grow as well, indicating an unstable system.
- If the exponential portion decays, indicating less than critical damping, the oscillations decrease in size until the system stabilizes (Figure 6.15(c) for $\zeta = 0.2$ and (d) for $\zeta = 0.4$).

Differentiating these equations and setting $t = 0$ will reveal that the slope of the response is zero. This means that unlike first-order systems where the initial slope is never zero, the initial slope of second-order transfer functions is always zero, indicating a slower initial response.

The steady-state gain (or the final value in response to a step function) is:

$$F_{ss} = \lim_{s \to 0} s \left(\frac{\omega_n^2}{s^2 + 2\zeta\omega_n s + \omega_n^2} \right) \frac{P}{s} = P$$

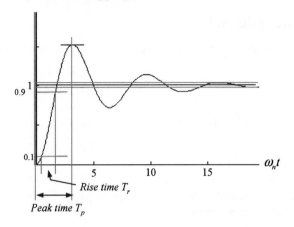

Figure 6.16 A typical response of a second-order transfer function and its characteristics.

Figure 6.16 shows a typical second-order response to a step function. The following characterize this response:

- The *peak time* T_p is the time to the maximum response value and can be found by taking the derivative of Equation (6.31) and setting it to zero as:

$$T_p = \pi/\omega_n\sqrt{1-\zeta^2} \tag{6.33}$$

- The *rise time* T_r is the time it takes to go from 10% of the response to 90%.
- Unlike for a first-order system, no time constant is defined for a second-order transfer function.
- Settling time T_s is reached when the response does not vary more than $\pm2\%$ or

$$T_s = 4/\zeta\omega_n \tag{6.34}$$

- Percent overshoot (%OS) is the ratio of the overshoot to the steady-state value, or

$$\%OS = \frac{F(\text{max}) - F_{ss}}{F_{ss}} \times 100\% = e^{\left(-\zeta\pi/\sqrt{1-\zeta^2}\right)} \times 100\% \tag{6.35}$$

For the response in Figure 6.16 with $\zeta = 0.2$, %OS = 53%. Similarly, the %OS for zero damping results in 100% overshoot, as shown in Figure 6.15(a).

6.11 Characteristic Equation: Pole/Zero Mapping

When the denominator of the transfer function is set to zero, the resulting equation is called the *characteristic equation*. The roots of the characteristic equation are called *poles*, whereas the roots of the numerator of the transfer function are called *zeros*. Pole/zero mapping is the graphical representation of the locations of the poles and zeros in a real-imaginary plane.

Example 6.13

Draw the pole-zero map of the following transfer function:

$$TF = \frac{s(s+3)}{(s+5)(s+2)(s^2+4s+5)}$$

Solution: The characteristic equation yields the poles as follows:

$$(s+5)(s+2)(s^2+4s+5) = 0$$

Therefore, $s = -5$, $s = -2$, $s = -2 \pm j$. The zeros are at $s = 0$, $s = -3$. Note that complex conjugate roots are always in pairs. Figure 6.17 shows the poles (shown as ×) and zeros (shown as o).

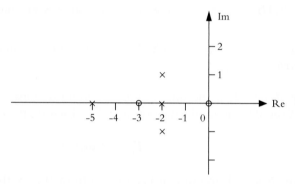

Figure 6.17   Pole–zero mapping of the roots of Example 6.13.

■

The loci of the poles and zeros reveal much information about the system and how it behaves. From the graph, we can identify the transfer function, the order of the system, and whether it is underdamped, critically damped, overdamped, stable, or unstable.

For a first-order transfer function of Equation (6.26), the only pole is:

$$\tau s + 1 = 0 \quad \rightarrow \quad s = -\frac{1}{\tau} \tag{6.36}$$

Therefore, the reciprocal of the pole location is the time constant. Clearly, as s increases, indicating a smaller time constant, the response of the system is faster. When the pole moves to the right (closer to the origin), the time constant is larger with a slower response. As long as the pole is on the left side of the imaginary axis, the system is stable. A pole at the origin is a pure integrator, which we will study later.

For second-order transfer functions, the solution for the characteristic equation is:

$$s^2 + 2\zeta\omega_n s + \omega_n^2 = 0$$

$$s = \frac{-b \pm \sqrt{b^2 - 4ac}}{2a} = \frac{-2\zeta\omega_n \pm \sqrt{(2\zeta\omega_n)^2 - 4\omega_n^2}}{2} \tag{6.37}$$

$$s = -\zeta\omega_n \pm \omega_n\sqrt{\zeta^2 - 1}$$

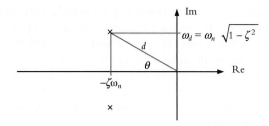

Figure 6.18   An underdamped system with its pair of complex conjugate poles and its characteristics.

Four possibilities exist:

- $\zeta = 1$, therefore $s = -\omega_n$, repeated twice. This means there will be two poles at the same location. This system is critically damped, and the response is as shown in Figure 6.15(b).
- $\zeta > 1$, therefore $\sqrt{\zeta^2-1}$ is positive, resulting in a pair of real and distinct roots and a system that is overdamped.
- $\zeta < 1$, therefore $\sqrt{\zeta^2-1}$ is negative. In this case, the roots are complex conjugate pair $s = -\zeta\omega_n \pm \omega_n\sqrt{1-\zeta^2}j$. The system is underdamped as shown in Figure 6.15(c) and (d).
- $\zeta = 0$, therefore $s = \pm\omega_n j$, which is an undamped system with complex conjugate roots on the imaginary axis. The response is as shown in Figure 6.15(a).

Figure 6.18 shows the mapping of the poles of an underdamped system. The following relationships hold true:

- $d = \sqrt{\left(-\zeta\omega_n\right)^2 + \omega_n^2(1 - \zeta^2)} = \omega_n$
- $\cos\theta = \dfrac{\zeta\omega_n}{\omega_n} = \zeta \quad \rightarrow \quad \theta = \cos^{-1}\zeta$ and hence, when $\zeta = 0, \theta = 90°$, and as we saw

before, the system is undamped and the roots are on the imaginary axis. When $\zeta = 1$, $\theta = 0$ and the system is critically damped with real poles.

Figure 6.19 shows how the response of the system changes as the poles move in different directions. These conclusions are based on the preceding results.

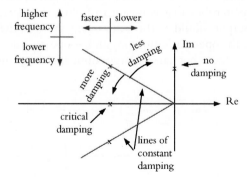

Figure 6.19   The response of the system changes as the poles move in different directions.

Higher-order transfer functions may be analyzed similarly through forming the time-domain response of the system based on partial fraction expansion and plotting the result. Usually, the result includes exponential portions, oscillatory sections, as well as step functions. Depending on the magnitudes of the poles and zeros, it may be possible to assume that certain portions of the time response are negligible while others are dominant. Regardless, the response can be plotted and analyzed either through the time-domain plotting or the pole/zero plotting. For further reading on higher-order systems, refer to related books and journal articles.

6.12 Steady-State Error

Figure 6.20 shows a typical control loop. The transfer function for this system is:

$$\frac{Y(s)}{R(s)} = \frac{k_1 k_2}{k_2 H + s(\tau s + 1)}$$

The steady-state error signal E_{ss} is a function of both the transfer function and the type of input to the system. It can be written as:

$$E_{ss} = R k_1 - Y_{ss} H \tag{6.38}$$

where:

$$Y_{ss} = \lim_{s \to 0} s\left(\frac{k_1 k_2}{k_2 H + s(\tau s + 1)}\right) R(s) \tag{6.39}$$

For a step input, the steady-state output and the steady-state error signal are:

$$Y_{ss} = \lim_{s \to 0} s\left(\frac{k_1 k_2}{k_2 H + s(\tau s + 1)}\right) \frac{1}{s} = \frac{k_1}{H} \tag{6.40}$$

$$E_{ss} = 1 \times k_1 - \frac{k_1}{H} H = 0 \tag{6.41}$$

As Equations (6.40) and (6.41) show, although the input to the system was a unit step, the output is k_1/H (unless $k_1 = H$) and the steady-state error signal is zero. This can be very handy depending on our application. For example, in a telerobot (such as a surgical robot or a space repair-robot that is meant to follow the operator's motions accurately), the steady-state output should be the same as the input. In that case, the robot's motions will be the same as the operator's input joystick—one-to-one mapping—with no steady-state error signal. On the other hand, a large robot with large motions—for example, the space

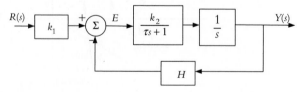

Figure 6.20   A typical control loop.

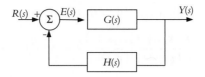

Figure 6.21 A typical control loop.

shuttle robot that handles satellites—must make large motions for small motions of the joystick, and therefore, must have a large gain, even though no steady-state error signal is desired. In this case, an appropriate gain may be selected to provide the desired motions while the steady-state error signal remains zero. Note that it is assumed that the dynamics of the robot are included in the model; therefore, the above example must be modified to represent a robot appropriately.

In order to see how the steady-state error can be found for any system, let's consider a typical feedback loop as shown in Figure 6.21.

The error signal and the transfer function for the system are:

$$E(s) = R(s) - Y(s)H(s)$$

$$Y(s) = \frac{G(s)}{1 + G(s)H(s)} R(s) \quad \rightarrow \quad E(s) = \left(\frac{1}{1 + G(s)H(s)}\right) R(s) \tag{6.42}$$

$$E_{ss} = \lim_{s \to 0} s \left(\frac{1}{1 + G(s)H(s)}\right) R(s)$$

Assume the open-loop transfer function $G(s)H(s)$ can be represented as:

$$G(s)H(s) = \frac{K(\tau_a s + 1)(\tau_b s + 1) \cdots}{s^n(\tau_1 s + 1)(\tau_2 s + 1) \cdots} \tag{6.43}$$

The value of n in Equation (6.43) determines the *type* of the system and is an indication of how many pure integrators are present in the feed-forward path. For $n = 0$, the system is type-0, for $n = 1$, the system is type-1, etc. Substituting different inputs for R for different system types in Equation (6.42) will yield the steady-state error signal.

Step inputs For step inputs, we define *static position error coefficient* $K_p = \lim_{s \to 0} G(s)H(s)$. Therefore:

$$E_{ss} = \frac{1}{1 + K_p}$$

For a type-0 system with $n = 0$:

$$G(s)H(s) = \frac{K(\tau_a s + 1)(\tau_b s + 1) \cdots}{(\tau_1 s + 1)(\tau_2 s + 1) \cdots} \quad \rightarrow \quad K_p = K \quad \text{and} \quad E_{ss} = \frac{1}{1 + K} \tag{6.44}$$

For a type-1 or higher system with $n \geq 1$:

$$G(s)H(s) = \frac{K(\tau_a s + 1)(\tau_b s + 1) \cdots}{s^n(\tau_1 s + 1)(\tau_2 s + 1) \cdots} \quad \rightarrow \quad K_p = \infty \quad \text{and} \quad E_{ss} = 0 \qquad (6.45)$$

Notice how a type-1 system has zero steady-state error compared to a type-0 system. As we will see later, having an additional pole in the denominator is equivalent to adding an integrator to a control system, bringing the error to zero.

Ramp input   For ramp inputs, we define *static velocity error coefficient* $K_v = \lim_{s \to 0} sG(s)H(s)$. Therefore:

$$E_{ss} = \lim_{s \to 0} s\left(\frac{1}{1 + G(s)H(s)}\right)\frac{1}{s^2} = \lim_{s \to 0}\frac{1}{sG(s)H(s)} = \frac{1}{K_v}$$

For a type-0 system with $n = 0$:

$$G(s)H(s) = \frac{K(\tau_a s + 1)(\tau_b s + 1) \cdots}{(\tau_1 s + 1)(\tau_2 s + 1) \cdots} \quad \rightarrow \quad K_v = 0 \quad \text{and} \quad E_{ss} = \infty \qquad (6.46)$$

For a type-1 system with $n = 1$:

$$G(s)H(s) = \frac{K(\tau_a s + 1)(\tau_b s + 1) \cdots}{s(\tau_1 s + 1)(\tau_2 s + 1) \cdots} \quad \rightarrow \quad K_p = K \quad \text{and} \quad E_{ss} = \frac{1}{K} \qquad (6.47)$$

Therefore, for a ramp input, the steady-state error in a type-0 system is infinite, whereas for type-1 system, it is finite. A similar analysis may be applied for higher-order inputs.

6.13 Root Locus Method

The root locus is the collection of the loci of the roots of the characteristic equation plotted on a real-imaginary plane as parameters vary. Root locus is a powerful tool for both analysis of the system—whether or not it is stable, system sensitivity, whether or not it is underdamped, critically damped, or overdamped, and so on—as well as system design (determining the location of roots or the magnitude of the gains for specific system behavior).

To see how root locus is formed and what it means, let's consider the system shown in Figure 6.22.

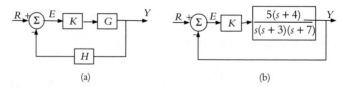

(a)   (b)

Figure 6.22   A typical control system.

The transfer function and characteristic equations for the system in part (a) are:

$$TF = \frac{Y}{R} = \frac{KG}{KGH + 1} \quad \text{and} \quad KGH + 1 = 0 \tag{6.48}$$

If we write the open-loop transfer function in polynomial form as $GH = \dfrac{N(s)}{D(s)}$, the characteristic equation can be written as:

$$K\frac{N(s)}{D(s)} + 1 = 0 \quad \rightarrow \quad KN(s) + D(s) = 0 \quad \text{or} \quad -K = \frac{D(s)}{N(s)} \tag{6.49}$$

Therefore, using Equation (6.49), the characteristic equation for part (b) is:

$$5K(s + 4) + s(s + 3)(s + 7) = 0 \tag{6.50}$$

The root locus is the loci of the roots of Equation (6.49) as K varies. If $K = 0$, the roots are the poles ($s = 0$, -3, and -7, for part (b)). As K increases, the location of the roots change until K approaches ∞, at which time the roots converge to the zeros of the open-loop transfer function ($s = -4$ for part (b)). For every value of K, the roots will be at a different location, yielding a different behavior. Plotting these roots for all values of K (root locus) allows us to both analyze and predict the behavior of the system.

Figure 6.23 shows the root locus for the system of Figure 6.22(b). As you can see, the roots start at the poles (where $K = 0$) and move in the direction of the arrows as K increases.

One of the conclusions we can make right away is that since the poles and the zero are all to the left of the imaginary axis, this system can never be unstable. Conclusions such as this make root locus a very powerful and useful technique.

In the next sections, we will study the root locus technique and its applications in the design of control systems.

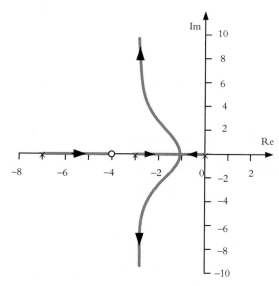

Figure 6.23 The root locus for the system of Figure 6.22(b).

Start and End of the Root Locus The start of root locus is where K in Equation (6.49) is zero. This corresponds to the poles of the open–loop transfer function. Therefore, by plotting the poles in the real-imaginary plane, we have the start of all portions of the root locus.

Each portion of the root locus ends at a zero or at ∞, where K in Equation (6.49) approaches ∞. Therefore, by plotting the open-loop transfer function zeros, the ends of root loci can be marked off. Each portion starts at a pole and ends at a zero or ∞.

If all the roots are numbered sequentially from right to left, the root locus exists to the left of the odd-numbered roots only. Each section starts at a pole and ends at a zero or continues to ∞.

Root Locus between the Start and the End Points The location of each point on the root locus relative to a pole or zero is represented by a vector with real or real-imaginary components. The magnitude of the overall transfer function at this point is the ratio of the products of all vectors from this point to each zero and pole, or:

$$M_{TF} = \frac{\prod M_{z_i}}{\prod M_{p_i}} = \frac{M_{z_1} M_{z_2} \cdots}{M_{p_1} M_{p_2} \cdots} \tag{6.51}$$

where M_{z_i} and M_{p_i} are the magnitudes of each vector between the zeros or the poles to the point of interest. Similarly, the corresponding angles of vectors are added as:

$$\theta = \sum \theta_{z_i} - \sum \theta_{p_i} = \pm 180 \tag{6.52}$$

Example 6.14

Calculate the magnitude and the angle of the vectors for the characteristic equation $K(s-1)(s+1) + (s+4)(s+6) = 0$, shown in Figure 6.24, for point $s = 0 + 4j$

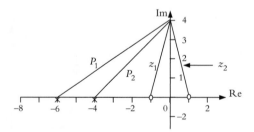

Figure 6.24   The vectors between a point in the Re-Im plane and the poles and zeros.

Solution: Using Equations (6.51) and (6.52), we get:

$$M = \frac{\sqrt{1^2 + 4^2} \times \sqrt{1^2 + 4^2}}{\sqrt{4^2 + 4^2} \times \sqrt{6^2 + 4^2}} = \frac{17}{\sqrt{32} \times \sqrt{52}} = 0.4167$$

$$\theta_{z_1} = \tan^{-1}\frac{4}{1} = 76°, \quad \theta_{z_2} = \tan^{-1}\frac{4}{-1} = 104°,$$

$$\theta_{p_1} = \tan^{-1}\frac{4}{6} = 33.7°, \quad \theta_{p_2} = \tan^{-1}\frac{4}{4} = 45°$$

$$\theta = 76 + 104 - 33.7 - 45 = 101.3°$$

■

Magnitude Criterion   Equation (6.51) can be used as a criterion for both determining root locus as well as for design purposes. Based on Equation (6.48), if the following magnitude criterion is satisfied, the closed–loop characteristic equation is also satisfied and the chosen point is on the root locus:

$$KGH = 1\angle 180° \tag{6.53}$$

Angle Criterion   Similarly, based on Equation (6.53), since K is a real value, the angle criterion is satisfied when $\angle GH = \pm 180°$.

Example 6.15

Based on the magnitude and angle criteria, point $s = 0 + 4j$ of Example 6.15 is not on the root locus because neither its magnitude ($0.4167 \neq 1$) nor its angle ($101.3 \neq \pm 180$) satisfy the requirements. ∎

Asymptotes   The total number of asymptotes is:

$$\alpha = \#poles - \#zeros \tag{6.54}$$

Asymptote Angles   The angles of asymptotes are:

$$\theta = \frac{\pi, \, 3\pi, \, 5\pi, \, \cdots}{\alpha} \tag{6.55}$$

This can be summarized as in Table 6.5.

Table 6.5   *The Angles of Asymptotes Based on Their Number.*

α	Angles of Asymptotes
1	180
2	90, 270
3	60, 180, 300
4	45, 135, 225, 315

Asymptote Center   Designating the real components of the poles and zeros as σ_p and σ_z, the center of the asymptotes (where they intersect the real axis) is:

$$\sigma_A = \frac{\sum \sigma_p - \sum \sigma_Z}{\alpha} \tag{6.56}$$

Example 6.16

As shown in Figure 6.23, based on Equation (6.56), the center of the asymptotes is:

$$\sigma_A = \frac{\sum \sigma_p - \sum \sigma_Z}{\alpha} = \frac{(0 - 3 - 7) - (-4)}{2} = -3$$

∎

Breakaway and Break-in Points   These are the points where the value of K is the largest on the real axis; therefore, at these points, the system is critically damped with the fastest rise time without any overshoot or any oscillations. No imaginary components are present and, consequently, the system will not oscillate. To find these points, we may take the derivative of the closed-loop characteristic equations and set it to zero as follows:

$$KG(s)H(s) + 1 = 0 \quad \rightarrow \quad K = \frac{-1}{G(s)H(s)}$$

$$\frac{dK}{ds} = \frac{d}{ds}\left(\frac{-1}{G(s)H(s)}\right) = 0 \tag{6.57}$$

Except for simple cases, calculation of breakaway and break-in points require solving higher-order polynomials, which is not always readily possible. Therefore, an estimate can be made for drawing the root locus, followed by an iterative process for exact location, if necessary.

Example 6.17

The root locus for the following characteristic equation can be drawn as follows:

$$GH = \frac{1}{(s-1)(s+4)(s+6)} \quad \rightarrow \quad p = 1, -4, -6$$

Number of asymptotes: $\alpha = 3 - 0 = 3$, angles of asymptotes: 60, 180, 300.

Asymptote center: $\sigma_A = \dfrac{\sum \sigma_p - \sum \sigma_z}{\alpha} = \dfrac{(1-4-6)-0}{3} = -3$

Breakaway point: $\dfrac{dK}{ds} = \dfrac{d}{ds}(-(s-1)(s+4)(s+6)) = 0$

Solve to get $s = -0.9$ and -5.08. Figure 6.25 shows the root locus.

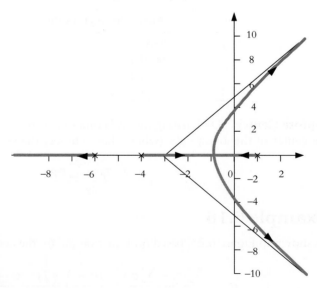

Figure 6.25   The root locus for Example 6.17.

Example 6.18

The root locus for the following characteristic equation can be drawn as follows:

$$s^2 + Ks + \omega^2 = 0 \quad \rightarrow \quad (s + j\omega)(s - j\omega) = -Ks \text{ (see Equation (6.49))}$$

Therefore, $p = \pm j\omega, \quad z = 0$

Number of asymptotes: $\alpha = 2 - 1 = 1$, angle of asymptote: 180.

Asymptote center is at $\sigma_A = 0$.

Breakaway point: $\dfrac{dK}{ds} = \dfrac{d}{ds}\left[-\dfrac{(s^2 + \omega^2)}{s}\right] = 0 \quad \rightarrow \quad s = \pm\omega$

Figure 6.26 shows the root locus.

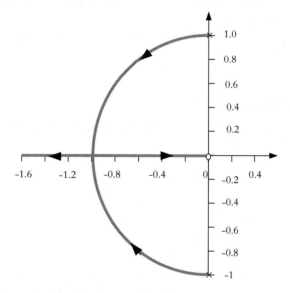

Figure 6.26 The root locus for Example 6.18.

Root Locus with MATLAB® Although it is crucial that the details of root locus be learned to be able to draw it and use it as a design tool, it is also convenient to use programs such as MATLAB to draw the root locus. To do so, if it is available to you, first enter the characteristic equation in the form g = zpk($[z_1, z_2, \cdots]$, $[p_1, p_2, \cdots]$, K). Then type *rlocus*. MATLAB draws the root locus. Please refer to Appendix C for more information.

Additionally, the MATLAB command *rltools* can assist in the controller design process. We will use both analytical tools and MATLAB in the next sections.

6.14 Proportional Controllers

Figure 6.22(a) shows a system with a proportional gain in the feed-forward loop. As we discussed earlier, when gain K varies, the locations of the poles and zeros of the system change as shown on a root locus. Therefore, the system's behavior is dependent on the value of this gain K. Consequently, it is possible to select (design) a value for the gain that makes the system behave in a particular way. For example, we may desire a system with an overshoot less than a certain percent, an overdamped system, or a system whose rise time is less than a certain value. This process, called *pole placement*, allows the designer to select poles and calculate the value of K that yields the particular pole locations.

Proportional controllers are the simplest controllers and are very common. We only need to change the amplification value of a controller that already exists without having to add anything to the system. However, it is not always possible to find appropriate pole locations with proportional controllers that yield satisfactory results. In that case, as will be seen next, other types of controllers may be used.

Please review Figure 6.19 before we continue with the next subject. Note the lines of constant damping, the directions of faster or slower response, directions of less or more damping, and so on.

In order to see how root locus may be used to design a proportional controller, we will use the following example.

Example 6.19

Find a proper value for the proportional gain that yields the following requirements for the system:

- Settling time ≤ 1 sec.
- % overshoot $\leq 5\%$
- $GH = \dfrac{K}{(s+1)(s+8)}$

Solution: Figure 6.27 shows the root locus for the system. As expected, there are two poles at -1 and -8 and two asymptotes located at $s = -4.5$. Since the system is second-order, we use Equations (6.34) and (6.35) to get:

$$T_s = 4/\zeta\omega_n \;\rightarrow\; \zeta\omega_n = \frac{4}{T_s} = \frac{4}{1} = 4$$

$$\%OS = e^{\left(-\zeta\pi/\sqrt{1-\zeta^2}\right)} \times 100\% \;\rightarrow\; \zeta = 0.69 \text{ (by trial and error)}$$

$$\theta = \cos^{-1}\zeta = 46.5°$$

$$\omega_n = 5.8 \text{ rad/sec}$$

We were able to use Equation (6.34) only because this is a second-order system. Otherwise, we need to either use approximation based on which pole is dominant, or use MATLAB, as we will discuss.

Applying the minimum requirements for the given specifications ($\zeta\omega_n \geq 4$ and $\theta \leq 46.5°$) to the root locus of Figure 6.27, we find all the acceptable possible root locations; left of the vertical line and between the constant damping lines. For

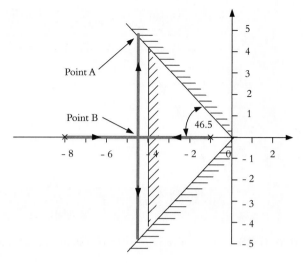

Figure 6.27 The root locus for Example 6.19.

example, point A and its conjugate, or point B, with their corresponding K values, can be used as possible roots.

Now we will find the steady-state error to see if that is acceptable. Note that this particular system is type-0, and with a step input, its steady-state error is finite. Conforming the given GH to the form in Equation (6.44), we get:

$$G(s)H(s) = \frac{K}{(s+1)(s+8)} \quad \rightarrow \quad K_p = \frac{K}{8} \quad \text{and} \quad E_{ss} = \frac{1}{1 + \frac{K}{8}}$$

Higher K yields lower steady-state error, and, therefore, we should select point A and its conjugate at $s = -4.5 \pm 4.75j$. We can use Equation (6.51) to calculate K as follows:

$$K = \frac{\prod M_{p_i}}{\prod M_{z_i}} = \frac{\sqrt{3.5^2 + 4.75^2} \times \sqrt{3.5^2 + 4.75^2}}{1} = 34.8$$

$$E_{ss} = \frac{1}{1 + \frac{K}{8}} = 0.187$$

Figure 6.28 shows the response of this system to a unit step function plotted by MATLAB. Notice how the steady-state error matches the analytical results. Also notice that the overshoot is about 5% above the final value (not the desired value of 1). The settling time is also about 1 second.

Figure 6.29 shows the response of the system if the roots were selected at point B of Figure 6.27. As you can see, the settling time is still about 1 second, but there is no overshoot (critical damping) and the steady-state error is much bigger at about 0.4. We will discuss how to reduce this error in the next section.

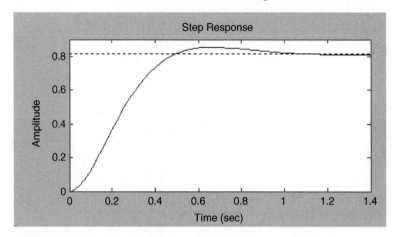

Figure 6.28   The step response of the system of Example 6.19.

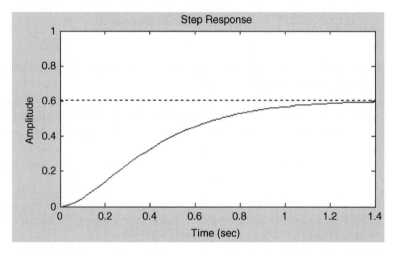

Figure 6.29   The response of the system of Example 6.19 with critical damping.

■

Example 6.20

In Example 6.2 (shown in Figure 6.6), we studied a hydraulic lift. Adding a floating lever to the system, as shown in Figure 6.30(a), makes this system a proportional servovalve. As the lever arm is pushed up, the supply fluid will move the power piston down, which in turn, brings down the spool in the valve and eventually closes it. Figure 6.30(b) shows the feedback loop. The system may be represented as follows. Note how the transfer function is type-0 (proportional controller):

$$\frac{Y(s)}{R(s)} = \frac{\dfrac{l_2}{l_1 + l_2}\dfrac{K}{s}}{1 + \dfrac{l_1}{l_1 + l_2}\dfrac{K}{s}} = \frac{l_2 K}{(l_1 + l_2)s + l_1 K}$$

■

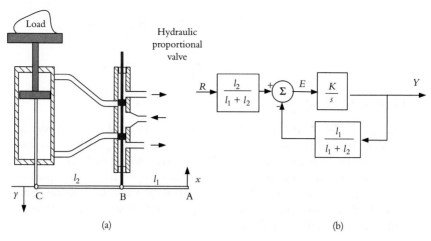

Figure 6.30 A proportional hydraulic servovalve.

6.15 Proportional-plus-Integral Controllers

Integral controllers provide a means for eliminating steady-state error in a system. This is because an integrator adds an additional s to the denominator of the transfer function, therefore raising its type. Referring to Section 6.12, recall that for type-0 systems, the steady-state error for a step function is a finite value; however, for type-1 systems, it is zero. Similarly, the steady-state error with a ramp input for a type-0 system is infinite, but finite for a type-1 system. Each integrator within the system raises its type, reducing the error boundary.

Now imagine that we are designing a control system for a robot. It should be clear that (1) the response of the robot actuators to a step function (to go from one location to another) should not overshoot, (2) it should rise to the value of the input signal as quickly as possible, and (3) it should not have any steady-state error. Obviously, if the response has an error, all our estimations of acceleration, velocity, and positional accuracy will be wrong. Therefore, we need to create a controller that delivers all these requirements simultaneously. However, as we discussed in Example 6.19, when a proportional controller was used, even allowing an overshoot resulted in a significant steady-state error. When we placed the poles on the real-axis, making it critically damped and eliminating the overshoot, the steady-state error was further increased. In order to have faster response, a high gain is needed, but that creates overshoot and error. When overshoot is reduced, error increases further. Consequently, a proportional controller alone cannot simultaneously provide for fast response, no overshoot, and zero steady-state error. However, a system with both proportional and integral elements will improve system response.

A proportional-plus-integral (PI) controller with gains K_P and K_I can be represented as shown in Figure 6.31 and derived as follows (please note that K_P used here is different from K_p, the static position error coefficient in Section 6.12):

$$G = \frac{V_a}{E} = K_P + \frac{K_I}{s} = \frac{K_P\left(s + \frac{K_I}{K_P}\right)}{s} = \frac{K(s + z_I)}{s} \qquad (6.58)$$

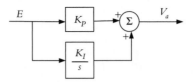

Figure 6.31 A proportional-plus-integral controller.

where $z_I = \frac{K_I}{K_P}$. As Equation (6.58) shows, the controller adds a pole at the origin as well as a zero at z_I, which is influenced by our choice of K_I and K_P. To not severely affect the shape of the root locus by this addition, we should pick z_I to be close to the origin. Therefore, the integral gain should be small compared to the proportional gain. With this choice, the pole at the origin and the zero near it add an additional small part to the root locus without changing its general shape. The following example demonstrates how this can affect the system's behavior.

Example 6.21

The system of Example 6.19 can be modified into a proportional-plus-integral controller to eliminate the steady-state error. We modify the system by adding an integrator pole at the origin and a zero at $z_I = -0.1$. Figure 6.32 shows the root locus for the system:

$$GH = \frac{K(s+0.1)}{s(s+1)(s+8)}$$

As shown, the root locus looks similar to Example 6.19, except that it has a small portion between the origin and the zero. Since the system is no longer second-order,

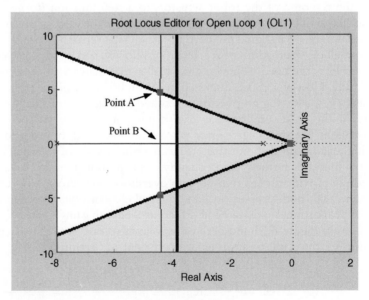

Figure 6.32 The root locus for Example 6.21.

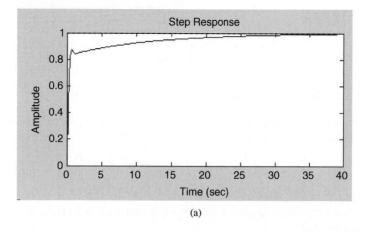

(a)

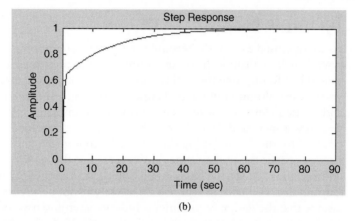

(b)

Figure 6.33 The step responses for the system of Example 6.21.

the equation for settling time no longer holds, but the acceptable range for roots remains about the same.

Figure 6.33(a) is the step response for the system when the roots are selected at point A and its conjugate $(-4.46 \pm 4.75j)$. As you can see, there is a 5% overshoot at the beginning, but the integrator reduces the steady-state error to zero, although taking a long time. Figure 6.33(b) shows the step response when the roots are selected at point B. Note that since this corresponds to critical damping, there is no overshoot, but the response is slower with zero steady-state error. Therefore, the addition of the integrator (and a zero near it) has improved the system's steady-state performance without changing the characteristics of the overall system. ∎

6.16 Proportional-plus-Derivative Controllers

Sometimes it is impossible to meet the design requirements with proportional or proportional-plus-integral controllers. In these cases, the dynamic behavior of the system must be altered in order to achieve the design requirements. This may be achieved by a

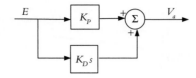

Figure 6.34   A proportional-plus-derivative controller.

proportional-plus-derivative (PD) controller. A PD controller with gains K_P and K_D can be represented as shown in Figure 6.34 and derived as follows:

$$G = \frac{V_a}{E} = K_P + K_D s = K_D\left(s + \frac{K_P}{K_D}\right) = K(s + z_D) \tag{6.59}$$

where $z = \frac{K_P}{K_D}$. The controller adds a zero to the root locus, and therefore, changes its characteristics. In order to see how this may affect the system, we continue with Example 6.21.

Example 6.22

As we discussed earlier, the settling time may be an important issue in the design of a system such as a robot. In order to improve the positional accuracy of a robot, we would like to improve the settling time of the system of Examples 6.19 and 6.21. As you can see, although an integral controller was added to the system of Example 6.19 and consequently its steady-state error was eliminated, the settling time for the system was increased. However, in both cases, the root locus limitation was set at or near -4 for the given settling time. In order to improve this design requirement, we need to reduce the settling time. However, since the center of the asymptotes is near -4.5 on the real axis, the best settling time we can get is $T_s = 4/\zeta\omega_n = 4/4.5 \approx 0.9$. Beyond that, as shown in Figure 6.35, there will not be any roots available. Now let's assume that the design requirements indicate a settling time of 0.6 sec with the same 5% overshoot. The limits are shown in Figure 6.35. Let's (arbitrarily) choose point A and its conjugate at $s = -6.5 \pm 7j$ as desired points for achieving the design requirements. We need to find the location of the derivative portion of a controller that will yield a root locus that includes these points.

To find the location of this zero, we will do the following:

1. Calculate the angle deficiency. For the original transfer function of $GH = \dfrac{K}{(s+1)(s+8)}$, the roots are at $s = -1$ and $s = -8$. From Equation (6.52), the complex vector angles to points $s = -6.5 \pm 7j$ are:

$$\theta_{p(-1)} = 180° - \tan^{-1}\left(\frac{7}{-6.5 - (-1)}\right) = 128.2°$$

$$\theta_{p(-8)} = \tan^{-1}\left(\frac{7}{-6.5 - (-8)}\right) = 77.9°$$

$$\sum \theta_{z_i} - \sum \theta_{p_i} = \theta_z - (128.2 + 77.9) = \pm 180 \quad \rightarrow \quad \theta_z = 26.1°$$

2. With this deficiency angle, the zero should be located at:

$$\tan 26.1 = \frac{7}{-6.5 - (z)} \quad \rightarrow \quad z = -20.8$$

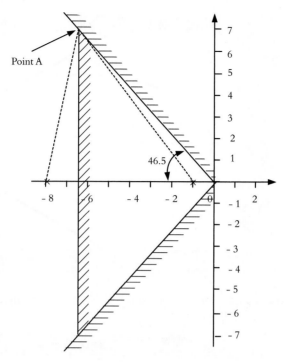

Figure 6.35 No roots are available for settling time less than 0.9.

3. The gain for these points can also be calculated from Equation (6.51) as:

$$M_{TF} = \frac{\prod M_{z_i}}{\prod M_{p_i}} = \frac{\sqrt{5.5^2 + 7^2}\sqrt{1.5^2 + 7^2}}{\sqrt{14.3^2 + 7^2}} = 4$$

4. The overall transfer function for the system is:

$$GH = \frac{4(s + 20.8)}{(s + 1)(s + 8)}$$

Figure 6.36 shows the root locus for this system. Figure 6.37 shows the response to a step input. As indicated, both %OS and settling time requirements are met. However, the steady-state error is not zero because we did not include an integrator in the system. ■

It is important to note here that a derivative controller, especially at high gains, is very susceptible to high frequency noise. Since this type of controller differentiates the signal, when high frequency (or sharply changing) signals are present, the derivative of the signal may become excessively large. Therefore, it may be necessary to use filters to reduce high frequency noise in the system. This can be easily demonstrated as follows:

$$e(t) = 1\sin(10t) + 0.01\sin(1000t)$$

$$\frac{d}{dt}e(t) = 10\cos(10t) + (10)\cos(1000t)$$

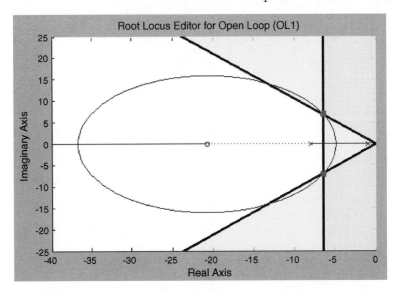

Figure 6.36   The root locus for the system of Example 6.22.

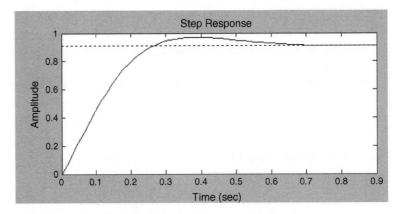

Figure 6.37   The response of the system of Example 6.22 to a step function.
$\zeta = 0.68$, $s = -6.5 \pm 7j$, gain $= 4$.

Although the amplitude of the noise is very small, after differentiation, it may have a significant effect on the system.

6.17 Proportional-Integral-Derivative Controller (PID)

As we saw, the addition of a derivative component to a controller changes its behavior, allowing the placement of the roots in desirable locations and thereby achieving design requirements. However, a proportional and derivative controller may not result in zero steady-state error. Many systems, including robots, may require zero steady-state error in addition to other requirements. Therefore, it will be necessary to add an integrator to the

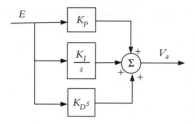

Figure 6.38 A proportional-integral-derivative (PID) controller.

system as well. However, care must be taken to ensure that the addition of the integrator does not otherwise change the behavior of the system.

Figure 6.38 shows how a PID controller may be constructed. The transfer function for this system is:

$$G = \frac{V_a}{E} = K_P + \frac{K_I}{s} + K_D s = \frac{K_D\left(s^2 + \frac{K_P}{K_D}s + \frac{K_I}{K_D}\right)}{s} = \frac{K_D(s + z_1)(s + z_2)}{s} \quad (6.60)$$

In order to maintain the behavior of the system and the general shape of the root locus, we may place one of the zeros of Equation (6.60) near the origin, and therefore, cancel the dynamic effect of the integrator pole at the origin (called zero-pole cancellation). With this, although the system behavior remains almost unchanged, its steady-state error will go to zero because the system type is raised. Remember that zeros must be real and distinct.

Example 6.23

To eliminate the steady-state error of the system of Example 6.22, we add a pole at the origin and a zero near it. We can express the system as:

$$GH = \frac{K(s + 20.8)(s + 0.5)}{s(s + 1)(s + 8)}$$

The root locus of this system is shown in Figure 6.39. Notice the similarity of this root locus to the one in Figure 6.36, even though a pole and a zero were added to the transfer function. Please also note that the location of the zero was arbitrarily chosen. Other values may also be acceptable. Figure 6.40 shows the response of this system to a step function. Notice how the design requirements are mostly met.

The selected poles are at $s = -14.3 \pm 14.8j$ with $\zeta = 0.697$. The loop gain is 20.2. From Equation (6.60), we can calculate the gains of the system as follows:

$$G = \frac{K_D\left(s^2 + \frac{K_P}{K_D}s + \frac{K_I}{K_D}\right)}{s} = \frac{20.2(s + 20.8)(s + 0.5)}{s}$$

$$K_D\left(s^2 + \frac{K_P}{K_D}s + \frac{K_I}{K_D}\right) = 20.2(s^2 + 21.3s + 10.4)$$

$$K_D = 20.2$$

$$K_P = 430.3$$

$$K_I = 210.1$$

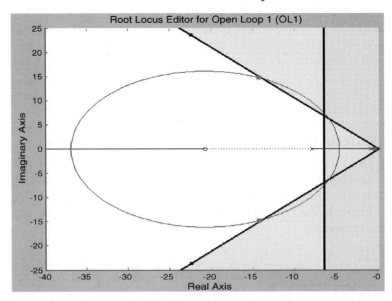

Figure 6.39   The root locus of the system of Example 6.23 with a proportional-integral-derivative controller.

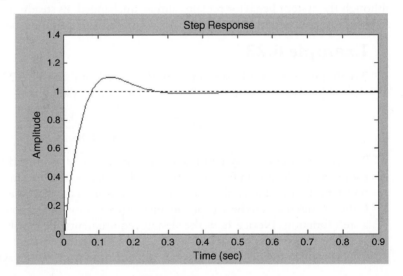

Figure 6.40   The response of the system of Example 6.23 to a step function. ∎

6.18 Lead and Lag Compensators

Ideal integral and derivative controllers are used to change the response of a plant according to the required design specifications such as the settling time, speed of the response, percent overshoot, and steady-state error elimination. However, they are both active systems and require power. In addition, a derivative controller has a wide bandwidth; therefore, although it can differentiate high frequencies in the system, it can also create problems when noise is present.

Alternately, a lead compensator or a lag compensator may be used. In each case, the circuits for lead and lag compensators are passive, basically consisting of resistors, capacitors, and inductors. A lead compensator has a limited bandwidth, and therefore, may be even better for high frequency noise reduction.

Lead and lag compensation is usually performed along with frequency domain analysis of systems (such as the Bode diagram).

Lag Compensators   A lag compensator consists of a zero placed near a pole close to the origin. The addition of the pole near the origin (and not exactly at the origin which makes it a pure integrator) acts similar to an integrator, but over time the system loses its accuracy as the steady-state error increases. Therefore, lag compensators are assumed to be *leaky*. The addition of the zero near the pole keeps the root locus about the same.

Lead Compensators   A lead compensator consists of a zero near the origin that acts similar to a derivative controller, plus a pole near it. A lead compensator causes little change in the overall shape of the root locus, but provides for passive derivative compensation with limited bandwidth.

6.19 Bode Diagram and Frequency Domain Analysis

The analysis and the design techniques associated with root locus are based on the time (or Laplace) domain. However, many systems function with inputs that vary continuously; therefore, it is better to analyze them in the frequency domain. Bode diagram is a graphical representation of the open-loop transfer function in frequency domain when s is replaced with $j\omega$. It consists of two graphs—one for the magnitude of GH in the logarithmic scale, another for the phase angle as ω varies. Figure 6.41 shows a typical Bode diagram plotted for $G(s) = 1/(s^2 + s + 10)$ by MATLAB. As you can see, the magnitude is drawn on a log/log scale in dB, where $dB = (20)\log|G(j\omega)|$. The phase is also drawn. Notice how the magnitude (output/input ratio) increases at the natural frequency of the system ($\omega_n = 10$). Both graphs also vary significantly depending on the damping ratio. However, it is possible to draw the Bode diagram, albeit with some error at the corner frequencies, by drawing asymptotes whose slopes are functions of the order of the system. For example, for the system of Figure 6.41, since the transfer function is second-order, the magnitude at higher frequencies decreases at $-40\ dB$/decade while the phase change is $-90°$/decade. Since the magnitude and phase are in logarithmic scale, the Bode plots of different parts of the characteristic equation can simply be added together to get the Bode diagram of the whole system. For further reading and for design techniques using the Bode diagram, refer to related books and journal articles.

6.20 Open-Loop versus Closed-Loop Applications

You may have noticed that both open-loop and closed-loop representations have been used for different applications. The following is a summary of which one is used for which application. Remember that the closed-loop transfer function

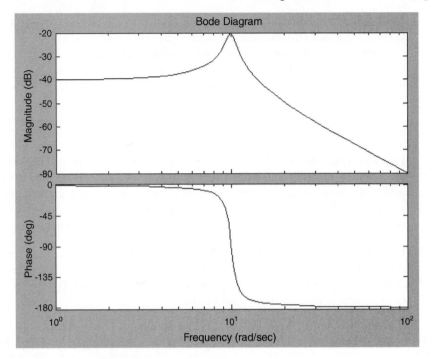

Figure 6.41   A typical Bode diagram for a second-order system.

and characteristic equation and the open-loop transfer function for the system of Figure 6.42 are:

$$CLTF = \frac{Y(s)}{R(s)} = \frac{KG}{KGH + 1}$$

$$OLTF = KGH$$

Characteristic Equation $= KGH + 1$

The open-loop transfer function represents the output of the system as measured by the feedback sensor.

- **Stability:** The closed-loop transfer function poles are plotted. They must be in the ½–left plane.

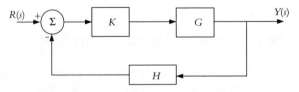

Figure 6.42   A typical feedback control system.

- **Steady-state error *Ess*:** This is related to the system type as well as the type of input. The open-loop transfer function with unity feedback is used to calculate the steady-state error.
- **Root locus:** Open-loop transfer function is used. The root locus starts at the poles and ends at zeros or ∞.

6.21 Multiple-Input and Multiple-Output Systems

Most systems we have considered so far are single-input, single-output (SISO) systems, where there is one input and one output. For example, when a voltage is supplied to a motor, the motor rotates and its angular velocity (output) can be measured. However, many systems have multiple DOF, where more than one variable controls the systems. In this case, there are multiple inputs and multiple outputs (MIMO) present. The following are simple examples of how linear MIMO systems may be analyzed.

Example 6.24

Figure 6.43 shows a multiple-input, single-output system. Derive the relationship between the inputs and the output.

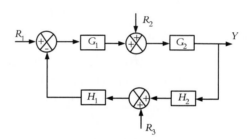

Figure 6.43   Multiple-input, single-output system of Example 6.24.

Solution: There are multiple ways of solving this problem. However, we simply write:

$$\{[R_1 - ((YH_2 + R_3)H_1)]G_1 + R_2\}G_2 = Y$$
$$R_1G_1G_2 - YH_1H_2G_1G_2 - R_3H_1G_1G_2 + R_2G_2 = Y$$
$$G_2(R_1G_1 - R_3H_1G_1 + R_2) = Y(1 + H_1H_2G_1G_2)$$
$$Y = \frac{G_2(R_1G_1 + R_2 - R_3H_1G_1)}{1 + H_1H_2G_1G_2}$$

■

Example 6.25

Derive the equations that represent each output of the multiple-input, multiple-output system of Figure 6.44.

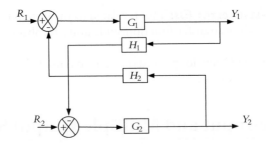

Figure 6.44 Multiple-input, multiple-output system for Example 6.25.

Solution: We write the equations relating to each input and output as follows:

$$\begin{cases} (R_1 - Y_2 H_2)G_1 = Y_1 \\ (R_2 - Y_1 H_1)G_2 = Y_2 \end{cases}$$

Substitute Y_2 into Y_1 to get:

$$(R_1 - (R_2 - Y_1 H_1)G_2 H_2)G_1 = Y_1$$
$$R_1 G_1 - R_2 G_1 G_2 H_2 = Y_1(1 - G_1 G_2 H_1 H_2)$$
$$Y_1 = \frac{R_1 G_1 - R_2 G_1 G_2 H_2}{(1 - G_1 G_2 H_1 H_2)}$$

Similarly, substitute Y_1 into Y_2 to get:

$$(R_2 - (R_1 - Y_2 H_2)G_1 H_1)G_2 = Y_2$$
$$R_2 G_2 - R_1 G_1 G_2 H_1 = Y_2(1 - G_1 G_2 H_1 H_2)$$
$$Y_2 = \frac{R_2 G_2 - R_1 G_1 G_2 H_1}{(1 - G_1 G_2 H_1 H_2)}$$

These two equations may be written in matrix form as:

$$\begin{bmatrix} Y_1 \\ Y_2 \end{bmatrix} = \begin{bmatrix} \dfrac{G_1}{\mathbf{K}} & \dfrac{-G_1 G_2 H_2}{\mathbf{K}} \\ \dfrac{-G_1 G_2 H_1}{\mathbf{K}} & \dfrac{G_2}{\mathbf{K}} \end{bmatrix} \times \begin{bmatrix} R_1 \\ R_2 \end{bmatrix} \quad \text{where } \mathbf{K} = (1 - G_1 G_2 H_1 H_2)$$

■

A multi-axis robot has multiple inputs and multiple outputs that must be controlled simultaneously. However, in most robots, each axis is controlled individually as a single-input, single-output unit. Although this introduces some error, the error is small for most practical purposes. The analysis of these systems is beyond the scope of this introduction to control theory. For further reading, refer to related books and journal articles on this topic.

6.22 State-Space Control Methodology

The transfer function that describes the relationship between the input and output of a system is only applicable when the system is initially relaxed (no initial conditions; otherwise, the Laplace transform cannot be applied) and only relates the inputs and output,

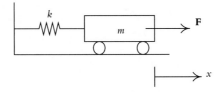

Figure 6.4 Repeated.

but not the internal signals within the system. An alternative to this method of representation is *state-space*, where different signals within the system may be linked together to create a set of first-order linear equations that are easy to solve and provide information about internal signals. Consider the mechanical system of Figure 6.4, repeated here.

The equations describing the motions of the mass as a result of the application of force **F** are given in Equation (6.3) as:

$$m\frac{d^2x}{dt^2} + b\frac{dx}{dt} + kx = \mathbf{F}$$

The actual motion is, of course, a function of the initial conditions of the location and velocity of the mass. In other words, the succeeding motion (location and velocity of the mass) depends on where the mass is located and its velocity when the force is applied. Each one of these variables is a *state*. Therefore, we can express the equation in a different form, as follows. We choose the position of the mass as one state, $y_1 = x$. We also choose the velocity as the second state; therefore $y_2 = \dot{x} = \dot{y}_1$. As you can see, both the initial conditions and the states within the system (position and velocity) are included in our representations. Equation (6.3) can be written as:

$$m\ddot{y}_1 + b\dot{y}_1 + ky_1 = \mathbf{F} \quad \text{or} \quad m\dot{y}_2 + by_2 + ky_1 = \mathbf{F}$$

$$\begin{cases} \dot{y}_1 = y_2 \\ m\dot{y}_2 = -by_2 - ky_1 + \mathbf{F} \end{cases} \quad \rightarrow \quad \begin{bmatrix} \dot{y}_1 \\ \dot{y}_2 \end{bmatrix} = \begin{bmatrix} 0 & 1 \\ -k/m & -b/m \end{bmatrix} \begin{bmatrix} y_1 \\ y_2 \end{bmatrix} + \begin{bmatrix} 0 \\ 1/m \end{bmatrix} \mathbf{F}$$

The output can be represented as: $x = \begin{bmatrix} 1 & 0 \end{bmatrix} \begin{bmatrix} y_1 \\ y_2 \end{bmatrix}$

This is the state-space equation describing the system. As you can see, this is a two-dimensional, linear, time-invariant equation, where the second-order part of the equation is converted into first-order by the introduction of the velocity state variable. y_1 and y_2 are states of the system that can be measured, say, with sensors, if necessary.

Now consider the system shown in Figure 6.45 with three states x_1, x_2, x_3, and its transfer function derived as:

$$R - E_1\left(\frac{a_1}{s} + \frac{a_2}{s^2} + \frac{a_3}{s^3}\right) = E_1 \quad \rightarrow \quad R = E_1\left(1 + \frac{a_1}{s} + \frac{a_2}{s^2} + \frac{a_3}{s^3}\right)$$

$$Y = \frac{E_1}{s^3}$$

$$\frac{Y}{R} = \frac{1}{s^3 + a_1s^2 + a_2s + a_3}$$

(6.61)

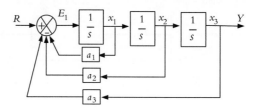

Figure 6.45   A system with three states.

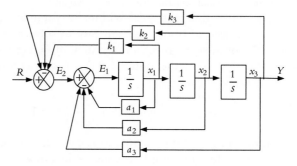

Figure 6.46   The representation of a state-space system.

Now consider the system in Figure 6.46, where the input to the system is augmented by feedback from the states, as shown, such that $E_2 = R - k_1x_1 - k_2x_2 - k_3x_3$. Notice how each state is the integral of the previous state and how these states are available at any given time.

The transfer function of the system can be derived as:

$$E_2 - E_1\left(\frac{a_1}{s} + \frac{a_2}{s^2} + \frac{a_3}{s^3}\right) = E_1 \quad \rightarrow \quad E_2 = E_1\left(1 + \frac{a_1}{s} + \frac{a_2}{s^2} + \frac{a_3}{s^3}\right)$$

$$R - E_1\left(\frac{k_1}{s} + \frac{k_2}{s^2} + \frac{k_3}{s^3}\right) = E_2 \quad \rightarrow \quad R = E_1\left(1 + \frac{a_1 + k_1}{s} + \frac{a_2 + k_2}{s^2} + \frac{a_3 + k_1}{s^3}\right)$$

$$Y = \frac{E_1}{s^3}$$

$$\frac{Y}{R} = \frac{1}{s^3 + (a_1 + k_1)s^2 + (a_2 + k_2)s + (a_3 + k_3)}$$

(6.62)

From Equations (6.61) and (6.62), it is clear that the two systems are the same, except that each root is supplemented with a k value. This provides for a very convenient way to place the roots of the characteristic equation at any desired place; therefore, it is a powerful technique to design a control system.

Example 6.26

A plant is modeled as $\frac{Y}{R} = \dfrac{1}{s(s+1)(s+5)}$ with poles at $s = 0$, $s = -1$, and $s = -5$. However, to change the behavior of the plant, we wish to place the poles at $s = -10$ and $s = -1 \pm j\sqrt{3}$. Find the appropriate values of k to accomplish this.

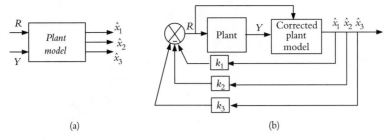

Figure 6.47 The application of estimators in control systems.

Solution: The original plant is represented by:

$$R = \ddot{y} + 6\ddot{y} + 5\dot{y} \quad \text{and} \quad s^3 + 6s^2 + 5s = 0.$$

The desired plant may be represented as:

$$(s + 10)(s^2 + 2s + 4) = (s^3 + 12s^2 + 24s + 40) = 0.$$

The augmented characteristic equation may be set equal to the desired one as:

$$(s^3 + 12s^2 + 24s + 40) = [s^3 + (6 + k_1)s^2 + (5 + k_2)s + k_3]$$

$$\text{and} \begin{cases} 6 + k_1 = 12 \\ 5 + k_2 = 24 \\ k_3 = 40 \end{cases} \rightarrow \begin{cases} k_1 = 6 \\ k_2 = 19 \\ k_3 = 40 \end{cases}$$

Therefore, the poles can easily be located at a desired location. ■

One additional benefit of this system is that in cases where the states of the system are not available or the cost of measuring the states may be high, it is possible to estimate the expected value of the state from the dynamics of the system. In other words, if the dynamics of the system are known, the states of the system may be estimated, as shown in Figure 6.47(a). Therefore, a system may be devised with *estimators* that provide values for each state, the values are fed into the control system, and the response is measured and corrected as needed, as shown in Figure 6.47(b). Estimators are readily used in many systems, from rockets to simple devices.

Example 6.27

Using state-space methodology, derive the equations describing a DC motor as shown in Figure 6.48.

Solution: For the electrical circuit portion of the system, where $v_{bemf} = K_B\dot{\theta}$ (K_B is a constant), we may write:

$$Ri + L\frac{di}{dt} = e(t) - v_{bemf} = e(t) - K_B\dot{\theta}$$

For the mechanical side of the system, with its inertia (of the armature and load) and damping, and $T_{bemf} = K_t i$ (where K_t is a constant), we may write:

$$T_{bemf} = K_t i = J\ddot{\theta} + b\dot{\theta}$$

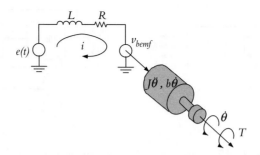

Figure 6.48   The electromechanical system for Example 6.27.

The states of the motor are the current i, angular position θ, and angular velocity $\dot\theta$. Let's select state variables as $x_1 = \theta$, $x_2 = \dot\theta = \dot{x}_1$, and $x_3 = i$. Therefore, these equations can be written as:

$$\begin{cases} L\dot{x}_3 = e(t) - K_B x_2 - R x_3 \\ J\dot{x}_2 = K_t x_3 - b x_2 \qquad\qquad \text{or} \\ \dot{x}_1 = x_2 \end{cases}$$

$$\begin{bmatrix} \dot{x}_1 \\ \dot{x}_2 \\ \dot{x}_3 \end{bmatrix} = \begin{bmatrix} 0 & 1 & 0 \\ 0 & -\dfrac{b}{J} & \dfrac{K_t}{J} \\ 0 & -\dfrac{K_B}{L} & -\dfrac{R}{L} \end{bmatrix} \begin{bmatrix} x_1 \\ x_2 \\ x_3 \end{bmatrix} + \begin{bmatrix} 0 \\ 0 \\ 1/L \end{bmatrix} e(t) \quad \text{and} \quad \theta = \begin{bmatrix} 1 & 0 & 0 \end{bmatrix} \begin{bmatrix} x_1 \\ x_2 \\ x_3 \end{bmatrix}$$

Once again, the second-order equations are transformed into first-order linear time-invariant equations, where the state variables can be measured and used in controls. ■

Obviously, there is much more to the state-space control methodology than is covered in this book. For further reading, refer to related books and journal articles on this topic.

6.23 Digital Control

Digital control is used in systems where microprocessors are used for controlling the system and in which signals are sampled. Many of the techniques used in analog control are also used in digital control systems, including root locus, lead-lag, proportional, integral, and derivative control, Bode diagrams, and others. However, one essential difference is that digital systems are discrete, not continuous.

Principally, a digital system can first be designed in the s-plane as an analog system, and subsequently through digital filtering, be converted to digital domain (called z-plane), or it can be designed in the digital domain. So long as the sampling rate in the system is relatively high, both techniques are acceptable. Otherwise, the system should be designed in digital domain (we will discuss the sampling theorem in Chapter 9). Suffice it to say

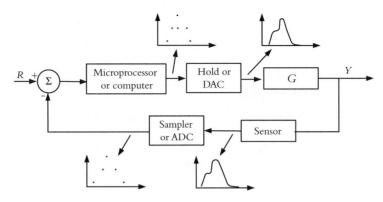

Figure 6.49 A typical digital system.

here that the sampling rate ω_s for a system with a maximum frequency-of-interest of ω should be $\omega_s \geq 2\omega$.

Figure 6.49 shows a general representation for a digital system. A microprocessor (or computer) generates a control signal that must be converted to an analog signal (by a "hold" circuit or a digital-to-analog converter). The plant's response is read by a sensor. If the sensor is not digital, the signal must be sampled and converted to digital form with an analog-to-digital converter before it can be used by the microprocessor or computer.

Now consider a differential equation $\dot{y} + ay = u$. Since in digital systems an analog signal is sampled and held until the next sample is taken, the signal is discrete; it changes when the sample is taken but remains the same during the interval. Therefore, $\dot{y}$ must be represented by a finite difference as:

$$\dot{y} = \frac{\Delta y}{\Delta t} = \frac{y(n) - y(n-1)}{T}$$

Substituting into the differential equation, we get:

$$\frac{y(n) - y(n-1)}{T} + ay(n) = u(n)$$
$$y(n) = \frac{1}{1 + aT}[y(n-1) + Tu(n)] \tag{6.63}$$

where T is the sampling period. Therefore, we need to be able to relate each value of the sample to the previous ones. This is done using the z-transform. As we saw, the Laplace transform for functions and their derivatives in continuous domain are defined as:

$$\mathcal{L}[f(t)] = F(s) = \int_0^\infty f(t)e^{-st}dt \quad \text{and} \quad \mathcal{L}(f'(t)) = sF(s) - f(0)$$

Similarly, a z-transform for a discrete domain is defined as:

$$F(z) \triangleq \sum_{n=0}^{\infty} f(n)Z^{-n} \tag{6.64}$$

and

$$Z(f(n-1)) = z^{-1}F(z) \tag{6.65}$$

Example 6.28

If $f(n) = 1$ for $n = 0, 1, 2, \cdots$, the z-transform of the function is:

$$F(z) = \sum_{n=0}^{\infty} z^{-n} = \sum_{0}^{\infty} \frac{1}{z^n} = 1 + \frac{1}{z} + \frac{1}{z^2} + \cdots = \frac{1}{1 - z^{-1}} = \frac{z}{z - 1}$$

∎

Example 6.29

Derive the z-transform of the following equation.

$$y(u) = -a_1 y(n-1) - a_2 y(n-2) + \cdots + b_0 u(n) + b_1 u(n-1) + b_2 u(n-2) + \cdots$$

Solution: Using Equations (6.64) and (6.65), we get:

$$Y(z) = [-a_1 z^{-1} - a_2 z^{-2} - \cdots]Y(z) + [b_0 + b_1 z^{-1} + b_2 z^{-2} + \cdots]U(z)$$
$$Y(z)[1 + a_1 z^{-1} + a_2 z^{-2} + \cdots] = [b_0 + b_1 z^{-1} + b_2 z^{-2} + \cdots]U(z)$$
$$\frac{Y(z)}{U(z)} = \frac{b_0 + b_1 z^{-1} + b_2 z^{-2} + \cdots}{1 + a_1 z^{-1} + a_2 z^{-2} + \cdots}$$

∎

There is much more to this subject than is presented here as an introduction. For more information about digital control systems, refer to related books and journal articles.

6.24 Nonlinear Control Systems

A system is considered linear if the differential equation describing it is linear and if the components of the system behave in a linear fashion. This means that if a system responds with outputs $y_1(t)$ and $y_2(t)$ for inputs $x_1(t)$ and $x_2(t)$ respectively, its response to an input $a_1 x_1(t) + a_2 x_2(t)$ will be $a_1 y_1(t) + a_2 y_2(t)$. Otherwise, the system is nonlinear.

Most systems are inherently nonlinear. For example, a spring's constant is not really constant. However, for a small range of displacements, we may assume the response is linear. Other examples of nonlinearity are saturation, backlash, hysteresis, or piecewise behavior such as in a relay. In robotic applications too, many elements of the system are inherently nonlinear but they may be assumed linear or they may be linearized for small ranges in order to simplify the analysis. In other cases, it is also possible to multiply the response of an element or system by the inverse of the component that makes the system nonlinear in order to eliminate its effect. For example, if the output of a component is a function of $\sin\theta$, multiplying it by $1/\sin\theta$ will linearize the output. Alternately, since $\sin\theta$ may be represented by a Taylor series function, ignoring the higher-order terms also linearizes the function, albeit for small angles. Figure 6.50 shows examples of nonlinear behavior of system elements.

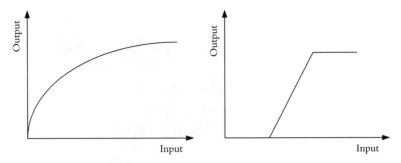

Figure 6.50 Examples of nonlinear behavior of system elements.

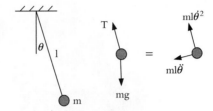

Figure 6.51 The pendulum of Example 6.30.

Example 6.30

The nonlinear equation describing the motion of a pendulum shown in Figure 6.51 may be linearized for small motions close to $\theta = 0$ as follows:

$$ml\ddot{\theta} = mg\sin\theta \quad \rightarrow \quad l\frac{d^2\theta}{dt^2} = g\sin\theta$$

$$\frac{d^2\theta}{dt^2} + \left(\frac{g}{l}\right)\sin\theta = 0$$

$$\sin\theta = \sum_{n=0}^{\infty}\frac{\theta^n}{n!}\left(\frac{d^n}{d\theta^n}(\sin\theta)|_{\theta=0}\right) = \theta - \frac{\theta^3}{3!} + \frac{\theta^5}{5!} - \cdots$$

Ignoring the higher-order terms of the Taylor series simplifies (and linearizes) the equation to:

$$\frac{d^2\theta}{dt^2} + \left(\frac{g}{l}\right)\theta = 0 \text{ (for small } \theta).$$ ∎

6.25 Electromechanical Systems Dynamics: Robot Actuation and Control

Figure 6.52 shows a robot and its actuators. Although hydraulic and pneumatic actuators are used in certain applications, most common industrial robots are electromechanical. We will study all these systems in Chapter 7. However, in this section, we will model electromechanical systems as they relate to robotic actuation.

Figure 6.52 A typical robot and its actuators. (HP3 robot, printed with permission from Motoman, Inc.)

The role of a robot actuator—prismatic or revolute—is to move a joint or a link and change its position. The role of the (feedback) control system is to ensure that the position is achieved in a manner that is satisfactory, as planned. A system whose role is to control the position of a system and track its motions is called a *servomechanism*.

Figure 6.53 shows a simplified depiction of a control system for a robot. As we discussed in Chapters 2, 3, 4, and 5, the joint values (displacement, velocity, acceleration, and applied forces and torques) are calculated from kinematic, dynamic, and trajectory analyses. These values are sent to the controller which, in turn, applies appropriate actuating signals to the actuators to run the joints to their destination in a controlled manner. The sensors measure the outputs and feed the signals back to the controller, which, in turn, controls the actuating signals accordingly. Figure 6.54 shows the schematic of the control system for a Quattro robot by Adept technology, Inc.

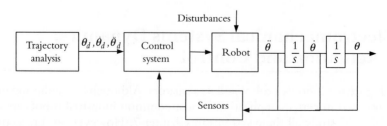

Figure 6.53 The feedback loop of a robot controller.

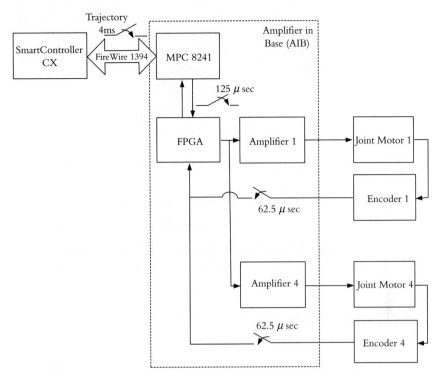

Figure 6.54   The control system schematic for Adept Technology's Quattro robot.
(Printed with permission from Adept technology, Inc.)

A multi-axis robot has multiple inputs and multiple outputs for each joint that must be controlled simultaneously. However, in most robots, each axis is controlled individually (called *independent joint control*) as a single-input, single-output unit. The coupling effects from other joints are usually treated as disturbances and are taken care of by the controller. Although this introduces some error, the error is small for most practical purposes. Additionally, robot dynamics equations have nonlinearities that require more sophisticated control schemes that are beyond the scope of this text. For more information on nonlinear control theory, see related books and journal articles. However, the following section shows how a robot actuator may be modeled for control purposes.

A robot's actuator consists of a motor, sensors, a controller through which a position reference signal is issued and which provides an actuating signal to the motor, and the external load. These elements form a system as shown in Figure 6.55. The motor model contains both an electric circuit as well as mechanical elements such as inertia and damping. These two systems are coupled together through the back-emf torque-voltage.

For the electrical circuit portion of the system, where $v_{bemf} = K_B\dot{\theta}$ (K_B is a constant), we may write:

$$Ri + L\frac{di}{dt} = e(t) - v_{bemf} = e(t) - K_B\dot{\theta}$$

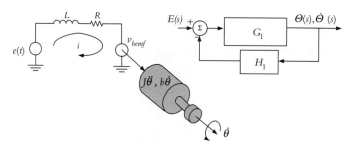

Figure 6.55   An electromechanical actuating system and its model.

This equation may be written in the Laplace form as:

$$E(s) - RI(s) - LsI(s) - K_{BS}\Theta(s) = 0 \qquad (6.66)$$

For the mechanical side of the system, with its inertia (of the armature and load) and damping, and $T_{bemf} = K_t i$ (where K_t is a constant), we may write:

$$T_{bemf} = K_t i = J\ddot{\theta} + b\dot{\theta}$$

This equation may be written in Laplace form as:

$$K_t I(s) = Js^2\Theta(s) + bs\Theta(s) \qquad (6.67)$$

Combining Equations (6.66) and (6.67) and rearranging the terms, we get:

$$E(s) = \left[\frac{R(Js^2 + bs)}{K_t} + \frac{Ls(Js^2 + bs)}{K_t} + K_{BS} \right] \Theta(s) \qquad (6.68)$$

In practice, the inductance of the motor L is usually much smaller than the inertia of the rotor and the load combined and can therefore easily be ignored for analysis. Consequently, Equation (6.68) may be simplified to:

$$E(s) = \left[\frac{R(Js^2 + bs)}{K_t} + K_{BS} \right] \Theta(s)$$

The transfer function between the output $\Theta(s)$ and input $E(s)$ is:

$$TF = \frac{\Theta(s)}{E(s)} = \frac{K_t}{R(Js^2 + bs) + K_t K_{BS}} = \frac{K_t/RJ}{s\left(s + \dfrac{b}{J} + \dfrac{K_t K_B}{RJ} \right)} \qquad (6.69)$$

If we are interested in the velocity of the motor (robot's arm) in response to the input voltage, we may multiply the s in the denominator and $\Theta(s)$ to get $\Omega(s)$. Therefore, the transfer function may be written as:

$$TF = \frac{\Omega(s)}{E(s)} = \frac{K}{s + a} \quad \text{where} \quad K = \frac{K_t}{RJ} \quad \text{and} \quad a = \frac{1}{J}\left(b + \frac{K_t K_B}{R} \right) \qquad (6.70)$$

This transfer function is a first-order differential equation, relating the motor angular velocity to the input voltage. We may use this equation to analyze the response of the motor. For example, when a particular input voltage is applied to the motor, we may study the response of the motor, its steady-state operation, how fast the response is, and much more.

Example 6.31

Assume that the input voltage to the system of Figure 6.55 is a step function $Pu(t)$. Determine the response of the motor and its steady-state value.

Solution: Using Equation (6.70) as the transfer function and referring to Table 6.3, we get:

$$\Omega(s) = \frac{K}{s+a}\frac{P}{s} = \frac{KP}{s(s+a)} = \frac{a_1}{s} + \frac{a_2}{(s+a)}$$

$$\text{where } a_1 = \left|s\left(\frac{KP}{s(s+a)}\right)\right|_{s=0} = \frac{KP}{a} \quad \text{and} \quad a_2 = \left|(s+a)\left(\frac{KP}{s(s+a)}\right)\right|_{s=-a} = \frac{KP}{-a}$$

$$\text{Hence, } \Omega(s) = \frac{KP}{sa} - \frac{KP}{(s+a)a} = \frac{KP}{a}\left(\frac{1}{s} - \frac{1}{s+a}\right)$$

The inverse Laplace transform of the equation is $\omega(t) = \frac{KP}{a}\left(1 - e^{-at}\right)$ and is shown in Figure 6.56. The steady-state velocity output of the motor, using the final value theorem is:

$$\omega_{ss} = \lim_{s \to 0} s\frac{KP}{s(s+a)} = \frac{KP}{a}$$

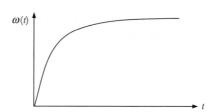

Figure 6.56   The approximate response of the motor of Example 6.31. ∎

Now let's add a tachometer to the system as a feedback sensor. The tachometer measures the angular speed of the motor in response to the actuating signal. Figure 6.57 shows the system of Figure 6.55 with the added tachometer.

For the tachometer, $v_b = K_f\dot\theta$. The circuit representing the tachometer can be expressed in Laplace domain as:

$$I(s) \times (R_a + R_L + Ls) = V_b(s) = K_f s\Theta(s)$$

$$V_o(s) = I(s)R_L = \frac{K_f s\Theta(s)R_L}{R_a + R_L + Ls}$$

The transfer function for the tachometer is:

$$TF = \frac{V_0(s)}{s\Theta(s)} = \frac{V_0(s)}{\Omega(s)} = \frac{K_f R_L}{(R_a + R_L + Ls)} = \frac{m}{s+n} \tag{6.71}$$

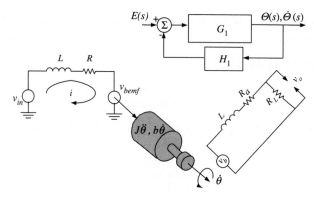

Figure 6.57   An electromechanical system with a tachometer sensor.

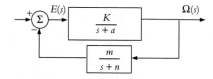

Figure 6.58   Completed block diagram for the robot's actuating motor.

where $m = \dfrac{K_f R_L}{L}$ and $n = \dfrac{R_a + R_L}{L}$. Figure 6.58 shows the completed block diagram of the system of Figure 6.57.

Equation (6.68) can also be used to calculate the natural frequency and damping ratio of the system. It can be rewritten as:

$$E(s) = \left[\frac{LJs^2 + s(RJ + Lb) + Rb + K_t K_B}{K_t}\right] s\Theta(s)$$

Therefore, the transfer function between the input voltage and output angular velocity is:

$$\frac{\Omega(s)}{E(s)} = \left[\frac{K_t/LJ}{s^2 + s\left(\dfrac{RJ + Lb}{LJ}\right) + \dfrac{Rb + K_t K_B}{LJ}}\right]$$

Since the characteristic equation is second-order in the form of $s^2 + 2\zeta\omega_n s + \omega^2$, the damping coefficient and natural frequency of the joint (and the connected load) can be calculated.

6.26 Design Projects

You may now decide how your robot(s) will be controlled and how you will implement the controller for your robot(s). Many different specifications and characteristics must be defined for a proper controller, both in terms of what the controller will include as well as how it will influence the behavior of the robot. You may try to decide these specifications for your robot(s), including % overshoots, damping, settling and rise times, and others. However, remember that this was an introduction to control systems design.

We will learn about robot actuators and sensors in the next chapters, where many of these decisions will be made when the type of actuators, loads, and other factors are specified. Many student projects include relatively simple microprocessors, small loads, slow motions, and simple control schemes. It will be your decision as to what level of sophistication is needed for your particular project.

Summary

In this chapter, we studied the basic principles of control systems, how they are analyzed, and how they may be designed. We also studied some introductory robot actuator modeling techniques. You should have learned enough material to be able to follow the design methodology for robot controllers. However, there is much more to control systems beyond what we can cover in one chapter. You should learn more from other references if you expect to design a working controller.

In the next chapters, we discuss actuators, sensors, and applications, where you will better understand how control systems play a role in the overall robot design.

References

1. Nise, Norman, "Control Systems Engineering," 4th Edition, John Wiley and Sons, 2004.
2. Dorf, Richard, Robert Bishop, "Modern Control Systems," 11th Edition, Prentice Hall, 2008.
3. Bateson, Robert, "Introduction to Control System Technology," 7th Edition, Prentice Hall, 2002.
4. Sciavicco, Lorenzo, Bruno Siciliano, "Modeling and Control of Robot Manipulators," McGraw-Hill, 1996.
5. Spong, Mark W., Seth Hutchinson, M. Vidyasagar, "Robot Modeling and Control," John Wiley and Sons, 2006.
6. Craig, John J., "Introduction to Robotics: Mechanics and Control," 3rd Edition, Prentice Hall, 2005.
7. Ogata, Katsushito, "System Dynamics," 4th Edition, Prentice Hall, 2004.

Problems

6.1. Derive the inverse Laplace transform of the following equation:

$$F(s) = \frac{3}{(s^2 + 5s + 4)}$$

6.2. Derive the inverse Laplace transform of the following equation:

$$F(s) = \frac{(s + 6)}{s(s^2 + 5s + 6)}$$

6.3. Derive the inverse Laplace transform of the following equation:

$$F(s) = \frac{1}{(s+1)^2(s+2)}$$

6.4. Derive the inverse Laplace transform of the following equation:

$$F(s) = \frac{10}{(s+4)(s+2)^3}$$

6.5. Simplify the block diagram of Figure P.6.5.

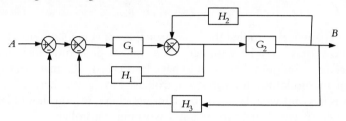

Figure P.6.5

6.6. Simplify the block diagram of Figure P.6.6.

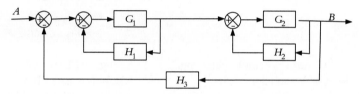

Figure P.6.6

6.7. Simplify the block diagram of Figure P.6.7.

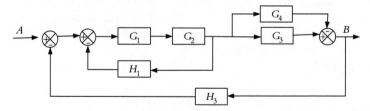

Figure P.6.7

6.8. Write an equation that describes the output of the system of Figure P.6.8.

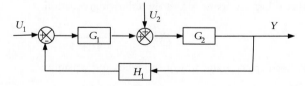

Figure P.6.8

6.9. Write the equations that describe the input–output relationships for Figure P.6.9.

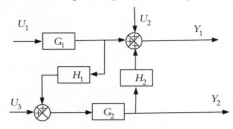

Figure P.6.9

6.10. Sketch the root locus for the following:

$$GH = \frac{K}{s(s+1)(s+3)(s+4)}$$

6.11. Sketch the root locus for the following:

$$GH = \frac{K(s+6)}{s(s+4)}$$

6.12. Sketch the root locus for the following:

$$GH = \frac{K(s+6)}{s(s+10-j10)(s+10+j10)(s+12)}$$

6.13. For the system of Problem 6.10, assume that two of the roots are chosen at $s = -5 \pm 2.55j$. Find the system's gain, damping ratio, and natural frequency. Show that the angle criterion is met. Can you determine from the root locus whether or not the system is stable?

6.14. For the system of Problem 6.10, assume that two of the roots are chosen at $s = -4 \pm 1.24j$. Find the system's gain, damping ratio, and natural frequency.

6.15. For the system of Problem 6.11, assume that the roots are chosen at $s = -3 \pm 1.73j$. Find the system's gain, damping ratio, and natural frequency. Show that the angle criterion is met. Can you determine whether or not the system may become unstable as the gain changes?

6.16. For the system of Problem 6.11, find the roots, the gain, and the steady-state error for a settling time of less that 1 second and overshoot of 4% or less.

6.17. For the following system, find the roots, the gain, and steady-state error for the fastest response and a settling time of less than 2 seconds and an overshoot of less than 4%.

$$GH = \frac{K}{(s+1)(s+3)}$$

6.18. For the system of Problem 6.17, select the locations and the proportional and integral gains to change it to a proportional-plus-integral system with zero steady-state error.

6.19. For the system of Problem 6.17, we would like to improve the settling time to 1 second by adding a zero to the system (proportional-plus-derivative). Find a proper location for the zero and the loop gain.

6.20. For the system of Problem 6.19, add an integrator to the system to make it into a PID system in order to achieve a zero steady-state error. Find the location of an additional zero and proportional, derivative, and integral gains.

CHAPTER 7

....

Actuators and Drive Systems

7.1 Introduction

Actuators are the muscles of robots. If the links and the joints are the skeleton of the robot, the actuators act as muscles that move or rotate the links to change the configuration of robots. The actuator must have enough power to accelerate and decelerate the links and to carry the loads and yet be light, economical, accurate, responsive, reliable, and easy to maintain.

There are many types of actuators available and, undoubtedly, there will be more varieties available in the future. At least, the following types are worth mentioning:

- Electric motors
  - Servomotors
  - Stepper motors
  - Direct drive electric motors
- Hydraulic actuators
- Pneumatic actuators
- Novelty actuators

Electric motors, and specially servomotors, are the most commonly used robotic actuators. Hydraulic systems were very popular for large robots in the past and are still around in many places but are no longer as popular, except for large applications. Pneumatic actuators are used in robots that have 1/2 degree of freedom, on-off type joints, as well as for insertion purposes. Novelty actuators, including direct drive electric motors, electroactive polymer actuators, muscle-wire actuators, and piezoelectric actuators, are mostly used in research and development work or as specialty items for specific purposes but may become more useful in the near future.

In the following section, we will compare the common characteristics of different types of actuators, followed by the individual study of each type.

7.2 Characteristics of Actuating Systems

The following characteristics may be used to compare different actuating systems. In addition to these, depending on the special circumstances in which they will be used, other characteristics may play a role in the design of robots. Examples include underwater systems, where waterproof operation of a system is very important, and space systems, where the lift-off weight and reliability are of absolute importance.

7.2.1 Nominal Characteristics—Weight, Power to Weight Ratio, Operating Pressure, Voltage, and Others

It is important to consider the nominal characteristics of actuators. These include weight, power and power to weight ratio, operating pressure, operating voltage, temperatures, and others. For example, since in many robotic systems the actuators are placed directly at the joints, and therefore, move with them, the weight of the actuator acts as a load on the preceding actuators and must be accelerated and decelerated by them. A heavier actuator downstream requires more torque upstream, resulting in larger power requirements and heavier actuators. Consequently, it is very important to consider the weight and placement of actuators.

Another important characteristic is power to weight ratio. For example, the power to weight ratio of electric systems is average. Stepper motors are generally heavier than servomotors for the same power, and therefore, have a lower power to weight ratio. Hydraulic systems have the highest power to weight ratio. However, it is important to realize that in these systems, the weight is actually composed of two portions. One is the hydraulic actuator, and the other is the hydraulic power unit. The system's power unit consists of a pump that generates the high pressure needed to operate the actuator, a reservoir, filters, electric drive motors to drive the pump, cooling units, valves, and so on. However, the power unit is normally stationary, somewhere away from the robot itself. It does not move with the actuator. The power is brought to the robot via an umbilical tether hose. Consequently, the actual power to weight ratio of the actuator is very high for the moving parts. If the power unit must also move with the robot (e.g., a hydraulic transportation robot), the total power to weight ratio will be much less. Pneumatic actuators deliver the lowest power to weight ratio. The power the hydraulic system delivers is very high due to high operating pressures. This may range from 50 psi to 5000 psi. Pneumatic cylinders normally operate around 100 to 120 psi. The higher pressures in hydraulic systems create higher power concentrations, but also require higher maintenance; if a leak occurs, they can be more dangerous.

Electric motors that operate at a higher voltage have a better power to weight ratio, too. Additionally, as we will see later, for the same power output, as the voltage to an electric motor increases the required current will decrease, reducing the size of required wires. The heat generated in the motor is a second-order function of the current; therefore, as the current decreases, the generated heat decreases and efficiency is increased.

7.2.2 Stiffness versus Compliance

Stiffness is the resistance of a material against deformation. It may be the stiffness of a beam against bending under the load, the resistance of a gas against compression in a cylinder under load, or resistance of wine against compression in a bottle during corking operation. A stiffer system requires a larger load to deform. Conversely, a more compliant system requires a smaller load to deform.

Stiffness is directly related to the modulus of elasticity of the material. The modulus of elasticity of fluids can be around 1×10^6 psi, which is very high. As a result, hydraulic systems are very stiff and noncompliant. Conversely, because air is compressible, pnuematic systems are compliant.

Stiff systems have a more rapid response to changing loads and pressures and are more accurate. Obviously, if a system is compliant, it can easily deform (or compress) under changing load or changing driving force; consequently, it is inaccurate. Similarly, if a small driving force is applied to a hydraulic ram, due to its stiffness, it responds more rapidly and more accurately than a pneumatic system that deforms under the load. Additionally, a stiff system resists deformation under the load, and therefore, holds its position more accurately. Now consider a robot that is used to insert an IC chip into a circuit board. If the system is not stiff enough, it will not be able to push the chip into the board because the actuator may deform under the resistive force. On the other hand, if the part and the holes are not perfectly aligned, a stiff system does not deform enough to prevent damage to the robot or the part, whereas a compliant system gives to prevent damage. So, although stiffness causes a more responsive and more accurate system, it also creates a danger if all things are not always perfect. Consequently, a working balance is needed between these two competing characteristics. We will discuss a particular solution to this problem called Remote-Center Compliance (RCC) device later.

7.2.3 Use of Reduction Gears

Some systems, such as hydraulic actuators and direct-drive electric motors, produce very large forces or torques with short strokes. This means that the actuator may be moved very slightly while delivering its full force or torque. As a result, there is no need to use reduction gear trains to increase the torque it produces and to slow it down to manageable speeds. For this reason, hydraulic actuators can be directly attached to the links, which simplify the design, reduce the weight, cost, and rotating inertia of joints, reduce backlash, increase reliability of the system due to simpler design and fewer parts, and reduce noise. On the other hand, electric motors rotate at high speeds, up to many thousands of revolutions per minute, and must be used in conjunction with reduction gears to increase their torque and decrease their speed, as no one would want a robot arm to be rotating at such high speeds. This, of course, increases the cost, number of parts, backlash, inertia of the rotating body, and so on, as we discussed earlier, but also increases the resolution of the system, as it is possible to rotate the link a very small angle.

Now suppose that, through a set of reduction gears with a reduction ratio of N, a load with inertia I_l, is connected to a motor with inertia I_m (including the inertia of the reduction gears), as shown in Figure 7.1. The torque and speed ratio between the motor and the load are:

$$T_l = NT_m$$
$$\dot{\theta}_l = \frac{1}{N}\dot{\theta}_m \quad \text{and} \quad \ddot{\theta}_l = \frac{1}{N}\ddot{\theta}_m \tag{7.1}$$

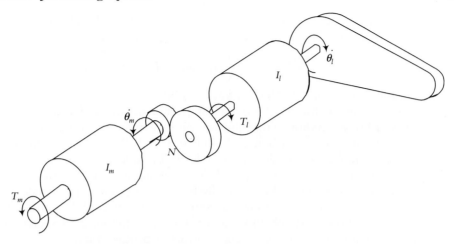

Figure 7.1 Inertia and torque relationship between a motor and a load.

If we write the torque balance equation for the system from free-body-diagrams of Figure 7.2, and substitute from Equation (7.1), we get:

$$T_m - \frac{1}{N} T_l = I_m \ddot{\theta}_m + b_m \dot{\theta}_m \quad \text{and} \quad T_l = I_l \ddot{\theta}_l + b_l \dot{\theta}_l$$

$$T_m = I_m \ddot{\theta}_m + b_m \dot{\theta}_m + \frac{1}{N} \left(I_l \ddot{\theta}_l + b_l \dot{\theta}_l \right) \tag{7.2}$$

$$T_m = I_m \ddot{\theta}_m + b_m \dot{\theta}_m + \frac{1}{N^2} \left(I_l \ddot{\theta}_m + b_l \dot{\theta}_m \right)$$

where b_m and b_l are viscous coefficients of friction for the motor and the load.

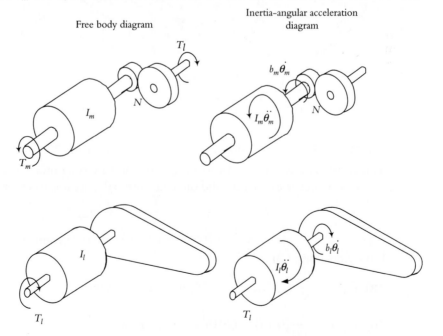

Figure 7.2 Free body diagrams of the motor and the load.

As Equation (7.2) indicates, the effective inertia of the load felt by the motor is conversely proportional to the square of the reduction gear ratio, or:

$$I_{Effective} = \frac{1}{N^2}I_l \quad \text{and} \quad I_{Total} = \frac{1}{N^2}I_l + I_m \tag{7.3}$$

So, the motor will only "feel" a fraction of the actual inertia of the load, which, in the case of a robot, constitutes both the manipulator and the load it carries. Reduction ratios of 20–100 are common in manipulator robots. Therefore, the total inertia applied to the motor from the load may only be $1/400^{th}$ to $1/10,000^{th}$ of the actual inertia, allowing the motor to accelerate and decelerate quickly. In direct drive systems, both electric and hydraulic, the actuators are exposed to the full inertial loads. With high gear ratios, the inertial effects of the load can actually be ignored in the control system of the robot.

   Please note that the opposite is also true, that the effect of the inertia of the motor on the load is also 400–10,000 times larger. To reduce this effect, designers opt to use low-inertia motors with long and slender rotors or pancake motors.

Example 7.1

A motor with rotor inertia of 0.015 Kgm^2 and maximum torque of 8 Nm is connected to a uniformly distributed arm with a concentrated mass at its end, as shown in Figure 7.3. Ignoring the inertia of a pair of reduction gears and viscous friction in the system, calculate the total inertia felt by the motor and the maximum angular acceleration it can develop if the gear ratio is (a) 3 or (b) 30.

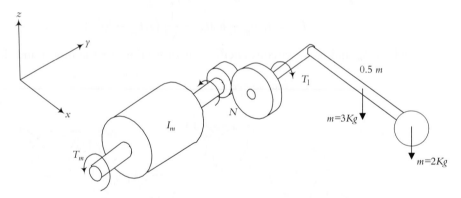

Figure 7.3 Schematic drawing of the system of Example 7.1.

Solution: This is very similar to a robot arm and a servomotor actuator. The total moment of inertia of the arm and the concentrated mass at the center of rotation is:

$$I_l = I_{arm} + I_{mass} = \frac{1}{3}m_{arm}l^2 + m_{mass}l^2$$
$$= \frac{1}{3}(3)(0.5)^2 + (2)(0.5)^2 = 0.75 \text{ Kgm}^2$$

From Equation (7.3):

(a) $I_{Total} = \dfrac{1}{N^2}I_l + I_m = \dfrac{1}{9}(0.75) + 0.015 = 0.098 \text{ Kgm}^2$

(b) $I_{Total} = \dfrac{1}{900}(0.75) + 0.015 = 0.0158 \text{ Kgm}^2$

As you can see, the total inertia with the higher gear reduction ratio is practically the same as the rotor inertia of the motor. The maximum angular accelerations will be:

(a) $\ddot{\theta}_m = \dfrac{T_m}{I_{total}} = \dfrac{8}{0.098} = 82 \, \text{rad/sec}^2$

(b) $\ddot{\theta}_m = \dfrac{T_m}{I_{total}} = \dfrac{8}{0.0158} = 506 \, \text{rad/sec}^2$

The no-load maximum angular acceleration of the motor would be about 530 rad/sec^2. ■

7.3 Comparison of Actuating Systems

Table 7.1 is a summary of actuator characteristics. We will refer to, and discuss, these characteristics throughout this chapter.

Table 7.1 *Summary of Actuator Characteristics.*

Hydraulic	Electric	Pneumatic
+ Good for large robots and heavy payload	+ Good for all sizes of robots	+ Many components are usually off-the-shelf
+ Highest power/weight ratio	+ Better control, good for high precision robots	+ Reliable components
+ Stiff system, high accuracy, better response	+ Higher compliance than hydraulics	+ No leaks or sparks
+ No reduction gear needed	+ Reduction gears reduce inertia on the motor	+ Inexpensive and simple
+ Can work in wide range of speeds without difficulty	+ Does not leak, good for clean room	+ Low pressure compared to hydraulics
+ Can be left in position without any damage	+ Reliable, low maintenance	+ Good for on-off applications and for pick and place
− May leak; not fit for clean room applications	+ Can be spark-free; good for explosive environments	+ Compliant systems
− Requires pump, reservoir, motor, hoses, and so on	− Low stiffness	− Noisy
− Can be expensive and noisy; requires maintenance	− Needs reduction gears, increased backlash, cost, weight, and so on	− Require pressurized air, filter, and so on
− Viscosity of oil changes with temperature	− Motor needs braking device when not powered; otherwise, the arm will fall	− Difficult to control their linear position
− Very susceptible to dirt and other foreign material in oil		− Deform under load constantly
− Low compliance		− Very low stiffness; inaccurate response
− High torque, high pressure, large inertia on the actuator		− Lowest power to weight ratio

7.4 Hydraulic Actuators

Hydraulic systems and actuators offer high power to weight ratio, large forces at low speeds, both linear and rotary actuation, compatibility with microprocessor and electronic controls, and tolerance of extreme hazardous environments. They can hold a load without need for a brake, generate less heat at the actuator, and apply a torque without the need for gearing. Many large robots of the past decades, mostly used in automobile production, were Cincinnati Milacron™ T3 hydraulic robots or other brands with similar characteristics. The T3 robot offered a payload of over 220 lb at 7 feet high, an impressive value. However, due to the leakage problem that is almost inevitable in hydraulic systems, and due to their power unit weight and cost, they are no longer common. Today, most robots are electric; however, still many robots in the industry have hydraulic actuators. Additionally, for special applications such as very large robots in civil and military service, hydraulic actuators may be the appropriate choice.

One important difference between hydraulic actuators and electric motors is that a hydraulic pump (not the actuator) may be sized for average load, whereas electric actuators must be sized for maximum load. This is because a hydraulic system uses an accumulator that stores the constant energy of the pump and can deliver larger loads when necessary. Therefore, if there are breaks in the motion, the accumulator stores energy needed for maximum loads. Another important consideration is that in general, an electric actuator must be located at or near a joint, adding to the mass and inertia of the robot. However, in hydraulic systems, only the actuator, the control valves, and the accumulators are near the joints. The hydraulic power unit may be located remotely, reducing the mass and inertia, especially when many joints are present.

The total force that a linear cylinder can deliver can be tremendously large for its size. A hydraulic cylinder can deliver a force of $F = p \times A$ lb, where A is the effective area of the piston or ram, and p is the working pressure. For example, for a 1000 psi pressure, every square inch of the cylinder develops 1000 lb of force. In rotary cylinders, the same principle is true, except that the output is a torque where:

$$dA = t \cdot dr$$

$$T = \int_{r_1}^{r_2} p \cdot r \cdot dA = \int_{r_1}^{r_2} p \cdot r \cdot t \cdot dr = pt \int_{r_1}^{r_2} r \cdot dr = \frac{1}{2} pt \left(r_2^2 - r_1^2 \right) \qquad (7.4)$$

where p is the fluid pressure, t is the thickness or width of the rotary cylinder, and r_1 and r_2 are the inner and outer radii of the rotary cylinder, as shown in Figure 7.4.

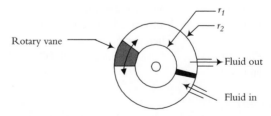

Figure 7.4   A rotary hydraulic actuator. This actuator can be directly attached to a revolute joint without any need for gear reduction.

The flow rate and volume of fluid needed in a hydraulic system are:

$$d(Vol) = \frac{\pi d^2}{4} dx \tag{7.5}$$

$$Q = \frac{d(Vol)}{dt} = \frac{\pi d^2}{4} \frac{dx}{dt} = \frac{\pi d^2}{4} \dot{x} \tag{7.6}$$

where dx is the desired displacement, and $\dot{x}$ is the desired velocity of the piston. By controlling the volume of the fluid going into the cylinder, the total displacement can be controlled. By controlling the rate at which the fluid is delivered to the cylinder, the velocity can be controlled. This is done through a servovalve that controls both the volume of the fluid as well its rate. We will discuss the servovalves later.

A hydraulic system generally consists of the following components or subsystems:

- Hydraulic linear or rotary actuators that provide the force or torque needed to move the joints and are controlled by the servovalves or manual valves.
- Hydraulic pump that provides high-pressure fluid to the system.
- Electric motor (or in cases such as in a mobile unit, an engine) that operates the hydraulic pump.
- Cooling system, which rids the system of the generated heat. In some systems, in addition to cooling fans, radiators and cooled air are used.
- Reservoir, which keeps the fluid supply available to the system. Since the pump is usually constantly supplying the system, whether the system is using it or not, the extra pressurized fluid and the returned fluid from the cylinders flow back into the reservoir.
- Accumulators that are used to store some extra energy for maximum loads, especially when the pump is designed for average use.
- Servovalves that control the amount and the rate of fluid to the cylinders. The servovalve is generally driven by a hydraulic servomotor.
- Check valves that are used for safety and controlling maximum pressure.
- Holding valves that are used to prevent the actuator from moving when the system is turned off or power is lost. Holding valves fulfill the need for a brake to prevent accidental motions due to power loss or when the system is shut down.
- Connecting hoses that are used to transport the pressurized fluid to the cylinders and back to the reservoir.
- Filtering system that maintains the quality and purity of the fluid. Due to its nature, moisture in the fluid damages hydraulic actuators and must be separated from the fluid.
- Sensors that are used as feedback to control the motion of the actuators. They include position, velocity, magnetic, touch, and other sensors. Additionally, other sensors are used for safety. For example, in case a servovalve is missing, the pump should not turn on. Otherwise, dangerous high-pressure fluid will squirt out of the ports.

Figure 7.5 is a schematic drawing of a typical hydraulic system. Refer to Chapter 6 for more about hydraulic controllers.

Figure 7.6 is a schematic drawing of a position control pilot valve for a hydraulic cylinder (also called a spool valve). This is a balanced valve, which means that the pressures on the two sides of the spool are equal. Therefore, it takes very little force to move the spool (due to friction), even though it may be under very high pressures. A

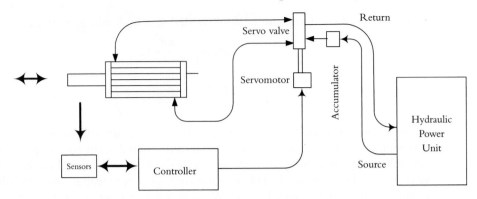

Figure 7.5   Schematic of a hydraulic system and its components.

spool valve may be operated manually or by a servomotor, called a servovalve. The servovalve and the cylinder together form a hydraulic actuator. As the spool moves up or down, it opens the supply and return ports through which, the fluid travels to the cylinder or is returned to the reservoir. Depending on the size of the opening of the port as the spool moves, the supply fluid flow rate is controlled, and so is the velocity of the cylinder. Depending on the length of time that the port is kept open, the total amount of the fluid to the cylinder, and consequently its total travel, is controlled. This can be written as:

$$q = Cx \qquad (7.7)$$

$$\text{and} \quad q(dt) = d(vol) = A(dy) \qquad (7.8)$$

where q is the flow rate, C is a constant, x is the spool's displacement, A is the area of the piston, and y is piston's displacement. Combining Equations (7.7) and (7.8), and

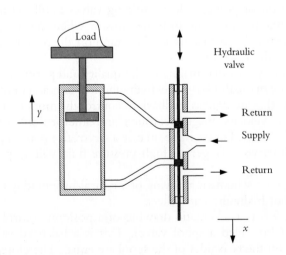

Figure 7.6   Schematic of a spool valve in neutral position.

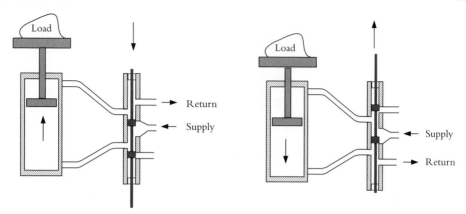

Figure 7.7 Schematic of a spool valve in open position. Depending on which ports are opened, the direction of motion of the piston will change.

designating d/dt as D, we get:

$$Cx(dt) = A(dy)$$

$$\text{and} \qquad y = \frac{C}{AD}x \qquad\qquad (7.9)$$

which shows that the hydraulic servomotor is an integrator. (Refer to Chapter 6 for more detail.)

The command to the servomotor controlling the spool valve comes from the controller. The controller sets the current to the servomotor as well as the duration the current is applied, which in turn, controls the position of the spool. Therefore, for a robot, when the controller has calculated how much and how fast a joint must move, it sets the current and its duration to the servomotor, which in turn, controls the position and rate of movement of the spool valve, controlling the flow of the fluid and its rate to the cylinder and moving the joint. The sensors provide feedback to the controller for accurate and continued control. Figure 7.7 shows the flow of the fluid as the spool valve moves up or down. As you can see, a simple motion of the spool controls the motion of the cylinder.

To provide feedback to the servovalve (otherwise it will not be a servovalve, but a manual spool valve), either electronic or mechanical feedback can be added (also refer to Figure 6.30). Figure 7.8 shows a simple mechanical feedback loop. A similar design is used in a two-stage spool valve to provide feedback. As you can see in Figure 7.8, a simple lever is added between the output and the input to provide an error signal to the system. As the desired position for the load is set by the setpoint lever (moved up, which will move down the load), the spool valve is opened, which will operate the cylinder. However, in reality, this lever arm is providing an error signal to the system, and therefore, the system responds by moving the cylinder. The error signal is integrated by the integrator (in this case, the cylinder), and as the error approaches zero, the signal to the system goes to zero. As the cylinder starts to move in the direction of the desired position (in this case, downward), the error signal becomes smaller which in turn reduces the output signal by closing the spool valve. By the time the load is at the desired position, the spool valve is closed and the load stops. Figure 7.9 is the schematic of the block diagram

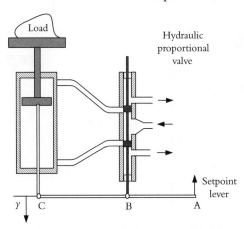

Figure 7.8   Schematic of a simple control device with proportional feedback.

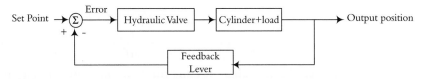

Figure 7.9   Block diagram of the hydraulic system with proportional feedback control scheme.

for this feedback loop. In reality, the feedback mechanism shown here is integrated into the servovalve.

There are many intricate and small passageways inside a hydraulic valve, especially in servovalves with feedback control integrated into the valve. In fact, this is why these systems are so susceptible to dirt or viscosity changes due to temperature. The smallest of foreign particles can affect the servovalve by restricting its passageways or ports. Similarly, viscosity changes can change the response of the valve as the fluid becomes less or more viscous. To visualize how these valves are constructed, imagine that you slice the valve into thin layers. Each layer will have corresponding portions of the passageways and ports and openings in it. We may manufacture the slices, which is easy to do by different methods such as a press, assemble the layers, and then either diffuse the layers together into one piece (under proper pressure and temperature), or connect them into one piece by through-bolts. This method is extensively used in industry to manufacture valves and many other products such as camera bodies. Figure 7.10 shows a simple example of this technique. When the thin slices are put together, a three-dimensional object results. Many rapid prototyping machines use the same technique to create 3-D prototypes of products.

Other designs have also been used for hydraulic actuation. For example, IBM 7565 robotic manufacturing system consisted of a gantry, 6-axis, hydraulic robot that was actuated at each of its three linear joints by a linear hydraulic motor. Each linear motor consisted of a set of four small hydraulic cylinders, which sequentially moved against a

Figure 7.10 An object can be sliced into thin layers. When the layers are cut out individually and put together later, they will recreate the object. This method is used for rapid prototyping and for manufacturing complicated parts with intricate internal openings and passageways. (b) In this simple case, the fluid enters in the right, travels up one layer, and exits on lower left.

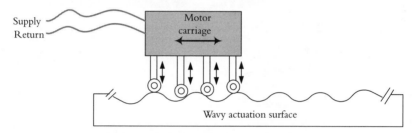

Figure 7.11 IBM 7565 linear hydraulic motor. As the four cylinders moved against the wavy actuation surface, the motor carriage moved sideways.

wavy actuation surface, as shown in Figure 7.11. The four cylinders were forced into a position that corresponded to the desired position against the actuation surface. This in turn forced the motor carriage to move sideways. The four cylinders were controlled by a single servovalve. The advantage of this system was that by adding simple sections of the waved surface, the actuation could be made as long as desired. Figure 7.12 shows this actuator on the IBM 7565 robot.

Another hydraulic actuator is modeled after biological muscles. A muscle actuates a bone by contracting and, consequently, becoming shorter. In the hydraulic version, a similar design is used, where an oval shaped bladder is placed inside a shear sheath, as shown in Figure 7.13. When the pressure in the bladder increases, it becomes more spherical, causing the shear sheath to bulge out and shorten, just like a muscle. This design is promising, but due to nonlinearities inherent in the system and technical difficulties, it has not become practical yet. However, since it looks like a biological muscle, it can be very useful in humanoid robots.

Refer to our previous discussion in Chapter 6 regarding PID or other schemes for controlling a servovalve. All applicable theory and practice discussed there applies to servovalves, too.

Figure 7.12   The linear hydraulic motor of IBM 7565 robot.

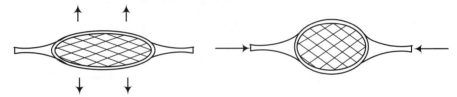

Figure 7.13   Schematic of a muscle type hydraulic actuator.

7.5 Pneumatic Devices

Pneumatic devices are principally very similar to hydraulic systems. A source of pressurized air is used to power and drive linear or rotary cylinders, controlled by manual or electrically controlled solenoid valves. Since the source of pressurized air is separate from the moving actuators, these systems have lower inertial loads. However, since pneumatic devices operate at a much lower air pressure, usually up to 100–120 psi max, their power to weight ratio is much lower than hydraulic systems.

The major problem with pneumatic devices is that air is compressible and, as a result, its volume changes under load. Consequently, pneumatic actuators are usually only used for insertion purposes, where the actuator is all the way forward or all the way backward, or they are used with 1/2-DOF joints that are fully on or fully off. Otherwise, controlling the exact position of pneumatic cylinders is very difficult.

One way to control the displacement of the pneumatic cylinders is called differential dithering. In this system, the exact location of the piston is sensed by a feedback sensor such as a linear encoder or potentiometer. This information is used in a controller that controls the air pressure on the two sides of the cylinder through a servovalve to control the exact position.[1]

Pneumatic actuators are simple, rugged, and safe. Even if they leak, the air is not a contaminant. Most components are off the shelf, and therefore, easy to use and inexpensive. They are mostly used either as on/off devices or as accessories in a robotic cell in conjunction with robots for material handling and similar purposes.

7.6 Electric Motors

When a wire carrying a current is placed within a magnetic field, it will experience a force normal to the plane formed by the magnetic field and the current. Therefore, each side of the wire (coil) experiences a force as shown in Figure 7.14. If the wire is attached to a center of rotation, the resulting torque will cause it to rotate about the center of rotation. Changing the direction of the magnetic field or the current will cause a change in the direction of the force and the wire will continue to rotate about the center of rotation. As long as this change continues, the coil will continue to rotate. In practice, in order to accomplish this change in the current, either a set of commutators and brushes or slip rings are used for DC motors, the current is electronically switched for DC brushless motors, or AC current is used for AC motors, although in this case the permanent magnets and the coil are switched.[2] This is the basic principle behind all electric motors. Similarly, if a conductor is moved within a magnetic field crossing the flux, a current is induced in the conductor like a generator. This is called back-electromotive-force, or back-emf.

There are many types of motors, including AC induction motors, AC synchronous motors, DC brushed motors, DC brushless motors, stepper motors, direct drive DC motors, switched reluctance motors, AC/DC universal motors, and other varieties such as three-phase AC motors, disk motors, and others. Although we will discuss a number of issues about electric motors, the assumption is that you have already studied about different motors and how they operate. So, the discussion here will be at a minimum and only about the subjects directly related to robotic actuation. Please refer to other sources for additional detail about motors and their drive circuitry.

Except for stepper motors, all other types of motors can be used as a servomotor, which will be discussed later. In each case, the torque or power output of the motor is a function of the strength of the magnetic fields and the current in the windings as well as the length of the conductors of the coils. Some motors have permanent magnets (PM). These motors generate less heat, since the field is always present and no current is needed to build it.

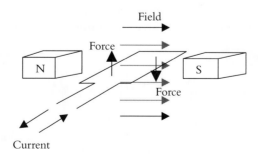

Figure 7.14 A wire carrying a current, placed within a magnetic field, will experience a force in a direction normal to a plane formed by the current and the field.

Others have a soft iron core and windings, where an electric current creates the magnetic field. In this case, more heat is generated, but when needed, the magnetic field can be varied by changing the current, whereas in permanent magnet motors, the field is constant. Additionally, under certain conditions, it is possible that the permanent magnet may get damaged and lose its field strength, in which case, the motor becomes useless.

7.6.1 Fundamental Differences between AC and DC-Type Motors

There are some fundamental differences between AC- and DC-type motors that dictate their power range, control, and applications. In the following sections, we will discuss these differences and their effects.

The first major difference between these motors is whether or not their speed can be controlled. This will be discussed later with servomotors, but suffice it to say that the speed of rotation of a DC motor can be controlled by changing the current to the windings; as the current increases or decreases, for the same load on the rotor, its speed also increases or decreases. However, the speed of an AC motor, among other things, is a function of the frequency of the supplied AC power. Since the frequency of the AC power is constant, the speed of an AC motor is generally constant.

The second major factor in the difference between brushed and brushless motors is the life of the brushes and commutators as well as the physical limitation of mechanical switching by brushes. Brushless DC motors, AC motors, and stepper motors are all brushless; therefore, they are sturdy and generally have a long life (only limited by the life of rotor bearings). Since brushes wear out, the life of brushed motors is limited and they require more maintenance.

The third important issue in the design and operation of all motors is heat dissipation. Like many other devices, the generated heat in motors eventually becomes the deciding factor about its size and power. The heat is generated primarily from the resistance of the wiring to electric current (load related), but includes heat due to iron losses including eddy current losses and hysteresis losses, friction losses, brush losses, and short-out circuit losses (speed related). As we will see in more detail later, the generated heat is a function of current ($W = i^2 R$). Thicker wires generate less heat but are more expensive, heavier (more inertia), and require more space. All motors generate this heat. However, what is important is the rate at which the heat is dissipated, and therefore, the path that the heat must take to leave the motor. The heat path is more important than the amount of heat, since if the dissipation is faster, more generated heat can be dissipated before damage occurs.

Figure 7.15 shows the heat dissipation path to the environment for an AC type motor and a DC-type motor. In DC-type motors, the rotor contains the winding and carries the

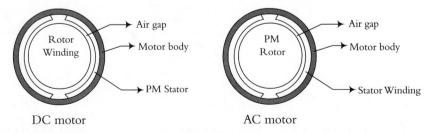

Figure 7.15 Heat dissipation path of motors.

current and, consequently, the heat is generated in the rotor. This heat must travel from the rotor, through the air gap, through the permanent magnets, through the motor's body, and be dissipated into the environment (it may also go through the shaft to the bearings and out). As you know, air is a good insulator. Therefore, the total heat transfer coefficient for the DC motor is relatively low. On the other hand, in an AC-type motor, the rotor is a permanent magnet and the winding is in the stator. The generated heat in the stator is dissipated to the air by conduction through the motor's body. As a result, the total heat transfer coefficient is relatively high, especially because no air gap exists. As a result, AC-type motors can be exposed to relatively higher currents without damage and, as a result, they are generally more powerful for the same size. Stepper motors and brushless DC motors, although not AC motors, have similar construction; the rotor is a permanent magnet and the stator contains the windings. Consequently, they also have better heat dissipation capability.

Overheating may be the most common cause of failure for electric motors. It can lead to:

- Failure of the winding isolation, causing shorts or burnouts,
- Failure of the bearings, resulting in jamming the rotor shaft,
- Degradation of the magnets, permanently reducing the motor's torque.

Heat development in a motor is not linear and is a function of the following parameters:[3]

- The basic heat generated in the winding.
- As the winding heats up, the resistance of the wiring increases, further increasing heat generation.
- The strength of the magnetic flux is negatively affected by heat, reducing the torque and requiring additional current.
- Heat dissipation increases as temperature rises.

Therefore, it is possible that the motor may achieve equilibrium at an acceptable temperature. However, there is also a danger that the temperature may continue to rise to failure. The heat generated in a motor is:

$$P_{electric} = i^2 R = \frac{T^2}{K_t^2} R_t \tag{7.10}$$

where i is the current in the winding, R and R_t are the nominal electrical resistance of the winding and the resistance at temperature t, T is the torque, K_t is the *torque constant* at temperature t and is a function of the magnetic field, number of turns of the windings, the effective area of the air gap, the radius of the rotor, and material properties. The variations in the winding electrical resistance can be described by:

$$R_t = R_{ref} \left[1 + \left(t_{winding} - t_{ref} \right) \cdot \alpha \right] \tag{7.11}$$

where R_t and R_{ref} are the electrical resistance at the temperature of interest $t_{winding}$ and room temperature t_{ref}, and α is a material constant. $\alpha_{copper} = 0.00393 \, (\text{K}^{-1})$. The torque constant, representing the magnetic field of the motor, varies with temperature as:

$$K_t = K_{ref} \left[1 + \left(t_{magnet} - t_{ref} \right) \cdot \beta \right] \tag{7.12}$$

Here, K_t and K_{ref} represent torque constants at the temperature of interest t_{magnet} and the reference temperature ($20°C$), and β represents the decay of magnetic flux-density, a material dependent property. Since this decay is negative, magnetic flux decreases as temperature increases. A simplified model for the final temperature of the motor can be written as:

$$t_{motor} = \left(P_{electric} + P_{friction}\right)R_{thermal} + t_{ref} \qquad (7.13)$$

where $R_{thermal}$ is the thermal resistance between the motor and the ambient. Through these equations, we can estimate the elevated temperature of the motor and find out whether or not it is at equilibrium.

Example 7.2

A motor develops a torque of 1.2 Nm at its nominal speed, with reference electrical resistance of 8Ω. The reference torque constant for the motor is 0.5 Nm/A and the thermal resistance for the motor is 1.05 K/W. Ambient temperature is assumed to be $20°C$. For the material used in the motor, $\beta = -0.002$ (K). Friction is ignored.

(a) Find whether or not the motor temperature will stabilize, and if so, find the converged value.
(b) Repeat the same assuming that due to improvements in heat dissipation (for example, by using a fan), thermal resistance is reduced to 1.02 K/W.

Solution:

(a) Substituting these values in Equations (7.10) through (7.13), we get the following power based on initial resistance:

$$P_{electric} = \frac{T^2}{K_t^2}R_t = \frac{1.2^2}{0.5^2}(8) = 46 \text{ W}$$

and $\quad t_{motor} = \left(P_{electric} + P_{friction}\right)R_{thermal} + t_{ref} = (46 + 0)(1.05) + 20 = 68°C$

At this temperature, both the resistance and the torque constant change and we get:

$$R_t = R_{ref}\left[1 + \left(t_{winding} - t_{ref}\right)\cdot\alpha\right] = 8[1 + (68-20)(0.00393)] = 9.5\Omega$$

$$K_t = K_{ref}\left[1 + \left(t_{magnet} - t_{ref}\right)\cdot\beta\right] = 0.5[1 + (68-20)(-0.002)] = 0.452 \text{ Nm/A}$$

Resubstituting these values into Equations (7.10) and (7.13), we get a new temperature of:

$$P_{electric} = \frac{1.2^2}{0.452^2}(9.5) = 67 \text{ W}$$

$$t_{motor} = (67 + 0)(1.05) + 20 = 90°C$$

The next iterations of the same process show that the temperature continues to rise. At iteration #20, the temperature is 180 degrees and rising uncontrollably, indicating that the motor will overheat.

(b) Repeating the same with the new thermal resistance will converge after about 30 iterations to near $130°C$. Therefore, it is clear that the proper dissipation of heat is a major issue in motors. ■

It should be clear from the preceding discussion that AC-type operation is preferable when considering heat generation and brushless operation, whereas DC-type operation is preferred for speed control or when only DC power is available. A combination of both characteristics would be ideal. As we will see in more detail later, brushless DC motors and stepper motors possess these characteristics, and therefore, are more robust and last longer.

In the following sections, we will briefly discuss DC motors, servomotors, brushless DC motors, and stepper motors.

7.6.2 DC Motors

DC motors are very common in industry, and have been used for a long time. As a result, they are reliable, sturdy, and relatively powerful.

In DC motors, the stator is a set of fixed permanent magnets, creating a fixed magnetic field, while the rotor carries a current. Through brushes and commutators, the direction of the current is changed continuously, causing the rotor to rotate continuously. Conversely, if the rotor is rotated within the magnetic field, a DC current will develop, and the motor will act as a generator (the output is DC, but not constant). Figure 7.16 shows the construction of a DC motor, its commutators and brushes, and the permanent magnet stator.

If permanent magnets are used to generate the magnetic field, the output torque T is proportional to the magnetic flux ϕ and the current in the rotor windings i. Then:

$$T = \alpha \cdot \phi \cdot i = k_t \cdot i \tag{7.14}$$

where k_t is called the *torque constant*. Since in permanent magnets, the flux is constant, the output torque becomes a function of i, and to control the output torque, i (or corresponding voltage) must be changed. If, instead of permanent magnets, soft iron cores with windings are used for the stator as well, then the output torque is a function of currents in both the rotor and the stator windings, as:

$$T = k_t k_f i_{rotor} i_{stator} \tag{7.15}$$

Figure 7.16 The stator, rotor, commutators, and brushes of a DC motor.

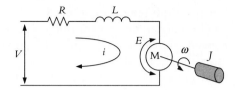

Figure 7.17   Schematic diagram showing a DC motor armature circuit.

where both k_t and k_f are constants. Assuming no power loss during this energy conversion, the total input must be equal to the output. Therefore:

$$P = T \cdot \omega = E \cdot i \rightarrow E = \frac{T \cdot \omega}{i} = k_t \cdot \omega \tag{7.16}$$

which indicates that voltage E is proportional to the angular velocity of the motor ω. This voltage is called the back-emf voltage and is generated across the motor because the windings cross the magnetic field. Therefore, as the rotor speed increases, so does the back-emf voltage. Since in reality the rotor windings have both resistance and inductance, we can write (see Figure 7.17):

$$V = Ri + L\frac{di}{dt} + E \tag{7.17}$$

Substituting Equations (7.14) and (7.16) into (7.17) and rearranging it, we get:

$$\frac{k_t}{R} V = T + \frac{L}{R}\frac{dT}{dt} + \frac{k_t^2}{R}\omega \tag{7.18}$$

$\frac{L}{R}$ is called *motor reactance* and is usually small, and therefore, the differential term may be ignored for simplicity of analysis at this time. Consequently, we get:

$$T = \frac{k_t}{R} V - \frac{k_t^2}{R}\omega \tag{7.19}$$

Equation (7.19) shows that as input voltage V increases, the output torque of the motor increases as well. It also shows that as the angular velocity increases, the torque decreases due to back-emf. Therefore, when $\omega = 0$, torque is the greatest (stalled motor), and when ω is at its nominal maximum value, $T = 0$ and the motor does not produce any useful torque (see Figure 7.18).

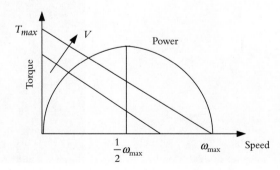

Figure 7.18   The output torque and power of a DC motor versus its angular velocity.

Figure 7.19 Eliminating the soft iron from a rotor makes it lightweight with low inertia.

Referring to Equation (7.16), the output power of the motor is zero when either the angular velocity of the motor (at stall condition) or the torque (at maximum speed) are zero (see Figure 7.18).

Through the use of powerful magnets made of rare-earth materials and alloys such as neodymium, the performance of motors has improved significantly. As a result, the power to weight ratio of motors is much better than before, and they have replaced almost all other types of actuators. However, these motors are also more expensive.

To overcome the problem of high inertia and the large size of many electric motors, a disk or hollow rotor can be used. In these motors, the iron core of the rotor winding is eliminated to reduce its weight and inertia; as a result, these motors are capable of producing very large accelerations (zero to 2000 rpm in one ms[5]), and they respond very favorably to changing currents for control purposes. Figure 7.19 shows a motor in which the hollow rotor has no soft iron. In a disk motor, the rotor is a flat, thin plate with windings pressed (etched) into it, as if we would flatten a rotor into a disk. The permanent magnets are generally small, short, cylindrical magnets that are placed on the two sides of the disk. As a result, disk motors are very thin and are used in many applications where both space and acceleration requirements are important (Figure 7.20).

7.6.3 AC Motors

Electric AC motors are similar to DC motors, except that the rotor is permanent magnet, the stator houses the windings, and all commutators and brushes are eliminated. This is possible because the changing flux is provided by the AC current, and not by

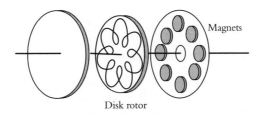

Magnets

Disk rotor

Figure 7.20 Schematic of a disk motor. The rotor has no iron core and consequently, has very little inertia. As a result, it can accelerate and decelerate very quickly.

Figure 7.21   A center-tapped AC motor winding.

commutation. As the flux generated by the AC current changes, the rotor follows it and rotates. As a result, AC motors have fixed nominal speeds, a function of the number of poles on their rotor, and the line frequency (e.g., 60 Hz). Since AC motors can dissipate heat more favorably than DC motors, they can be more powerful. The same principles of back-emf (see section 7.6.6) hold for AC motors as well.

Nowadays, with the available power electronics, it is possible to use methodologies such as *flux vector control* to generate a sinusoidal current with desired frequency and amplitude that enable us to control the speed and torque of an AC motor. This is accomplished by converting the AC line voltage to a DC form and creating an approximate pulsed sinusoidal DC current at the desired frequency and magnitude to drive the motor. Stepper motors (driven in microstepping mode) and brushless DC motors are similar attempts at making DC motors operate similar to an AC motor.

There are also reversible AC motors available. In this case, the motor winding is center-tapped; therefore, as the current flows in each half of the winding, the direction of the flux, and consequently the rotation of the rotor, changes (Figure 7.21). However, since the current flows in only half of the winding, the generated torque is half as much.

7.6.4 Brushless DC Motors

Brushless DC motors are a hybrid between AC motors and DC motors. Although not exactly the same, their construction is very similar to an AC motor. The major difference is that brushless DC motors are operated with an electronically switched DC waveform that is similar to an AC current (either sine wave or trapezoidal waveform) but is not necessarily at 60 Hz. As a result, unlike AC motors, DC brushless motors can be operated at any speeds, including very low speeds. To operate, a feedback signal is necessary to determine when to switch the direction of the current. In practice, a resolver, an optical encoder, or Hall-effect sensors send signals to a controller, which switches the current to the rotor. For smooth operation and almost constant torque, the rotor usually has three phases in it.[2,4] Therefore, three currents, with 120° phase shift, are fed into the rotor. Brushless DC motors are operated by a controller circuit. They will not operate if you connect them directly to a DC power source. Figure 7.22(a) shows a radial-type brushless DC motor with the rotor and stator side by side. Figure 7.22(b) is the stator of an axial-type brushless motor.

7.6.5 Direct Drive Electric Motors

Direct drive electric motors are very similar in construction to brushless DC motors or stepper motors. The major difference is that they are designed to deliver a very large torque at very low speeds, with very high resolution. These motors are intended to be used directly with a joint without any gear reduction. Direct drive motors can be very expensive and very heavy, but have impressive characteristics. In one model by

(a)   (b)

Figure 7.22 Brushless DC motors.

NSK^(TM) Corporation, a 40 Kg motor produces a continuous torque of 150 Nm at 3 rps maximum, with a resolution of 30 arc-sec.

A voice coil may be used as a direct drive actuator for low torque but high resolution applications. Voice coils are commonly used in disk drives and deliver impressive characteristics in reliability and resolution. Figure 7.23 shows a disk drive voice coil actuator.

7.6.6 Servomotors

An important issue in all electric motors is the back electromotive force, or back-emf. As discussed earlier, a conductor carrying a current within a magnetic field will experience a force, which causes it to move. Similarly, if a conductor moves within a magnetic field such that it crosses the field lines, a current will be induced into the conductor. This is the basic principle of electric power generation. However, it also means that when the wires of the windings in a motor are rotating within the magnetic field of the magnets, a current (or voltage) will be induced in them in the opposite direction of the input current, called

Figure 7.23 Disk drive voice coil actuator.

back-emf, that tends to reduce the effective current of the motor. The faster the motor rotates, the larger the back-emf. Back-emf current is usually expressed as a function of rotor speed as shown in Equation (7.16) and repeated here:

$$E = k_t \cdot \omega \tag{7.16}$$

where k_t is typically given in volts per 1000 rpm. As the motor approaches its nominal no-load speed, the back-emf is large enough, such that the motor speed will stabilize at the nominal no-load speed with its corresponding effective current. However, at this nominal speed, the output torque of the motor is essentially zero. The motor's velocity is governed by Equation (7.19), repeated here:

$$T = \frac{k_t}{R} V - \frac{k_t^2}{R} \omega \tag{7.19}$$

As shown in Figure 7.18, at maximum ω, the output torque is zero. For constant input voltage V, if a load is applied to the motor, it will slow down, resulting in smaller back-emf, larger effective current, and consequently, a positive net torque. The larger the load, the slower the motor will rotate in order to develop a larger torque. If the load becomes increasingly larger, there comes a time when the motor stalls, there is no back-emf, the effective current is at its maximum, and the torque is at its maximum. Unfortunately, in each case, when the back-emf is smaller, although the output torque is larger, since the net current is larger, so is the generated heat. Under stall or near-stall conditions, the generated heat may be large enough to damage the motor.

To increase the motor torque while maintaining a desired speed, the input voltage V (or current) to the rotor, stator, or both if soft iron magnets are used, must be increased. In such a case, although the motor rotates at the same speed, and although the back-emf is still the same, the larger voltage will increase the net effective current, and consequently, the torque. By varying the voltage (or corresponding current), the speed-torque balance can be maintained as desired. This system is called a servomotor.

A servomotor is a DC, AC, brushless, or even stepper motor with feedback that can be controlled to move at a desired speed and torque for a desired angle of rotation. To do this, a feedback device sends signals to the servocontroller circuit reporting its angular position and velocity. If, as a result of higher loads, the velocity is lower than the desired value, the voltage (or current) is increased until the speed is equal to the desired value. If the speed signal shows that the velocity is larger than desired, the voltage is reduced accordingly.[6] If position feedback is used as well, the feedback signal is used to shut off the motor as the rotor approaches the desired angular position.

We will discuss sensors later. Here, suffice it to say that many different types of sensors may be used for this purpose, including encoders, resolvers, potentiometers, and tachometers. If a position sensor such as a potentiometer or encoder is used, its signal can be differentiated to produce a velocity signal. Figure 7.24 is a schematic of a simple control block diagram for a servomotor. Refer to Chapter 6 for more detail.

7.6.7 Stepper Motors

Stepper motors are versatile, robust, simple motors that can be used in many applications. In most applications, stepper motors are used without feedback. This is because, unless a

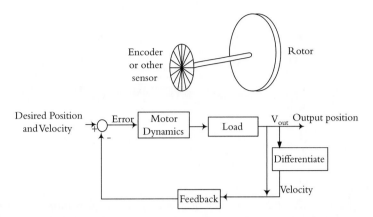

Figure 7.24 Schematic of a servomotor controller. A sensor sends velocity and position signals to the controller, which controls the output velocity and position of the servomotor.

step is missed, the motor steps a known angle each time it is moved. Therefore, its angular position is always known and no feedback is necessary. Stepper motors come in many different forms and principles of operation. Each type has certain characteristics unique to it, yielding it an appropriate choice for particular applications. Most stepper motors can be used in different modes by wiring them differently.

Unlike regular DC or AC motors (but like brushless DC motors) if you connect a stepper motor to power, it will not rotate. Steppers rotate only when the magnetic field is rotated through its different windings. In fact, their maximum torque is developed when they do not turn. Even when not powered, steppers have a residual torque called *detent* or *residual torque*. An external torque must be applied to turn a stepper motor even when not powered. Stepper motors need a microprocessor or driver/controller (indexer) circuit for rotation. You may either create your own driver, or you may purchase a device called an indexer that drives the stepper motor for you. Similar to servomotors, which need feedback circuitry, stepper motors need drive circuitry. So, in each application, the designer must decide which type of motor is more appropriate. For industrial robotic actuation, except in small tabletop robots, stepper motors are hardly ever used. However, stepper motors are used extensively in nonindustrial robots and robotic devices as well as in machines used in conjunction with robots, from material handling machines to peripheral devices and from automatic manufacturing to control devices.

Structure of Stepper Motors   Generally, stepper motors have soft iron or permanent magnet rotors, while their stators house multiple windings. Based on the discussion in section 7.6.1, since the heat generated in the coils can more easily dissipate through the motor's body, stepper motors are less susceptible to heat damage, and since there are no brushes or commutators, they have long life.

Rotors of stepper motors are not all alike. We will discuss two types of rotors later. In each case though, the rotor follows a moving magnetic field generated by the coils. As a result, somewhat similar to both AC motors and brushless DC motors, a rotor follows a moving flux under the control of a controller or driver. In the next sections, we will study how stepper motors operate.

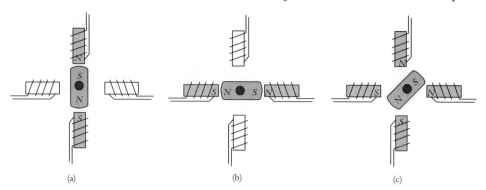

Figure 7.25   Basic principle of operation of a stepper motor. As the coils in the stator are turned on and off, the rotor will rotate to align itself with the magnetic field.

Principle of Operation   Generally, there are three types of stepper motors: variable reluctance, permanent magnet (also called *canstack*), and hybrid. Variable reluctance motors use a soft iron, toothed, rotor. Permanent magnet canstack-type motors use a permanent magnet rotor without teeth. Hybrid motors have a permanent magnet, but toothed, rotor. These simple differences result in somewhat different principles of operation. In the following sections, we will discuss canstack- and hybrid-type motors. The variable reluctance motors operate with the same principle as hybrid motors.

Imagine a stepper motor with two coils in its stator and a permanent magnet as its rotor, as in Figure 7.25. When one of the coils of the stator is energized, the permanently magnetized rotor (or soft iron in variable reluctance motors) will rotate to align itself with the stator magnetic field (a). The rotor will stay at this position unless the field rotates. As the power to the present coil is discontinued and is directed to the next coil, the rotor will rotate to once again align itself with the field in the new position (b). Each rotation is equal to the step angle, which may vary from 180 degrees to as little as a fraction of a degree (in this example, 90°). Next, the first coil will once again be turned on, but in the opposite polarity, while the second is turned off. This will cause the rotor to rotate another step in the same direction. The process continues as one coil is turned off and another is turned on. A sequence of four steps will bring the rotor back to exactly the same state it was at the beginning of the sequence.

Now imagine that at the conclusion of the first step, instead of turning off one coil and turning on the second coil, both would be turned on. In that case, the rotor would only rotate 45° to align itself to the path of least reluctance (c). Later, if the first coil is turned off while the second remains on, the rotor will rotate another 45°. This is called half-step operation and includes a sequence of eight movements.

Of course, with the opposite on-off sequence, the rotor will rotate in the opposite direction. Typical steppers run between 1.8 to 7.5 degrees in full-step mode. Obviously, one way to reduce the size of the steps is to increase the number of coils. However, there is a physical limit to how many coils may be used. To further increase the number of steps per revolution, different numbers of teeth can be built into the stator and rotor creating an effect similar to a caliper. For instance, 50 teeth on the rotor and 40 teeth on the stator will result in a 1.8 degree step angle with 200 steps per revolution, as will be discussed later.

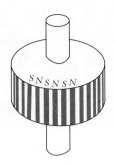 

Figure 7.26 The rotor of a canstack stepper motor.

Canstack Motors These motors are very common, are usually 7.5° steppers or similar, and are used in many different applications. Due to their construction, they are relatively short and lend themselves to applications with low vertical clearance.

The rotor is a cylinder with alternate strips of north-south polarities, usually made of resins embedded with ferrite particles (similar to refrigerator magnets), as shown in Figure 7.26. A typical rotor for a 7.5° stepper motor will have 24 pairs of poles on it.

A refrigerator magnet sticks to ferrite materials on one side only; the side that is meant to have a message or advertisement on it does not stick. Is it because of the additional printed layer? Not really. It is because there is no magnetic flux on that side, but only on the side that sticks. These magnets are made of a resin mixed with a ferrite powder that is magnetized as a series of small horseshoe magnets next to each other called *Halback Array*, as shown in Figure 7.27.

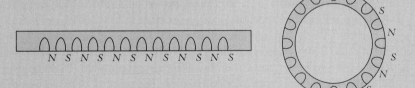

Figure 7.27 Magnetic flux pattern on a refrigerator magnet and the rotor of a stepper motor or brushless DC motor.

The permanent magnet rotors of many stepper motors and brushless DC motors are made the same way, and therefore, the rotor's cross-section is made up of a series of horseshoe magnets next to each other. This arrangement creates a unique rotor that allows small steps in a brushless DC motors (see Figure 7.22) and canstack-type stepper motor (see Figure 7.26).

Figure 7.28 The stator of a canstack stepper motor.

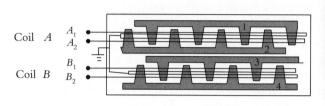

Coil A A_1
         A_2

Coil B B_1
         B_2

Cross-section of 4 plates staggered   A single plate with tabs (teeth)

Figure 7.29 Canstack stator windings and plates.

The stator is made up of four plates, each with 12 teeth, stacked on top of each other, with the teeth staggered in 1-3-2-4 order as shown in Figures 7.28 and 7.29. One coil is wrapped around plates 1 and 2, as is one around plates 3 and 4. The four plates create two independent magnets on top of each other, each with a winding that is center-tapped (called bifilar) and grounded at the center. A current going through one coil and exiting at the center will create a certain magnetic polarity which is the opposite of the polarity if the current goes in from the other end and exits at the center wire. As a result, each coil may be energized at either polarity. Each winding will cause all the teeth on each plate of a pair to be of similar polarity: all north or all south. Consequently, energizing both windings will create repeating patterns of north and south in all four plates.

Figure 7.29 is a schematic of a canstack stator, linearized for better visualization. Plates 1 and 2 are related to coil $A1/A2$ (two halves of coil A), as are plates 3 and 4 to coil $B1/B2$ (two halves of coil B). If coil $A1$ is turned on, plate 1 will be north and plate 2 will be south. However, if coil $A2$ is turned on, the polarity of plates 1 and 2 will be reversed; plate 1 will be south and plate 2 will be north. The same will happen to plates 3 and 4 with coils $B1$ and $B2$. This is called center-tapping and is shown in Figure 7.30. The advantage

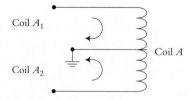

Coil A_1

Coil A_2

Coil A

Figure 7.30 Center tapping of a coil allows changing the polarity of the magnetic field by having the current flow in either half of the coil.

of center tapping is that during manufacturing, a set of two wires is simultaneously wrapped around the pole. At the conclusion, the two heads of the wires are connected and form the ground, while the other two heads will be contact points for input current. As a result, the magnet may easily be energized with either polarity by simply having current in each half of the double winding.

During the operation of the motor, two coils will be turned on in the following sequence resulting in the indicated polarities:

Step	A1	A2	B1	B2	Plate 1	Plate 2	Plate 3	Plate 4
1.	on	off	on	off	N	S	N	S
2.	off	on	on	off	S	N	N	S
3.	off	on	off	on	S	N	S	N
4.	on	off	off	on	N	S	S	N

Notice that $A1$ and $A2$ or $B1$ and $B2$ are never turned on simultaneously (as they would cancel each other's fields). During each step of the sequence, the poles on the rotor will align themselves between the stator poles with the least reluctance such that any south pole on the rotor will be between two north poles on the stator, and any north pole on the rotor will be between two south poles on the stator, as in Figure 7.31. In this Figure, "arcs" of S-N show the polarities of the poles at each step of the sequence. At the end of the four-step sequence, the rotor has moved four steps, which will bring it to exactly the same situation as in the beginning of the sequence. Therefore, repeating the four-step sequence will rotate the rotor continuously. The faster the sequencing of steps, the faster the rotor will rotate. As a result, by carefully controlling how many sequences

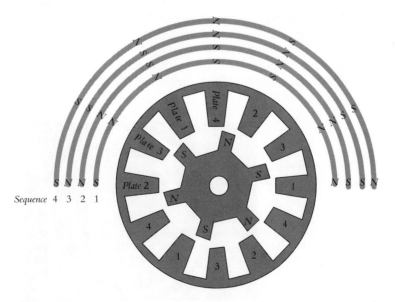

Figure 7.31 The cross-section of a canstack stepper motor. Each S-N arc shows the polarities of each pole during one step of the sequence of four.

Table 7.2   (a) Full Step and (b) Half Step Sequence for Stepper Motors.

(a)					(b)			
Step	**A1**	**A2**	**B1**	**B2**	**Step**	**A1**	**A2**	**B1**	**B2**
1	On	Off	On	Off	1	On	Off	On	Off
2	Off	On	On	Off	2	Off	Off	On	Off
3	Off	On	Off	On	3	Off	On	On	Off
4	On	Off	Off	On	4	Off	On	Off	Off
					5	Off	On	Off	On
					6	Off	Off	Off	On
					7	On	Off	Off	On
					8	On	Off	Off	Off

are provided and at what speed, the rotational displacement as well as angular velocity of the motor can be controlled. Note that Figure 7.31 is drawn with only 1/4 of all stator and rotor poles. In actual stepper motors, there are a total of 48 poles, providing a 7.5° step angle.

Instead of always energizing two coils simultaneously, if either one or two coils are energized alternately, rendering an eight-step sequence, the stepper will step at half the angle, therefore half stepping. However, this is not common in canstack motors. Table 7.2 shows the on-off sequence for full stepping as well as for half stepping the stepper motor. Reversing the sequence will cause the stepper motor to rotate in the opposite direction.

Hybrid Stepper Motors   These steppers are usually made with two coils—either center tapped, or two independent windings in opposite directions—each with four poles. The rotor is made of two collinear cylinders mounted on a stainless steel shaft, such that one end of the rotor is north and the other end is south (Figures 7.32 and 7.33). The rotor and the stator poles are all cut to have teeth, where the teeth on one half of the rotor are offset by a half tooth from the other half of the rotor. However, to understand the role

Figure 7.32   Stator of a hybrid stepper motor.

Figure 7.33 Rotor of a hybrid stepper motor.

of the teeth on the rotor and stator of these motors, we first review the concept behind a caliper.

Let's take two parallel lines capable of sliding relative to each other, each one unit of length long. Next, divide each line into 10 equal divisions. In order to have the division lines aligned with each other when moving one step, the sliding line will have to move one whole division to the next line (Figure 7.34).

Similarly, take two other lines, divide one into ten equal divisions and the other line into 11 equal divisions. In this case, the divisions will be 0.1 and approximately 0.09, respectively. If two of the division lines are aligned and one of the lines slides, it will only take a distance of $0.1 - 0.09 = 0.01$ units to align the next pair of division lines. Calipers do exactly the same. By dividing the lines into different lengths, we can easily measure a distance at a fraction of the smallest division.

The teeth on the stepper motor stator and rotor are also the same. Since the number of teeth on the stator and rotor are different, for each step, the rotor only rotates an angle equal to the difference between the two divisions. With 40 teeth on the stator poles and 50 teeth on the rotor, the step angle will be 1.8 degrees at full steps, because:

$$\frac{360°}{40} - \frac{360°}{50} = 9° - 7.2° = 1.8° \tag{7.20}$$

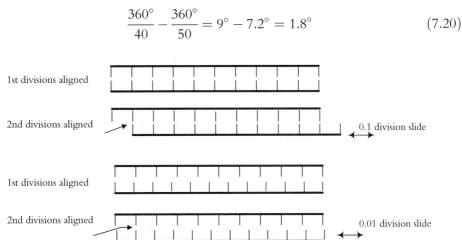

Figure 7.34 Application of unequal divisions for measuring lengths as in a caliper.

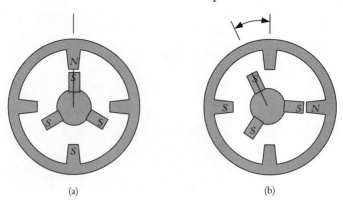

(a)   (b)

Figure 7.35 Basic operation of a hybrid stepper motor.

The two rotor cylinders (Figure 7.33) are a half-step out of phase and the stator poles run across the whole length of the stepper. Although the construction of hybrid and canstack steppers are different, driving the motors is similar.

Figure 7.35 is a simplified hybrid stepper motor with only two coils and a rotor with three teeth. As we discussed earlier, in actual motors, many teeth are cut into the poles to create smaller steps, but to see how the motor works this example only uses three teeth. As long as the number of teeth in the stator and the rotor are different, the above-mentioned effect will be true. Now suppose that one coil is energized as shown in (a). The rotor, with all its teeth as south on one side and north on the other will align itself to the path of least reluctance, as shown. If the coil is turned off and the second coil is turned on, although the flux (or field) has turned 90°, the rotor will rotate only 30° to the new location in (b). The same sequence will continue similarly to a canstack motor and the rotor will rotate accordingly. Changing the sequence backwards will cause the rotor to rotate backwards. Applying the sequence faster will cause the rotor to move faster. Therefore, by controlling the sequence, its direction, and its speed, the rotor's motion and angular velocity can be controlled. Of course, the same eight-step sequence can also be applied to these motors for half-step operation.

Unipolar, Bipolar, and Bifilar Stepper Motors   Unipolar stepper motors are designed to work with one power source. Generally, it is desirable to have one power source that powers both the motor windings as well as the drive circuits. Unless other techniques of switching are used, with one power source, it is impossible to simply change the polarity of the coils by changing the polarity of the power from the power source as this would ruin the electronic drive circuitry. Since the polarity of the coils cannot be changed, these motors may not be run in half-step mode. However, there is only one power source used, which reduces the cost, and the motor will develop it full power.

For bipolar motors, the assumption is that the polarity of the power source can be changed. As a result, there will either be two power sources, where one powers the motor windings and its polarity can be switched, and one is used to power the drive circuitry, or that more sophisticated switching is used with one power supply to prevent damage to the circuits. In this case, the motor will require more power supplies or

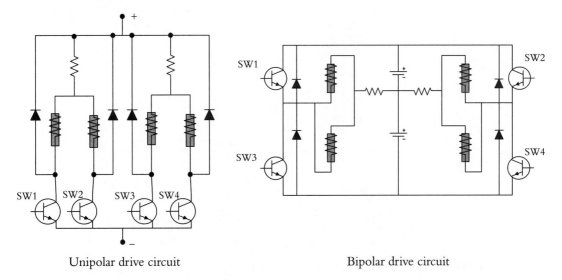

Unipolar drive circuit   Bipolar drive circuit

Figure 7.36 Schematic drawing for unipolar and bipolar drive circuits.

electronics, but can be run either at full or half-step modes, and will develop its full power.

In bifilar motors, the coils are center-tapped, as was previously discussed. In this case, the polarity of the coils can be changed simply by flowing the current in each half of the coil. Therefore, with simple circuits and one power supply, the motor can be run in either full or half-step mode, but since only half of the coil is energized, the motor develops only half of its full power. Figure 7.36 is a schematic drawing for unipolar and bipolar drive circuits. The switches are connected to the microprocessor ports and are turned on and off by the microprocessor.

Most motors are wired to be used in any of the three modes. Some motors are not. Figure 7.37 shows how stepper motors may be wired. In 8-lead configuration, you may wire the motor in any mode, as the coils are completely detached from each other. In this case, it is also easy to find which two wires are connected to which coil. The 6-lead motors are connected such that each coil is center tapped, but the two coils are detached. In 5-lead motors, the two center taps are connected together. 4-lead motors may not be used in bifilar mode. By measuring the resistance between different wires, we may find which wire is connected to which coil.

Figure 7.38 shows how a 5-phase "pentagon" stepper motor may be connected.

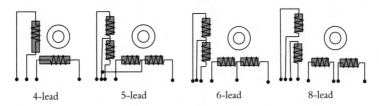

4-lead   5-lead   6-lead   8-lead

Figure 7.37 Stepper motor lead configurations.

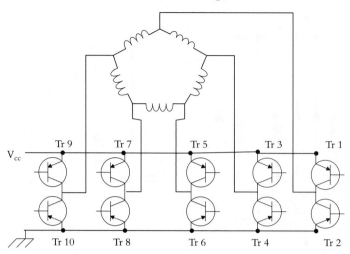

Figure 7.38   Connection schematic for a bipolar drive mode with pentagon connection for a 5-phase stepper motor. (Adapted by permission from the Oriental Motor USA Corp.)

Stepper Motor Speed-Torque Characteristics   Stepper motors are very useful in many applications. They do not require any feedback, as it is assumed that stepper motors advance a known angle every time a signal is sent. As long as the load on the motor is less than the torque it can deliver, the steps will not be missed, and this assumption is correct. However, if the load is too large, or in fact, if the speed is more than what the motor is capable of rotating, steps may be missed, and since there is no feedback, all subsequent positions will be incorrect.

Stepper motors develop their maximum torque, called *holding torque*, at zero angular velocity, when the rotor is stationary (the torque developed with no power is called *detent* or *residual torque*). Like all motors, as the speed of the stepper motor increases, the torque it develops reduces, but more significantly than others. Therefore, it is very important that the designer checks the speed-torque characteristic of stepper motors from manufacturers before a choice is made. Figure 7.39 is a typical speed-torque characteristic. The useful torque (also called *pull-out torque*) depends on the way the stepper is externally wired and the drive power signals used.

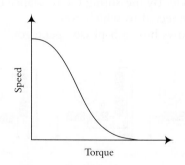

Figure 7.39   A typical speed-torque curve for a stepper motor.

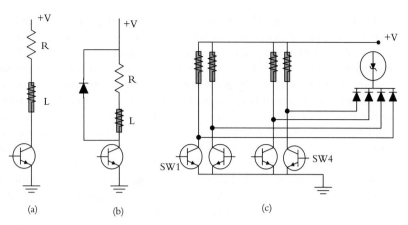

Figure 7.40 Application of (a) a resistor, (b) a diode, and (c) a zener diode with stepper motors for increasing the maximum velocity.

One reason for this poor performance at higher speeds is that since at each step, the rotor must accelerate, cruise, decelerate, and stop at the next step, and that this process must be repeated for every single step, stepper motors cannot rotate quickly, especially if the rotor is heavy. If the signals coming to the motor are too fast, the rotor will not have the time to accelerate/decelerate and will miss steps. So, one reason for losing output torque at higher speeds is the inertia load. However, even more important is the alternating magnetic field of the stator. As was discussed earlier, every time a coil is turned off, there will be a varying (in this case a decaying) flux. The wires of the coils, in the presence of a changing flux, will develop a back-emf that will slow down the decay of the flux as a new one builds up due to the back-emf current. The generated flux will tend to "hold" the rotor from rotating and will slow it down. As a result, the rotor is pulled back and is not able to rotate freely.

To remedy this problem and to prevent high-current "sparking" across the switches, a freewheeling diode may be added to the circuit, as in Figure 7.40(b). The diode will allow the current to continue flowing through the coil and be dissipated through conversion to heat in the resistor. The effectiveness of the process can be increased by adding a zener diode to the circuit, as shown in Figure 7.40(c). It is important to realize that the zener diode's breakdown voltage must be near the transistor's breakdown voltage rating. Otherwise, it will have no effect.

Microstepping Stepper Motors In microstepping, instead of turning the coils on and off abruptly, the power-up or power-down for each coil is done gradually by dividing the changes into smaller divisions, usually up to 250 steps. As an example, suppose coil A is on and coil B is off. In full stepping, the next sequence will be for coil A to be off and for coil B to be on. In microstepping, this is divided, let's say, into 100 divisions. As a result, in the next microstep, coil A will be 99% on and coil B will be 1% on. Therefore, the rotor will turn slightly to the point of least reluctance, which is a microstep away from the previous original step. In the next step, coil A would become 98% strong and coil B would be at 2%, still microstepping the rotor a little further. The process continues until both are equal for the half-step, and then, until coil A is turned off and coil *B* is 100% on. This would divide the full step into 100 smaller steps. For a 200-step 1.8° stepper motor, this means that with microstepping, the motor will have 20,000 steps per revolution. In practice, each step is usually divided into 125 or 250 microsteps, requiring 25,000 or 50,000 microsteps per revolution for a 1.8° stepper.

This dividing of the voltages is done with electronic circuits that resolve the voltage into smaller divisions and corresponding steps. Since, unlike full-stepping, the dissipation of the flux is not abrupt, the back-emf is much smaller. As a result, with microstepping, stepper motors' performance is improved significantly. The motor develops a higher torque, and its maximum speed without missed steps is much higher. Additionally, because microsteps are very small, the motor's movement is not as piecewise and, consequently, vibration of the motor is smaller and less pronounced. Of course, the disadvantage is that microstepping requires a more sophisticated driver that is much more accurate with better current resolution, and is more expensive.

In practice, instead of dividing the steps into linear and equally divided divisions (like 1% divisions), the changes follow a sine wave. In other words, the voltage to each coil is a piecewise, digital voltage resolved into up to 250 steps. Then, principally, in this mode, the stepper is in fact similar to an AC synchronous motor, except that its angular velocity is not fixed with line frequency but is driven by the microstepper driver circuit. As a result, the motor is driven at a desired angular velocity, for any desired angular displacement, and can be stopped at any instant.

Stepper Motor Control   Stepper motors may be driven by microprocessors (or microcontrollers) either directly or through driver circuits. They can also be driven by a dedicated stepper driver/translator that motor manufacturers offer.

To drive a stepper motor directly with a microprocessor, the power to the motor coils must be directly controlled by the processor. This is accomplished in two ways, depending on the microprocessor. If the output port of the processor is low power (mA), it will not be able to provide a high enough current to the motor windings (the output port of a personal computer is an example of this situation). As a result, the output port must turn on and off power transistors that control the power flow into the motor coils at higher currents. If the output ports of the microprocessor can provide high current flow, so long as the current requirement of the motor is below the level of the microprocessor's, the motor coils can be directly connected to the output ports. In either case, the microprocessor controls the current flow to the coils of the stepper motor by sequentially turning on and off the output ports, as discussed previously. The stepper will move in one direction or another depending on the order of the sequence. In this situation, four output ports are needed to drive one stepper motor, but the stepper's displacement, velocity, and direction are all under control. Figure 7.41 shows a schematic arrangement in which a microprocessor with a low power output port is used to drive a stepper motor with transistors (a) as well as a microprocessor with high current capability, which is used without transistors (b).

Alternatively, it is possible to use a dedicated IC chip to control the motion of a stepper motor.[7,8] These IC chips, called stepper driver with translator, are designed to sequence the stepper motor based on the information they receive. In most cases, the information needed is a stream of pulses. Every time the driver receives a pulse (the input to the chip changes from low to high), it will sequence the stepper by one step. If "n" signals are received, the motor will be sequenced for "n" steps. When a driver is used, the

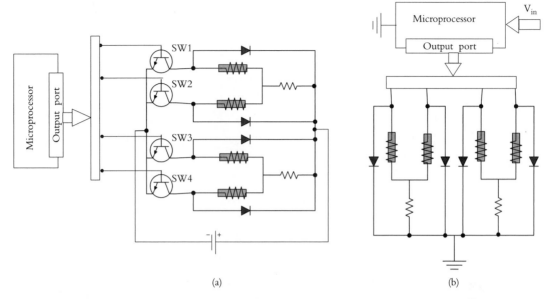

Figure 7.41 Schematic drawing of the application of a microcontroller in driving a stepper motor (a) with low current output, and (b) with high current output. (Schematics adapted by permission from Oriental Motor USA Corp.)

microprocessor only provides the stream of pulses, which are always the same, and not the sequences. A second input pulse to the driver determines the direction of motion. Consequently, only 2 output ports are needed to drive a stepper motor for any displacement, at either direction, at any speed. Therefore, using stepper drivers, the same number of output ports as before (four outputs) can drive twice as many stepper motors. Most drivers provide a choice of full or half stepping by making one of their pins high or low. In this case, three output lines will be necessary. Drivers are very easy to use, are very inexpensive, and simplify programming of the processors. Relatively in-expensive microstepping stepper driver IC chips are also commercially available and may be used for limited microstepping.

Drivers, like microprocessors, may be low current or high current. If they are low current, it is necessary to use power transistors as switches, where the driver turns the transistors on and off, thus driving the stepper motor. If the driver is high current (such as Allegro MicroSystems Inc. A3982 with a current rating of 2 amps), the output port of the driver may directly be connected to the stepper motor, simplifying the circuit. Figure 7.42 is a schematic of a typical stepper motor driver (Allegro Micro Systems A3982 chip). Figure 7.43 shows how it can be used. As shown, the microprocessor provides a pulse train to pin 16 of the driver that automatically sequences the stepper motor for each pulse. The faster the pulses, the faster the stepper motor's rate of rotation. To control the direction of rotation, pin 20 of the driver must change from high to low or from low to high.

Another way to run a stepper motor with a driver is to provide the pulse train to the driver by means other than a microprocessor. As an example, the timer circuit in Figure 7.44 can provide a steady stream of pulses to the driver, causing it to run the

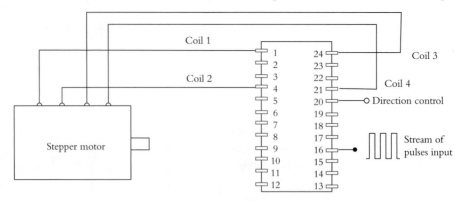

Figure 7.42 Schematic of a typical stepper motor driver.

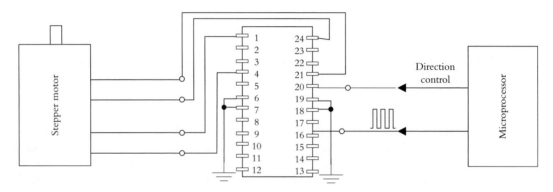

Figure 7.43 Schematic drawing of the application of a driver to run a stepper motor with a microprocessor.

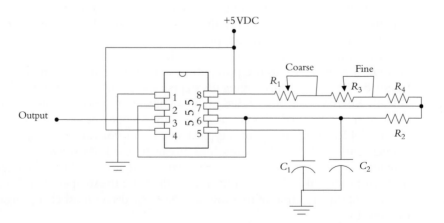

Figure 7.44 Schematic drawing of a timer circuit that creates a pulse train, which can be used to drive a stepper driver. Adjusting the potentiometers will change the period of the output pulses.

stepper motor. However, to change the velocity of the stepper, or to control the total displacement, other means of control are needed (for example, by adjusting the potentiometers in the circuit). Still, in certain applications where the velocity remains constant, or where the total displacement can be controlled by other means such as a microswitch, the timer circuit may be very useful. For $R_H = R_1 + R_3 + R_4$, the following equations apply to this circuit:

$$t_{high} = 0.693C_2(R_H + R_2)$$
$$t_{low} = 0.693C_2(R_2)$$
$$t_{total} = 0.693C_2(R_H + 2R_2)$$

(7.21)

while for practical purposes, $\max(R_H + R_2) \leq 3.3\,M\Omega$, $\min(R_H, R_2) = 1\,K\Omega$, and $500\,pF \leq C_2 \leq 10\,\mu F$. With an appropriate choice of resistors and capacitors, a wide range of periods may be achieved. Refer to References 9 through 13 for more information on mechatronics applications and design.

7.7 Microprocessor Control of Electric Motors

As we discussed in Chapter 1, a robot is supposed to be controlled by computers or microprocessors. Therefore, it is important to be able to control the motions of the actuators with microprocessors. We have already discussed in some detail how a stepper motor or a brushless DC motor may be controlled by a microprocessor. In the following sections, a few other common techniques for controlling electric motors will be explored.

A microprocessor is a digital device. It only deals with digital inputs and digital outputs. Any voltage lower than about 0.8 volts is considered low (off, 0). Any voltage above approximately 2.4 volts is considered high (on, 1). The microprocessor can only read the 0s and 1s. It can also only output 0s and 1s. All analog or continuous input signals or information must be digitized for use by a microprocessor. All desired analog or continuous output signals or information must be converted from digital to analog as well. Analog to digital converters (ADC) and digital to analog converters (DAC) are used for this purpose. However, a key element in both DAC and ADCs is their resolution.

Digital devices handle numbers (and all other information) by bits, which can only be 0 or 1. A 1-bit piece of information can only have two states, 0 or 1 (off or on). To add to this capability, we may use two bits. Then, there are four possibilities: 00, 01, 10, 11. As the number of bits increases, the variations increase with 2^n. Therefore, a 4-bit set will have 16 distinct possibilities, and an 8-bit set would have 2^8 or 256 possibilities. Every 4 bits is called a nibble, and every 8 bits is a byte.

Suppose you want to read a variable voltage between 0–5 volts into your microprocessor and be able to use it for running a device. Of course, the processor can only read high or low, not a continuous number. If you use a 1-bit input port, it can only recognize whether the voltage is high or low, hardly a continuous process. Now suppose you use two bits to read the voltage. There are four distinct possibilities for 2 bits. Consequently, the voltage can be divided into four different values, perhaps 0, 1.67, 3.34, and 5, which correspond to 00, 01, 10, 11 bit mapping. Then we may distinguish four different levels of voltage with the processor. If this resolution is not enough, we would have to increase the number of bits. With a 4-bit input, the 5-volt voltage can be divided into 16 portions, corresponding to 0000, 0001, 0010, 0011, . . . mapping. Then the smallest change in voltage we could read (resolution) is 0.33 volts. As you can see, the larger the number of bits, the better the resolution. However, what this means is that in order to read one input voltage, four input ports of the processor would have to be dedicated to it. Additionally, you would have to use an ADC to convert the analog voltage signal to a digital form, and feed the information from ADC into the processor.

Conversely, suppose you want to control a servomotor with a microprocessor. In order to have a variable voltage to control the speed of the motor, the digital information must be converted to analog form through a DAC. The resolution of this information is also limited to the number of bits used. For better resolution, more bits are necessary. Now you can imagine that for a 6-axis robot with multiple servomotors, inputs, and sensors, how many input and output ports would be necessary to have an accurate robot.

To control the motions of a robot, or to move from one point to another, the robot controller would have to calculate the magnitude of the change in each joint based on the kinematic equations governing the motions of the robot. If it is desired to also have velocity control, the speed at which each joint must move is also calculated. This information determines how much and how fast each joint must move, which alternately, determines how much and how fast the servomotors of each joint must rotate (based on the gear ratio of the joint, and so on). This information is converted to a set of voltages and voltage profiles. The voltage is fed to the servomotor, and the feedback signals going back to the controller are checked. The voltage is adjusted accordingly for desired velocity until the joint is moved a desired amount. This process continues until the robot's movements are finished. Consequently, it is necessary to have control over the voltages that go to each servomotor, and to be able to read the feedback signals from each joint.

It should be mentioned here that a new category of Intelligent Motor Controllers (IMC) is evolving.[14] These dedicated, high-speed controllers can be programmed for the high performance motor control of brushless motors and AC induction motors.

7.7.1 Pulse Width Modulation

As we discussed earlier, to be able to run a servomotor with almost-analog voltage, we would have to dedicate a large number of output ports or bits to it to have a good resolution. This is expensive, and at times—where large numbers of inputs and outputs are needed—prohibitive. Additionally, since this voltage is low power and cannot directly run the motor, it has to be used as input to a power transistor that controls the motor. Now let's see what happens when a transistor is not run in full power mode.

Figure 7.45(a) shows a simple circuit to control a motor. V_{CE} of the ideal transistor is controlled by the microprocessor, which in turn, controls the voltage across the motor, and therefore, the current. The power loss in the transistor can be found as:

$$i = \frac{V_m}{R}$$
$$P_{trans} = V_{CE} \cdot i = (V_{in} - V_m) \cdot i = \frac{(V_{in} - V_m) \cdot V_m}{R} \tag{7.22}$$

Figure 7.45(b) shows the power loss of the transistor. As Equation (7.22) indicates, when either $V_m = 0$ or $V_m = V_{in}$, power loss is zero (this assumes that the internal resistance of

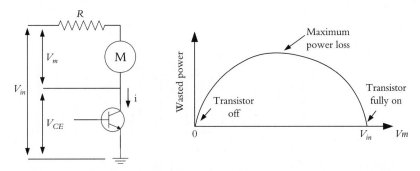

Figure 7.45 (a) Application of a power transistor to control the supplied voltage to a motor, (b) the power loss of the transistor at different voltages.

an ideal transistor is very low, and therefore, when fully on, the voltage drop across it is very little). Otherwise, the power loss is a second-order function of the V_m motor voltage as shown in Figure 7.45(b). Therefore, if the transistor is fully on or off, there is no power loss. Otherwise, it generates (and wastes) a lot of power. Therefore, it is beneficial to run the transistor fully on or fully off.

To overcome both preceding problems, a method called *Pulse Width Modulation* (PWM) is used. PWM can create variable voltages with only one output port of the microprocessor without any loss in the power transistors. To do this, the voltage on the output port of the processor is turned on and off repeatedly, many times a second, such that by varying the length of time that the voltage is on or off, the average effective voltage will vary. In other words, as in Figure 7.46, as t_1 versus t changes, the average voltage at the motor changes accordingly. The average output voltage in PWM is:

$$V_{out} = V \frac{t_1}{t} \tag{7.23}$$

The pulse rate of PWM may be 2-20 kHz, while the natural frequency of a motor is much smaller. If the rate of PWM switching remains many times larger than natural frequency of the motor's rotor, the switching will have little effect on the performance of the motor. The motor effectively acts as a low pass filter and does not respond to on/off signals except through the average value of the input voltage due to PWM.

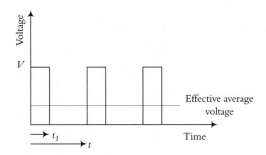

Figure 7.46 Pulse Width Modulation Timing.

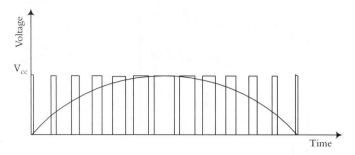

Figure 7.47 Sine wave generation with pulse width modulation.

PWM can create an audible noise in a motor, which becomes inaudible as the frequency increases beyond the hearing threshold of humans. On the other hand, theoretically, the power loss of the transistor is zero when it is off or fully on. However, in reality, every time the transistor is turned on or off, it takes a finite time for the voltage to build up or break down, causing heat generation. As the frequency increases, heat generation increases as well. Therefore, it is crucial to use transistors that have very fast switching capability (e.g., MOSFETs for low power and IGBTs for higher powers).

Additionally, as the PWM rate increases, the back–emf voltage in the motor due to $L\dfrac{di}{dt}$ in Equation (7.17) can also increase. Therefore, it may be necessary to insert a diode across the motor armature to protect the system.

By varying the PWM timing continuously, a variable voltage can be created which can be used in brushless DC motors or similar applications. Figure 7.47 depicts a sine wave generation with PWM.

7.7.2 Direction Control of DC Motors with an H-Bridge

Another troublesome issue in controlling with a microprocessor is the changing of polarity for direction change. In microprocessor control of motors, it is desirable to change the direction of current flow in a motor for changing its direction of rotation with only two bits of information. In other words, instead of actually changing the polarity, we may change the direction of the flow by changing bit information from the microprocessor. A simple circuit called an H-bridge can accomplish this, as shown in Figure 7.48. As

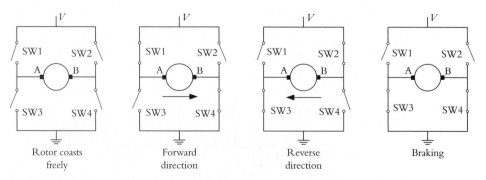

Figure 7.48 Directional control of a motor using switches in an H-bridge.

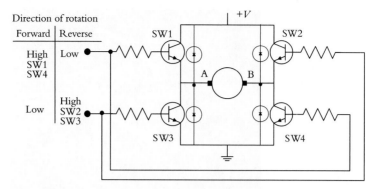

Figure 7.49 Application of H-bridge in controlling the direction of rotation of a motor.

shown, if all four switches are off, the rotor coasts freely. If SW1 and SW4 are connected, the current flows from A to B and the rotor rotates in one direction, whereas if SW2 and SW3 are connected, the current flows from B to A and the rotor rotates in the opposite direction. In fact, if SW3 and SW4 are connected, due to back-emf, a braking effect is created on the rotor.

Figure 7.49 shows how an H-bridge is wired. The diodes are necessary to prevent damage to the circuitry during switching. Additionally, if two switches on the same side are turned on together, a short circuit occurs. The same is true if one switch on one side does not turn off before the other is turned on. Most commercial H-bridges have internal protection built into them. The switches are turned on and off by the microprocessor; therefore, only two bits are needed to control the H-bridge.

7.8 Magnetostrictive Actuators

When a piece of a material called Terfenol-D is placed near a magnet, this special rare-earth-iron material will change its shape slightly. This phenomenon, called *magneto-striction* effect, is used to make linear motors with μ-inch displacement capabilities.[15,16] To run the actuator, a magnetostrictive rod, covered by a magnetic coil, is attached to two chassis. As the magnetic field changes, causing the rod to contract and expand, one chassis moves relative to the other. A similar concept is used with piezo crystals to create a linear motor with nano-inch displacements, including flexible actuators for robots.[17–20]

7.9 Shape-Memory Type Metals

One particular type of shape-memory Titanium-Nickel alloy, called Biometal[TM] (muscle wire), a patented alloy, shortens by about 4% when it reaches a certain temperature. The transition temperature can be designed into the material by changing the composition of the alloy, but standard samples are set for about 90°C. At around this temperature, the crystalline structure of the alloy makes a transition from martensitic to austenitic state, and consequently, it shortens. However, unlike many other shape-memory alloys, it will once

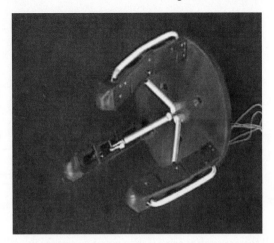

Figure 7.50   A 3-fingered end effector with Biometal-wire actuators.

again switch back to martensitic state when it is cooled down. This process can continue for hundreds of thousands of cycles if the loading on the wire is low. The common source of heat for this transition is an electric current flowing through the metal itself, which heats due to its electrical resistance. As a result, a piece of Biometal wire can easily be shortened by an electric current from a battery or other power sources. The major disadvantage of the wire is that the total strain happens within a very small temperature range, and therefore, except in on-off situations, it is very difficult to accurately control the strain, and thus, the displacement.[21]

Although Biometal wires are not yet fit to act as actuators, like other novel systems, it is possible to expect that someday they will become useful. In that case, a robot arm could be developed with muscles similar to those of humans or animals, which would only require a current to operate, as muscles do. Figure 7.50 shows a 3-fingered end effector with Biometal wires as actuators. Because of its small size, the wire is not visible. However, the wire was looped multiple times around the digits to create large deflections. Nevertheless, no solenoid, motor, or pneumatic actuators were needed to operate the gripper; therefore, the gripper is simple and small. The springs return the fingers to their neutral position.

7.10 Electroactive Polymer Actuators (EAP)

When exposed to high-voltage electric fields, due to a phenomenon called Maxwell stress, dielectric elastomers such as silicones and acrylics contract in the direction of the field and expand in the direction normal to it. By laminating thin films of the material with layers of electrodes on two sides, the elastomer can contract in the direction of the electric field formed between the electrodes (that act as a capacitor) and expand perpendicularly to it.[22] To act as an actuator, the laminate can be formed in many shapes, in series, parallel, or series and parallel, as well as in cylindrical configuration.[23] When in the form of a sheet, the material reacts to a voltage between any two points on it and bends locally. Although no real robot has been made with these actuators, no doubt they can be used for many other purposes, including in future robots.

7.11 Speed Reduction

Any means of speed reduction common in industry may be used in robotics as well. This includes fixed axis and planetary gear trains with many types of gears. Additionally, particular types of planetary gear trains called Harmonic Drive[TM] and Orbidrive[TM] as well as a novel design with nutating gears may be used. The following sections describe the Harmonic Drive concept as well as the nutating gears concept.

In planetary gear trains, there are normally four basic elements: the sun gear, the ring gear, the arm, and the planets. For calculation purposes, the sun and the ring gears are called *first* and *last* gears. Although we will not develop the following equation here, it can be stated that for a gear train, with corresponding angular velocities and number of teeth for the first and last gears, the following is true:[24]

$$\frac{\omega_{LA}}{\omega_{FA}} = \frac{\omega_L - \omega_A}{\omega_F - \omega_A} = \frac{N_F \times N_3}{N_2 \times N_L} \tag{7.24}$$

where ω_F, ω_L and ω_A are the angular velocities of the first and last gears and the arm, ω_{LA} and ω_{FA} are the angular velocities of the first and last gears relative to the arm, and N_L, N_F, N_2, and N_3 are the number of teeth of the first, last, and the planet gears as in Figure 7.51.

In this gear train example, it is assumed that the first gear is stationary and its angular velocity is zero. The input to the system is the arm, and the output is the last gear. There are two planets designated as gears 2 and 3. For $\omega_F = 0$, from Equation (7.24), we get:

$$-N_F N_3 \omega_A = N_2 N_L \omega_L - N_2 N_L \omega_A$$
$$\omega_A(N_2 N_L - N_F N_3) = N_2 N_L \omega_L$$

and
$$e = \frac{\omega_L}{\omega_A} = \frac{N_2 N_L - N_F N_3}{N_2 N_L} \tag{7.25}$$

where e is the gear ratio for the system. This can be used to calculate the gear ratio based on the number of teeth of the four gears.

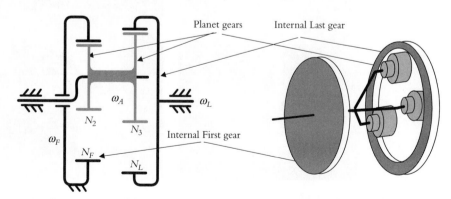

Figure 7.51 Schematic drawing of a planetary gear train.

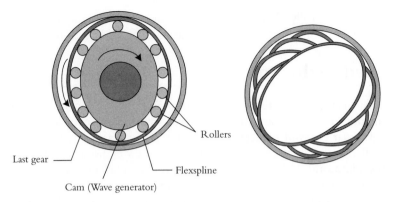

Rollers

Last gear

Flexspline

Cam (Wave generator)

Figure 7.52   Schematic of the strain wave gearing train.

In Harmonic drives, the number of teeth of the gears are chosen such as to render very large gear ratios with very few gears. To do this, suppose the teeth are chosen such that $N_F = N_2 + 1$ and $N_L = N_3 + 1$. In that case, Equation (7.25) will reduce to:

$$\frac{\omega_A}{\omega_L} = \frac{(N_F - 1)N_L}{N_F - N_L} \tag{7.26}$$

For example, with $N_L = 50$ teeth and $N_F = 45$ teeth, the gear ratio will be 440, which is huge (the negative number of the answer only indicates a direction change between the input and output of the gear train). The problem is that although theoretically we can pick the gears such that $N_F = N_2 + 1$ and $N_L = N_3 + 1$, in practice this will not work, because a pair of internal-external gears engaged with each other and having this many teeth will not rotate relative to each other. To overcome this problem, the planet in the strain wave gearing system is a flexible metal band with teeth (flexspline) that rolls over a cam (wave generator). As the wave generator rotates, it changes the shape of the flexspline planet such that it remains in contact with the internal gear at two opposing points only, while the remaining teeth are disengaged (Figures 7.52 and 7.53). For more information, refer to other references.[25–27]

Figure 7.53   A Harmonic Drive[TM] strain wave gear. (Reprinted with permission from Harmonic drive, LLC. "Harmonic Drive" is a trademark of Harmonic Drive LLC.)

Figure 7.54 Nutating gears train.

   The nutating gear train concept is somewhat similar. Gears that are very close in size wobble relative to each other at their edges.[28] As a result, after each revolution, one gear falls slightly behind the other, thus creating a large gear ratio. The equations governing the systems are the same as Equation (7.25). A simple rendition of the nutating gears is shown in Figure 7.54.

7.12 Other Systems

Many other novel systems can be used in robotic actuation, and there will be yet others available in the future. For example, although not used in any industrial robot, there is a spherical actuator made up of 80 magnets attached to the inside of a hollow sphere. The sphere is placed inside a cone made up of 16 circular electromagnets. By controlling which magnets are turned on in relation to the permanent magnets on the sphere, it can be forced to rotate in any direction.[29] Piezo-electrically generated traveling waves can drive a harmonic drive,[30] or a planetary speed reducer may use balls instead of gears.[31] The same principle used in harmonic drives may be used to create a linear drive, where slightly different pulley sizes create a differential motion between the carriage and the base (Figure 7.55). In this case, there is no need for gear reduction either.[32] And finally, instead of using standard brakes for electrically-actuated joints, a fail-safe electromagnetic system may be used that requires little power.[33] Look around for other innovations.

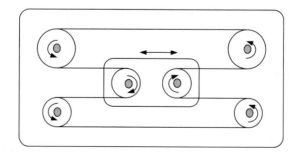

Figure 7.55 The concept behind a harmonic linear differential drive. (Adapted from OEM-dynamics.com.)

7.13 Design Projects

In the following section, you will find a few suggestions for possible projects. These are presented only as a starting point. Obviously, countless other possibilities exist. Be creative and think of other projects as well.

7.13.1 Design Project 1

Considering the advantages, disadvantages, capabilities, and limitations of each type of actuator, and depending on what may be available to you, design actuators for your robot. Whatever system you pick, you will have to consider what components are needed, how you can later run and control the actuators, their cost, weight, robustness, and availability. You must also consider how you will connect the actuators to the joints and links, the gear ratios needed, and so on. Additionally, make sure the actuator you pick can deliver the desired torque or force.

You must also design and program a controller for controlling the actuators. Stepper motors can be easily programmed and controlled by a simple microprocessor, with or without driver chips, commercial stepper drivers, or with a pulse generating circuit and a driver chip. Depending on the capability of your driver or controller, you may be able to control the speed of your stepper motors as well as their displacements. For servomotors, you will need a servo controller. Commercial servo controllers are expensive. However, with an appreciable amount of effort, one may design and build a servo controller. Please note that it is relatively easy to make a controller for controlling the position in a servomotor. This can be done with a simple feedback device like a potentiometer or an encoder. However, the design of a servo controller becomes much more complicated if you decide to control the velocity of the robot as well (which is why a servomotor would be used). Inexpensive, simple, geared servomotors, available commercially, that are designed for use with remote control airplanes are a good choice for inexpensive robots or for other actuation. These servomotors rotate a specific angle depending on the length of the signal they receive. Therefore, a simple command signal from your microprocessor can be used to easily control the servomotor. Speed control is usually possible, too. However, the accuracy and the power of these motors are limited.

You must also pick an appropriate microprocessor that has enough computing power to calculate the kinematic equations, but also has enough input and output ports for adequate communication with the motors. Integrating the robot manipulator with the actuators and the microprocessor will complete your robot. When you have completed programming the robot, you will have a functioning robot. In the next chapters, we will learn more about sensors, at which time you may integrate desired sensory information into the robot as well.

7.13.2 Design Project 2

This project involves the design and manufacturing of a rolling-cylinder robot rover. It is meant to consist of two colinear hollow cylinders that can rotate independently, and therefore, navigate. Its power source, the electronic drive circuitry, the actuators, and the

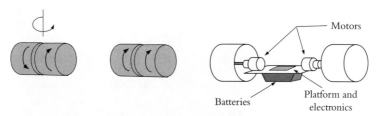

Figure 7.56 A schematic depiction of a possible arrangement for the cylinder rover.

sensors are all supposed to be inside the two cylinders. As a result, from outside, the rover should look like a cylinder. However, you should be able to program the rover to follow a predetermined path, or using its sensors, navigate through hallways, a maze, or similar environments. Figure 7.56 is a schematic depiction of the design. As shown, if the two cylinders rotate relative to each other, the rover will rotate about a vertical axis. If they rotate together in the same direction, the rover will move forward. The exploded view shows how the circuitry and the drive motors may be mounted on a platform and assembled inside the cylinders. Stepper motors as well as servomotors may be used for this purpose, each with its own advantages and disadvantages. The motors, mounted on a platform, may be attached to the cylinders by shafts or by rotating friction wheels. The center of gravity of the assembly, consisting of the platform, the motors, the power source, and the circuitry must be below the centerline of the rotating cylinders (bottom heavy). This creates a downward gravitational force that keeps the platform from rotating. As a result, as the motors rotate the cylinders, the whole assembly will move. You must realize that this is similar to a pendulum attached to a rolling wheel. If you write the equations of motion, you will see that the system will oscillate every time the torque on the system changes. As a result, you should expect to have rolling cylinders that oscillate every time they start, stop, or rotate. However, increasing the mass of the platform as well as increasing the distance of the center of mass of the platform from the center of the cylinders will reduce the frequency of the oscillations. Note that since a stepper motor's rotation involves stop-and-go steps, it increases the oscillations of the platform. Therefore, a servomotor may be a better choice. After you have built a prototype rover, you may add damping to the system, create additional points of support between the platform and the cylinders, or improve your drive programs (by controlling accelerations and decelerations) to reduce or eliminate the oscillations.

The motors, the sensors (as will be discussed in Chapter 8), and the added intelligence can all be controlled by a microprocessor or other control circuitry. This design project is intentionally left open-ended in order to allow you to be as creative as you wish. As an example, you may use light sensors to start and stop the rover and to navigate it (the light can be projected onto a sensor between the two cylinders). Alternately, you may use a microprocessor that responds to sensed information to do the same, or use a wireless communication link.

To complete the design project, first choose the cylinders, the platform size, the motors, the control system, and the sensors you want to incorporate in the design. You must make sure the arrangement of the subsystems will render a bottom heavy system. Consider how the motors will be attached to the cylinders. After the design details are completed, you may proceed with the manufacturing of the design, testing, and

Figure 7.57   An example of a cylinder robot rover (made by Nick Supat, 2008).

modifying until all problems are solved. When the rover is completed, you may proceed with programming it to do more sophisticated navigation and sensory information processing. Figure 7.57 is one example of a cylinder robot rover.

7.13.3 Design Project 3

The purpose of this project is to design and build a sphere robot rover. The basic idea is similar to design project 2. However, the rover is placed inside a hollow sphere. When the robot moves inside the sphere, it will force the sphere to move forward. It may appear that as the robot turns sideways inside the sphere, it will force the sphere to turn sideways, too. However, since the sphere already has a forward velocity, turning the two-wheeled robot sideways inside the sphere will push up the wheels and interfere with the motion. Instead, design and build a simple gyroscopic steering mechanism. In this case, one motor turns both wheels together (or even three wheels), moving the sphere forward, but the turning results from a gyroscope. To do this, imagine that you attach a flywheel to a motor that constantly rotates about the x-axis. Install the flywheel and the motor in gimbals that can rotate about the y-axis on the platform. When the flywheel is rotated about this axis, it will turn the assembly about the z-axis, forcing the sphere to change direction. Therefore, you can control the motions of the sphere by a microprocessor that sends signals to the motors running the wheels and the gyro. Figure 7.58 shows one rendition of this idea.

Placed inside a frictionless cone (or similar support), the sphere robot rover can in fact function as a spherical motor too.

7.13.4 Design Project 4

Design and build a snake robot. One way to do this is to build a series of links, attached in series by joints, that act as the snake's body. Attach pieces of Biometal between the links (e.g., as shown in Figure 7.59). By sending signals from your microprocessor to these simple actuators and contracting them sequentially, you can create a slithering motion. Just remember that you might need to create a one-directional frictional force in the forward direction at the underbelly of the snake to force the slithering motion forward. The simplicity of Biometal muscle wires is a good choice for this project, although other means may also be used.

Figure 7.58   A sphere robot. (Cal Poly Robotics Club.)

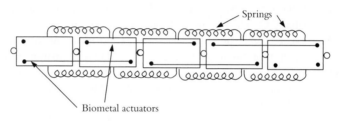

Figure 7.59   A possible design for a snake robot. (Design by Bryan Terry.)

You may also design and build other insect robots, other animal robots, and walking machines. A sophisticated toy called Pleo life form is in the form of a baby dinosaur that starts as a baby and grows in about 10 hours (not in size, but in behavior). Its final behavior is dependent on how it is treated by the user.

Summary

In this chapter, a variety of different actuating systems were presented. Each system has its advantages and disadvantages that make it useful for particular applications. In addition to their application in actuating robots, they can also be used in other devices used with robotic systems.

Although hydraulic systems are no longer common for industrial robots, they are still used for heavy payload robots and for devices that require a high power to weight ratio. Most industrial robots are actuated by servomotors. Stepper motors are very common in many other peripheral devices and in small robots. Other novel actuators can also be used for specific purposes in robotics. The design engineer must consider the best application for each actuator based on the design specifications.

In the next chapter, we will discuss a variety of sensors that are used in conjunction with robots and robotic applications.

References

1. "Servopneumatic Positioning System," Festo AG & Co., 2000.
2. "Step Motors and servomotors Control Catalog," Parker Hannifin, 1996–7.
3. Hage, Edward, "Size Indeed Matters," Power Transmission Engineering, February 2009, pp. 34–37.
4. McCormik, Malcolm, "A Primer on Brushless DC Motors," *Mechanical Engineering*, February 1988, pp. 52–57.
5. Mazurkiewicz, John, "From Dead Stop to 2,000 rpm in One Millisecond," Motion Control, September/October 1990, pp 41–44.
6. Shaum, Loren, "Actuators," International Encyclopedia of Robotics: Applications and Automation, Richard C. Dorf, Editor, John Wiley and Sons, N.Y., 1988, pp. 12–18.
7. "Technical Information on Stepping Motors," Oriental Motors U.S.A. Corporation.
8. "Application of Integrated Circuits to Stepping Motors," Oriental Motor U.S.A. Corporation, 1987.
9. Shetty, Devdas, R. Kolk, "Mechatronics System Design," PWS Publishing, MA, 1997.
10. Auslander, David, C. J. Kempf, "Mechatronics: Mechanical System Interfacing," Prentice Hall, NJ, 1996.
11. Stiffler, Kent A., "Design with Microprocessors for Mechanical Engineers," McGraw-Hill, NY, 1992.
12. "Mechatronics '98, proceedings of the 6th UK Mechatronics Forum International Conference, Skovde, Sweden," J. Adolfsson, J. Karlsen, Editors, Pergamon Press, Amsterdam, 1998.
13. Bolton, W., "Mechatronics: Electronic Control Systems in Mechanical and Electrical Engineering," 2nd Edition, Addison Wesley Longman, NY, 1999.
14. Lewin, C., "Intelligent Motor Control ICs Simplify System Design", Motion Control Technology Tech Briefs, December 2009, pp. 53–54.
15. Ashley, S., "Magnetostrictive Actuators," *Mechanical Engineering*, June 1998, pp. 68–70.
16. "Push/Pull Magnetostrictive Linear Actuators," NASA Tech Briefs, August 1999, pp. 47–48.
17. "Tiny Steps for a Big Job," NASA Motion Control Tech Briefs, August 1999, pp. 1b–4b.
18. "MEMS-Based Piezoelectric/Electrostatic Inchworm Actuator," NASA Motion Control Tech Briefs, June 2003, p. 68.
19. "Flexible Piezoelectric Actuators," NASA Tech Briefs, July 2002, p. 27.
20. "Magnetostrictive Motor and Circuits for Robotic Applications," NASA Motion Control Tech Briefs, August 2002, p. 62.
21. Toki America Technologies Biometal Reference.
22. Ashley, Steven, "Artificial Muscles," Scientific American, October 2003, pp. 53–59.
23. "Electroactive-Polymer Actuators with Selectable Deformations," NASA Tech Briefs, July 2002, p. 32.
24. Erdman, Arhtur, G. N. Sandor, "Mechanism Design: Analysis and Synthesis," Prentice Hall, New Jersey, 1984.
25. "Hollow Shaft Actuators with Harmonic Drive Gearing," NASA Motion Control Tech Briefs, December 1998, pp. 10b–14b.
26. Orbidrive Catalog, Compudrive Corporation.
27. "Planetary Speed Reducer with Balls Instead of Gears," NASA Tech Briefs, October 1997, p. 15b.
28. Kedrowski, D., Scott Slimak, "Nutating Gear Drivetrain for a Cordless Screwdriver," *Mechanical Engineering*, January 1994, pp. 70–74.
29. Stein, David, Gregory S. Chirikjian, "Experiments in the Commutation and Motion Planning of a Spherical Stepper Motor," Proceedings of DETC'00, ASME 2000 Design Engineering Technical Conferences and Computers and Information in Engineering Conference, Baltimore, Maryland, September 2000, pp. 1–7.
30. "Travelling Wave Rotary Actuators: Piezoelectrically Generated Travelling Waves Drive Harmonic Gears," NASA Tech Briefs, October 1997, p. 10b.
31. "Planetary Speed Reducer with Balls Instead of Gears," NASA Tech Briefs, October 1997, p. 15b.
32. "Harmonic Goes Linear," *Mechanical Engineering*, July 2008, p. 20.
33. "Fail-Safe Electromagnetic Motor Brakes," NASA Motion Control Tech Briefs, December 2002, p. 54.

Problems

7.1. A motor with a rotor inertia of 0.030 Kgm^2 and maximum torque of 12 Nm is connected to a uniformly distributed arm with a concentrated mass at its end, as shown in Figure P.7.1. Ignoring the inertia of a pair of reduction gears and viscous friction in the system, calculate the total inertia felt by the motor and the maximum angular acceleration it can develop if the gear ratio is a) 5, b) 50, c) 100. Compare the results.

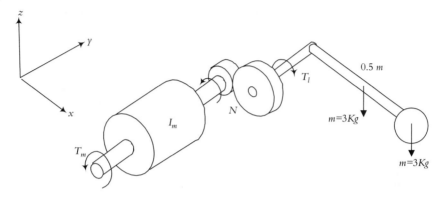

Figure P.7.1

7.2. Repeat Problem 7.1, but assume that the two gears have 0.002 Kgm^2 and 0.005 Kgm^2 inertias respectively.

7.3. The three-axis robot shown in Figure P.7.3. is powered by geared servomotors attached to the joints by worm gears. Each link is 22 cm long, made of hollow aluminum bars, each weighing 0.5 Kg. The center of mass of the second motor is 20 cm from the center of rotation. The gear ratio is 1/3 in the servomotor and 1/5 in the worm gear set. The worst case scenario for the elbow joint is when the arm is fully extended, as shown. Calculate the torque needed to accelerate both arms together, fully extended, at a rate of 90 rad/s^2. Assume the inertias of the worm gears are negligible.

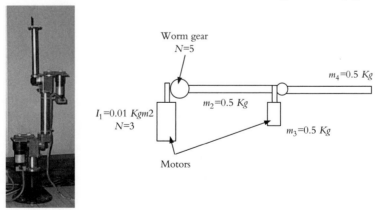

Figure P.7.3

7.4. Repeat Problem 7.3, but suppose the maximum torque this motor can provide is 0.9 Nm. Therefore, a new motor must be picked. Two other motors are available, one with the inertia of 0.009 Kgm^2 and torque of 0.85 Nm, one with inertia of 0.012 Kgm^2 and torque of 1 Nm. Which one would you use?

7.5. Estimate how much the torque/inertia ratio of a disk motor might be if it can go from zero to 2000 rpm in one millisecond, and compare it to the motor of Problem 7.1.

7.6. Using a timer circuit, design a pulse generating circuit that will deliver a range of 5–500 pulses per second to a stepper motor driver.

7.7. Calculate the gear ratio for a Harmonic drive if $N_L = 100$, $N_F = 95$, $N_2 = 90$, $N_3 = 95$.

7.8. Write a program to generate a variable pulse stream to drive a motor with pulse-width-modulated voltages of 1, 2, 3, 4, and 5 volts for a 5-volt input.

7.9. Write a program to generate a sinusoidal pulse-width-modulated output for a constant input voltage.

7.10. If you have access to a microprocessor and electronic components such as transistors, make an H-bridge and write a control program to drive a motor in either direction or to brake it. Be mindful of the problems associated with an H-bridge's transistors turning on and off at inappropriate times.

CHAPTER 8

Sensors

8.1 Introduction

In robotics, sensors are used for both internal feedback control and external interaction with the outside environment. Animals and humans have similar distinct sensors. For example, when you wake up, even before you open your eyes, you know where your extremities are; you do not have to look to know that your arm is beside you, or that your leg is bent. This is because neurons in the muscles send signals to the brain, and as they are stretched or relaxed with the contracting, stretching, or relaxing muscles, the signal changes and the brain determines the state of each muscle. Similarly, in a robot, as the links and joints move, sensors such as potentiometers, encoders, and resolvers send signals to the controller, allowing it to determine joint values. Additionally, as humans and animals possess senses of smell, touch, taste, hearing, vision, and speech to communicate with the outside world, robots may possess similar sensors that allow them to communicate with the environment. In certain cases, the sensors may be similar in function to that of humans such as vision, touch, and smell. In other cases, the sensors may be something humans lack such as a radioactive sensor.

There is a huge array of sensors available for measuring almost any phenomenon. However, in this chapter, we will only discuss sensors used in conjunction with robotics and automatic manufacturing.

8.2 Sensor Characteristics

To choose an appropriate sensor for a particular need, we have to consider a number of different characteristics. These characteristics determine the performance, economy, ease of application, and applicability of the sensor. In certain situations, different types of sensors may be available for the same purpose. Therefore, the following may be considered before a sensor is chosen:

- **Cost:** The cost of a sensor is an important consideration, especially when many sensors are needed for one machine. However, the cost must be balanced with other requirements of the design such as reliability, importance of the data they provide, accuracy, life, and so on.
- **Size:** Depending on the application of the sensor, the size may be of primary importance. For example, the joint displacement sensors have to be adapted into the design of the joints and move with the robot's body elements. The available space around the joint may be limited. Additionally, a large sensor may limit the joint's range. Therefore, it is important to ensure that enough room exists for the joint sensors.
- **Weight:** Since robots are dynamic machines, the weight of a sensor is very important. A heavy sensor adds to the inertia of the arm and reduces its overall payload. Similarly, a heavy camera mounted on a robotic insect airplane will severely limit its flying capabilities.
- **Type of output (digital or analog):** The output of a sensor may be digital or analog and, depending on the application, this output may be used directly or have to be converted. For example, the output of a potentiometer is analog, whereas that of an encoder is digital. If an encoder is used in conjunction with a microprocessor, the output may be directly routed to the input port of the processor, while the output of a potentiometer has to be converted to digital signal with an analog-to-digital converter (ADC). The appropriateness of the type of output must be balanced with other requirements.
- **Interfacing:** Sensors must be interfaced with other devices such as microprocessors and controllers. The interfacing between the sensor and the device can become an important issue if they do not match or if other add-on components and circuits become necessary (including resistors, transistor switches, power source, and length of wires involved).
- **Resolution:** Resolution is the minimum step size within the range of measurement of the sensor. In a wire-wound potentiometer, it will be equal to the resistance of one turn of the wire. In a digital device with n bits, the resolution will be:

$$Resolution = \frac{Full\ Range}{2^n} \qquad (8.1)$$

As an example, an absolute encoder with 4 bits can report positions up to $2^4 = 16$ different levels. Therefore, its resolution is $360/16 = 22.5°$.
- **Sensitivity:** Sensitivity is the ratio of a change in output in response to a change in input. Highly sensitive sensors will show larger fluctuations in output as a result of fluctuations in input, including noise.
- **Linearity:** Linearity represents the relationship between input variations and output variations. This means that in a sensor with linear output, the same change in input at any level within the range will produce a similar change in output. Almost all devices in nature are somewhat nonlinear, with varying degrees of nonlinearity. Some devices may be assumed to be linear within a certain range of their operation. Others may be linearized through assumptions. A known nonlinearity in a system may be overcome by proper modeling, equations, or additional electronics. For example, suppose a displacement sensor has an output that varies as a second-order equation. Using the square root of the signal, either through programming or by a simple electronic circuit, will yield a linear output proportional to the displacement. Therefore, the output will be as if the sensor were linear.

- **Range:** Range is the difference between the smallest and the largest outputs the sensor can produce, or the difference between the smallest and largest inputs with which it can operate properly.

- **Response time:** Response time is the time that a sensor's output requires to reach a certain percentage of the total change. It is usually expressed in percentage of total change, such as 95%. It is also defined as the time required to observe the change in output as a result of a change in input. For example, the response time of a simple mercury thermometer is long, whereas a digital thermometer's response time, which measures temperature based on radiated heat, is short. A special response time of 63.2% is called *time constant* τ. Similarly, rise time is the time required between 10% and 90% of the final value and settling time is the time between 0% and 98% rise.

- **Frequency response:** Suppose you attach a very high-quality radio tuner to a small, cheap speaker. Although the speaker will reproduce the sound, its quality will be very low, whereas a high-quality speaker system with a woofer and tweeter can reproduce the same signal with much better quality. This is because the frequency response of the two-speaker system is very different from the single, cheap speaker. The natural frequency of a small speaker is high, and therefore, it can only reproduce high frequency sounds. On the other hand, the speaker system with at least two speakers will run the signal into both the tweeter and woofer speakers, one with high natural frequency and one with low natural frequency.

  The summation of the two frequency responses allows the speaker system to reproduce the sound signal with much better quality (in reality, the signals are filtered for each speaker). All systems can resonate at around their natural frequency with little effort. As the input frequency deviates from the natural value, the response falls off. The frequency response is the range in which the system's ability to resonate (respond) to the input remains relatively high. The larger the range of the frequency response, the better the ability of the system to respond to varying input. Otherwise, the phenomenon measured may vary quickly, before the sensor has a chance to respond and send a signal. Therefore, it is important to consider the frequency response of a sensor and determine whether or not the sensor's response is fast enough under all operating conditions (we will discuss this in more detail in Chapter 9).

- **Reliability:** Reliability is the ratio of how many times a system operates properly, divided by how many times it is used. For continuous, satisfactory operation it is necessary to choose reliable sensors that last a long time while considering the cost and other requirements.

- **Accuracy:** Accuracy is defined as how close the output of the sensor is to the expected value. If for a given input, the output is expected to be a certain value, accuracy is related to how close the sensor's output is to this value. For example, a thermometer should read $100°C$ when placed in pure boiling water at sea level.

- **Repeatability:** If the sensor's output is measured a number of times in response to the same input, the output may be different each time. Repeatability is a measure of how varied the different outputs are relative to each other. Generally, if a sufficient number of tries are made, a range can be defined that includes all results around the nominal value (the radius of a circle that encompasses all results). This range is defined as repeatability. In general, repeatability is more important than accuracy, since in most cases, inaccuracies are systematic and can be corrected or compensated because they

Accurate and   Inaccurate and   Inaccurate but
repeatable   not repeatable   repeatable

Figure 8.1 Accuracy versus repeatability.

can be predicted and measured. Repeatability is generally random and cannot be easily compensated (see Figure 8.1).

The following is a review of some sensors used in robotics, mechatronics, and automation.

8.3 Sensor Utilization

Figure 8.2(a) shows a basic sensor circuit with a voltage source. As the sensor turns on and off, due to the back-emf principle, the wires act as inductors and, consequently, a voltage spike is generated in the wires that can create false readouts. To prevent this, it is advisable to add a monolithic-type capacitor to the circuit, as shown in Figure 8.2(b). The capacitor should be placed as close to the sensor as possible.

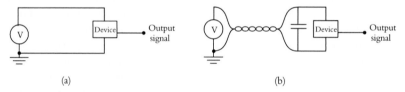

(a)   (b)

Figure 8.2 Application of a capacitor, added to prevent voltage spikes in reading sensors.

Similarly, if long wires (longer than a few inches) are used to connect a sensor to a voltage source or to where the signal is read, the wires can act as antennae and interfere with the signal. The solution is to use shielded or coaxial wires or to twist the wires together.

By the way, the above is true in other cases too. For example, long wires that connect a motor to a voltage source can also act as antennae, and therefore, it is better to twist the wires together. Similarly, voltage spikes can create problems with integrated circuit chips. Therefore, it is advisable to place a capacitor between the voltage-in and ground pins of an IC chip as close to it as possible (for example, next to, or under, the chip).

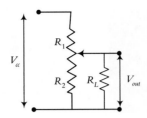

Figure 8.3 A potentiometer as a position sensor.

8.4 Position Sensors

Position sensors are used to measure displacements, both angular and linear, as well as movements. In many cases, such as in encoders, the position information may also be used to calculate velocities. The following are common position sensors used in robotics:

8.4.1 Potentiometers

A potentiometer converts position information into a variable voltage through a resistor. As the sliding contact (wiper) slides on the resistor due to a change in position, the proportion of the resistance before or after the point of contact with the wiper compared to the total resistance varies (Figure 8.3). The resistive external load R_L is in parallel with R_2, and both are in series with R_1. Since in this capacity, the potentiometer acts as a voltage divider, the output will be proportional to the resistance as:

$$V_{out} = \frac{R_2 R_L}{R_1 R_L + R_2 R_L + R_1 R_2} \cdot V_{cc} \qquad (8.2)$$

Assuming that R_L is large, the quantity $R_1 R_2$ can be ignored, and the equation simplifies to:

$$V_{out} = V_{cc} \frac{R_2}{R_1 + R_2} \qquad (8.3)$$

Example 8.1

Assume that $R_1 = R_2 = 1\ k\Omega$. Calculate the difference between the values of V_{out} based on Equations (8.2) and (8.3) if (a) $R_L = 10\ k\Omega$ and (b) $R_L = 100\ k\Omega$.

Solution:

a. $V_{out} = \dfrac{10}{10 + 10 + 1} \cdot V_{cc} = \dfrac{10}{21} V_{cc} = 0.476 V_{cc}$ and $V_{out} = \dfrac{1}{2} \cdot V_{cc} = 0.5 V_{cc}$

b. $V_{out} = \dfrac{100}{100 + 100 + 1} \cdot V_{cc} = \dfrac{100}{201} V_{cc} = 0.498 V_{cc}$ and $V_{out} = \dfrac{1}{2} \cdot V_{cc} = 0.5 V_{cc}$

Clearly, it is crucial that the resistive load be large enough for acceptable accuracy. ■

Potentiometers can be rotary or linear, and therefore, can measure linear or angular motions. Rotary potentiometers can also be multiple-turn, enabling the user to measure many revolutions of motion.

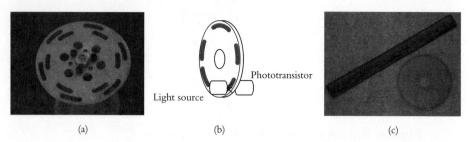

(a)   (b)   (c)

Figure 8.4   (a) A simple rotary incremental encoder disk mounted on a motor shaft. This
encoder measures angular rotations. (b) Schematic of a rotary encoder arrangement. (c) A
reflective-type linear absolute encoder that can measure linear movements and a rotary
incremental encoder disk with 1024 slots.

Potentiometers are either wire-wound or use a conductive polymer resistor paste—a
deposit of a thin film of resistive carbon particles in a polymer or ceramic and metal mix
called *cermet* on a phenolic substrate. The major benefit of conductive polymers is that
their output is continuous, and therefore, less noisy. As a result, it is possible to
electronically differentiate the output of this type of resistor to find velocity. However,
since the output of a wire-wound potentiometer is stepwise, it cannot be differentiated.

Potentiometers are generally used as internal feedback sensors in order to report the
position of joints and links. Potentiometers are used both alone as well as together with
other sensors such as encoders. In this case, the encoder reports the current position of
joints and links, whereas the potentiometer reports the startup positions. As a result, the
combination of the sensors allows minimal input requirement with maximum accuracy.
This will be discussed in more detail later.

8.4.2 Encoders

An encoder is a simple device that can output a digital signal for each small portion of a
movement. To do this, the encoder disk or strip is divided into small sections, as in
Figure 8.4. Each section is either opaque or transparent (it can also be either reflective or
nonreflective). A light source, such as an LED on one side, provides a beam of light to
the other side of the encoder disk or strip, where it is seen by a light-sensitive sensor,
such as a phototransistor. If the disk's angular position (or in the case of a strip, the linear
position) is such that the light is revealed, the sensor on the opposite side will be turned
on and will have a high signal. If the angular position of the disk is such that the light is
occluded, the sensor will be off and its output will be low (therefore, a digital output).
As the disk rotates, it can continuously send signals. If the signals are counted, the
approximate total displacement of the disk can be measured at any time.

Incremental Encoders   There are two basic types of encoders: incremental and
absolute. Figures 8.4(a) and 8.4(b) are incremental encoders. In this type of encoder, the
areas (arcs) of opaque and transparent sections are all equal and repeating. Since all arcs are
the same size, each represents an equal angle of rotation. If the disk is divided into only
two portions, each portion is 180 degrees, its resolution will also be 180 degrees, and

within this arc, the system is incapable of reporting any more accurate information about the displacement or position. If the number of divisions increases, the accuracy increases as well. Therefore, the resolution of an optical encoder is related to the number of arcs of transparent/opaque areas. Typical incremental encoders can have 512 to 1024 arcs, reporting angular displacements with a resolution of 0.7 to 0.35 degrees. High resolution encoders with thousands of pulses per revolution (PPR) are also available.

Optical encoders are either opaque disks with the material removed for transparent areas (Figure 8.4(a) and 8.4(c)) or are clear material like glass with printed opaque areas. Many encoder disks are also etched, such that they either reflect the light or do not reflect the light. In that case, the light source and the pick-up sensor are both on the same side of the disk.

An incremental encoder is like an integrator. It only reports changes to angular position (it reports the change in location, which is the displacement). However, it cannot report or indicate directly the actual value of the position. In other words, an incremental encoder can only tell how much movement is made. But unless the initial location is known, the actual position cannot be discerned from the sensor. An incremental encoder acts as an integrator, because the controller actually counts the number of signals the encoder sends, determining the total positional change, and consequently, it is integrating the position signal. Unless the controller knows the start-up position, it can never determine where the robot is. In all systems that track positions with incremental encoders, it is necessary to reset the system at the beginning of operations (or at wake-up). The controller will subsequently know the displacements at all times, so long as the reset position is known. (In some Adept robots, a 16-bit encoder is used together with a Hall-effect sensor to provide $\pm20\mu$ accuracy).

Most photodetectors are analog devices. This means that as the magnitude of the light varies, their output varies too. Therefore, as one section on the encoder disk approaches the detector and the projected light intensity increases to a maximum, the output of the detector rises before falling again as it departs. Consequently, a squaring circuit is used to condition the signal. Figure 8.5 shows the output of an incremental encoder. If only one

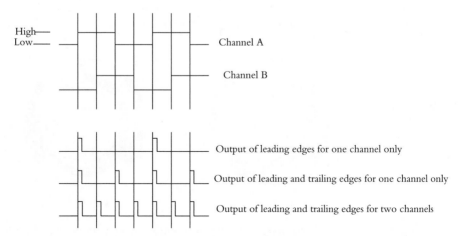

Figure 8.5 Output signals of an incremental encoder.

set of slots is used, it will be impossible to determine whether the disk is rotating clockwise or counterclockwise. To remedy this, encoder disks have two sets of slots (two channels), 1/2 step out of phase with each other (Figure 8.4(a)). As a result, the output signals of the two sets of slots are also a 1/2 step out of phase with each other. The controller can compare the two signals and determine which one changes from high to low or vice versa before the other signal. Through this comparison, it is possible to determine the direction of rotation of the disk.

By counting both the leading edges as well as the trailing edges of the output signals of the encoders on both channels, it is actually possible to increase resolution of the output of incremental encoders without increasing the number of slots.

Note that it is crucial to set up your system to look for *changes* in the signal, not whether the signal is high or low. If your circuit keeps counting when the signal is high, it may register a significantly high false count, especially if your system is fast or if the shaft rotates slowly. Only counting when there is a change (high to low or low to high) ensures that a correct number of signals are registered and counted.

Absolute Encoders   An alternative to incremental optical encoders is an absolute encoder. Each portion of the encoder disk's angular displacement has a unique combination of clear/opaque sections that give it a unique signature. Through this unique signature, it is possible to determine the exact position of the disk at any time, without the need for a starting position. In other words, even at start time, the controller can determine the position of the disk by considering the unique signature of the disk at that location. As shown in Figure 8.6, there is a multiple row of sections, each one different from the others. The first row may have only one clear and one opaque section (one on, one off). The next row has 4 (or 2^2), followed by 8 (or 2^3), and so on. Each row must have its own light source and light detector assembly. Each sensor assembly sends out one signal. Therefore, two rows require two inputs to the controller (2 bits), three rows require 3 bits, and so on.

As shown in Figure 8.6, an encoder with 4 rows can have $2^4=16$ distinct combinations, each section covering an angle of 22.5°. This means that within this section of 22.5°, the controller cannot determine where the encoder is. Therefore, the resolution is only 22.5°. To increase the resolution, there would have to be more sections, or bits. An encoder with 1024 divisions on one row has 10 ($1024 = 2^{10}$) bits of information that must

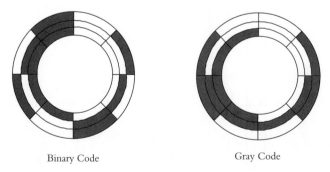

Binary Code   Gray Code

Figure 8.6   Each portion of the angular position of an absolute encoder has a unique signature. Through this signature, the angular position of the encoder can be determined.

Table 8.1 *Binary and Gray Codes.*

#	Gray Code	Binary Code		#	Gray Code	Binary Code
0	0000	0000		6	0101	0110
1	0001	0001		7	0100	0111
2	0011	0010		8	1100	1000
3	0010	0011		9	1101	1001
4	0110	0100		10	1111	1010
5	0111	0101		11	1110	1011

be communicated to the controller. With 10-bit resolution, a robot with 6 joints would require 60 input lines to the controller. Consequently, it is necessary to consider the advantages and disadvantages of incremental and absolute encoders. Commercial encoders with as high as 15–16 bits are available.

Another method to increase the resolution of encoders is to add a supplemental light sensing device to it. In one rendition,[1] a faceted mirror was attached to the encoder shaft, which reflected a laser light onto a low-line-density diffraction grating. The diffracted light was projected onto an array of 200–8000 photodiodes. Depending on the angle of the shaft, the light reflected by the mirror onto the grating would change, therefore changing the output signals from the array. A combination of signals from the encoder and the photodiode array increases the resolution significantly, but at a great cost.

Figure 8.6 also shows the difference between a binary code and a gray code. In the binary code system, there are many instances where more than one set of bits change sign simultaneously, whereas in gray code, at any particular location, there is always only one bit-change to go back or forth. The importance of this difference is that in digital measurements, unlike popular perception, the values of signals are not constantly read, but the signal is measured (sampled) and held until the next sample reading. In binary code, where multiple bits change simultaneously, if all changes do not happen exactly at the same time, they may not all register. In gray code, since there is only one change, the system will always find it. Table 8.1 lists the gray code for numbers 0–11.

8.4.3 Linear Variable Differential Transformer (LVDT)

A linear variable differential transformer (or transducer) is actually a transformer whose core moves with the distance measured and that outputs a variable analog voltage as a result of this displacement. In general, a transformer is an electric-to-electric energy converter that changes the voltage/current ratio. Except for losses, the total input energy to the device is the same as the total output energy. As a transformer increases or decreases a voltage in proportion to the number of turns in its coils, the corresponding current changes inversely with it. This occurs because there are two coils with different numbers of turns. The electrical energy into one coil creates a flux, which induces a voltage in the second coil proportional to the ratio of the number of turns in the windings. As the number of turns in the secondary coil increases, the voltage increases proportionally, and consequently, the current decreases proportionally. However, the induction of voltage in the secondary is very much a function of the strength of the flux. If no iron core is present, the flux lines can disperse, reducing the strength of the magnetic field. As a result, the induction of voltage in the secondary will be minimal. In the presence of an iron core, the flux lines are gathered

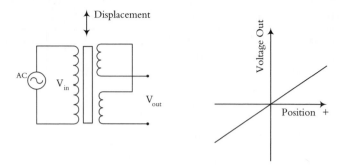

Figure 8.7 Linear Variable Differential Transformer.

inward, increasing the density of the field, and consequently, the induced voltage. This is used to create the variable output voltage in the LVDT, as in Figure 8.7. The output of an LVDT is very linear and proportional to the input position of the core.

8.4.4 Resolvers

Resolvers are very similar to LVDTs in principle, but are used to measure an angular motion. A resolver is also a transformer, where the primary coil is connected to the rotating shaft and carries an alternating current, either through slip rings or from a brushless transformer within it (Figure 8.8). There are two secondary coils, placed 90° apart from each other. As the rotor rotates, the flux it develops rotates with it. When the primary coil in the rotor is parallel to either of the two secondary coils, the voltage induced in that coil is maximum, while the other secondary coil perpendicular to it does not develop any voltage. As the rotor rotates, eventually the voltage in the first secondary coil goes to zero, while the second coil develops its maximum voltage. For all other angles in between, the two secondary coils develop a voltage proportional to the sine and cosine of the angle between the primary and the two secondary coils. Although the output of a resolver is analog, it is equal to the sine and cosine of the angle, eliminating the necessity to calculate these values later. Resolvers are reliable, robust, and accurate.

8.4.5 (Linear) Magnetostrictive Displacement Transducers (LMDT or MDT)

In this sensor, a pulse is sent through a conductor, which bounces back as it reaches a magnet. The time of travel to the magnet and back is converted to a distance if the speed of travel is known. By attaching the moving part to either the magnet or the conductor,

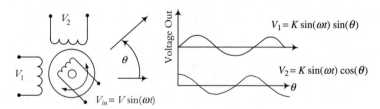

Figure 8.8 Schematic of a resolver.

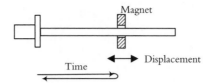

Figure 8.9 Schematic drawing of a magnetostrictive displacement sensor.

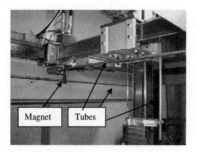

Figure 8.10 The IBM 7565 magnetostrictive displacement transducers. A pulse generated at one end of a long tube travels within the tube until it reaches a magnet and bounces back. The time of travel of the signal is converted to position information.

the displacement can be measured. A simple schematic of the sensor is shown in Figure 8.9. The IBM 7565 hydraulic gantry robot displacement sensors were of this type, as shown in Figure 8.10.

8.4.6 Hall-effect Sensors

A Hall-effect sensor works on the Hall-effect principle, where the output voltage of a conductor that carries a current changes when in the presence of a magnetic field. Therefore, the output voltage of the sensor changes when a permanent magnet or a coil that produces a magnetic flux is close to the sensor. A Hall-effect transducer's output is analog and must be converted for digital applications. It is used in many applications, including the sensing of the position of the permanent magnet rotors of brushless DC motors.

8.4.7 Other Devices

Many other devices can be used as position sensors, some novel and hi-tech, some simple and old. For example, in order to measure the angles of finger joints in a glove (such as in a virtual-reality glove), conductive elastomer strips were attached to the glove above the fingers. Conductive elastomer is a urethane-based synthetic rubber filled with conductive carbon particles. Its electrical resistance decreases as the tension on it increases. Therefore, as the finger bends within the glove, it stretches the strip, changing its resistance, which can be measured and converted to a position signal.[2]

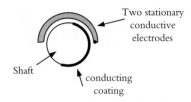

Figure 8.11 Shaft-angle measuring device based on a tunnel-diode oscillator and capacitance between a shaft and stationary electrodes.[3]

In another device, one-half side of a nonconductive shaft is coated with a conducting material. Two $\frac{1}{2}$-cylinder conductive electrodes with radii slightly larger than the shaft's are mounted concentrically over the shaft, creating a capacitor between the shaft and the stationary electrodes (Figure 8.11). As the shaft rotates, the capacitance changes, too. Used as a capacitor within a tunnel-diode oscillator circuit, the output frequency varies as the capacitance varies relative to the shaft position. Therefore, by measuring the frequency of the oscillation, the position of the shaft can be measured.[3]

8.5 Velocity Sensors

The following are the more common velocity sensors used in robotics. Their application is very much related to the type of position sensor used. Depending on the type of position sensor used, there may not even be a need to use a velocity sensor.

8.5.1 Encoders

If an encoder is used for displacement measurement, there is in fact no need to use a velocity sensor. Since encoders send a known number of signals for any given angular displacement, by counting the number of signals received in a given length of time dt velocity can be calculated. A typical number for dt may be 10 ms. However, if the encoder shaft rotates slowly, the number of signals received may be too small for an accurate calculation of velocity. On the other hand, if the time is increased in order to increase the total number of signals per cycle, the rate at which velocity is updated and sent to the controller will decrease. This will diminish the accuracy and effectiveness of the controller. In some systems, the cycle time dt is varied depending on the angular velocity of the encoder shaft. A smaller number is used if it rotates fast, increasing the effectiveness of the controller, and a larger number is used otherwise to gather enough data.

8.5.2 Tachometers

A tachometer is in fact a generator that converts mechanical energy into electrical energy. Its output is an analog voltage proportional to the input angular speed. It may be used along with potentiometers to estimate velocity. Tachometers are generally inaccurate at very low speeds.

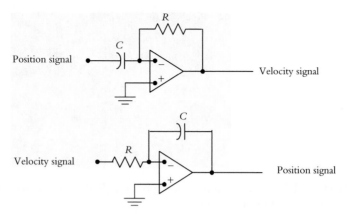

Figure 8.12 Schematics of differentiating and integrating R-C circuits with an op-amp.

8.5.3 Differentiation of Position Signal

If the position signal is clean, it is actually possible, and simple, to differentiate the position signal and convert it to velocity signal. To do this, it is necessary that the signal be as continuous as possible to prevent the creation of large impulses in the velocity signal. Therefore, it is recommended that a resistor with conductive polymer film be used for position measurement, as a wire-wound potentiometer's output is piecewise and unfit for differentiation. However, differentiation of a signal is always noisy and should be done very carefully. Figure 8.12 shows a simple R–C circuit with an op-amp that can be used for differentiation, where the velocity signal is:

$$V_{out} = -RC\frac{dV_{in}}{dt} \tag{8.4}$$

Similarly, the velocity (or acceleration) signal can be integrated to yield position (or velocity) signals as:

$$V_{out} = -\frac{1}{RC}\int V_{in}dt \tag{8.5}$$

8.6 Acceleration Sensors

Accelerometers are very common sensors for measuring accelerations. However, in general, accelerometers are not used with industrial robots. Recently, acceleration measurements have been used for high precision control of linear actuators[4] and for joint feedback control of robots.[5]

8.7 Force and Pressure Sensors

8.7.1 Piezoelectric

Piezoelectric material compresses if exposed to a voltage and produces a voltage if compressed. This was used in devices such as the phonograph to create a voltage from the

Figure 8.13   A typical force sensing resistor (FSR). The resistance of this sensor decreases as the force acting on it increases.

variable pressure caused by the grooves in the record. Similarly, a piece of piezoelectric can be used to measure pressures, or forces, in robotics. The analog output voltage must be conditioned and amplified for use.

8.7.2 Force Sensing Resistor

The Force Sensing Resistor (FSR) is a polymer thick-film device that exhibits a decreasing resistance with increasing force applied perpendicular to its surface. In one particular model, the resistance changes from about 500 $k\Omega$ to about 1 $k\Omega$ for forces of 10 to 10,000 gr (refer to References 6, 7, and 8 for more information about UniForce^{TM} sensors and others). Figure 8.13 shows a typical force sensing resistor.

8.7.3 Strain Gauge

A strain gauge can also be used to measure force. The output of the strain gauge is a variable resistance, proportional to the strain, which itself is a function of applied forces. Therefore, measuring the resistance, we can determine the applied force. Strain gauges are used to determine the forces at the end effector and the wrist of a robot. Strain gauges can also be used for measuring the loads on the joints and links of the robot, but this is not very common. Figure 8.14(a) is a simple schematic drawing of a strain gauge. Strain gauges are used within a Wheatstone bridge, as shown in Figure 8.14(b). A balanced Wheatstone bridge would have similar potentials at points A and B. If the resistance in any of the four resistors changes, there will be a current

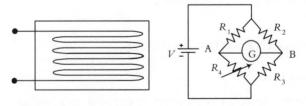

Figure 8.14   (a) A strain gauge and (b) a Wheatstone bridge.

flow between these two junctions. Consequently, it is necessary to first calibrate the bridge for zero flow in the galvanometer. Assuming that R_1 is the strain gauge, when under stress, its value will change, causing an imbalance in the Wheatstone bridge and a current flow between A and B. By carefully adjusting the resistance of one of the other resistors until the current flow becomes zero, the change in the resistance of the strain gauge can be determined from:

$$\frac{R_1}{R_4} = \frac{R_2}{R_3} \tag{8.6}$$

Strain gauges are sensitive to changes in temperature. To remedy this problem, a dummy strain gauge can be used as one of the four resistors in the bridge to compensate for temperature changes.

8.7.4 Antistatic Foam

The antistatic foam used for transporting IC chips is conductive and its resistance changes due to an applied force. It can function as a crude and simple, yet inexpensive, force and touch sensor. To use a piece of antistatic foam, insert a pair of wires into two sides of it and measure the voltage or resistance across it.

8.8 Torque Sensors

Torque can be measured by a pair of strategically placed force sensors. Suppose that two force sensors are placed on a shaft, opposite of each other, on opposite sides. If a torque is applied to the shaft, it generates two opposing forces on the shaft's body, causing strains in opposite directions. The two force sensors can measure the forces, which can be converted to a torque. To measure torques about different axes, three pairs of mutually perpendicular sensors must be used. However, since forces can also be measured with the same sensors, a total of six force sensors can generally report forces and torques about three axes, independent of each other, as depicted in Figure 8.15. Pure forces generate

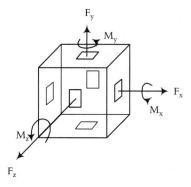

Figure 8.15 Arrangement of three pairs of strain gauges along the three major axes for force and torque measurements.

Figure 8.16 Typical industrial force/torque sensors. (IP65 Gamma and Mini 85, printed with permission from ATI Industrial Automation.)

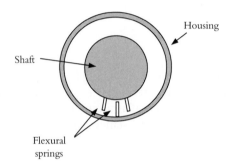

Figure 8.17 The torque can be measured by measuring the changes in the frequency of oscillation of a tunnel-diode oscillator when the capacitance of the flexural springs changes due to the applied torque.

similar signals in a pair, while torques generate pairs of signals with opposite signs. Figure 8.16 shows typical industrial force-torque sensors.

A miniature load sensor, designed to be used as fingertips for anthropomorphic robot hands, uses a spring instrumented with at least six strain gauges. The wires are attached to a small interface board at the base of the spring. The sensor is attached to an A/D converter as close to the sensor as possible. The data is transmitted to the controller by wires, routed at the neutral axis of the fingers.[9]

Figure 8.17 shows a schematic depiction of a system in which flexural springs, attached to a shaft, form a pair of capacitors used as part of a tunnel-diode oscillator circuit. As the shaft rotates slightly under the load, the capacitance of each pair changes, causing a change in the oscillation frequency of the circuit. By measuring the frequency of oscillations, the torque can be determined.[10]

8.9 Microswitches

Microswitches, though extremely simple, are very useful and common in all robotic systems. They cut off the electrical current, and therefore, can be used for safety purposes, for determining contact, for sending signals based on displacements, and many other uses. Microswitches are robust, simple, and inexpensive.

8.10 Visible Light and Infrared Sensors

These sensors react to the intensity of light projected onto them by changing their electrical resistance. If the intensity of light is zero, the resistance is at maximum. As the light intensity increases, the resistance decreases, and consequently, the current increases. These sensors are inexpensive and very useful. They can be used for making optical encoders and other devices as well. They are also used in tactile sensors, as will be discussed later.

A phototransistor can also be used as a light sensor, where in the presence of a certain intensity of light, it will turn on; otherwise, it will be off. Phototransistors are usually used in conjunction with an LED light source.

A light sensor array can be used with a moving light source to measure displacements as well. This has been used to measure deflections and small movements in robots and other machinery.[11] Light sensors are sensitive to the visible light range. Infrared sensors are sensitive to infrared range. Since infrared is invisible to human eyes, it will not disturb humans. For example, if a device needs light to measure a large distance for navigation purposes, infrared can be used without attracting attention or disturbing anyone. Simple infrared remote control devices are also available that can be used to establish remote control communication links between devices and robots. Refer to Reference 8 for specifications.

8.11 Touch and Tactile Sensors

Touch sensors are devices that send a signal when physical contact has been made. The simplest form of a touch sensor is a microswitch, which either turns on or off as contact is made. The microswitch can be set up for different sensitivities and ranges of motion. As an example, a strategically placed microswitch can send a signal to the controller if a mobile robot reaches an obstacle during navigation. More sophisticated touch sensors may send additional information. For example, a force sensor used as a touch sensor may not only send touch information, but also report the magnitude of the contact force.

A tactile sensor is a collection of touch sensors which, in addition to determining contact, can also provide additional information about the object. This additional information may be about the shape, size, or type of material. In most cases, a number of touch sensors are arranged in an array or matrix form, as shown in Figure 8.18. In this design, an array of six touch sensors is arranged on each side of a tactile sensor. Each touch

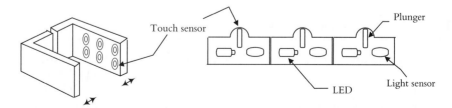

Figure 8.18 Tactile sensors are generally a collection of simple touch sensors arranged in an array form with a specific order to relay contact and shape information to the controller.

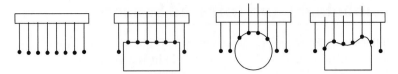

Figure 8.19   A tactile sensor can provide information about the object.

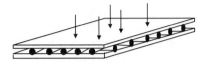

Figure 8.20   Skin-like tactile sensor.

sensor is made up of a plunger, an LED, and a light sensor. As the tactile sensor closes and the plunger moves in or out, it blocks the light from the LED projecting onto the light detector. The output of the light sensor is proportional to the displacement of the plunger. As you can see, these touch sensors are in fact displacement sensors. Similarly, other types of displacement sensors may be used for this purpose, from microswitches to LVDTs, pressure sensors, magnetic sensors, and so on.

As the tactile sensor comes in contact with an object, depending on the shape and size of the object, different touch sensors react differently at a different order. This information is then used by the controller to determine the size and the shape of the object. Figure 8.19 shows three simple set-ups, one touching a cube, one touching a cylinder, and one touching an arbitrary object. As can be seen, each object creates a different unique signature that can be used for detection.

Attempts have also been made to create somewhat of a continuous skin-like tactile sensor that could function similarly to human skin. In most cases, the design revolves around a matrix of sensors embedded between two polymer-type layers, separated by a mesh, as shown schematically in Figure 8.20. As a force is applied to the polymer, it is distributed between a few surrounding sensors, where each one sends a signal proportional to the force applied to it. For low resolution, these tactile sensors work satisfactorily.[12] Other designs include a similar polymer-type substrate populated with capacitive sensors. A microprocessor reads the sensors sequentially in order to determine the shape of the object and the contact force at each location. In another design, a flexible circuit board, populated with proximity sensors (see Section 8.12) provides a skin-like covering to help robots avoid collisions.[13]

8.12 Proximity Sensors

A proximity sensor is used to determine that an object is close to another object before contact is made. This noncontact sensing can be useful in many situations, from measuring the speed of a rotor to navigating a robot. There are many different types of proximity sensors, such as magnetic, eddy current and Hall-effect, optical, ultrasonic, inductive, and capacitive. The following is a short discussion of some of these sensors.

8.12.1 Magnetic Proximity Sensors

These sensors are activated when they are close to a magnet. They can be used for measuring rotor speeds (and the number of rotations) and turning a circuit on or off.[8] Magnetic proximity sensors may also be used to count the number of rotations of wheels and motors, and therefore, be used as position sensors. Imagine a mobile robot, where the total displacement of the robot is calculated by counting the number of times a particular wheel rotates, multiplied by the circumference of the wheel. A magnetic proximity sensor can be used to track wheel rotations by mounting a magnet on the wheel (or its shaft) and having the sensor stationary on the chassis. Similarly, the sensor can be used for other applications, including safety. For example, many devices have a magnetic proximity sensor that sends a signal when the door is open to stop the rotating or moving parts.

8.12.2 Optical Proximity Sensors

Optical proximity sensors consist of a light source called emitter (either internal to the sensor, or external to it), and a receiver, which senses the presence or the absence of light. The receiver is usually a phototransistor and the emitter is usually an LED. The combination of the two creates a light sensor, and is used in many applications, including optical encoders.

As a proximity sensor, it is set up such that the light, emitted by the emitter, is not reflected to the receiver unless an object is within range. Figure 8.21 is a schematic drawing of an optical proximity sensor. Unless a reflective object is within the range of the switch, the light is not seen by the receiver; therefore, there will be no signal.

Figure 8.22 shows another variation of an optical proximity sensor. In this simple system that can determine both proximity as well as short–range distance (and therefore

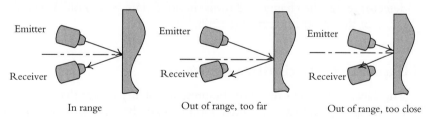

Figure 8.21 Optical proximity sensor.

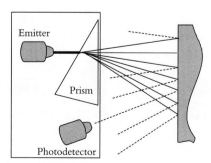

Figure 8.22 An alternative optical proximity sensor.

act as a range finder for short distances), a beam of light is passed through a prism that refracts the light into its constituent primary colors. Depending on the distance of the object from the sensor, one particular color of light is reflected back to the sensor's photodetector. By measuring the energy of the reflected light, the distance can be determined and reported.

8.12.3 Ultrasonic Proximity Sensors

In ultrasonic proximity sensors, an ultrasonic emitter emits frequent bursts of high frequency sound waves (usually in the 200 kHz range). There are two modes of operation for ultrasonic sensors, namely, opposed mode and echo (diffused) mode. In opposed mode, a receiver is placed in front of the emitter; in echo mode, the receiver is either next to, or integrated into, the emitter and receives the reflected sound wave. If the receiver is within range, or if the sound is reflected by a surface close to the sensor, it is sensed and a signal is produced. Otherwise, the receiver will not sense the wave and there is no signal. All ultrasonic sensors have a blind zone near the surface of the emitter in which the distance and presence of an object cannot be detected. Ultrasonic sensors cannot be used with surfaces such as rubber and foam that do not reflect the soundwaves in echo mode. For more information about ultrasonic sensors, refer to section 8.13.1. Figure 8.23 is a schematic drawing of this type of sensor.

8.12.4 Inductive Proximity Sensors

Inductive proximity sensors are used to detect metal surfaces. The sensor is a coil with a ferrite core, an oscillator/detector, and a solid state switch. In the presence of a metal object in the close vicinity of the sensor, the amplitude of the oscillation diminishes. The detector senses the change and turns the solid state switch off. When the part leaves the range of the sensor, it turns on again.

8.12.5 Capacitive Proximity Sensors

The capacitive sensor reacts to the presence of any object that has a dielectric constant more than 1.2. In that case, when within range, the material's capacitance raises the total

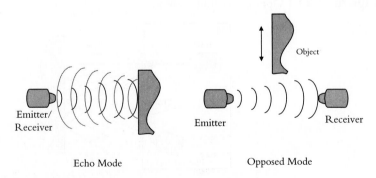

Figure 8.23 Ultrasonic proximity sensors.

Table 8.2 *Dielectric Constants for Select Materials.*

Air	1.000	Porcelain	4.4–7
Aqueous solutions	50–80	Cardboard	2–5
Epoxy resin	2.5–6	Rubber	2.5–3.5
Flour	1.5–1.7	Water	80
Glass	3.7–10	Wood, dry	2–7
Nylon	4–5	Wood, wet	10–30

capacitance of the circuit. This triggers an internal oscillator to turn on the output unit which will send out an output signal. Consequently, the sensor can detect the presence of an object within a range. Capacitive sensors can detect nonmetal materials such as wood, liquids, and chemicals. Table 8.2 shows dielectric constants for select materials.

8.12.6 Eddy Current Proximity Sensors

As we discussed in Chapter 7, when a conductor is placed within a changing magnetic field, an electromotive force (emf) is induced in it that causes a current to flow in the material. This current is called eddy current. An eddy current sensor typically has two coils, where one coil generates a changing magnetic flux as reference. In the close proximity of conducting materials, an eddy current is induced in the material, which in turn creates a magnetic flux opposite of the first coil's, effectively reducing the total flux. The change in the total flux is proportional to the proximity of the conducting material and is measured by the second coil. Eddy current sensors are used to detect the presence of conductive materials as well as the nondestructive testing of voids and cracks, thickness of materials, and so on.

8.13 Range Finders

Unlike proximity sensors, range finders are used to find larger distances, to detect obstacles, and to map the surfaces of objects. Range finders are meant to provide advance information to the system. Range finders are generally based on light—visible light, infrared light, or laser—and ultrasonics. Two common methods of measurement are triangulation and time-of-flight or lapsed time.

Triangulation involves illuminating the object by a single ray of light that forms a spot on the object. The spot is seen by a receiver such as a camera or photodetector. The range or depth is calculated from the triangle formed between the receiver, the light source, and the spot on the object, as shown in Figure 8.24.

As is evident from Figure 8.24(a), the particular arrangement between the object, the light source, and the receiver only occurs at one instant. At this point, the distance d can be calculated by:

$$\tan \beta = \frac{d}{l_1}, \quad \tan \alpha = \frac{d}{l_2} \quad \text{and} \quad L = l_1 + l_2$$

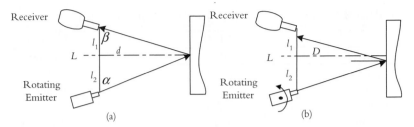

Figure 8.24   Triangulation method for range measurement. The receiver will only detect the spot on the object when the emitter is at a particular angle, which is used to calculate the range.

Substituting and manipulating the equation will yield:

$$d = \frac{L \tan \alpha \tan \beta}{\tan \alpha + \tan \beta} \tag{8.7}$$

Since L and β are known, if α is measured, d can be calculated. You can see from Figure 8.24(b) that except at that instant, the receiver will not see the reflected light. Consequently, it is necessary to rotate the emitter and, as soon as the reflected light is observed by the receiver, record the angle of the emitter and use it to calculate range. In practice, the emitter's light (such as laser) is rotated continuously by a rotating mirror and the receiver is checked for signal. As soon as the light is observed, the angle of the mirror is recorded.

Time of flight or **lapsed time** ranging consists of sending a signal from a transmitter that bounces back from an object and is received by a receiver. The distance between the object and the sensor is half the distance traveled by the signal, which can be calculated by measuring the time of flight of the signal and by knowing its speed of travel. This time measurement must be very fast to be accurate. For small distance measurements, the wavelength of the signal must be very small.

8.13.1 Ultrasonic Range Finders

Ultrasonic systems are rugged, simple, inexpensive, and low powered. They are readily used in cameras for focusing, in alarm systems for motion detection, and in robots for navigation and range measurement. Their disadvantage is in their limited resolution which is due to the wavelength of the sound and natural variations of temperature and velocity in the medium, and in their maximum range which is limited by the absorption of the ultrasound energy in the medium. Typical ultrasonic devices have a frequency range of 20 kHz to above 2 MHz.

Most ultrasonic devices measure the distance using the time-of-flight technique, in which, a transducer emits a pulse of high-frequency ultrasound that is reflected back when it encounters a separation in the medium and a receiver that receives the reflected signal. The distance between the transducer and the object is half the distance traveled, which is equal to the time-of-flight times the speed of sound. Of course, the accuracy of the measurement depends on the wavelength of the signal and the accuracy of the time measurement and the speed of sound. The speed of sound in a medium is dependent on the frequency of the wave (at above 2 MHz level) and the density and temperature of the medium. To increase the accuracy of the measurement, a calibration bar is usually placed about an inch in front of the transducer, which is supposed to calibrate the system for varying temperatures. This is only good if the temperature is uniform throughout the traveled distance, which may or may not be true.

Time measurement accuracy is also very important for accurately measuring distance. Usually, the worst case error in time measurement is $\pm 1/2$ wavelength if the clock is stopped as soon as the receiver receives the returned signal at a minimum threshold. Therefore, higher frequency ultrasound devices yield better accuracy. For example, for 20 kHz and 200 kHz systems, the wavelengths will respectively be about 0.67 and 0.067 inches (17 and 1.7 mm) yielding a minimum worst case accuracy of 0.34 and 0.034 inches (8.5 and 0.85 mm). Cross correlation, phase comparison, frequency modulation, and signal integration methods have been used to increase the resolution and accuracy of ultrasonic devices. However, although higher frequencies yield a better resolution, they attenuate much faster than the lower frequency signals, which severely limits their range. On the other hand, the lower frequency transducers have wide beam angles and a severely deteriorated lateral resolution. Consequently, there is a tradeoff between the lateral resolution and signal attenuation in relation with the beam frequency.

Background noise is another problem with ultrasonic sensors. Many different industrial and manufacturing operations and techniques produce soundwaves that contain ultrasonics as high as 100 kHz, which can interfere with the ultrasonic device operation. As a result, it is recommended that frequencies above 100 kHz be utilized in industrial environments.

Ultrasonics can be used for distance measurement, mapping, and flaw detection. A single-point distance measurement is called *spot checking*, versus *range array acquisition* for multiple data point acquisition techniques used for 3-D mapping. In this case, a large number of distances to different locations on an object are measured. The collection of distance data provides a 3-D map of the surface of the object. It should be noted that since only half the surface area of a 3-D object can be ranged, these measurements are also referred to as 2.5-D. The backside of the object or areas obscured by other parts cannot be ranged.

8.13.2 Light-Based Range Finders

Light (including infrared and laser)-based range finders measure the distance from an object by three different methods: direct time delay measurement, indirect amplitude modulation, and triangulation. The direct time delay measurement method measures the time required for a collimated beam of light (usually laser, since it does not divert) to travel to an object and back, similar to an ultrasonic sensor. Since the speed of light in air is 186,000 miles/sec (300,000 km/sec), it travels about 1 ft (30 cm) in 1 ns. Therefore, extremely high speed electronics and high resolutions are required to use this method.

In one indirect method, the time delay is measured by modulating a long burst of light with a low-frequency sinusoidal wave (Time-to-Amplitude Converter, TAC) and measuring the phase difference between the modulations between the emitted light and the backscattered light. This, in effect, is slowing down the wave speed to measurable scales by substituting the speed of light with low-speed modulations, but still taking advantage of the long travel range of laser lights.

Triangulation is the common technique used in range finding using light beams. For shorter distances encountered in navigation, triangulation yields the most accurate and best resolution among the three different techniques.

Another technique for measuring range with light sources is stereo imaging, which we will discuss in Chapter 9. A variation to this technique involves the use of a small laser pointer along with a single camera.[14] In this technique, the location of the laser light within the camera image is measured relative to the center of the image. Since the laser light and the axis of the camera are not parallel, the location of the laser dot within the image is a function of the distance between the object and the camera.

LIDAR (Light Detection and Ranging) is similar to radar, but uses light instead of radio waves. A beam of light (laser or infrared) is fired toward the target, and the properties of scattered light are measured to find the range and/or other information about a distant target. To gather information on a continuous basis, thousands of pulses of light are reflected by a rotating mirror. In a system developed by Velodyne Lidar, Inc., a set of 64 laser emitters fire thousands of pulses per second while the unit rotates between 5–15 Hz. It can collect data about the environment at 360 degrees azimuth and 27 degrees elevation, with a range of 120 meters.[15] Another time-of-flight laser-based sensor that measures distances up to 30 meters at a resolution of 0.25 degrees costs several thousands of dollars.

8.13.3 Global Positioning System (GPS)

This positioning system is based on a radio-navigation system for civilian use, freely available to anyone. With a GPS receiver, we can determine a global position and time that can be used for navigation and mapping. The system includes 29 satellites orbiting the Earth, a control and monitoring station on Earth, and the GPS receivers. The receiver uses the transmitted data from the satellites to calculate its position. This information can be sent directly to the control system of a mobile robot for positioning purposes and navigation.

Each satellite sends signals at precise intervals with information about the time the signal was sent and location of the satellite. The GPS unit reads the signals sent by four satellites and, using the difference between the current time and the time at which each signal was sent (which is contained in the message received), calculates the distance to the satellite. Each distance forms a sphere centered at the satellite, on which the GPS unit resides. The intersection between these spheres is the location of the GPS unit.

In theory, signals from only three satellites should suffice; the GPS unit should be able to determine its location relative to three satellites (two spheres intersect at a circle, and the circle generally intersects the third sphere at two points; the one closer to the Earth's surface is the desired location). However, because the signals move at the speed of light, the accuracy of the system is greatly dependent on the accuracy of the GPS unit's clock. The commercially mass-produced GPS clocks are not accurate enough to yield precise positioning. Therefore, the signal from a fourth satellite is also used to increase the accuracy of the system from about 100 meters to about 20 meters. Military devices use a more accurate clock and high performance signals for improved positional accuracy.

A GPS unit can be integrated into a robotic system for navigation and positioning. The position information is fed into the microprocessor which uses it to decide the succeeding actions or motions. A 3-D roll–pitch–yaw compass may also be used for global direction and navigation. Although this compass is not a GPS system, it can provide directional information about the three axes of motion, and therefore, aid in controlling a robot's position and orientation.

8.14 Sniff Sensors

Sniff sensors are similar to smoke detectors. They are sensitive to particular gases and send a signal when they detect the gas. They are used for safety purposes as well as for search and detection purposes.[16,17]

8.15 Taste Sensors

A taste sensor is a device that determines the composition of particles in a medium. One device uses an array of potentiometric sensors to evaluate the five basic tastes of sweetness, bitterness, sourness, saltiness, and umami (although no smell has been integrated into the system yet). To distinguish different varieties of wine, a wine-tasting artificial tongue uses an array of ion-sensitive field-effect transistors within a single chip to measure relative levels of ions of sodium, potassium, calcium, copper, and silver. These are used to evaluate and classify samples of wine.[18] Another sensor uses ion-specific electrodes, oxidation/reduction sensor pairs, an electrical conductivity sensor, and an array of galvanic cells to measure the presence of contaminants such as copper, zinc, lead, and iron ions in water as low as 10 ppm.[19] This information can either be used directly, or in combination with other data, in robotic systems and automated activities.

8.16 Vision Systems

Vision systems are perhaps the most sophisticated sensors used in robotics. Due to their importance and complexity, they will be discussed separately in Chapter 9. However, note that vision systems are, in fact, sensors, and that they relate the function of a robot to its environment as do all other sensors.

8.17 Voice Recognition Devices

Voice recognition involves determining what is said and taking an action based on the perceived information. Voice recognition systems generally work on the frequency content of spoken words. As you may remember from other courses, any signal may be decomposed into a series of sines and cosines of different frequencies at different amplitudes, which will reconstruct the original signal if combined. (We will discuss this in more detail in Chapter 9.) However, it is useful to realize that all signals have certain major frequencies that constitute a particular spectrum and that this spectrum is generally different for other signals. In voice recognition systems, it is assumed that every word (or letter or sentence), when decomposed into its constituent frequencies, will have a unique signature composed of its major frequencies, which allow the system to recognize the word.

To do this, the user must train the system by speaking the words *a priori* to allow the system to create a look-up table of the major frequencies that represent the spoken words. Later, when a word is spoken and its frequencies are determined, the result is compared to the look-up table. If a close match is found, the word is recognized. For better accuracy, it is necessary to train the system with more repetitions. On the other hand, a more accurate list of frequencies will reduce allowable variations. This means that if the system tries to match many frequencies for better accuracy, in the presence of any noise or any variations in the spoken words, the system will not be able to recognize the word. On the other hand, if a limited number of frequencies is matched in order to allow for variations, the system may recognize a similar, but incorrect, word. A universal system that recognizes all accents and variations in speaking may not be either possible or useful. Many robots have been equipped with voice recognition systems in order to communicate with the users. In most cases, the robot is trained by the user and can recognize words that trigger a certain action in response. For example, a particular word may be programmed to relate to a certain position and orientation. When the voice recognition system recognizes the word, it will send a signal to the controller which, in turn, will run the robot to the desired location and orientation. This has been particularly useful with robots that aid the disabled as well as for medical robots.

8.18 Voice Synthesizers

Voice synthesis is accomplished in two different ways. One way is to re-create the words by combining phonemes and vowels. In this case, each word is recreated when the phonemes and vowels are combined. This can be accomplished with commercially available phoneme chips and a corresponding program. Although this type of system can reproduce any word, it sounds unnatural and machine-like. As an example of the difficulty encountered by this kind of system, consider the two words "power" and "mower." Although both words are written very similarly, they are pronounced differently. This kind of a system will not be able to recognize this (unless every conceivable exception is programmed into the chip).

The alternative is to record the words that the system may need to synthesize and to access them from memory as needed. Telephone announcements, video games, and many other machine voices are pre-recorded and accessed as needed. Although this system sounds very natural, it is limited. As long as all the words the machine needs to say are known *a priori*, this system can be used. With advances in computer technology, voice recognition and synthesis will be advanced significantly in the future.

8.19 Remote Center Compliance (RCC) Device

Although this device is not an actual sensor, it is discussed here because it acts as a sensing device for misalignments and provides a means of correction for robots. However, remote center compliance (RCC) devices, also called *compensators*, are completely passive and there are no input or output signals.

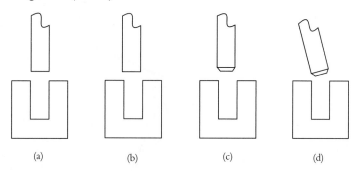

Figure 8.25 Misalignment of assembling elements.

An RCC device is an attachment added to the robot between the wrist and the end effector. It is designed to provide a means of correction for misalignments between the end effector and a part.

Suppose a robot is to push a peg into a hole in a part, as shown in Figure 8.25. If the hole and the peg are exactly the right sizes, and if they are exactly aligned, both laterally and axially, the robot may push the peg into the hole. However, this is often impossible to achieve. Imagine the hole is slightly off such that the centerline of the hole and the peg are a small distance apart, as in Figure 8.25(b).

If the robot is in position-control mode, it will attempt to push the peg into the hole even if there is a misalignment. As a result, either the robot or the part will deflect or break. A stiffer robot, a sign of a "good" robot, worsens this problem. If the robot has some compliance, it is actually possible to cut a chamfer (Figure 8.25(c)) around the hole (or the peg, or both) to allow the robot to move laterally to align itself with the hole and prevent deflections or breakage. Alternately, it is possible to allow the part to move to align itself with the robot.

Now assume that instead of an axial misalignment, there is an angular (cocking) misalignment between the two centerlines (as in Figure 8.25(d)). In this case, even if the peg and the hole are exactly aligned at the mouth of the hole, if the peg is pushed in, one of the two will have to either deflect or break, unless one is allowed to move. However, a compliant robot that "gives" enough to prevent breakage will probably have unacceptable accuracy.

Imagine that in order to resolve these problems, a spring is used to connect the end effector to the robot wrist. In this case, the misalignment can be overcome, but the compliant connection between the robot and the part does not allow insertion of the peg into the hole; the spring simply compresses instead. Therefore, a device is needed that can provide selective compliance to the end effector to allow the robot to correct itself in directions where correction is needed but without affecting its accuracy in other directions. A remote center compliance device provides this selective compliance through a simple 4-bar mechanism.

To understand how the RCC device works, consider a simple 4-bar mechanism as shown in Figure 8.26. In a mechanism, there are a total of $M = n \times (n-1)/2$ instantaneous centers of zero velocity, where n is the number of links, including the ground. Each instantaneous center of zero velocity is a point where the instantaneous velocity of one body *relative* to another is zero. In a 4-bar mechanism, there will be a total

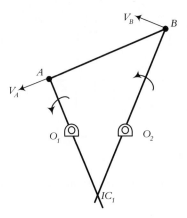

Figure 8.26   Instantaneous centers of zero velocity for a 4-bar mechanism.

of six such centers. Each of the two pin joints O_1 and O_2 attached to the ground is a center of rotation for the two links attached to the ground. The other two pin joints A and B are centers of rotation (or zero velocity) of the coupler AB relative to links O_1A and O_2B, and vice versa. However, in addition to these, there are two more centers of instantaneous zero velocity, one between the ground and the coupler and one between the two links O_1A and O_2B.

To find the instantaneous center of zero velocity for the coupler (in which we are interested for this subject), we need to find the velocities of two arbitrary points on it. The instantaneous center of zero velocity for the coupler will be at the intersection of two lines perpendicular to the velocities of the two points on the coupler, such as points A and B. This is true because, since $\bar{V} = \bar{\omega} \times \bar{\rho}$, the velocity of any point is normal to its radius of curvature $\bar{\rho}$. As a result, the instantaneous center of zero velocity must be somewhere on the normal-to-velocity line (which is along the length of each link), where two such lines intersect. Since this point has a zero instantaneous velocity, it means that at this instant, it is not moving, and consequently, the body *must* be rotating about it. Therefore, at the instant shown, the coupler link AB is rotating about point IC_1. This point will be at another location in the next instant, and as a result, its acceleration cannot be zero.

Now consider a parallelogram 4-bar mechanism as shown in Figure 8.27(a). Since the two normals to the two velocities at A and B are parallel, the instantaneous center of zero

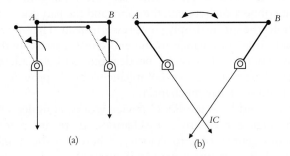

Figure 8.27   Special 4-bar mechanisms, the basis for a remote-center compliance device.

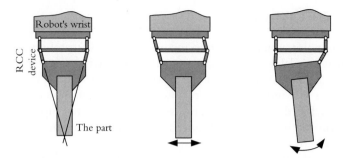

Figure 8.28 Schematic depiction of how an RCC device operates.

velocity for the coupler will be at infinity, indicating that the coupler is not rotating but is in pure translation. This means that the coupler will always translate to the left or right without any rotation (although its motion is curvilinear). Figure 8.27(b) shows a 4–bar mechanism with two links of equal lengths and the instantaneous center of zero velocity for its coupler which allows an instantaneous rotation of the coupler link about the *IC*. These two mechanisms can provide simple translation or rotation about a remote center when needed. An RCC device is a combination of these two mechanisms such that when needed, it can provide slight translation or rotation of the object about a distant point (therefore remote-center compliance). The distant point is the point of contact between the two parts, such as the peg and the hole, which is remote from the robot. However, you realize that this compliance is only lateral (or angular), where it is needed. The robot is still axially stiff, since the mechanism does not provide any motion in the direction normal to the coupler. As a result, it provides a selective compliance in the direction needed, without reducing the robot's stiffness, and consequently, its accuracy.

Figure 8.28 is a schematic drawing of how an RCC device works. In reality, each device provides a certain stiffness (or compliance) in lateral and axial directions, or in bending and cocking directions, and must be picked based on need. Each device also has a given center-to-center distance, which determines its remote center location relative to the center of the device. Therefore, there may be a need for multiple RCC devices if more than one part or operation is performed, and it must be picked accordingly.[20] Figure 8.29 shows a commercial RCC device.

Figure 8.29 A commercial RCC device. (Printed with permission from ATI Industrial Automation Corporation.)

8.20 Design Project

At this point, you may want to incorporate into your robots as many sensors as you want or have available to you. Some of the sensors will be necessary for feedback, which are essential if you are to control the robots. Others are added based on need and availability. This is a very interesting part of any robotic project. You may experiment with different sensors for different applications—and even come up with your own. You may experiment with other sensors that have not been mentioned here but are available from electronic warehouses.

Similarly, you may integrate sensors to the rolling-cylinder or sphere robot rovers for control and added intelligence. For example, visible light and infrared sensors located in the center of their platform will allow you to communicate with the rover by projecting a visible or infrared light beam through the gap between the cylinders. Similarly, you may route a sensor's wires through the central shaft of the robot and connect to the microprocessor. This way, without the need for slip-rings, you can have sensors outside of the robots. Proximity sensors and range finders can also be used to determine proximity or distance to walls and other obstacles and for navigating in different environments.

Summary

In this chapter, we discussed a variety of different sensors used in conjunction with robots and robotic applications. Some of these sensors are used for internal feedback. Others are used for communication between the robot and the environment. Some sensors are easy to use and inexpensive while others are expensive, difficult to use, and require a lot of support circuitry. Each sensor has its own advantages and disadvantages. As an example, an incremental encoder can provide simple, digital, position, and velocity information with minimum input requirements. However, the absolute position cannot be measured with it. An absolute encoder provides absolute position information in digital form but requires many lines of input that may not be available. A potentiometer can also provide absolute position information, is very simple to use, and is very inexpensive, but its output is in analog form and must be digitized before a microprocessor can use it. However, in some applications, an encoder and a potentiometer are used together, one to report the absolute position at wake-up and one to accurately report the changes in the position. Together, they provide all the information needed to run the system. It is the role of the design engineer to decide what type of sensor is needed or is best suited for a particular application.

References

1. "Higher Resolution Optoelectronic Shaft-Angle Encoder," NASA Tech Briefs, March 2000, pp. 46–48.

2. "Glove Senses Angle of Finger Joints," NASA Tech Briefs, April 1998.

3. "Shaft-Angle Sensor Based on Tunnel-Diode Oscillator," NASA Tech Briefs, July 2008, pp. 22–24.

4. Tan, K. K., S. Y. Lim, T. H. Lee, H. Dou, "High Precision Control of Linear Actuators

Incorporating Acceleration Sensing," Journal of Robotics and Computer Integrated Manufacturing, Vol. 16, No. 5, October 2000, pp. 295–305.

5. Xu, W. L., J. D. Han, "Joint Acceleration Feedback Control for Robots: Analysis, Sensing, and Experiments," Journal of Robotics and Computer Integrated Manufacturing, Vol. 16, No. 5, October 2000, pp. 307–320.

6. Interlink Electronics, Santa Barbara, California.

7. Force Imaging Technologies, Chicago, IL.

8. Jameco Electronics Catalog, Belmont California.

9. "Miniature Six-Axis Load Sensor for Robotic Fingertips," NASA Tech Briefs, July 2009, p. 25.

10. "Torque Sensor Based on Tunnel-Diode Oscillator," NASA Tech Briefs, July 2008, p. 22.

11. Puopolo, Michael G., Saeed B. Niku, "Robot Arm Positional Deflection Control with a Laser Light," Proceedings of the Mechatronics '98 Conference, Skovde, Sweden, Adolfsson and Karlsen Editors, Pergamon Press, Sep. 98, pp. 281–286.

12. Hillis, Daniel, "A High Resolution Imaging Touch Sensor," Robotics Research, 1:2, MIT press, Cambridge, MA.

13. "Flexible Circuit Boards for Modular Proximity-Sensor Arrays," NASA Tech Briefs, January 1997, p. 36.

14. Niku, S. B., "Active Distance Measurement and Mapping Using Non Stereo Vision Systems," Proceedings of Automation '94 Conference, July, 1994, Taipei, Taiwan, R.O.C., Vol. 5, pp. 147–150.

15. www.velodyne.com/lidar.

16. "Sensors that Sniff," High Technology, February 1985, p. 74.

17. "Electronic Noses Made From Conductive Polymer Films," NASA Tech Briefs, July 1997 pp. 60–61.

18. "Robotic Taster," Mechanical Engineering, October 2008, p. 20.

19. "Electronic Tongue for Quantization of Contaminants in Water," NASA Tech Briefs, February 2004, pp. 31–32.

20. ATI Industrial Automation Catalogs for Remote Center Compliance Devices.

CHAPTER 9

Image Processing and Analysis with Vision Systems

9.1 Introduction

A very large body of work is associated with vision systems, image processing, and pattern recognition that addresses many different hardware and software related topics on these subjects. This information has been accumulated since the 1950s, and with the added interest in the subject from different sectors of the industry and economy, it is rapidly growing. The enormous number of papers published every year indicates that many useful techniques constantly appear in the literature. At the same time, it also means that many of these techniques may be suitable for certain applications only. In this chapter, we will study and discuss some fundamental techniques for image processing and image analysis, with a few examples of routines developed for certain purposes. This chapter does not intend to be a complete survey of all possible vision routines, but only an introduction. If interested, it is recommended that you continue studying the subject through other references.

9.2 Basic Concepts

The following sections include some fundamental definitions of terms and basic concepts that we will use throughout the chapter.

9.2.1 Image Processing versus Image Analysis

Image processing relates to the preparation of an image for later analysis and use. Images, as captured by a camera or other similar techniques (such as a scanner), are not necessarily in a form that can be used by image analysis routines. Some may need improvement to reduce noise; some may need to be simplified; others may need to be enhanced, altered, segmented, filtered, and so on. Image processing is the collection of routines and techniques that improve, simplify, enhance, and otherwise alter an image.

Image analysis is the collection of processes by which a captured and processed image is analyzed to extract information about the content and to identify objects or other related facts about the objects within the image or the environment.

9.2.2 Two- and Three-Dimensional Image Types

Although all scenes are three-dimensional, images can either be two-dimensional (lacking depth information) or three-dimensional (containing depth information). Most images with which we normally deal, obtained by cameras, are two-dimensional. However, other systems such as Computed Tomography (CT) and CAT-scans create three-dimensional images that contain depth information. Therefore, these images can be rotated about different axes in order to better visualize the depth information. A two-dimensional image is extremely useful for many applications even though it has no depth information. This includes feature extraction, inspection, navigation, parts handling, and many more.

Three-dimensional images are used with applications that require motion detection, depth measurement, remote sensing, relative positioning, and navigation. CAD/CAM-related operations also require three-dimensional image processing, as do many inspection and object recognition applications. For three-dimensional images, either X-rays or ultrasonics are used to get images of one slice of the object at a time; later, all images are put together to create a three-dimensional image representation of the internal characteristics of the object.

All three-dimensional vision systems share the problem of coping with many-to-one mapping of scenes to images. To extract information from these scenes, image processing techniques are combined with artificial intelligence techniques. When the system is working in environments with known characteristics (e.g., controlled lighting), it functions with high accuracy and speed. On the contrary, when the environment is unknown or noisy and uncontrolled (e.g., underwater operations), the systems are not very accurate and require additional processing of the information, and therefore, operate at lower speeds.

9.2.3 The Nature of an Image

An image is a representation of a real scene, either in black and white (B/W) or in color, and either in print or in digital form. Printed images may be reproduced either by multiple colors and gray scales, such as CMYK in color print or halftone black and white print, or by a single ink source. For example, to reproduce a photograph with real halftones, we use multiple gray inks, which when combined, produce a somewhat realistic image. However, in most print applications, only one ink color is available (such as black ink on white paper in a newspaper or copier). In this case, all gray levels must be produced by changing the ratio of black versus white areas (the size of the black dot). Imagine that a picture to be printed is divided into small sections. In each section, if the ink portion of the section is smaller than the white, the section will look a lighter gray

Figure 9.1 Examples of how gray intensities are created in printed images. In print, only one color ink is used, while the ratio of black to white area of the pixel is changed to create different gray levels.

(Figure 9.1). If the black ink area is larger than the white area, it will look a darker gray. By changing the size of the printed dot, many gray levels may be produced and, collectively, a gray scale picture may be printed.

Similar to printed images, electronic and digital images are also divided into small sections called picture cells, or *pixels* (in three-dimensional images, it is called volume cell or *voxel*), where the size of all pixels is the same. To capture an image, the intensity of each pixel is measured and recorded; similarly, to recreate an image, the intensity of light at each pixel location is varied. Therefore, an image file is the collection of the data representing the light intensities of a large number of pixels. This file can be recreated, processed, modified, or analyzed. A color image is essentially the same, except that the original image is separated into three images of red, green, and blue before capturing and digitization. When the three colors with different intensities at each pixel location are superimposed, color images are recreated.

9.2.4 Acquisition of Images

There are two types of vision cameras: analog and digital. Analog cameras are no longer common but are still around and used to be the standard camera at television stations. Digital cameras are the current standard and are practically all the same. A digital moving-picture camera is similar to a digital still-camera plus a recording section; otherwise, the mechanism of image acquisition is the same. Whether the captured image is analog or digital, in vision systems, the image is eventually digitized. In digital systems, all data is in binary form and is stored in a computer file or memory device. Therefore, we ultimately deal with a file of numbers 0 and 1, from which we extract information and make decisions.

Appendix B presents a short discussion about analog and digital cameras and how the image is captured. The final outcome of these systems is a file that contains sequential pixel-location and pixel-intensity data that we use in our discussions. Refer to Appendix B for understanding the fundamental issues about image acquisition. For more information about details of these systems, refer to digital data acquisition references.

9.2.5 Digital Images

The light intensities at each pixel location are measured and converted to digital form regardless of the type of camera or image acquisition system. The data is either stored in

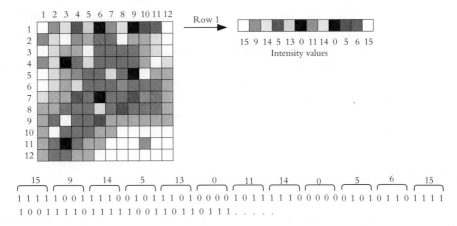

Figure 9.2 An image and the binary representation of its first row using 4 bits per pixel.

memory, in a file, or in recording devices with an image format such as TIFF, JPG, Bitmap, and so on, or is displayed on a monitor. Since it is digitized, the stored information is a collection of 0s and 1s that represent the intensity of light at each pixel; a digitized image is nothing more than a computer file that contains the collection of these 0s and 1s, sequentially stored to represent the intensity of light at each pixel. However, these files can be accessed and read by a program, duplicated and manipulated, or rewritten in a different form. Vision routines generally access this information and perform some function on the data and either display the result or store the manipulated result in a new file. A fundamental issue is to be able to extract information or manipulate this collection of 0 and 1 values in a meaningful way.

To understand this better, consider the simple low resolution image in Figure 9.2. Each pixel is referred to by row and column numbers. Assuming the system is digitized with only 4 bits (we will discuss this further shortly), there will be up to $2^4 = 16$ distinct light intensities possible. The sequence of 0 and 1 numbers representing the first row of the image will look as shown in the figure (all with only 4 bits per pixel). Different file formats list these numbers differently. In a simple Portable Gray Map (PGM) format, the intensities are listed sequentially as shown. A header at the beginning of the file will indicate the number of pixels in each row and column (in this case, it is 12 × 12). The program knows that every 4 bits is 1 pixel. Therefore, it can access each pixel intensity directly. However, as you notice, the file is reduced to a string of 0 and 1 values, to which, image processing routines are applied, and from which, information is extracted. Now imagine the size of the string of 0 and 1 values that represents a large image (sometimes in mega-pixel range) at up to 24 bits per pixel, for three primary colors.

An image with different gray levels at each pixel location is called a gray image. A color image results by superimposing three images of red, green, and blue hues (RGB), each with a varying intensity and each equivalent to a gray image (but in one of the three hues). Therefore, when the image is digitized, it will similarly have strings of 0s and 1s for each hue (an alternative way is to assign a number to each color, all declared in a header at the beginning of the image file. Then the number representing the pixel represents the color reference and intensity). A binary image is an image where each pixel is either fully light or fully dark, either a 0 or a 1. To achieve a binary image, in most cases, a gray image is converted using the histogram of the image and a cut-off value called a threshold. A histogram of the pixel gray levels will determine the distribution of the different gray levels. We can pick a value that best determines a cut-off level with the least distortion and use this value as a threshold to assign a 0 (or off) to all pixels whose gray levels are below the threshold value and to assign a 1 (or on) to all pixels whose gray values are above the threshold. Changing the threshold will change the binary image. The advantage of a binary image is that it requires far less memory, and it can be processed much faster than gray or colored images.

9.2.6 Frequency Domain versus Spatial Domain

Many processes used in image processing and analysis are either based on frequency domain or on spatial domain. In frequency domain processing, the frequency spectrum of the image is used to alter, analyze, or process the image. In this case, the individual pixels and their contents are not used. Instead, a frequency representation of the whole image is used for the process. In spatial domain processing, the process is applied to the individual pixels of the image. As a result, each pixel is affected directly by the process. Both techniques are equally important and powerful and are used for different purposes. It should be noted here that although spatial and frequency domain techniques are used differently, they are both related. For example, suppose a spatial filter is used to reduce noise in an image. As a result of this filter, noise level in the image will be reduced, but at the same time, the frequency spectrum of the image will also be affected due to this reduction in noise.

The following sections discuss some fundamental issues about frequency and spatial domains. This discussion, although general, will help us throughout the entire chapter.

9.3 Fourier Transform and Frequency Content of a Signal

As you may remember from your mathematics or other courses, any periodic signal may be decomposed into a collection of sines and cosines of different amplitudes and frequencies, called Fourier series, as follows:

$$f(t) = \frac{a_0}{2} + \sum_{n=1}^{\infty} a_n \cos n\omega t + \sum_{n=1}^{\infty} b_n \sin n\omega t \qquad (9.1)$$

When these sines and cosines are added together, the original signal is reconstructed. This conversion to frequency domain is called Fourier series, and the collection of different frequencies present in the equation is called *frequency spectrum* or *frequency content* of the signal. Of course, although the signal is in the amplitude-time domain, the frequency spectrum is in the amplitude-frequency domain. To understand this better, let's look at an example.

Consider a signal in the form of a simple sine function like $f(t) = \sin(t)$. Since this signal consists only of one frequency and a constant amplitude, the frequency spectrum representing it consists of a single value at the given frequency, as shown in Figure 9.3. Obviously, if we plot the function represented by the arrow in Figure 9.3(b) with the given frequency and amplitude, we will have the same sine function reconstructed. The

plots in Figure 9.4 are similar and represent $f(t) = \sum_{n=1,3\cdots15} \frac{1}{n} \sin(nt)$. The frequencies are

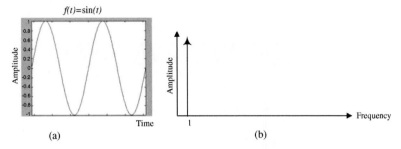

Figure 9.3 Time domain and frequency domain plots of a simple sine function.

also plotted in the frequency–amplitude domain. As you can see, when the number of frequencies contained in $f(t)$ increases, the summation gets closer to a square function.

Figure 9.5(a) shows a signal from a sensor and its frequency content. Although the signal is not a true sine function, the dominant frequency is 0.75 Hz. However, due to these discrepancies and the variations in the signal, the frequency spectrum contains many

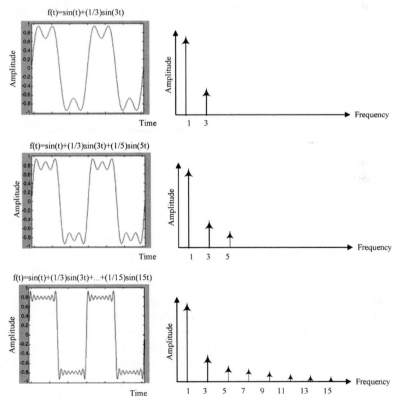

Figure 9.4 Sine functions in time and frequency domains for a successive set of frequencies. As the number of frequencies increases, the resulting signal gets closer to a square function.

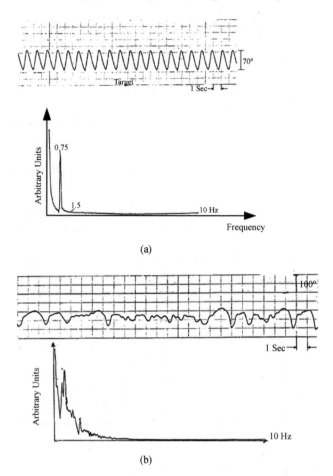

Figure 9.5   Two signals and their frequency spectrums.

other frequencies. Figure 9.5(b) shows a signal with more frequent variations and its frequency spectrum. Clearly, many more sine and cosine functions must be added in order to reconstruct this signal; therefore, the spectrum contains many more frequencies.

Theoretically, to reconstruct a square wave from sine functions, an infinite number of sines must be added together. Since a square wave function represents a sharp change, this means that rapid changes (such as an impulse, a pulse, square wave, or other similar functions) decompose into a large number of frequencies. The sharper the change, the higher the number of frequencies needed to reconstruct it. Therefore, any video or other signal that contains sharp changes (noise, edges, high contrasts, impulse, step function) or has detailed information (high resolution signals with fast, varying changes) will have a larger number of frequencies in its frequency spectrum.

A similar analysis can be made on nonrepeating signals too (called Fourier Transform, and particularly, *Fast Fourier Transform* or FFT). Although we will not discuss the details of the Fourier transform in this book, suffice it to say that an approximate frequency spectrum of any signal can be found. Although theoretically there will be infinite frequencies in the spectrum, generally there will be some major frequencies within the spectrum with larger amplitudes called *harmonics*. These major frequencies or harmonics are used in identifying and labeling a signal, including recognizing voices, shapes, objects, and the like.

9.4 Frequency Content of an Image; Noise, Edges

Figure 9.6 shows a low resolution artificial image and a graph of its pixel intensities versus their positions. A representation such as this may be obtained when an image is scanned by an analog camera or a frame grabber is used with a digital system to sample and hold the data (see Appendix B). The graph is a discrete representation of varying amplitudes showing the intensity of light at each pixel (or versus time). Let's say we are on the 9th row and are looking at pixel numbers 129–144. The intensity of pixel number 136 is very different from the ones around it and may be considered noise, which is generally information that does not belong to the surrounding environment. The intensities of pixels 134 and 141 are also different from the neighboring pixels and may indicate a transition between the object and the background, and therefore, can be construed as an edge of the object.

Although this is a discrete (digitized) signal, as discussed earlier, it may be transformed into a large number of sines and cosines with different amplitudes and frequencies which, if added, will reconstruct the signal. As discussed earlier, portions of the signal that change slowly, such as small changes between succeeding pixel gray values, will require fewer sines and cosines to be reconstructed, and consequently, contribute more low frequencies to the spectrum. On the other hand, parts of the signal that vary quickly or significantly, such as large differences between pixel gray levels, require a large number of higher frequencies to be reconstructed and, as a result, contribute more high frequencies to the spectrum. Both noises and edges are among cases where one pixel value is substantially different from the neighboring pixels. Therefore, noises and edges contribute to the higher frequencies of a typical frequency spectrum, whereas slowly varying gray level sets of pixels, representing the object, contribute to the lower frequencies of the spectrum.

If a high frequency signal is passed through a low-pass filter—a filter that allows lower frequencies through without much attenuation in amplitude, but which severely

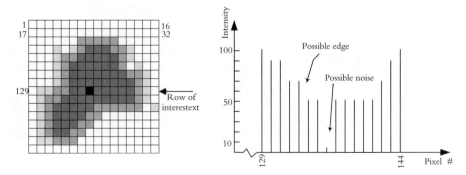

Figure 9.6 Noise and edge information in an intensity diagram of an image. The pixels with intensities much different from the neighboring pixels can be considered edges or noise.

attenuates the amplitudes of the higher frequencies in the signal—it will reduce the influence of all high frequencies, including the noises and edges. This means that although a low-pass filter reduces noise, it also reduces the clarity of an image by attenuating the edges and softening the image throughout. A high-pass filter, on the other hand, will increase the apparent effect of higher frequencies by severely attenuating the low frequency amplitudes. In such cases, noise and edges will be left alone, but slowly changing areas will disappear from the image. The application of different methods for noise reduction and edge detection will be discussed further in later sections of this chapter.

9.5 Resolution and Quantization

Two measures significantly affect the usefulness of an image and the data contained within it. The first one is resolution, which is affected by how often a signal is measured and read or sampled. Higher numbers of samples at equally spaced periodic times result in higher resolution, and therefore, more data. The resolution of an analog signal is a function of *sampling rate*. The resolution of a digital system is a function of how many pixels are present. Fundamentally, these two are the same measure; reading the light intensity of the image at more pixel locations is the same as sampling more often. Figure 9.7 shows an image sampled at (a) 432×576, (b) 108×144, (c) 54×72, and (d) 27×36 pixels. The clarity of the image is lost when the sampling rate decreases.

The second issue is how accurately the value of the signal at any given point is converted to digital form. This is called *quantization*—a function of how many bits are

(a)   (b)

(c)   (d)

Figure 9.7   Effect of different sampling rates on an image at (a) 432 × 576, (b) 108 × 144, (c) 54 × 72, and (d) 27 ×36 pixels. As the resolution decreases, the clarity of the image decreases accordingly.

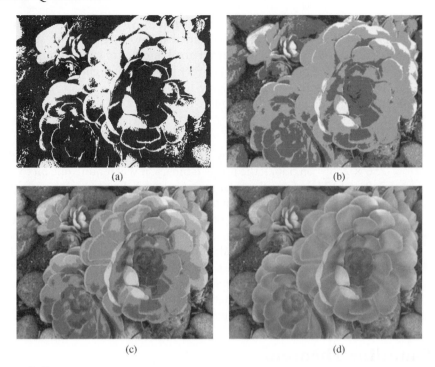

Figure 9.8 An image at different quantization levels of 2, 4, 8, and 44 gray levels. As the quantization resolution increases, the image becomes smoother.

used to represent the digitized magnitude of the sampled signal. Depending on the number of bits used for quantization, the grayness variations of the image will change. The total number of gray level possibilities is 2^n, where n is the number of bits. For a 1 bit analog to digital converter (ADC), there are only two possibilities, on or off, or 0 or 1 (called a binary image). For quantization with an 8-bit ADC, the maximum number of gray levels will be 256. Therefore, the image will have 256 different gray levels (0–255).

Quantization and resolution are completely independent of each other. For example, a high resolution image may be converted into a binary image, where there are only on and off pixels (0 and 1, or dark and light), or the same image may be quantized into 8 bits, which can yield a spectrum of 256 different shades of gray. Figure 9.8 shows the same image quantized at (a) 2 levels, (b) 4 levels, (c) 8 levels, and (d) the original at 44 levels.

Both the resolution and quantization must be sufficiently high in order to provide adequate information for a specific task. A low-resolution image may not be adequate for recognition of parts with high detail, but enough for distinguishing between a bolt and a nut. Low bit-count quantization may be enough for many applications where binary images are adequate, but not in others where different objects must be distinguished from each other. For example, a high-resolution image is necessary for reading the license plate of a car or recognition of faces with a security camera. However, because the license plate consists of primarily dark letters on a light background, even a binary image (only one bit per pixel) may be sufficient. A similar image must be quantized at a higher bit-count in order to allow face recognition. When choosing a camera, both these values must be considered.

The sampled light at a pixel, when quantized, yields a string of 0s and 1s representing the light at that pixel location. The total memory required to store an image is the product of the memory needed for the total number of samples (pixels) and the memory needed for each digitized sample. A larger image with higher resolution (total number of pixels) and a higher number of gray levels requires a larger memory size. The total memory requirement is a function of both values.

Example 9.1

Consider an image with 256 by 256 pixels. The total number of pixels in the image is $256 \times 256 = 65,536$. If the image is binary, it will require one bit to record each pixel as 0 or 1. Therefore, the total memory needed to record the image will be 65,536 bits, or with 8 bits to a byte, it will require 8192 bytes. If each pixel were to be digitized at the rate of 8 bits for 256 shades of gray, it would require $65,536 \times 8 = 524,288$ bits, or 65,536 bytes. For a video clip, changing at the rate of 30 images per second, the memory requirement will be $65,536 \times 30 = 1,966,080$ bytes per second. Of course, this is only the memory requirement for recording the image pixels, and does not include index information and other book-keeping requirements. The actual memory requirement may be less depending on the format in which the image is saved. ■

9.6 Sampling Theorem

Can you tell from Figure 9.9 what the image represents? Of course, since this is a very low-resolution 16×16 image, it is difficult to guess what the object is. This simple illustration signifies the relationship between sampling rate and the information obtained from it. To understand this, we will discuss some fundamental issues about sampling.

Consider a simple sinusoidal signal with frequency f as shown in Figure 9.10. Suppose the signal is sampled at the rate of f_s. The arrows in 9.10(b) show the corresponding sampled amplitudes.

Now suppose we want to use the sampled data to reconstruct the signal. This would be similar to sampling a sound source such as a CD and trying to reconstruct the sound signal from the sampled data through a speaker. One possibility would be that, by chance, the same signal might be reconstructed. However, as you can see in Figure 9.11, it is very

Figure 9.9 A low-resolution (16×16) image.

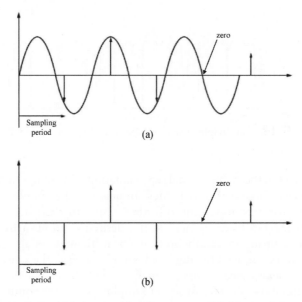

Figure 9.10 (a) Sinusoidal signal with a frequency of f, (b) sampled amplitudes at the rate of f_s.

possible that another signal may be reconstructed from the same data that is completely different from the original signal but yields the same sampled data. Both are valid, and many other signals can be valid and might be reconstructed from this sampled data too. This loss of information is called *aliasing* of the sampled data, and it can be a very serious problem.

In order to prevent aliasing, according to what is called *sampling theorem*, the sampling frequency must be at least twice as large as the largest frequency present in the signal. In that case, we can reconstruct the original signal without aliasing. The highest frequency present in the signal can be determined from the frequency spectrum of the signal. If a signal's frequency spectrum is found using the Fourier transform, it will contain many frequencies. However, as we have seen, the higher frequencies have smaller amplitudes. We can always pick a maximum frequency that may be of interest, while assuming that the frequencies with very low amplitudes beyond that point can be ignored without

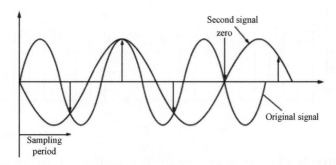

Figure 9.11 Reconstruction of signals from the sampled data. More than one signal may be reconstructed from the same sampled data.

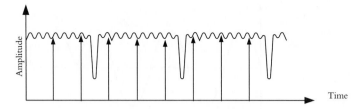

Figure 9.12   An inappropriate sampling rate may completely miss important data within a signal.

much effect in the system's total representation. The sampling rate of the signal must be at least twice as large as this frequency. In practice, the sampling rate is generally chosen to be larger than this minimum to further ensure that aliasing of the signal will not occur. Frequencies 4–5 times as large as the desired maximum frequency are common. For example, human ears can theoretically hear frequencies up to about 20,000 Hz. If a CD player is to reconstruct the digitized, sampled music, the sampling rate of the laser sensor must be at least twice as large, namely 40,000 Hz. In practice, CD players sample at the rate of about 44,100 Hz. At lower sampling rates, the sound may become distorted. In reality, if a signal changes more quickly than the sampling rate, the details of the change will be missed, and therefore, the sampled data will be inadequate. For example, it has been shown that the resulting vibration from a rotating gear with a broken tooth is distinctly different from a regular gear (Figure 9.12). However, at a low sampling rate, the sampled data may be completely void of this important information. Similarly, in Figure 9.13, the sampling rate is lower than the higher frequencies of the signal. As shown, although the lower frequencies of the signal are reconstructed, the signal will not have the higher frequencies of the original signal. The same is true with sound and image signals. If a sound signal is sampled at a low rate, the high frequency information will be lacking and the reconstructed sound will lack high frequency sounds. The output of the system, even if the best speakers are used, will be distorted and different from the real signal.

For images too, if the sampling rate is low, translating into a low resolution image, the sampled data may not have all the necessary detail; the information in the image is lost, and the image cannot be reconstructed like the original image. Figure 9.9 is sampled at a very low rate, and the information in it is lost. This is why you cannot decipher the

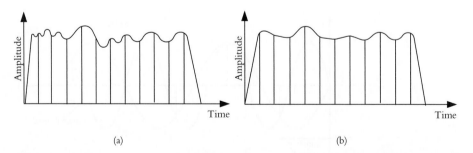

(a)   (b)

Figure 9.13   The original signal in (a) is sampled at a sampling rate lower than the higher frequencies of the signal. The reconstructed signal in (b) will not have the higher frequencies of the original signal.

Figure 9.14 The image of Figure 9.9, presented at higher resolutions of (a) 32 × 32, (b) 64 × 64, (c) 256 × 256.

image. However, when the sampling rate is increased, there will be a time when there is enough information to recognize the image. The still higher resolutions or sampling rates will transfer more information, and therefore, increasingly more detail can be recognized. Figure 9.14 is the same image as in Figure 9.9, but at 2, 4, and 16 times higher resolutions. Now suppose you need to recognize the difference between a bolt and a nut in a vision system in order to direct a robot to pick up the parts. Because the information representing a bolt and a nut is very different, a low-resolution image still enables you to determine what the part is. However, in order to recognize the license plate number of a car while moving in traffic, a high-resolution image is needed to extract enough information about the details such as the numbers on the license plate.

9.7 Image-Processing Techniques

As was mentioned earlier, image-processing techniques are used to enhance, improve, or otherwise alter an image and to prepare it for image analysis. Usually during image processing, information is not extracted from an image. Instead, the intention is to remove faults, trivial information, or information that may be important but not useful to improve the image. As an example, suppose an image was obtained while the object was moving, and as a result, the image is not clear. It would be desirable to see if the blurring in the image could be reduced or removed before the information about the object (such as its nature, shape, location, orientation, etc.) could be determined. Also consider an image corrupted by reflections due to direct lighting or an image that is noisy because of low light. In all these cases, it is desirable to improve the image and prepare it before image analysis routines are used. Similarly, consider the image of a section of a city fully detailed with streets, cars, shadows, and the like. It may actually be more difficult to extract information from this image than if all unnecessary detail, except for edges, were removed.

Image processing is divided into many sections, including histogram analysis, thresholding, masking, edge detection, segmentation, region growing, modeling, and many more. In the next sections, we will study some of these techniques and their applications.

9.8 Histogram of Images

A histogram is a representation of the total number of pixels of an image at each gray level. Histogram information is used in a number of different processes, including thresholding. For example, histogram information can help in determining a cutoff point for converting the image into binary form. It can also be used to decide if there are any prevalent gray levels in an image. For instance, suppose a systematic source of noise in an image causes many pixels to have one "noisy" gray level. A histogram can be used to determine the noisy gray level in order to attempt to remove or neutralize the noise. The same may be used to separate an object from the background so long as they have distinctly different colors or gray values.

Figure 9.15(a) shows a low-contrast image that has all its pixel gray levels clustered between two relatively close values. In this image, all pixel gray values are between 120 to 180 gray levels, at intervals of 4 (the image is quantized at 16 distinct levels between 0 to 256). Figure 9.15(c) shows the histogram of this image and, as you see, all pixel gray levels are between 120 to 180, a relatively low range. As a result, the image is not very clear and details are not visible. Now suppose we equalize the histogram such that the same 16 gray levels present in the image are spread out between 0 to 255 gray levels at intervals of 17, instead of the present 120–180 at intervals of 4. As a result of this histogram equalization, the image is vastly improved, as shown in Figure 9.15(b), with its corresponding histogram in (d). Notice that the number of pixels at each gray level are exactly the same in both cases, but the gray levels are spread out. The grayness values are given in Table 9.1.

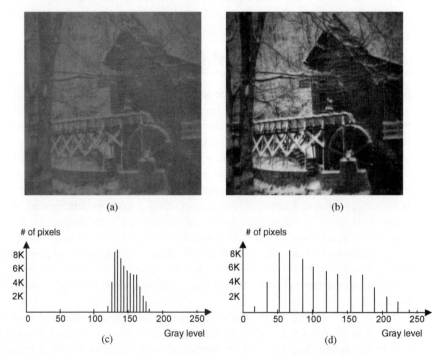

Figure 9.15   Effect of histogram equalization in improving an image.

Table 9.1   *The Actual Grayness Values and # of Pixels for Images in Figure 9.15(a) and 9.15(b).*

Levels	1	2	3	4	5	6	7	8	9	10	11	12	13	14	15	16
# of Pixels	0	750	5223	8147	8584	7769	6419	5839	5392	5179	5185	3451	2078	1692	341	0
For (b)	0	17	34	51	68	85	102	119	136	153	170	187	204	221	238	256
For (a)	120	124	128	132	136	140	144	148	152	156	160	164	168	172	176	180

Example 9.2

Assume the histogram of an image is spread between 100 and 150 out of the maximum grayness level of 255. What is the effect of multiplying the range by 1.5 or by 2? What is the effect of adding 50 to all gray values?

Solution: The two operations mentioned here are common in formatting images and in many vision systems. When all gray values are increased by the same amount, the image becomes brighter but the contrast does not change. So long as the added value does not increase the greyness level of any pixel beyond the 255 level, no information is lost and the original image may be regained by decreasing all pixel values by the same amount.

If the pixel greyness levels are multiplied by a number, so long as the maximum available gray levels are not exceeded, the histogram range is extended and contrast is increased. In this example, since the range is between 100 and 150, equalizing the histogram by 1.5 increases the range to 150 and 225. However, increasing the value to 2 will extend the histogram to 200 and 300, therefore saturating the image beyond 255 and changing its nature. Unless the original image is saved, dividing the pixel values by 2 will yield an image with a histogram between 100 and 127.

Figure 9.16(a) shows an original image that was later altered by an image-formatting routine for increased brightness (Figure 9.16(c)) and increased contrast (Figure 9.16(e)). As is evident in the histograms (b) and (d), when an image is brightened, its histogram distribution simply shifts, in this case by 30 points. When the contrast of the image is increased, in this case by 50%, the distribution of pixel gray levels is expanded, although the relationship remains the same. However, unlike the previous example, the distribution of gray levels is different because new gray values are introduced. ∎

9.9 Thresholding

Thresholding is the process of dividing an image into different portions, or levels, by picking a certain greyness level as a threshold, comparing each pixel with the threshold value, and assigning the pixel to the different portions (or levels) of interest depending on whether the pixel's greyness level is below the threshold (off, zero, or not belonging) or above the threshold (on, 1, or belonging). Thresholding can either be performed at a single level or with multiple thresholding values where the image is processed by dividing the image into layers, each layer with a selected threshold. To aid in choosing an appropriate threshold, many different techniques have been suggested. These techniques range from simple routines for binary images to sophisticated techniques for complicated images. Early routines were used for a binary image where the object was bright and the background was completely dark. This condition can be achieved in controlled lighting in industrial situations but may not be available in other environments. In binary images, the pixels are either on or off; therefore, choosing a threshold is simple and

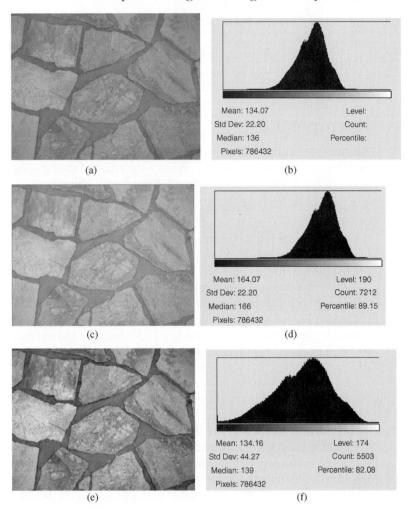

Mean: 134.07	Level:
Std Dev: 22.20	Count:
Median: 136	Percentile:
Pixels: 786432	

(a)   (b)

Mean: 164.07	Level: 190
Std Dev: 22.20	Count: 7212
Median: 166	Percentile: 89.15
Pixels: 786432	

(c)   (d)

Mean: 134.16	Level: 174
Std Dev: 44.27	Count: 5503
Median: 139	Percentile: 82.08
Pixels: 786432	

(e)   (f)

Figure 9.16   Increasing the contrast in an image expands the histogram to include new gray values.

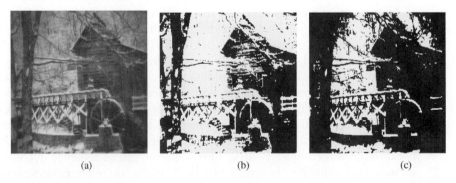

(a)   (b)   (c)

Figure 9.17   Thresholding an image with 256 gray levels at two different values of (b) 100 and (c) 150.

straightforward. In other situations, the histogram may be a multimodal distribution. In this case, the valley(s) are chosen as the threshold value. More advanced techniques use statistical information and distribution characteristics of the image pixels to develop a thresholding value. For example, the lowest value between two peaks, the midpoint between two peaks, the average of two peaks, and many other scenarios may be used. As the thresholding value changes, so does the image. Figure 9.17(a) shows an original image with 256 gray levels and the result of thresholding at greyness levels of (b) 100 and (c) 150.

Thresholding is used in many operations such as converting an image into binary form, filtering operations, masking, and edge detection.

Example 9.3

Figure 9.18(a) shows an image of a cutting board and its histogram. Due to the nature of this image, there are four peaks in the histogram. Figures 9.18(c), (e), and (g) show

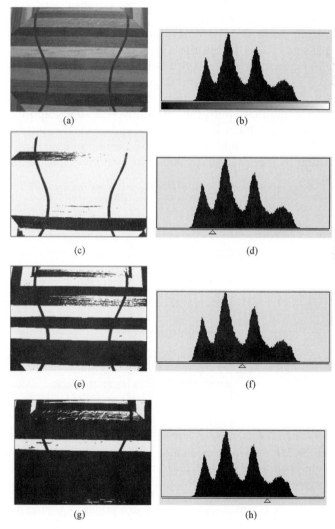

Figure 9.18 Images and histograms for Example 9.3.

the effect of thresholding at different levels. In fact, in this case, different types of wood can be identified and separated from each other due to their colors.   ■

9.10 Spatial Domain Operations: Convolution Mask

Spatial domain processes access and operate on the individual pixel information. As a result, the image is directly affected by the operation. Many processes used in vision systems are in spatial domain. One of the most popular and most common techniques in this domain is convolution, which can be adapted to many different activities such as filters, edge finders, morphology, and many more. Many processes in commercial vision systems and photography software are based on convolution, too. The following is a discussion of basic principles behind convolution. Later, we will apply the convolution idea to different purposes.

Imagine an image is composed of pixels, each with a particular gray level or color information that collectively constitute the image (in this example, the gray level is not digitized into 0s and 1s, but the actual value is indicated). As an example, let's say the image in Figure 9.19(a) is part of a larger image with pixel values shown symbolically as *A, B, C.* . . . Let's also assume there is a 3 × 3 kernel or mask, as shown, which has values in its cells as indicated by m_1 through m_9.

Applying the mask onto the image involves superimposing (convolving) the mask, first on the upper left corner of the image and taking the summation of the product of the value of each pixel multiplied by the corresponding mask value and dividing the summation by a normalizing value. This will yield (please follow carefully):

$$R = (A \times m_1 + B \times m_2 + C \times m_3 + E \times m_4 + F \times m_5 + G \times m_6 + I \times m_7 + J \times m_8 + K \times m_9)/S$$

$$(9.2)$$

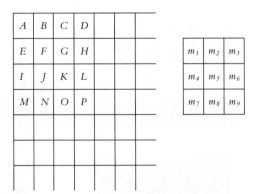

Figure 9.19   When a convolution mask (kernel) is superimposed on an image, it can change the image pixel by pixel. Each step consists of superimposing the cells in the mask onto the corresponding pixels, multiplying the values in the mask's cells by the pixel values, adding the numbers, and normalizing the result. The result is substituted for the pixel in the center of the area of interest. The mask is moved over pixel by pixel and the operation is repeated until the image is completely processed.

where S is the normalizing value. This is usually the summation of the values in the mask, or

$$S = |m_1 + m_2 + m_3 + \ldots + m_9| \tag{9.3}$$

If the summation is zero, substitute $S = 1$ or choose the largest number. The result R is substituted for the value of the pixel in the center of the block that was superimposed. In this case, R will replace the pixel value of F. Usually, the substitution takes place into a new file in order to not alter the original file $(R \to F_{new})$.

The mask is then moved one pixel to the right and the same is repeated for a new R which will replace G as follows:

$$R = G_{new} = (B \times m_1 + C \times m_2 + D \times m_3 + F \times m_4 + G \times m_5 + H \times m_6 + J \times m_7 + K \times m_8 + L \times m_9)/S$$

The result is, once again, substituted for G in a new file. The mask is then moved over one more pixel, and the operation is repeated until all the pixels in the row are changed. Then the operation continues in a raster scan fashion (see Appendix B) with the following rows until the image is completely affected. The resulting image will show characteristics that may be slightly or very severely affected by the operation, all depending on the m values in the mask. The first and last rows and columns are not affected by this operation, and therefore, are usually ignored. Some systems insert zeros for the first and last rows and columns or retain the original values. Another alternative is to copy the first and last rows and columns into an additional layer of rows and columns around the image in order to calculate new values for these pixels.

For an image $I_{R,C}$ with R rows and C columns of pixels, and for a mask $M_{n,n}$ with n rows and columns in the mask as shown in Figure 9.20, the value for the pixel $(I_{x,y})_{new}$ as the center of a block can be calculated by:

$$(I_{x,y})_{new} = \frac{1}{S} \left(\sum_{i=1}^{n} \sum_{j=1}^{n} M_{i,j} \times I \left[\left(x - \left(\frac{n+1}{2} \right) + i \right), \left(y - \left(\frac{n+1}{2} \right) + j \right) \right] \right) \tag{9.4}$$

$$S = \left| \sum_{i=1}^{n} \sum_{j=1}^{n} M_{i,j} \right| \qquad \text{if } S \neq 0$$

$$S = 1 \text{ or largest number} \qquad \text{if } S = 0 \tag{9.5}$$

$I_{1,1}$	$I_{1,2}$	$I_{1,3}$	$I_{1,4}$	$I_{1,5}$		
$I_{2,1}$	$I_{2,2}$	$I_{2,3}$	$I_{2,4}$	$I_{2,5}$		
$I_{3,1}$	$I_{3,2}$	$I_{3,3}$	$I_{3,4}$	$I_{3,5}$		
$I_{4,1}$	$I_{4,2}$	$I_{4,3}$	$I_{4,4}$	$I_{4,5}$		

$M_{1,1}$	$M_{1,2}$	$M_{1,3}$
$M_{2,1}$	$M_{2,2}$	$M_{2,3}$
$M_{3,1}$	$M_{3,2}$	$M_{3,3}$

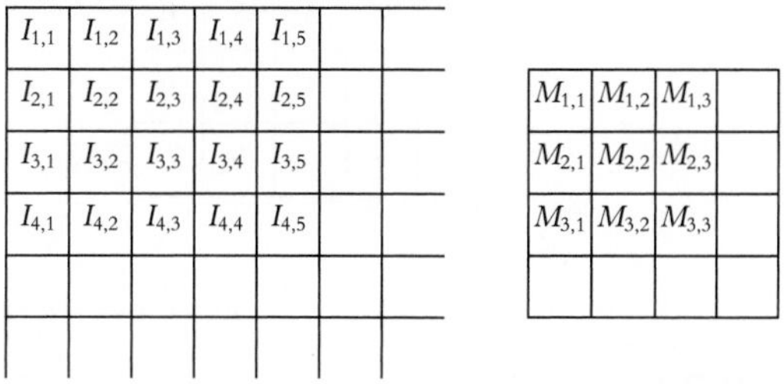

Figure 9.20 The representation of an image and a mask.

Note that the normalizing or scaling factor S is arbitrary and is used to prevent saturation of the image. As a result, the user can always adjust this number to get the best image without saturation.

Example 9.4

Consider the pixels of an image, with values as shown in Figure 9.21, as well as a convolution mask with the given values. Calculate the new values for the given pixels.

Figure 9.21   An example of a convolution mask.

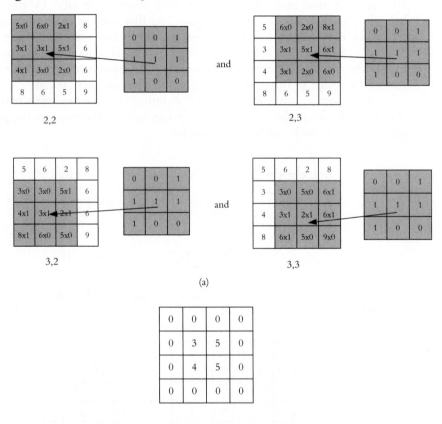

Figure 9.22   (a) Convolving the mask onto the cells of the image; (b) the result of the operation.

Solution: We substitute zeros for the first and last columns and rows because they are not affected by this process. For the remaining pixels, we superimpose the mask on the remaining cells of the image and use Equations (9.2) and (9.3) to calculate new pixel values, as shown in Figure 9.22(a), with the result shown in Figure 9.22(b). Superimposing the mask on the image as shown for each remaining element, we get:

2,2: $[5(0) + 6(0) + 2(1) + 3(1) + 3(1) + 5(1) + 4(1) + 3(0) + 2(0)]/5 = 3.4$
2,3: $[6(0) + 2(0) + 8(1) + 3(1) + 5(1) + 6(1) + 3(1) + 2(0) + 6(0)]/5 = 5$
3,2: $[3(0) + 3(0) + 5(1) + 4(1) + 3(1) + 2(1) + 8(1) + 6(0) + 5(0)]/5 = 4.4$
3,3: $[3(0) + 5(0) + 6(1) + 3(1) + 2(1) + 6(1) + 6(1) + 5(0) + 9(0)]/5 = 4.6$

In reality, greyness levels are integers, and therefore, all numbers are rounded to whole numbers. ∎

Example 9.5

Apply the 7×7 mask shown to the image of Figure 9.23.

Figure 9.23 The image and mask for Example 9.5.

Solution: Applying the mask to the image results in Figure 9.24. As you can see, the on-cells in the mask have convolved into the same shape within the image, albeit an upside down and mirror image, when applied to a single on-pixel in the image. This, in fact, demonstrates the real meaning of the convolution mask. Any set of numbers used in the mask will convolve into the image and will affect it accordingly. Therefore, the choice of numbers in the mask can have a significant effect on the image. Please also notice how the first and last three rows and columns are unaffected by the 7×7 mask.

0	0	0	0	0	0	0	0	0	0	0	0	0
0	0	0	0	0	0	0	0	0	0	0	0	0
0	0	0	0	0	0	0	0	0	0	0	0	0
0	0	0	0	0	0	1	1	0	0	0	0	0
0	0	0	0	0	0	1	0	0	0	0	0	0
0	0	0	0	0	0	1	0	0	0	0	0	0
0	0	0	0	0	0	1	0	0	0	0	0	0
0	0	0	0	0	0	1	0	0	0	0	0	0
0	0	0	0	0	0	1	0	0	0	0	0	0
0	0	0	0	1	1	1	1	1	0	0	0	0
0	0	0	0	0	0	0	0	0	0	0	0	0
0	0	0	0	0	0	0	0	0	0	0	0	0
0	0	0	0	0	0	0	0	0	0	0	0	0

Figure 9.24 The result of convolving the mask on the image of Example 9.5. ■

9.11 Connectivity

In a number of instances, we need to decide whether or not neighboring pixels are somehow "connected" or related to each other. This connectivity establishes whether they are of the same properties, such as being of the same region or object, similar textures or colors, and so on. To establish this connectivity of neighboring pixels, we first have to decide a connectivity path. For example, we need to decide whether only pixels on the same column and row are connected, or diagonally situated pixels are also accepted as connected.

There are three fundamental connectivity paths for 2D image processing and analysis: +4 or ×4-connectivity, H6 or V6 connectivity, and 8-connectivity. In 3D, connectivity between voxels (volume cells) can range from 6 to 26. Refer to Figure 9.25 for the following definitions:

+4-connectivity—a pixel *p*'s relationship is only analyzed with respect to the 4 pixels immediately above, below, to the left, and to the right of the pixel (*b,d,e,g*).
×4-connectivity—a pixel *p*'s relationship is only analyzed with respect to the 4 pixels diagonally across from it on 4 sides (*a,c,f,h*). For pixel *p(x,y)*, these are defined as:

$$\text{for} + 4\text{-connectivity}\ (x+1, y), (x-1, y), (x, y+1), (x, y-1) \tag{9.6}$$

$$\text{for}\ \times 4\text{-connectivity}\ (x+1, y+1), (x+1, y-1), (x-1, y+1), (x-1, y-1) \tag{9.7}$$

H6-connectivity—a pixel *p*'s relationship is only analyzed with respect to the 6 neighboring pixels on two rows immediately above and below the pixel (*a,b,c,f,g,h*).

a	*b*	*c*
d	*p*	*e*
f	*g*	*h*

Figure 9.25 Neighborhood connectivity of pixels.

V6-connectivity—a pixel p's relationship is only analyzed with respect to the 6 neighboring pixels on two columns immediately to the right and to the left of the pixel (a,d,f,c,e,h). For pixel $p(x,y)$, these are defined as:

for H6-connectivity

$$(x-1, y+1), (x, y+1), (x+1, y+1), (x-1, y-1), (x, y-1), (x+1, y-1)$$

$$(9.8)$$

for V6-connectivity

$$(x-1, y+1), (x-1, y), (x-1, y-1), (x+1, y+1), (x+1, y), (x+1, y-1)$$

$$(9.9)$$

8-connectivity—a pixel p's relationship is analyzed with respect to all 8 pixels surrounding it (a,b,c,d,e,f,g,h). For pixel $p(x,y)$, this is defined as:

$$(x-1, y-1), (x, y-1), (x+1, y-1), (x-1, y), (x+1, y), (x-1, y+1),$$
$$(x, y+1), (x+1, y+1)$$

$$(9.10)$$

Example 9.6

In Figure 9.26, starting with pixel (4,3), find all succeeding pixels that can be considered connected to each other based on +4, ×4, H6, V6, and 8-connectivity rules.

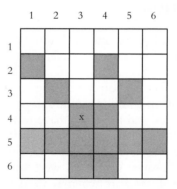

Figure 9.26 The image for Example 9.6.

Solution: Figure 9.27 shows the results of the connectivity search. Follow each one. You must take one pixel, find all others connected to it based on the applicable connectivity rule, and search the pixels found to be connected to the previous ones for additional connected pixels, until done. The remaining pixels are not connected. We will use the same rules later for other purposes such as region growing. The H6, V6, and 8-connectivity search is left for you to do as an exercise.

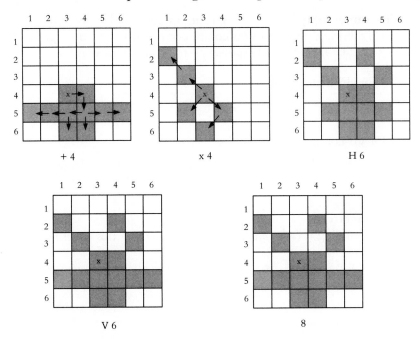

Figure 9.27   The results of the connectivity searches for Example 9.6.   ■

So far, we have studied some general issues and fundamental techniques used in image processing and analysis. Next, we will discuss particular techniques used for specific applications.

9.12 Noise Reduction

Similar to other signal-processing mediums, vision systems contain noise. Some noise is systematic and comes from dirty lenses, faulty electronic components, bad memory chips, and low resolution. Others are random and are caused by environmental effects or bad lighting. The net effect is a corrupted image that needs to be preprocessed to reduce or eliminate the noise. In addition, some images have low quality due to hardware and software inadequacies, and therefore, have to be enhanced and improved before other analyses can be performed on them. At the hardware level, in one attempt,[1] an on-chip correction scheme was devised for defective pixels in an image sensor. In this scheme, readouts from the nearest neighbors were substituted for identified defective pixels. However, in general, software schemes are used for most filtering operations.

Filtering techniques are divided into two categories of frequency domain and spatial domain. Frequency related techniques operate on the Fourier transform of the signal, whereas spatial domain techniques operate on the image at the pixel level, either locally or globally. The following is a summary of a number of different operations for reducing noise in an image.

9.12.1 Neighborhood Averaging with Convolution Masks

As discussed in section 9.10, a mask may be used for many different purposes, including filtering operations and noise reduction. In section 9.4, it was also discussed that noise,

20	20	20	20	20
20	100	20	20	20
20	20	20	20	20
20	20	20	20	20

1	1	1
1	1	1
1	1	1

Figure 9.28 Neighborhood averaging mask.

along with edges, creates higher frequencies in the spectrum. It is possible to create masks that behave like low-pass filters, such that the higher frequencies of an image are attenuated, while the lower frequencies are not changed much, and thereby, reduce the noise.

Neighborhood averaging with a convolution mask can be used to reduce the noise in images, but it also reduces the sharpness of an image. Consider the 3×3 mask in Figure 9.28 with its corresponding values, as well as a portion of an imaginary image, with its gray levels shown.

As you can see, all the pixels but one are at a gray value of 20. The pixel with a gray level of 100 may be considered noise since it is different from the pixels around it. Applying the mask over the corner of the image, with a normalizing value of 9 (summation of all values in the mask) will yield:

$$R = (20 \times 1 + 20 \times 1 + 20 \times 1 + 20 \times 1 + 100 \times 1 + 20 \times 1 + 20 \times 1$$
$$+ 20 \times 1 + 20 \times 1)/9 = 29$$

As a result of applying the mask on that corner, the pixel with the 100 value will change to 29. Consequently, the large difference between the noisy pixel and the surrounding pixels (100 versus 20) becomes much smaller (29 versus 20), thus reducing the noise. If the mask is applied to the set of pixels in columns 3, 4, 5, the average will be 20; therefore, the operation has no effect on the set. The difference between pixels remains low. With this characteristic, this mask acts as a low-pass filter because it attenuates the sharp differences between neighboring pixels but has little effect on pixels whose intensities are similar. Notice that this routine will introduce new gray levels in the image (29), and therefore, will change the histogram of the image. Similarly, this averaging low-pass filter will also reduce the sharpness of edges, making the resulting image softer and less focused. Figure 9.29 shows (a) an original image, (b) an image corrupted with noise, (c) the image after a 3×3 averaging filter application, and (d) the image after a 5×5 averaging filter application. As you can see, the 5×5 filter works even better than the 3×3 filter, but requires a bit more processing.

There are other averaging filters, such as Gaussian (also called Mild Isotropic Low-Pass), shown in Figure 9.30. This filter will similarly improve the image, but with a slightly different result.

9.12.2 Image Averaging

In this technique, a number of images of the exact same scene are averaged together. Since the camera has to acquire multiple images of the same scene, all actions in the scene must completely stop. As a result, in addition to being time consuming, this technique is

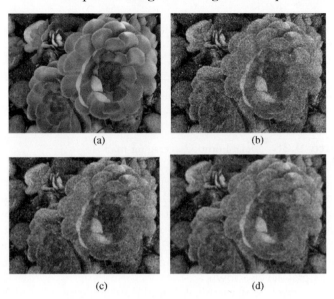

(a)   (b)

(c)   (d)

Figure 9.29   Neighborhood averaging of an image.

1	4	6	4	1
4	16	24	16	4
6	24	36	24	6
4	16	24	16	4
1	4	6	4	1

5×5

1	2	1
2	4	2
1	2	1

3×3

Figure 9.30   5×5 and 3×3 Gaussian averaging filters.

not suitable for operations that are dynamic and change rapidly. Image averaging is more effective at increased numbers of images and is fundamentally useful for random noise. If the noise is systematic, its effect on the image will be exactly the same for all multiple images and, as a result, averaging will not reduce the noise. If we assume that an acquired image $A(x,y)$ has random noise $N(x,y)$, then the desired image $I(x,y)$ can be found from this averaging because the summation of random noises will be zero, or:

$$A(x, y) = I(x, y) + N(x, y)$$

$$\frac{\sum_n A(x, y)}{n} = \frac{\sum_n I(x, y) + N(x, y)}{n} = \frac{\sum_n I(x, y)}{n} + \frac{\cancel{\sum_n N(x, y)}^{\,0}}{n} = I(x, y) \qquad (9.11)$$

Although image averaging reduces random noise, unlike neighborhood averaging, it will not blur the image or reduce its focus.

9.12.3 Frequency Domain

When the Fourier transform of an image is calculated, the frequency spectrum might show a clear frequency for the noise, which in many cases, can be selectively eliminated by proper filtering.

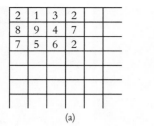 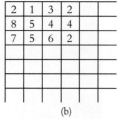

                      (a)   (b)

Figure 9.31 Application of a median filter.

9.12.4 Median Filters

One of the main problems in using neighborhood averaging is that along with removing noise, the filter will also blur the edges and reduce the sharpness of the image. A variation to this technique is to use a median filter in which the value of the pixel is replaced by the median of the values of the pixels in a mask around the pixel (the pixel plus the 8 surrounding pixels), sorted in ascending order. A median is the value where half of the values in the set are below and half are above the median (also called 50th percentile). Since, unlike an average, the median's final value is independent of the value of any single pixel in the set, the median filter will be much stronger in eliminating spike-like noises without blurring the object or decreasing the sharpness of the image.

Suppose we apply a median filter to the image in Figure 9.28. The sorted values in ascending order will be 20, 20, 20, 20, 20, 20, 20, 20, 100. The median is 20 (the fifth one from the left). Replacing the center pixel's value with 20 will completely eliminate the noise. Of course, noise is not always this easily removed. But this example shows how the effect of median filters can be very different from averaging. Notice that median filters do not create any new gray levels, but they do change the histogram of the image.

Median filters tend to make the image grainy, especially if applied more than once. Consider the image in Figure 9.31(a). The values in ascending order for the left-most corner are 1, 2, 3, 4, 5, 6, 7, 8, 9. The middle value is 5, resulting in the image in (b). The values for the second set of 9 pixels are 1, 2, 2, 3, 4, 5, 6, 7, 9, and the median is 4. As you can see, the image has become grainy because the pixel sets with similar values appear longer (as in 5 and 5 or 4 and 4).

Figure 9.32 shows (a) an original image, (b) the image corrupted with random Gaussian noise, (c) the image improved with a 3×3 median filter, and (d) a 7×7 median filter. Generally, larger size median filters are more effective.

9.13 Edge Detection

Edge detection is a general name for a class of routines and techniques that operate on an image and result in a line drawing of the image. The lines represent changes in values such as cross-sections of planes, textures, lines, and colors, differences in light intensities between parts and backgrounds or features such as holes and protrusions, as well as differences in shading and textures. Some techniques are mathematically oriented, some are heuristic, and some are descriptive techniques. They generally operate on the differences between the gray levels of pixels or groups of pixels through masks or

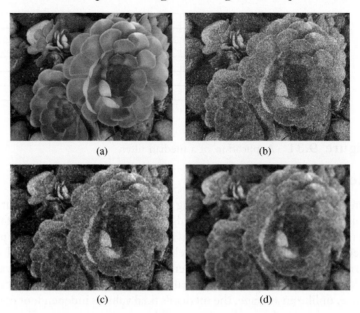

(a)   (b)

(c)   (d)

Figure 9.32   (a) is the original image, (b) is the same image corrupted with a random Gaussian noise, (c) is the image improved by a 3×3 median filter, and (d) is the same image improved with a 7×7 median filter.

thresholds. The final result is a line drawing or similar representation that requires much less memory, can be processed more easily, and saves in computational and storage costs. Edge detection is also necessary in subsequent processes such as segmentation and object recognition. Without edge detection, it may be impossible to find overlapping parts, calculate features such as diameter and area, or determine parts by region growing. Different techniques of edge detection yield slightly different results, and therefore, should be chosen carefully and used wisely.

Except in binary images, edges are generally not ideal. This means that instead of a clear distinction between two neighboring pixels' gray levels, the edge is spread over a number of pixels, as shown in Figure 9.33. A simple comparison between two pixels may be inadequate for edge detection. The first and second derivatives of the graph are also shown. It is possible to assume that the edge is at the peaks of the first derivative or at the zero crossing of the second derivative and to use these values to detect the edges. The problem is exacerbated when the image is noisy; therefore, the derivatives have excessive numbers of peaks or zero crossings.

Generally, the edges are at regions of rapid intensity change. Referring to Figure 9.34, the magnitude and direction of the gradient of image intensity can be calculated as:

$$\nabla I = \left(\frac{\partial I}{\partial x}, \frac{\partial I}{\partial y}\right)$$

$$(\nabla I)_{magnitude} = \sqrt{\left(\frac{\partial I}{\partial x}\right)^2 + \left(\frac{\partial I}{\partial y}\right)^2} \qquad (9.12)$$

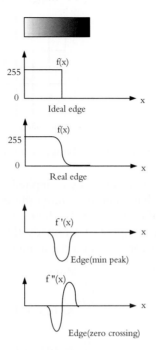

Figure 9.33   Edge detection with first and second derivatives.

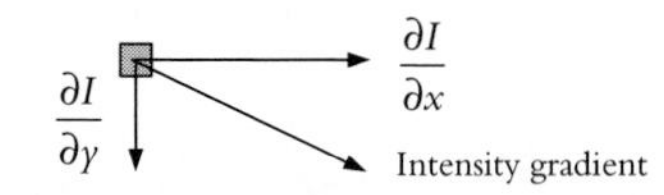

Figure 9.34   Gradient of image intensity.

$$(\nabla I)_{direction} = \tan^{-1}\frac{(\partial I/\partial y)}{(\partial I/\partial x)} \tag{9.13}$$

Similarly, the second gradient of the intensity, called *Laplacian*, is shown as Equation (9.14). The magnitude and orientation of the second gradient can be calculated in a similar fashion.

$$\nabla^2 I = \left(\frac{\partial^2 I}{\partial x^2}, \frac{\partial^2 I}{\partial y^2}\right) \tag{9.14}$$

Digital Implementation   Since images are discrete, a finite difference approach may be taken in order to calculate the gradients. For a one-dimensional system, the finite difference between successive elements is:

$$f'(x) = \lim_{dx \to 0}\frac{f(x+dx)-f(x)}{dx} \tag{9.15}$$

In an image, dx is 1 pixel wide. Therefore, the finite difference for an image can be simplified to $F'(x) = F(x+1) - F(x)$ and be implemented by kernel $\begin{bmatrix} -1 & 1 \end{bmatrix}$. For a

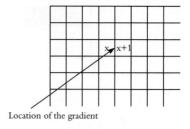

Location of the gradient

Figure 9.35   Intensity gradient between successive pixels.

two dimensional system, the same is applied in both x and y directions. Referring to Figure 9.35, notice that when the finite difference is calculated, it does not relate to the center of the pixel of interest; rather, there are two midpoints between successive pixels to which the gradients apply. To remedy this, the finite difference can be calculated between the pixels before and after the point of interest and averaged using the modified kernel (mask) $\frac{1}{2}[-1 \quad 0 \quad 1]$, yielding:

$$\frac{dF}{dx} \approx F(x+1) - F(x-1)$$
$$\frac{dF}{dy} \approx F(y+1) - F(y-1) \tag{9.16}$$

Similarly, the second derivative of the image intensities can be calculated with finite difference as:

$$\begin{aligned} F''(x) = \frac{\partial^2 F}{\partial x^2} &= [F(x+1) - F(x)]' \\ &= [F(x+1) - F(x)] - [F(x) - F(x-1)] \\ &= F(x-1) - 2F(x) + F(x+1) \end{aligned} \tag{9.17}$$

which can be implemented by a kernel $[1 \quad -2 \quad 1]$. Therefore, the approximate magnitude of the Laplacian for a two-dimensional image can be calculated by applying the following kernel (mask):

$$\text{Laplacian}(0°, 90°) = \begin{bmatrix} 0 & 1 & 0 \\ 1 & -4 & 1 \\ 0 & 1 & 0 \end{bmatrix}$$
$$\text{Laplacian}(45°) = \begin{bmatrix} 1 & 0 & 1 \\ 0 & -4 & 0 \\ 1 & 0 & 1 \end{bmatrix} \tag{9.18}$$

As we will soon see, this is a common way of detecting edges. In fact, many other common masks used for edge detection are variations of the gradient scheme.

As discussed earlier, like noise, edges are high frequency, and therefore, can be separated by high-pass filters. Masks can be designed to behave like a high-pass filter, reducing the amplitude of the lower frequencies, while not affecting the amplitudes of the higher frequencies as much, thereby separating noises and edges from the rest of the

Figure 9.36   The Laplacian-1 high-pass edge detector mask.

image. Consider the image and the Laplacian kernel (mask) in Figure 9.36. As you see, this mask has negative numbers. Applying the mask to the image at the corner yields:

$$R = (20 \times -1 + 20 \times 0 + 20 \times -1 + 20 \times 0 + 100 \times 4 + 20 \times 0 + 20 \times -1$$
$$+ 20 \times 0 + 20 \times -1)/1 = 320$$

The normalizing factor is 1, which results in the value of 100 replaced with 320, comparatively accentuating the original difference (from 100 versus 20 to 320 versus 20) while applying the mask to the set of pixels in columns 3, 4, 5 yields zero, indicating that the difference between pixels is not changed. Since this mask accentuates large intensity variations (higher frequencies) while ignoring similar intensities (lower frequencies), it is a high-pass filter. This also means that the noise and edges of objects in images will be shown more effectively. As a result, this mask acts as an edge detector. Some high-pass filters act as an image sharpener. Figure 9.37 shows some other high-pass filters.

The following three masks[2-6]—Sobel operator, Roberts edge, and Prewitt—shown in Figure 9.38 effectively do the same gradient differentiation with somewhat different results and are very common. When applied to an image, the two pairs of masks calculate the gradients in the x and y directions, which are added and compared to a threshold. Notice how these follow the gradient equations developed earlier.

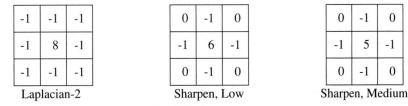

Figure 9.37   Other high-pass filters.

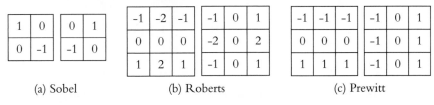

Figure 9.38   The Sobel, Roberts, and Prewitt edge detectors.

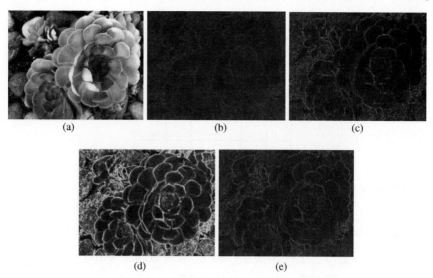

Figure 9.39 An image and its edges from Laplacian-1 (b), Laplacian-2 (c), Sobel operator (d), and Robert's edge (e).

Figure 9.39 is an original image (a) with its edges detected by a Laplacian-1 (b), Laplacian-2 (c), Sobel operator (d), and Robert's edge (e).

You must realize that although in this example the results are as shown, the result for other images may be different. This is because the histogram of the image and the chosen thresholds have great effects on the final outcome. Some routines allow the user to change the thresholding values, and some do not. In each case, the user must decide which routine performs the best.

Other simple methods can be used for binary images that are simple to implement and yield continuous edges. In one example,[7] a search technique, dubbed Left-Right (L-R) in this book, is used to quickly and efficiently detect edges in binary images of single objects that look like a blob. Imagine a binary image as shown in Figure 9.40. Let's assume that gray pixels are "on" (or the object) and white pixels are "off" (background).

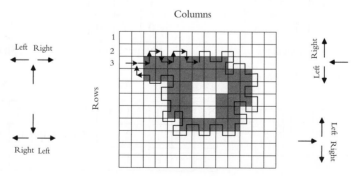

Figure 9.40 Left-Right search technique for edge detection.[7]

Table 9.2   *Possible Left-Right Schemes Based on the Direction of Search.*

If Vpresent-Vprevious>0		left	Unext=Upresent−1
		right	Unext=Upresent+1
if Vpresent-Vprevious<0		right	Unext=Upresent−1
		left	Unext=Upresent+1
If Upresent-Uprevious<0		right	Vnext=Vpresent+1
		left	Vnext=Vpresent−1
If Upresent-Uprevious>0		right	Vnext=Vpresent−1
		left	Vnext=Vpresent+1

Assume a pointer is moving from one pixel to another, in any direction (up, down, right, left). Any time the pointer reaches an "on" pixel, it will turn left. Any time it reaches an "off" pixel, it will turn right. Of course, as shown, depending on the direction of the pointer, the left and right might mean different directions. Starting at pixel 1,1, moving to 1,2, to the end, then row 2, and then row 3, the pointer will find the first "on" pixel at 3,3, will turn left, and encounter an "off" pixel, turn right twice, then left, and will go on. The process continues until the first pixel is reached. The collection of the pixels on the pointer's path is one continuous edge. Other edges can be found by continuing the process with a new pixel. In this example, the edge will be pixels 3,3-3,4-3,5-3,6 . . . 3,9-4,9-4,10-4,11 . . . .

Table 9.2 shows how a simple computer program can be developed to do the search. U and V are pixel coordinates.

Masks may also be used for the intentional emphasis of some characteristic of the image. For example, a mask may be designed to emphasize horizontal lines, vertical lines, or diagonal lines. Figure 9.41 shows three such masks. Figure 9.42 shows an original image (a), along with the effects of a vertical mask (b), a horizontal mask (c), and a diagonal mask (d).

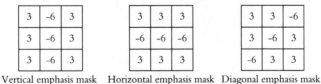

Vertical emphasis mask   Horizontal emphasis mask   Diagonal emphasis mask

Figure 9.41   These masks emphasize the vertical, horizontal, and diagonal lines of an image.

9.14 Sharpening an Image

Image sharpening can be accomplished in many different ways. The simplest is to apply a relatively high-pass filter to the image that increases the sharpness of the image by eliminating some of the lower frequencies from the edges. However, in sharpening operations, noise is increased too, and therefore, as the level of sharpening increases, so does the noise level. Figure 9.37, partially repeated here, shows two simple sharpening masks.

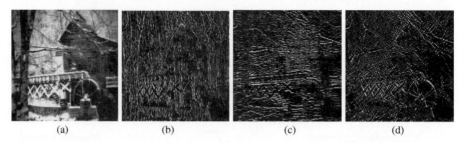

(a)   (b)   (c)   (d)

Figure 9.42   An original image (a) with effects of vertical emphasis mask (b), horizontal emphasis mask (c), and diagonal emphasis mask (d).

0	-1	0
-1	6	-1
0	-1	0

Sharpen, Low

0	-1	0
-1	5	-1
0	-1	0

Sharpen, Medium

Figure 9.37   Repeated

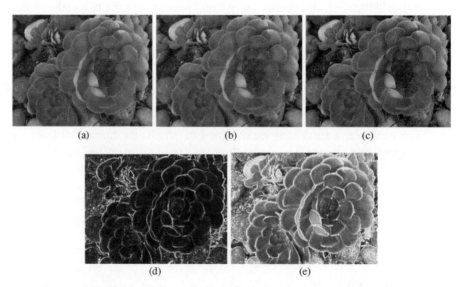

(a)   (b)   (c)

(d)   (e)

Figure 9.43   The original image (a), after an averaging mask was applied to it (b), the result of sharpening with a low sharpening mask (c), Sobel edge (d), the result of adding the Sobel edge to the original image.

Figure 9.43 shows a more sophisticated method to sharpen images. In this case, a 3×3 mask was applied to the original image to decrease noise (b), followed by a low sharpening mask (c), followed by a Sobel edge detector (d). The result was added to the original image (e). As you can see, the image shows more detail and is somewhat sharpened, but there is also more noise present.

9.15 Hough Transform

As you have probably noticed, in most edge detection techniques, the resulting edges are not continuous. However, there are many applications where continuous edges are either necessary or preferred. For example, as we will see later, in region growing, edges that define an area or region must be continuous and complete before a region growing routine can detect and label it. Additionally, it is desirable to be able to calculate the slope of detected edges in order to either complete a broken line, or to detect objects. *Hough transform*[8] is a technique used to determine the geometric relationship between different pixels on a line, including the slope of the line. For example, we can determine whether a cluster of points is on a straight line or not. This also aids in the further development of an image in preparation for object recognition since it relates individual pixels into recognizable forms.

Hough transform is based on transforming the image space (x,y) into either (r,θ) or (m,c) space. The normal from the origin to any line will have an angle of θ with respect to the x-axis and a distance of r from the origin. The transformation into the r,θ-plane (also called Hough plane) showing these values is called Hough transform (Figure 9.44(a)). Note that since all the points constituting the line in x,y-plane have the same r,θ values, they are all represented by the same point A in r,θ-plane. Therefore, all points on a straight line are represented by a single point in the Hough plane.

Similarly, a line in the x,y-plane with a slope m and intercept c can be transformed into a Hough plane of m,c with x and y as its slope and intercept (Figure 9.44(b)). Therefore, a line in the x,y-plane with a particular slope and intercept will transform into a point in the Hough plane. Since all points on this line have the same m and c, they are all represented by the same point in the Hough plane.

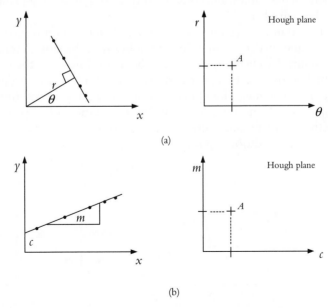

(a)

(b)

Figure 9.44 The Hough transform from x,y-plane into r,θ-plane or m,c-plane.

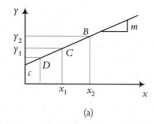
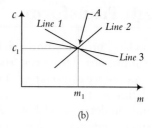

(a)   (b)

Figure 9.45   Hough transform.

Now consider the line in Figure 9.45(a), described by its slope m and intercept c as:

$$y = mx + c \tag{9.19}$$

Equation (9.19) can also be written in terms of m and c as variables as:

$$c = -xm + y \tag{9.20}$$

where, in the m,c-plane, the x and y will be the slope and the intercept.

As discussed earlier, the line of Equation (9.19) with m and c converts to a single point A in the m,c-plane. Whether the line is drawn with this equation, or in polar coordinates with (r, θ), the result is the same. Thus, a line (and all the points on it) are represented by a point in the Hough plane.

The opposite is also true. As shown in Figure 9.46, an infinite number of lines may go through a point in the x,y-plane, all intersecting at the same location. Although these lines have different slopes m and intercepts c, they all share the same point x,y which become the slope and intercept in the Hough plane. Therefore, the same x and y values represent all these lines, and consequently, a point in the x,y-plane is represented by a line in the Hough plane.

Hough transform converts the pixels (edges) within an image into lines in the Hough plane. If a group of points are colinear, their Hough transforms will all intersect at one point. By checking this, it can be determined whether a cluster of pixels is on a straight line or not. Hough transforms can also be used in determining the angle or orientation of a line. This application has found use in determining the orientation of an object in a plane by calculating the orientation of a particular line in the object. Since the intercept and slope of the line are now known, a broken line can easily be completed by additional points.

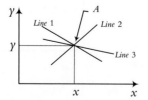
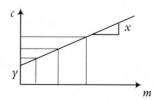

Figure 9.46   Transformation of a point in x,y-plane into a line in the Hough plane.

Example 9.7

The x and y coordinates of 5 points are given as (1,3), (2,2), (3,1.5), (4,1), and (5,0). Using the Hough transform, determine which points are on the same line. Find the slope and intercept of the line.

Solution: Of course, any two points form a line. So, we will look for at least three points that will be on the same line. Clearly, looking at the graph of the points, it is very easy to answer the questions, a trivial matter. However, in computer vision, since the computer does not have the intelligence to understand an image, it must be calculated. Imagine having thousands of points in a computer file representing an image. It is impossible, whether for a computer or a human, to tell which points are on the same line and which ones are not. We will perform a Hough transform to determine which points fall on the same line. The following table summarizes the lines formed in the m,c-plane that correspond to the points in the x,y plane:

y	x	x,y	m,c
3	1	$3 = m1+c$	$c = -1m+3$
2	2	$2 = m2+c$	$c = -2m+2$
1.5	3	$1.5 = m3+c$	$c = -3m+1.5$
1	4	$1 = m4+c$	$c = -4m+1$
0	5	$0 = m5+c$	$c = -5m+0$

Figure 9.47 shows the five corresponding lines drawn in the m,c-plane. As you see, three different lines intersect at two different places, while other intersections are just between two lines. These correspond to points (1,3), (3,1.5), (5,0) and to (2,2), (3,1.5), and (4,1). The slope and intercept of the first line are -0.75 and 3.75. The slope and intercept for the second line are -0.5 and 3 respectively. This shows how the Hough transform can be cluttered with an exceeding number of intersecting lines. Determining which lines are intersecting is the main issue in Hough transform analysis.

The equations representing these lines are $y = -0.75x + 3.75$ and $y = -0.5x + 3$. Using these equations, additional points lying on these lines can be assigned to the group, therefore completing broken lines.

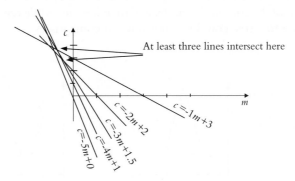

Figure 9.47 Hough transform for Example 9.7.

Coincidentally, the same analogy may be made for circles and points instead of lines and points. All points on a circle will correspond to intersecting circles in the Hough plane and vice versa. For more information, refer to Reference 3.

The Hough transform has many desirable features. For example, since each point in the image is treated independently, all points can be processed simultaneously with parallel processing methods. This makes the Hough transform a suitable candidate for real-time processing. It is also insensitive to random noise, since individual points do not greatly contribute to the final count of the part itself. However, Hough transform is computationally intensive. To reduce the number of calculations needed to determine whether lines are actually intersecting with each other at the same point, we may use a circle within which, if the lines approximately intersect with each other, they are assumed to be intersecting. Many variations to the Hough transform have been devised to increase its efficiency and utility for different tasks, including object recognition.[9]

9.16 Segmentation

Segmentation is a generic name for a number of different techniques that divide the image into segments or constituents. The purpose is to separate the information contained in the image into smaller entities that can be used for other purposes. For example, an image can be segmented by the edges in the scene, or by division into small areas (blobs), and so on. Each of these entities can then be used for further processing, representation, or identification. Segmentation includes, but is not limited to, edge detection, region growing, and texture analysis.

The early segmentation routines were all based on edge detection of simple geographic models such as polyhedrons. In three-dimensional analysis of objects, models such as cylinders, cones, spheres, and cubes were used as well. Although these shapes and figures did not necessarily match any real objects, they provided a means for early developmental work that evolved into more sophisticated routines and techniques. They also provided a means to develop schemes that could process complicated shapes and recognize objects. As an example, the routines could model a tree as a cone or sphere mounted on a cylinder (Figure 9.48) and could match it with a model of a tree, requiring very little processing power; the tree could be expressed with only a few pieces of information such as the diameters of the cone and cylinder and their heights, while representing all the information pertaining to a tree could be enormous in comparison.

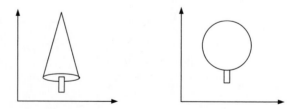

Figure 9.48 Representation of objects such as a tree with models such as a cone or sphere mounted on a cylinder can reduce processing requirements.

9.17 Segmentation by Region Growing and Region Splitting

In addition to edge detection routines, region growing and image splitting are other common techniques of segmentation. Through these techniques, an attempt is made to separate the different parts of an image into segments or components with similar characteristics that can be used in further analysis such as in object detection. Edges found by an edge detector are lines of textures, colors, planes, and gray levels, and therefore, may or may not be continuous. However, segmentation by regions naturally results in complete and closed boundaries. For a survey of other segmentation techniques, see Reference [10].

Two approaches are used for region segmentation. One is to grow regions by similar attributes such as a range of gray levels or other similarities. The other is region splitting which will split images into smaller areas using their finer differences.

One technique of region splitting is thresholding. The image is split into closed areas of neighboring pixels by comparing them to a thresholding value or range. Any pixel that falls below a threshold (or between a range of values) will belong to a region, and otherwise, to another. This will split the image into a series of regions or clusters of pixels that have common or similar attributes. Generally, although this is a very simple technique, it is not very effective since choosing an appropriate threshold is difficult. The results are also highly dependent on the threshold value and will change accordingly when the thresholds change. Still, it is a useful technique under certain conditions such as silhouettes and for images with relatively uniform regions.

In region growing, first nuclei pixels or regions are formed based on some specific selection law. Nuclei regions are the small clusters of pixels formed at the beginning of segmentation. They are usually small and act as a nucleus for subsequent growing and merging, as in alloys. The result is a large number of little regions. Successively, these regions are combined into larger regions based on some other attributes or rules. Although these rules will merge many smaller regions to create a smoother set of regions, they may unnecessarily combine certain features that should not be merged such as holes, smaller but distinct areas, or different distinct areas with similar intensities.

The following is a simple search technique for growing regions for a binary image (or with the application of thresholding, for gray images as well) that uses a bookkeeping approach to find all pixels that belong to the same region.[11] Figure 9.49 shows a binary image. Each pixel is referred to by a pair of index numbers. Assume a pointer starts at the top and searches for a nucleus to start a region. As soon as a nucleus is found (which does not already belong to another region), the program assigns a region number to it. All pixels connected to it receive the same region number and are placed in a stack. The search continues with all the pixels in the stack until the stack is emptied. The pointer will then continue searching for a new nucleus and a new region number.

It is important to decide what form of connectivity is to be used in growing regions, as this will change the final outcome. As discussed in section 9.11, +4-, ×4-, H6-, V6-, and 8-connectivity can be used for region growing. In Figure 9.49, the first nucleus is found at pixel 2d. Suppose we have chosen the +4-connectivity. The program will check the four corresponding pixels around the nucleus to determine connectivity. If there is an "on" pixel, its location index numbers will be placed in a stack, the cell is given the region number (n), and the pointer is moved down in the stack to the next cell, 3d. At this location, the connectivity of pixels around the cell is checked again, the "on" pixel index numbers are placed in the search stack, the cell is given the region-n designation, and the

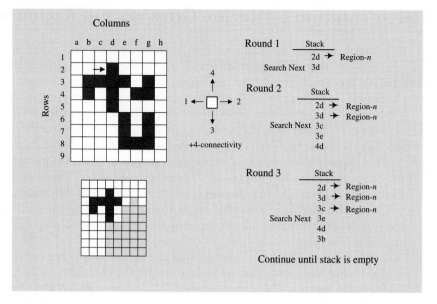

Figure 9.49 Region growing based on a search technique. With +4-conectivity search, region-*n* is as shown.

process is repeated for the next index number on the stack, 3c. The process continues until the stack is empty.

Notice that this is nothing more than a bookkeeping technique to make sure the computer program can find all connected pixels in the region without missing any. Otherwise, it is a simple search technique.

Example 9.8

Using ×4-, H6-, V6-, and 8-connectivity, determine the first region that results from a search of Figure 9.49.

Solution: The result is shown in Figure 9.50 for the ×4-connectivity. Please follow the routine for other connectivity rules.

Figure 9.50 The result of a search for ×4-connectivity for Example 9.8.   ■

There are many other segmentation schemes that apply to different situations. For example, in one technique, the following is done:

- Assign the image to *k* clusters.
- Calculate the mean of each cluster.

- If the mean of a token area (or pixel) is closer to its cluster's mean, keep it.
- If the mean of a token area (or pixel) is closer to another cluster's mean, reassign it to the other cluster.
- Continue until no changes are made.

Once again, as you notice, a bookkeeping and comparison routine is applied to the image in order to segment it based on a desired characteristic, in this case the mean of each area. Other schemes can be found in other references.[2,12–14]

9.18 Binary Morphology Operations

Morphology operations refer to a family of operations performed on the shape (therefore, morphology) of subjects in an image. They include many different operations, both for binary and gray images such as thickening, dilation, erosion, skeletonization, opening, closing, and filling. These operations are performed on an image in order to aid in image analysis as well as for reducing the "extra" information that may be present in the image. For example, consider the binary image in Figure 9.51(a) and the stick figure representing one of the bolts in (b). As we will see later, a moment equation may be used to calculate the orientation of the bolts. However, the same moment calculation can also be performed on the stick figure of the bolt, but with much less effort. As a result, it would be desirable to convert the bolt to its stick figure or skeleton. In the following sections, we will discuss a few of these operations.

Morphology operations are based on the set theory. For example, in Figure 9.52, the union between the two lines creates the parallelogram (apply the first line to the second line), while the union between the two smaller circles is the larger circle. In this case, the radius of the first circle is added to the second one, enlarging it. This is called *dilation*,

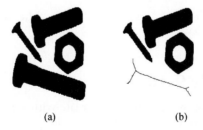

(a)   (b)

Figure 9.51   The binary image of a bolt and its stick (skeleton) representation.

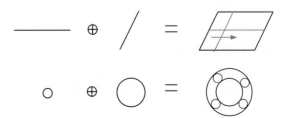

Figure 9.52   The union between two geometries creates *dilation*.

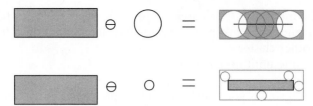

Figure 9.53   The subtraction of two geometries creates *erosion*.

shown with the symbol ⊕, and as we will soon discuss, it can be as little as one pixel added to the perimeter of a part in an image.

Similarly, in Figure 9.53, subtracting the second set from the first results in an eroded shape, therefore *erosion*, shown with the symbol ⊖, which as will be discussed later, may be as small as one pixel around the object. Similar combinations of dilation and erosion create other effects as follows.

Example 9.9

Figure 9.54 shows the effect of a union operation between two shapes. As shown, this union reduces the appearance of the peaks and valleys in the original shape. This is used for smoothing the jagged edges of shapes such as a bolt or gear.

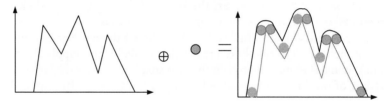

Figure 9.54   The result of the union of the two shapes reduces the appearance of the peaks and valley.   ∎

Example 9.10

Figure 9.55 shows the image of a plate with a small protrusion in it. In order to locate the protrusion, we may subtract a circle from the plate with a diameter slightly larger than the protrusion, add the circle to the result, and subtract the result from the original image. The remaining object is only the protrusion.   ∎

The following operations are all based on the previously mentioned operations.

9.18.1 Thickening Operation

A thickening operation fills the small holes and cracks on the boundary of an object and can be used to smooth the boundary. In the example shown in Figure 9.56, the thickening operation reduced the appearance of the threads of the bolts. This is a very useful operation when we try to apply other operations such as skeletonization to the object. The initial thickening will prevent the creation of whiskers caused by the threads,

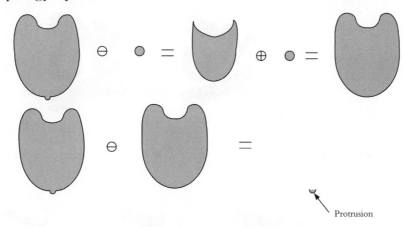

Figure 9.55 The application of union and subtraction operations for locating the protrusion.

Figure 9.56 The threads of the bolts are removed by a triple application of a thickening operation, resulting in smooth edges.

as we will see later. Figure 9.56 shows the effect of three rounds of thickening operations on the threads of the bolts.

9.18.2 Dilation

In dilation, the background pixels that are 8-connected to the foreground (object) are changed to foreground. As a result, effectively, a layer is added to the object every time the process is implemented. Due to the fact that dilation is performed on pixels that are 8-connected to the object, repeated dilations can change the shape of the object. Figure 9.57 (b) is the result of five dilation operations on the objects in (a). As you can see, due to this dilation, the four objects have bled into one piece. With additional applications of dilation, the four objects, as well as the disappearing hole, can become one solid piece, which can no longer be recognized.

9.18.3 Erosion

In this operation, foreground pixels that are 8-connected to a background pixel are eliminated. This effectively eats away a layer of the foreground (the object) each time it is performed. Figure 9.58(b) shows the effect of 3 repetitions of the erosion operation on

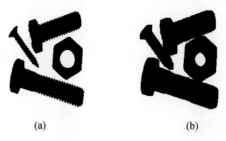

(a)   (b)

Figure 9.57   Effect of dilation operations. Here, the objects in (a) were subjected to five rounds of dilation (b).

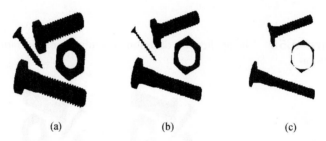

(a)   (b)   (c)

Figure 9.58   Effect of erosion operation on objects with (b) 3 and (c) 7 repetitions.

the binary image in (a). Since erosion removes one pixel from around the object, the object becomes increasingly thinner with each pass. However, erosion disregards all other requirements of shape representation. It will remove one pixel from the perimeter (and holes) of the object even if the shape of the object is eventually lost, as in (c) with 7 repetitions, where one bolt is completely lost and the nut will soon disappear. Erosion can eventually remove all objects. This means that if the reversing operation of dilation, which will add one pixel to the perimeter of the object with each pass, is used, the dilated object may not resemble the original object at all. In fact, if the object is totally eroded to one pixel, dilation will result in a square or circle. As a result, erosion can irreparably damage the image. However, it can also be successfully used to eliminate unwanted objects in an image. For example, if we want to identify the largest object in an image, successive erosions will eliminate all other smaller objects before the largest object is eliminated. Therefore, the object of interest can be identified.

9.18.4 Skeletonization

A skeleton is a stick representative of an object where all thicknesses have been reduced to one pixel at any location. Skeletonization is a variation of erosion. Whereas in erosion, the thickness may go to zero and the object may be totally lost, in skeletonization, as soon as the thickness of the object becomes one pixel, the operation at that location stops. Although in erosion the number of repetitions are chosen by the user, in skeletonization the process automatically continues until all thicknesses are 1 pixel (the program stops when no new changes are made as a result of the operation). The final result of

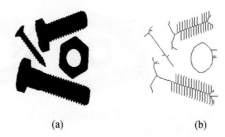

(a)   (b)

Figure 9.59   The effect of skeletonization on an image without thickening. The threads of the bolts have resulted in whiskers.

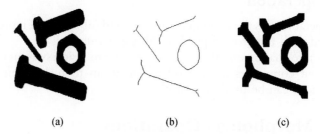

(a)   (b)   (c)

Figure 9.60   The skeleton of the objects in (a) after the application of thickening operation results in a clean skeleton (b). Part (c) is the dilated image of the skeletons.

skeletonization is a stick figure (skeleton) of the object, which is a good representation of the object, sometimes much better than the edges. Figure 9.59(b) shows the skeleton of the original objects in (a). The whiskers are created because the objects were not smoothed by thickening. As a result, all threads are reduced to one pixel, creating the whiskers. Figure 9.60 shows the same objects that are thickened to eliminate the threads, resulting in a clean skeleton. Figure 9.60(c) is the result of dilating the skeleton 7 times. As can be seen, the dilated objects are not the same as the original objects. Notice how the smaller screw appears as big as the bigger bolts.

Although dilation of a skeleton will also result in a shape different from the original object, skeletons are very useful in object recognition since they are generally a better representation of an object than others. The stick representation of an object can be compared to the available *a priori* knowledge of the object for matching.

9.18.5 Open Operation

Opening is an erosion operation followed by a dilation. This causes a limited smoothing of convex parts of the object and can be used as an intermediate operation before skeletonization.

9.18.6 Close Operation

Closing is a dilation operation followed by an erosion. This causes a limited smoothing of concave parts of the object and, like opening, can be used as an intermediate operation before skeletonization.

Figure 9.61 As a result of a fill operation, the hole in the nut is filled with foreground pixels, thus eliminating the hole.

9.18.7 Fill Operation

Fill operation fills the holes in the foreground (object). In Figure 9.61, the hole in the nut is filled with foreground pixels until it is eliminated.

For information on other operations, refer to vision systems manufacturers' references. Different companies include other operations to make their software unique. These operations can be used as available.

9.19 Gray Morphology Operations

Gray morphology operations are similar to binary morphology operations, except that they operate on a gray image. Usually, a 3 × 3 mask is used to apply the operations, where each cell in the mask may be either 0 or 1. Imagine a gray image is a multilayer, three-dimensional image, where the light areas are peaks and the dark areas are valleys. The mask will be applied to the image by moving it from pixel to pixel. Where the mask matches the gray values in the image, there are no changes made. If the gray values of the pixels do not match the mask, they will be changed according to the selected operation, as described in the following sections.

9.19.1 Erosion

In this case, each pixel will be replaced by the value of the darkest pixel in its 3 × 3 neighborhood, known as a Min Operator, effectively eroding the object. Of course, the result is dependent on which cells in the mask are 0 or 1. It removes light bridges between dark objects.

9.19.2 Dilation

In this case, each pixel will be replaced by the value of the lightest pixel in its 3 × 3 neighborhood, known as a Max Operator, effectively dilating the object. Of course, the result is dependent on which cells in the mask are 0 or 1. It removes dark bridges between light objects.

9.20 Image Analysis

Image analysis is a collection of operations and techniques used to extract information from images. This includes object recognition, feature extraction, analysis of position, size, orientation, and other properties of objects in images, and extraction of depth information. Some techniques may be used for multiple purposes, as we will see later. For example, moment equations may be used for object recognition as well as calculation of position and orientation of objects.

Generally, it is assumed that image processing routines have already been applied to the image or that they are available for further use when needed to improve and prepare the image for analysis. Image analysis routines and techniques may be used on both binary and gray images. In the following sections, some of these techniques are discussed.

9.21 Object Recognition by Features

Objects in an image may be recognized by their features. These features may include, but are not limited to: gray level histograms; morphological features such as area, perimeter, number of holes, and others; eccentricity; cord length; and moments. In many cases, the information extracted is compared to *a priori* information about the object, which may be in a look-up table. For example, suppose two objects are present in the image, one with two holes and one with one hole. Using previously discussed routines, it is possible to determine how many holes each part has and, by comparing the two parts (let's say they are assigned regions 1 and 2) to a look-up table, it is possible to determine what each of the two parts are. In another example, assume a moment analysis is performed for a known object and the moment, relative to an axis, is calculated at many angles and the data is stored in a look-up table. Later, when the moment of the part in the image is calculated relative to the same axis and is compared to the look-up table, the angle of the part in the image can be estimated.

The following is a discussion of a few techniques and different features that may be used for object recognition.

9.21.1 Basic Features Used for Object Identification

The following morphological features may be used for object recognition and identification:

(a) Gray Levels: Average, maximum, or minimum gray levels may be used to identify different parts or objects in an image. As an example, assume there are three parts in an image, each one with a different color or texture. The colors and textures will create different gray levels in the image. If the average, maximum, or minimum gray levels of the objects are found (e.g., through histograms mapping), the objects can be recognized by comparison of this information. In other cases, even the presence of one particular gray level may be enough to recognize a part.

(b) Perimeter, area, diameter, number of holes, and other similar morphological characteristics may be used for object identification. The perimeter of an object may be found by first applying an edge detection routine, and then, by counting the number of pixels on the perimeter. The Left-Right search technique of section 9.13 can also be used to calculate the perimeter by counting the pixels that are on the path in an accumulator. Area can be calculated by region growing techniques. Moment equations can also be used, as will be discussed later. Diameter for noncircular objects

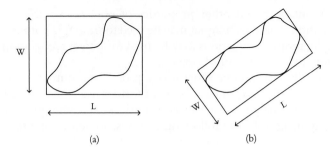

Figure 9.62   (a) Aspect ratio of an object, (b) minimum aspect ratio.

is the maximum distance between any two points on any line that crosses the identified area of the object.

(c) Aspect Ratio: Aspect ratio is the width to length ratio of an enclosing rectangle about the object, as shown in Figure 9.62. All aspect ratios are sensitive to orientation, except the minimum aspect ratio. Therefore, the minimum aspect ratio is usually used to identify objects.

(d) Thinness is defined as one of the two following ratios:

$$1. \text{Thinness} = \frac{(\text{perimeter})^2}{\text{area}} \tag{9.21}$$

$$2. \text{Thinness} = \frac{\text{diameter}}{\text{area}} \tag{9.22}$$

(e) Moments: Due to their importance, moments are discussed in the next section.

9.21.2 Moments

The moment of an object within an image is:

$$M_{a,b} = \sum_{x,y} x^a y^b I_{x,y} \tag{9.23}$$

where $M_{a,b}$ is the moment of the object with a and b indices, x *and* y are the coordinates of each pixel raised to the power of a and b, and $I_{x,y}$ is the intensity of the pixel, as in Figure 9.63. If the image is binary, the intensities are either 1 (or on) for the object and 0 (off) for the background, and therefore, only the pixels that are turned on are considered. In gray images, the intensities may vary greatly, and consequently, the value of the moment may be exceedingly influenced by gray values. As a result, although it is mathematically possible to apply moment equations to a gray image, it is not practical or useful unless other rules are engaged, too. For binary images, $I_{x,y}$ is either 0 or 1; therefore, considering only the on-pixels in the image, Equation (9.23) simplifies to:

$$M_{a,b} = \sum_{x,y} x^a y^b \tag{9.24}$$

To calculate the moments, first determine whether or not each pixel belongs to the object (is turned on); if so, raise the coordinates of the location of the pixel to the given

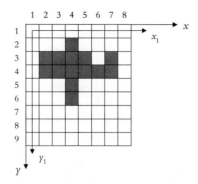

Figure 9.63 Calculation of the moment of an image. For each pixel that belongs to the object, the coordinates of the pixel are raised to the powers indicated by the moment's indices. The summation of the values thus calculated will be the particular moment of the image.

values of a and b. The summation of this operation over the entire image is the particular moment of the object with a and b indices.

$M_{0,0}$ is the moment of the object with $a = 0$ and $b = 0$. This means the x and y coordinate values of all on-pixels are raised to a power of 0. $M_{0,2}$ means all x values are raised to the power of 0 and all y values are raised to the power of 2, and so on. All combinations of values between 0 and 3 are common.

Distances x and y are measured either from a fictitious coordinate frame located at the edge of the image (x,y) or are measured from a coordinate frame formed by the first row and column of the image. Since the distances are measured by counting the number of pixels, the use of the first row and column as the coordinate frame is more logical. However, note that in this case, all distances should be measured to the centerline of the pixel row or column. As an example, the first on-pixel on the second row is pixel $(2,4)$. The x distance of the pixel from the x_1-y_1 coordinate frame will be 3, whereas the same coordinate from the x-y coordinate is 4 (or more accurately, 3.5 pixels). As long as the same distances are used consistently, the choice is not important.

Based on the above, since all numbers raised to the power of 0 are equal to 1, all x^0 and y^0s are equal to 1. Therefore, the $M_{0,0}$ moment is the summation of all on-pixels, which is the area of the object. This moment can be used to determine the nature of an object and to distinguish it from others that have a different area. Obviously, the $M_{0,0}$ moment can also be used to calculate the area of an object within an image.

Similarly, $M_{0,1}$ is $\sum x^0 y^1$ for all on-pixels, or the summation of $1 \times y$ values, which is the summation of the y-coordinates of all on-pixels from the x-axis. This is similar to the first moment of the area relative to the x-axis. Therefore, the location of the center of the area relative to the x-axis can be calculated by:

$$\bar{y} = \frac{\sum y}{area} = \frac{M_{0,1}}{M_{0,0}} \tag{9.25}$$

So, by simply dividing the two moments, you may calculate the $\bar{y}$ coordinate of the center of the area of the object. Similarly, the location of the center of the area relative to the y-axis is:

$$\bar{x} = \frac{\sum x}{area} = \frac{M_{1,0}}{M_{0,0}} \tag{9.26}$$

This way, an object may be located within an image regardless of its orientation (the orientation will not change the location of the center of an area). Of course, this information can be used to locate an object, say, for grabbing by a robot.

$M_{0,2}$ is $\sum x^0 y^2$ and represents the second moment of the area relative to the x-axis. Similarly, $M_{2,0}$ is the second moment of the area relative to the y-axis. As you can imagine, the moment of inertia of an object such as the one in Figure 9.63 varies significantly as the object rotates about its center. Suppose we calculate the moments of the area about an axis, say, the x-axis, at different orientations. Since each orientation creates a unique value, a look-up table that contains these values can later be used to identify the orientation of the object. Therefore, if a look-up table is prepared containing the values of the moments of inertia of the known object at different orientations, the subsequent orientation of the object can be estimated by comparing its second moment to the values in the look-up table. Of course, if the object translates within an image, its moments of inertia will also change, rendering it impossible to determine the orientation except in known locations. However, with a simple application of the parallel axes theorem, the second moments about the center of the area can be calculated. Since this measure is independent of the location, it can be used to determine the orientation of the object regardless of its location. Therefore, using the moment equations, the object, its location, and its orientation can be identified. In addition to identification of the part, the information can be used in conjunction with a robot controller to direct the robot to pick up the part and/or operate on it.

Other moments can also be used similarly. For example, $M_{1,1}$ represents the product of inertia of the area and can also be used for object identification. Higher-order moments such as $M_{0,3}, M_{3,0}, M_{1,2}$, and so on can also be used to identify objects and their orientation. Imagine two objects relatively similar in shape, as in Figure 9.64(a). It is possible that the second moments, areas, perimeters, or other morphological characteristics of the two objects may be similar or close to each other such that they may not be useful in object identification. In this case, a small difference between the two objects may be exaggerated through higher-order moments, making object identification possible. The same is true for an object with a small asymmetry (Figure 9.64(b)). The orientation of the object may be found by higher-order moments.

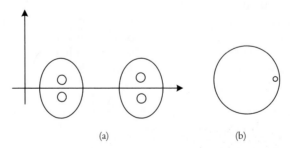

(a)   (b)

Figure 9.64 Small differences between objects or small asymmetry in an object may be detected using higher-order moments.

A moment invariant is a measure of an object based on its different moments, and is independent of its location, orientation, and scale factor. Therefore, the moment invariants may be used for object recognition and parts identification without regard to camera set-up, location, or orientation. There are seven different moment invariants, such as:

$$MI_1 = \frac{M_{0,0}M_{2,0} - M_{1,0}^2 + M_{0,0}M_{0,2} - M_{0,1}^2}{M_{0,0}^3} \tag{9.27}$$

Refer to Reference [2] for the other six moment invariants.

Example 9.11

For the simple object in a low-resolution image of Figure 9.65, calculate the area, center of the area, and second moments of area of the object relative to the x_1, y_1-axes.

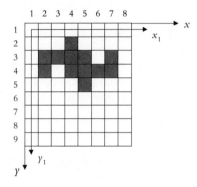

Figure 9.65 Image used for Example 9.11.

Solution: Measuring the distances of each on-pixel from the x_1, y_1-axes and substituting the measurements into the moment equations will yield the following results:

$$M_{0,0} = \sum x^0 y^0 = 12(1) = 12$$

$$M_{1,0} = \sum x^1 y^0 = \sum x = 2(1) + 1(2) + 3(3) + 3(4) + 1(5) + 2(6) = 42$$

$$M_{0,1} = \sum x^0 y^1 = \sum y = 1(1) + 5(2) + 5(3) + 1(4) = 30$$

$$\bar{x} = \frac{M_{1,0}}{M_{0,0}} = \frac{42}{12} = 3.5 \quad \text{and} \quad \bar{y} = \frac{M_{0,1}}{M_{0,0}} = \frac{30}{12} = 2.5$$

$$M_{2,0} = \sum x^2 y^0 = \sum x^2 = 2(1)^2 + 1(2)^2 + 3(3)^2 + 3(4)^2 + 1(5)^2 + 2(6)^2 = 178$$

$$M_{0,2} = \sum x^0 y^2 = \sum y^2 = 1(1)^2 + 5(2)^2 + 5(3)^2 + 1(4)^2 = 82$$

The same procedure may be used for an image with much higher resolution. There will just be many more pixels to deal with. However, a computer program can handle as many pixels as necessary without difficulty. ■

Example 9.12

In a certain application, a vision system looks at an 8×8 binary image of rectangles and squares. The squares are either 3×3-pixel solids, or 4×4 hollow, while the rectangles are 3×4 solids. Through guides, jigs, and brackets, we can be certain the objects are always parallel to the reference axes, as shown in Figure 9.66, and that

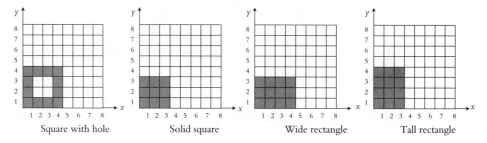

Square with hole   Solid square   Wide rectangle   Tall rectangle

Figure 9.66   Image used for Example 9.12.

the lower left corners of the objects are always at pixel 1,1. We want to only use the moment equations to distinguish the parts from each other. Find one set of lowest values a and b in the moment equation that would be able to do so, with corresponding values for each part. For this example, use the absolute coordinates of each pixel for distances from the corresponding axes.

Solution: Using the moment equations, we will calculate the different moments for all four until we find one set that are all unique for each object.

Square with hole	Solid square	Wide rectangle	Tall rectangle
$M_{0,0} = 12$	$M_{0,0} = 9$	$M_{0,0} = 12$	$M_{0,0} = 12$
$M_{0,1} = 30$	$M_{0,1} = 18$	$M_{0,1} = 24$	$M_{0,1} = 30$
$M_{1,0} = 30$	$M_{1,0} = 18$	$M_{1,0} = 30$	$M_{1,0} = 24$
$M_{1,1} = 75$	$M_{1,1} = 36$	$M_{1,1} = 60$	$M_{1,1} = 60$
$M_{0,2} = 94$	$M_{0,2} = 42$	$M_{0,2} = 56$	$M_{0,2} = 90$

As shown, the lowest set of moment indices that yields a unique solution for each object is $M_{0,2}$. Of course, $M_{2,0}$ would result in similar numbers. ∎

Example 9.13

For the image of the screw in Figure 9.67, calculate the area, $\bar{x}, \bar{y}, M_{0,2}, M_{2,0}, M_{1,1}$ $M_{2,0} @ \bar{x}$, $M_{0,2} @ \bar{y}$ and the moment invariant.

Solution: A macro program called moments.macro was written for the Optimas™ 6.2 vision software to calculate the moments. In this program, distances used for moments are all in terms of the number of pixels and not in units of length. The values were calculated for five separate cases, horizontal, 30°, 45°, 60°, and vertical. Small variations in the results are due to rotation operations. Every time a part of an image is rotated, since every point in the image must be converted with a sine or cosine function, it changes slightly. Otherwise, as you can see, the results are consistent. For example, as

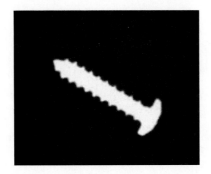

Figure 9.67 Image used for Example 9.13.

the part is rotated in place, the location of its center of area does not change. You also can see that the moment invariant is constant and, from the moment of inertia information about the centroid, the orientation can be estimated. This information can now be used to identify the object or to direct a robot controller to send the robot arm to the location, with proper orientation, to pick up the part.

	Horizontal	30°	45°	60°	Vertical
Area	3713	3747	3772	3724	3713
x-bar	127	123	121	118	113
y-bar	102	105	106	106	104
$M_{0,2}$	38.8 E6	43.6 E6	46.4 E6	47.6 E6	47.8 E6
$M_{2,0}$	67.6 E6	62.6 E6	59 E6	53.9 E6	47.8 E6
$M_{1,1}$	48.1 E6	51.8 E6	52 E6	49.75 E6	43.75 E6
Moment Invariant	7.48	7.5	7.4	7.3	7.48
$M_{2,0}@\bar{x}$	7.5 E6	5.7 E6	3.94 E6	2.07 E6	0.264 E6
$M_{0,2}@\bar{y}$	0.264 E6	2.09 E6	3.77 E6	5.7 E6	7.5 E6

■

The following is the listing of the moments.macro program, written for Optimas. Although this program cannot directly be used with other software, it is listed here to show how simply a program can be developed to do similar operations. The Excel part of the program is nothing more than a simple set of Excel equations that operate on the coordinates of all pixels, and later, are summed up.

```
/*MOMENTS.MAC PROGRAM Written by Saeed Niku, Copyright 1998
This macro checks an active image within Optimas vision system. It records the
coordinates of all pixels above the given threshold. It subsequently writes the
coordinates into an Excel worksheet, which determines the moments. Moments.mac will
then read back the data and display it. The DDE commands communicate the data between
Excel and Optimas macro program. If your number of coordinates is more than 20,000
pixels, you must change the command below. */
```

```
BinaryArray = GetPixelRect (ConvertCalibToPixels(ROI));
INTEGER NewArray[,]; Real MyArea; Real Xbar; Real Ybar;
Real Mymoment02; Real Mymoment20; Real Mymoment11; Real VariantM;
Real MyMXBar; Real MyMYBar;

For(Xcoordinate = 0; Xcoordinate<= (VectorLength(BinaryArray[0,])-1);
  Xcoordinate ++)
{
      For(Ycoordinate = 0; Ycoordinate <= (VectorLength(BinaryArray[,0])-1);
        Ycoordinate ++)
        {
            If (BinaryArray[Ycoordinate,Xcoordinate] > 100)
            {
                NewArray ::= Xcoordinate : Ycoordinate;
      } } }
hChanSheet1 = DDEInitiate ("Excel","Sheet1");
DDEPoke(hChanSheet1,"R1C1:R20000C2",NewArray);
DDETerminate(hChanSheet1);

Show("Please Enter to Show Values");
hChanSheet1 = DDEInitiate ("Excel","Sheet1");
DDERequest(hChanSheet1,"R1C14",MyArea);
DDERequest(hChanSheet1,"R2C14",Ybar);
DDERequest(hChanSheet1,"R3C14",Xbar);
DDERequest(hChanSheet1,"R4C14",Mymoment02);
DDERequest(hChanSheet1,"R5C14",Mymoment20);
DDERequest(hChanSheet1,"R6C14",Mymoment11);
DDETerminate(hChanSheet1);

VariantM=(MyArea*Mymoment20*1000000.0-Xbar*Xbar
          +MyArea*Mymoment02*1000000.0-Ybar*Ybar)
          /(MyArea*MyArea*MyArea);

MyMYBar=(Mymoment20*1000000.0-MyArea*Xbar*Xbar)/1000000.0;
MyMXBar=(Mymoment02*1000000.0-MyArea*Ybar*Ybar)/1000000.0;

MacroMessage("Area=",MyArea,"\n","Xbar=",Xbar,"\n",
            "Ybar=",Ybar,"\n","Moment02=",Mymoment02," x10^6",
            "\n","Moment20=",Mymoment20," x10^6","\n"
            ,"Moment11=" ,Mymoment11," x10^6","\n","Invariant 1="
            ,VariantM);
MacroMessage("Moment20@Xbar=",MyMXBar," x10^6","\n",
            "Moment02@Ybar=",MyMYBar," x10^6");
```

9.21.3 Template Matching

Another technique for object recognition is model or template matching. If a suitable line drawing of the scene is found, the topological or structural elements such as total number of lines (sides), vertices, and interconnections can be matched to a model. Coordinate transformations such as rotation, translation, and scaling can be performed to eliminate the differences between the model and the object resulted from position, orientation, or depth differences between them. This technique is limited by the fact that *a priori* knowledge of the object models is needed for matching. Therefore, if the object is different from the models, they will not match and the object will not be recognized. Another major limitation is that if one object is occluded by other objects, it will not match a model.

9.21.4 Discrete Fourier Descriptors

Similar to a Fourier transform calculated for an analog signal, a Discrete Fourier Transform (DFT) of a set of discrete points (such as pixels) can also be calculated. This means that if the contour of an object within an image is found (such as in edge detection), the discrete pixels of the contour can also be used for DFT calculations. The result of DFT calculation is a set of frequencies and amplitudes in frequency domain that describe the spatial relationship of the points in question.[14]

To calculate the DFT of a set of points in a plane, assume the plane is a real-imaginary plane, such that each point is described by an $x + iy$ relationship. If the contour is completely traced around, starting from any pixel, and the locations of the points are measured, the information can be used to calculate the corresponding frequency spectrum of the set. Matching these frequencies with the frequencies found for possible objects in a look-up table may be used to determine the nature of the object. In one unpublished experiment, matching 8 frequencies yielded enough information about the nature of the object (an airplane). Matching 16 frequencies could determine the type of an airplane from a large class of planes. An advantage of this technique is that the Fourier transform can very simply be normalized for size, position, and orientation. A disadvantage of the technique is that it requires a complete contour of the object. Of course, other techniques, such as the Hough transform, can be used to complete broken contours of the object.

9.21.5 Computed Tomography (CT)

Tomography is a technique of determining the distribution of material density in the examined part. In computed tomography (CT), a three-dimensional image of the object's density distribution is reconstructed from a large number of two-dimensional images of the density taken by different scanning techniques such as X-rays or ultrasonics. In computed tomography, it is assumed that the part consists of a sequence of overlaying slices. Images of density distribution of each slice are taken repeatedly around the object. Although partial coverage of the part has been used as well, a complete coverage of 360° is preferred. The data is stored in a computer, and subsequently, is reduced to a three-dimensional image of the part's density distribution that is shown on a monitor.

Although this technique is completely different from the other techniques discussed above, it is a viable technique for object recognition. In many situations, either alone or in conjunction with other techniques, computed tomography may be the only way to recognize an object or to differentiate it from other similar objects. Specifically, in medical situations, CT scan can be used in conjunction with medical robots where the three-dimensional mapping of the internal organs of the human body may be used to direct the robot for surgical operations.

9.22 Depth Measurement with Vision Systems

Extracting depth information from a scene is performed using two basic techniques. One is the use of range finders in conjunction with a vision system and image-processing techniques. In this combination, the scene is analyzed in relation to the information gathered by range finders about the distances of different portions of an environment or the location of particular objects or sections of the object. Second is the use of binocular or stereo vision similar to humans and animals. In this technique, either simultaneous images from multiple cameras or multiple images from one camera that moves on a track are used to extract depth information. As long as the scene does not change during this operation, the results will be the same as the use of multiple cameras. Since the location of the multiple (usually two) cameras in relation to any particular point in the scene is slightly different, each camera develops a slightly different image. By analyzing and measuring the differences between the two scenes, depth information can be extracted.

9.22.1 Scene Analysis versus Mapping

Scene analysis refers to the analysis of images developed by a camera or other similar devices in which a complete scene is analyzed. In other words, the image is a complete replica of the scene within the resolution limit of the device where all the details of the scene are included in the image. In this case, more processing is generally required to extract information from the image, but more information can be extracted. For instance, in order to identify an object within a scene, the image may have to be filtered and enhanced, segmented by edge detection or thresholding, the part isolated by region growing, and then identified by extracting its features and comparing them to a template or look-up table. On the other hand, mapping refers to drawing the surface topology of a scene or object where the image consists of a set of discrete distance measurements, usually at low resolutions. The final image is a collection of lines that relate to the relative position of points on the object at discrete locations. Since the image is already sliced, less processing is required in analysis of mapped images, but less information can be extracted from the scene as well. Each technique has its own merits, benefits, and limitations and is used for different purposes, including navigation.

9.22.2 Range Detection and Depth Analysis

Range measurement and depth analysis is performed using many different techniques such as active ranging,[20] stereo imaging, scene analysis, or specialized lighting. Humans use a combination of techniques to extract information about the depth and positional relationship between different elements of an image. Even in a two-dimensional image, humans can extract useful information using details such as the changing size of similar elements, vanishing lines, shadows, and changing intensity of textures and shades. Since many artificial intelligence techniques are based on, and studied for, understanding of the way humans do things, a number of depth measurement techniques are designed after similar human operations.[15]

9.22.3 Stereo Imaging

An image is the projection of a scene into the image plane through an ideal lens. Therefore, every point in the image will correspond to a certain point in the scene.

However, the depth information of the point is lost in this projection and cannot simply be retrieved from a single image. If two images of the same scene are taken, the relative depth of different points from the image plane can be extracted by comparing the two images; the differences represent the spatial relationship between different points.[16,17] Humans do the same, automatically, by combining the two images and forming a three-dimensional image.[18,19,21] The stereo image used for depth measurement is considered a 2.5-dimensional image. Many more images are required to form a true three-dimensional image.

Depth measurement using stereo images requires two operations:

1. To determine the point-pairs in the two images that correspond to the same point in the scene. This is called correspondence or disparity of the point-pair. This is a difficult operation since some points in one image may not be visible in another, or since due to perspective distortion, sizes and spatial relationships may be different in the two images.
2. To determine the depth or location of the point on the object or in the scene by triangulation or other techniques.

Generally, if the two cameras (or the relative locations of a single camera used twice to get two images of a nonmoving, static scene) are accurately calibrated, triangulation is relatively simple as long as enough corresponding points have been found.

Correspondence points can be determined by matching specific features, such as corners or small segments, from the two images. Depending on their locations, correspondence points can create matching problems. Consider the two marks A and B in Figure 9.68. In each case, the two cameras will see the marks as shown in (a) and (b). Although the locations of the two marks are different, the cameras will see them similarly. As a result, the marks may be located wrongly (unless additional information such as vanishing lengths are also considered).

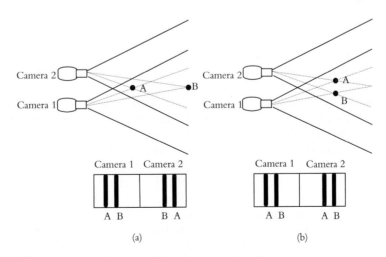

Figure 9.68   Correspondence problem in stereo imaging.

The accuracy of depth measurement in stereo imaging is dependent on the angle between the two images, and therefore, the disparity. However, larger disparities require more searching over larger areas. To improve the accuracy and reduce computation time, multiple images of the same scene can be used.[18] A similar technique was used in the Stanford Cart, where the navigation system would use a camera mounted on a shaft to take multiple images of the scene in order to calculate distances and find obstacles.[23]

9.22.4 Scene Analysis with Shading and Sizes

Humans use the details contained in a scene to extract information about the locations of objects, their sizes, and their orientations. One detail is the shading on different surfaces. Although the smoothly changing intensity of shades on surfaces is a source of difficulty in some other operations such as segmentation, it can be indirectly used in extracting information about the depth and shape of objects. Shading is the relationship between the orientation of the object and the reflected light. If this relationship is known, it can be used to derive information about the object's location and orientation. Depth measurement using shades requires *a priori* knowledge of the reflectance properties of the object and exact knowledge of the light source. As a result, its utility is limited.

Another source of information for depth analysis is the use of texture gradient, or the changes caused in textures as a result of depth changes. These variations are due to changes in the texture itself, which is assumed to be constant, or due to changes in the depth or distance (scaling gradient), or due to changes in the orientation of the plane (called foreshortening gradient). An example of this is the perceived change in the size of bricks on a wall. By calculating the gradient of the brick sizes on the wall, depth can be estimated.

9.23 Specialized Lighting

Another possibility for depth measurement is utilizing special lighting techniques that yield a specific result. The specialized result can be used for extracting depth information. Most of these techniques are designed for industrial applications, where specialized lighting is possible and the environment is controlled. The following is the theory behind one technique.

If a strip (narrow plane) of light is projected over a flat surface, it will generate a straight line in relation with the relative positions and orientations of the plane and light source. However, if the plane is not flat and an observer looks at the light strip in a plane other than the plane of light, a curved or broken line will be observed (Figure 9.69). By analyzing the reflected light, we can extract information about the shape of the object, its location, and orientation. In certain systems, two strips of light are used, such that in the absence of any object on the table, the two strips intersect exactly on the surface. When an object is present, the two strips of light develop two reflections. The reflections are picked up by a camera and depth information is calculated and reported. A commercial system based on this technique is called CONSIGHT™.

A disadvantage of this technique is that only information about the points that are lit can be extracted. Therefore, in order to have information about the complete image, it is necessary to scan the entire object or scene.

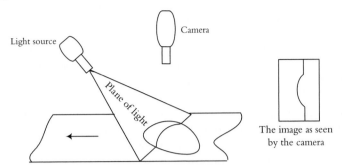

Figure 9.69 Application of specialized lighting in depth measurement. A plane of light strikes the object. The camera, located at a different angle than the plane of light, will see the reflection of the light plane on the object as a curved line. The curvature of the line is used to calculate depth.

9.24 Image Data Compression

Electronic images contain large amounts of information, and therefore, require data transmission lines with large bandwidth capacity. The requirements for spatial resolution, number of images per second, and number of gray levels (or colors for color images) are determined by the required quality of the images. Recent data transmission and storage techniques have significantly improved image transmission capability, including transmission over the Internet.

The following are some techniques that accomplish this task. Although there are many different techniques of data compression, only some of them directly relate to vision systems. The subject of data transmission in general is beyond the scope of this book and will not be discussed here. Image data compression techniques are divided into intraframe (within frame) and interframe (between frames) methods.

9.24.1 Intraframe Spatial Domain Techniques

Pulse Code Modulation (PCM) is a popular technique of data transmission in which the analog signal is sampled, usually at the Nyquist rate (a rate that will prevent aliasing), and quantized. The quantizer will have N levels where N is a power of 2. If N is 8, then 2^8 will yield a quantizer with 256 different gray levels (an 8-bit image quantizer). Certain applications (space and medical) use higher resolutions such as 2^{10} or 2^{12}.

In a technique called Pseudorandom Quantization Dithering,[22] random noise is added to the pixels' gray values in order to maintain the same quality while reducing the number of bits. This is done to prevent contouring, which happens when the number of bits of a quantizer is reduced (see section 9.5 and Figure 9.8). These contours can be broken up by adding a small amount of broadband pseudorandom, uniformly distributed noise called dither to the signal prior to sampling. The dither causes the pixel to oscillate about the original quantization level, removing the contours. In other words, the contours are forced to randomly make small oscillations about their average value. A proper amount of noise will enable the system to have the same apparent resolution while the number of bits is reduced significantly.

Another technique of data compression is halftoning. In halftoning, a pixel is effectively broken up into a number of pixels by increasing the number of samples per pixel. Instead, every sample is quantized by a simple 1–bit binary quantizer into a black or white pixel. Since the human eye will average the groups of pixels, the image will still look gray and not binary.

Predictive Coding refers to a class of techniques which are based on the theory that in highly repetitive images, only the new information (innovations) need be sampled, quantized, and transmitted. In these types of images, many pixels remain without change in multiple images (e.g., TV news sets). Therefore, the data transmission can be significantly reduced if only the changes between successive images are transmitted. To do this, a predictor is used to predict an optimum value for each pixel based on the information obtained from the previous images. The innovation is the difference between the actual value of the pixel and the predicted value. This value is transmitted by the system to update the previous image. If in an image many pixels remain the same, the innovations are few and transmission is reduced.

Example 9.14

In a similar attempt to reduce the amount of data transmission in space by the Voyager 2 Spacecraft, its computers were reprogrammed, while in space, to use a differential coding technique. At the beginning of its space travel, the Voyager's system was designed to transmit information about every pixel, at a 256 gray level scale. This took 5,120,000 bits to transmit one single image, not including error detection and correction codes, which were about the same length. Beginning with the Uranus fly-by, the system was reprogrammed to only send the difference between successive pixels rather than the absolute brightness of the pixels. Consequently, if there were no differences between successive pixels, no information would be transmitted. In scenes such as in space, where the background is essentially black, there are many pixels that are similar to their neighbors. This reduced data transmission by about 60%.[23] Other examples of fixed background information include theatrical sets and industrial images.   ■

In Constant Area Quantization (CAQ),[24,25] data transmission is reduced by transmitting fewer pulses at lower resolution in low contrast areas compared to high contrast areas. This, in effect, is taking advantage of the fact that higher contrast areas have higher frequency content and require more information transmission than the lower contrast areas.

9.24.2 Interframe Coding

These methods take advantage of the redundant information that exists between successive images. The difference between these and the intraframe methods is that rather than using the information within one image, a number of different images are used to reduce the amount of information to be transmitted.

A simple technique to achieve this is to use a frame memory at the receiver. The frame memory will hold an image and continually show it at the display. When information about any pixel is changed, the corresponding location in the frame memory is updated. As a result, the rate of transmission is significantly reduced. The disadvantage of this technique is that in the presence of rapidly moving elements, flickering may happen.

9.24.3 Compression Techniques

Two general methods are used for data compression. In one method (such as in zip archives) called *lossless compression*, codes are assigned to repetitive words, phrases, or values in order to reduce the size of the data file. As the name implies, in these methods no data is lost, and consequently, the original file may be reconstructed without any change or loss. However, the level of savings or compression in chromatic or achromatic gray images is not large because in these images pixels rarely have repeating patterns. These methods, however, are more useful in particular situations. For example, in a binary file, large areas (blobs) have similar values. If the data is presented line by line (such as in a facsimile file scanned line by line), the data may be compressed by coding the length of on-off sets of pixels rather than the value of each individual pixel. Therefore, many pixels with similar values can be represented by only specifying the starting point of each section and its length.

The second category covers methods that compress image data by reducing the information, and therefore, are called *lossy compression*, including the popular JPEG (Joint Photographers Expert Group) compression.[13] Although we will not discuss the elaborate sequence of steps taken in order to compress the data, it should be mentioned that a lot of information is lost during this process, although a much smaller file is generated when an image is saved or converted to JPEG format. However, unless the original detailed data is needed for other purposes or a picture must be zoomed in to extract information, the human eye may not recognize the difference as much.

9.25 Color Images

White light can be decomposed into a rainbow of colors that span the range of 400 to 700 nm wavelengths. Although it is rather difficult to subscribe an exact value to any particular hue, the primary colors of light are thought to be red, green, and blue (RGB). Theoretically, all other hues and color intensities can be recreated by mixing varying levels of the primary color lights, although in reality, the recreations are not truly accurate. However, in images, most colors can be recreated using RGB.

To capture color images, filters are used to separate the light into these three subimages and each one is captured, sampled, and quantized individually, therefore creating three image files. To recreate color images, the screen is composed of three sets of pixels, interlaced sequentially (RGBRGB . . .). Each set of pixels is recreated individually, but simultaneously. Due to the limited spatial resolution of our eyes, we tend to mix the three images together and perceive color images. However, as far as image processing is concerned, a color image is in fact a set of three images, each representing the intensities of the three primary colors of the original image.

To convert a color (chromatic) image into a black and white (achromatic) image, the intensities of the individual colored files must be converted into gray values. One method to do this is to take the average values of the three files for the same pixel location and to use that as a substitute for the gray value. Therefore, the histogram of a gray image shows the same exact values for all three channels of RGB. For more information about the image processing of colored images please refer to other resources such as References 2 and 13.

9.26 Heuristics

Heuristics is a collection of rules of thumb developed for semi-intelligent systems in order to enable them to select a predetermined decision from a list based on the current situation. Heuristics is used in conjunction with mobile robots, but has applications in many fields.

Consider a mobile robot that is supposed to navigate through a maze. Imagine the robot starts at a point and is equipped with a sensor that alerts its controller that the robot has reached an obstacle such as a wall. At this point, the controller has to decide what to do next. Let's say the first rule is that when encountering an obstacle, the robot should turn left. As the robot continues, it may reach another wall, turn left again, and continue. Suppose that after three left turns, the robot reaches the starting point. In this case, should it continue to turn left? Obviously, this will result in a never-ending loop. The second rule may be to turn right if the first point is encountered. Now imagine that after a left turn, the robot gets to a dead end. Then what? A third rule may be to trace back the path until an alternate route can be found. As you can see, there are many different situations the robot may encounter. Each one of these situations must be considered by the designer, and a decision must be provided. The collection of these rules is the heuristics rule base for the controller to "intelligently" decide how to control the motions of the robot. However, it is important to realize that this is not true intelligence since the controller is not really making decisions but merely selecting from a set of decisions that have already been made. If a new situation is encountered that is not in the rules base, the controller will not know how to respond.[26]

9.27 Applications of Vision Systems

Vision systems may be used for many different applications, including in conjunction with robotic operations and robots. Vision systems are commonly used for operations that require information from the work environment and include inspection, navigation, part identification, assembly operations, surveillance, control, and communication.

Suppose that in an automatic manufacturing setting, a circuit board is to be manufactured. One important part in this operation is the inspection of the board at different states before and after certain operations. A common method is to set up a cell where an image of the part is taken and subsequently modified, improved, and altered. The processed image is compared to a look-up image. If there is a match, the part is accepted. Otherwise, either the part is rejected or is repaired. These image-processing and analysis operations are generally made up of the processes discussed earlier. Most commercial vision systems have embedded routines that can be called from a macro program, making it very easy to set up a system.

Vision systems have been used for many applications, for example, for locating radioactive pucks,[27] random bin picking,[28] creating an automated brake inspection system,[29] measuring robotic motions and external objects,[30] food inspection such as texture of cookies and consistency of packaging,[31] creating adaptive behavior for mobile robots,[32] analyzing the health of agricultural crops,[33] and many others.

In navigation, the scene is usually analyzed for finding acceptable pathways, obstacles, and other elements that confront the robot.[34] In some operations, the vision system sends its information to an operator, who controls the motions from a distance. This is very common in telerobotics as well as in space applications.[35] In some medical applications, the surgeon guides the device, whether a surgical robot or a small investigative, exploratory device such as an angiogram, through its operations.[36] Autonomous navigation requires the integration of depth measurement with the vision system, either by stereo vision analysis, or by range finders. It also requires heuristic rules of behavior for the robotic device to navigate around an environment.

In another application,[37,38] an inexpensive laser diode was mounted next to a camera. The projected laser light was captured by the camera and was used to both measure the depth of a scene as well as to calibrate the camera. In both cases, due to the brightness of the laser light and bleeding effects, the image contained a large, bright circular spot. To identify the dot and separate it from the rest of the scene, a histogram and thresholding operation was used. Subsequently, the circle was identified and skeletonized until only the center of the circle remained. The location of the pixel representing the center of the circle was then used in a triangulation method to calculate the depth of the image, or to calibrate the camera.

These simple examples are all related to what we have discussed. Although many other routines are available, the fundamental knowledge about vision systems enables you to proceed with an application and adapt to your application what vision systems have to offer.

9.28 Design Project

There are many inexpensive digital cameras on the market that can be used to create a simple vision system. They are simple, small, and lightweight and provide a simple image that can be captured by computers and be used to develop a vision system. In fact, many cameras come with the software to capture and digitize an image. Standard still and video cameras can also be used for capturing images. In this case, although you can capture an image for later analysis, due to the additional steps of downloading the image from the camera to the computer, the image is not available for immediate use.

Additionally, many simple programs such as Adobe's Photoshop^{TM} have many routines similar to what we have discussed in this chapter. Additional routines may be developed using common computer languages such as C. Many other routines may be downloaded from the public domain. The final product will be a simple vision system that can be used to perform vision tasks. This may be done independently, or in conjunction with your 3-axis robot, and can include routines for parts identification and pick up, the development of mobile robots, and many other similar devices.

Most images shown in this chapter were captured and processed by standard digital cameras and the vision systems in the Mechanical Engineering Robotics laboratory at Cal Poly, including MVS909^{TM} and Optimas^{TM} 6.2 vision systems and Photoshop^{TM}.

You may also develop your own simple vision system using other programming languages and systems that can handle an image file. This includes Excel™, LabView™, and other development systems.

Summary

In this chapter, we studied the fundamentals of image processing to modify, alter, improve, or enhance an image as well as image analysis through which data can be extracted from an image for subsequent applications. This information may be used for a variety of applications, including manufacturing, surveillance, navigation, and robotics. Vision systems are very powerful tools that can be used with ease. They are flexible and inexpensive.

There are countless routines that can be used for a variety of different purposes. Most of these types of routines are created for specific operations and applications. However, certain fundamental techniques such as convolution masks can be applied to many classes of routines. We have mostly concentrated on these types of techniques, which enable you to adopt, develop, and use other routines and techniques for other applications. The advances in technology have also created tremendous opportunities in this area. There is no doubt that this trend will continue in the future as well.

References

1. Doudoumopoulos, Roger, "On-Chip Correction for Defective Pixels in an Image Sensor," NASA Tech Briefs, May 2000, p. 34.
2. Gonzalez, R. C., Richard Woods, "Digital Image Processing," Prentice Hall, New Jersey, 2002.
3. Low, Adrian, "Introductory Computer Vision and Image Processing," McGraw-Hill, 1991.
4. Horn, B. K. P., "Robot Vision," McGraw-Hill, 1986.
5. Hildreth, Ellen, "Edge Detection for Computer Vision System," *Mechanical Engineering*, August 1982, pp. 48–53.
6. Olson, Clark, "Image Smoothing and Edge Detection Guided by Stereoscopy," NASA Tech Briefs, September 1999, pp. 68–69.
7. Groover, M. P., et al. "Industrial Robotics, Technology, Programming, and Applications," McGraw-Hill, 1986, p. 177.
8. Hough, P. V. C., *A Method and Means for Recognizing Complex Patterns*, U.S. Patent 3,069,654, 1962.
9. Illingworth, J., J. Kittler, "A Survey of the Hough Transform," Computer Vision, Graphics, and Image Processing, Vol. 44, 1988, pp. 87–116.
10. Kanade, T., "Survey; Region Segmentation: Signal vs. Semantics," Computer Graphics and Image Processing, Vol. 13, 1980, pp. 279–297.
11. Snyder, Wesley, "Industrial Robots: Computer Interfacing and Control," Prentice Hall, 1985.
12. Haralick, Robert M., L. G. Shapiro, "Computer and Robot Vision," Volume I, Addison Wesley, MA, 1992.
13. Russ, John C., J. C. Russ, "Introduction to Image Processing and Analysis," CRC Press, 2008.
14. Gonzalez, Rafael, P. Wintz, "Digital Image Processing," 2nd Edition, Addison-Wesley, Reading, Mass., 1987.
15. Liou, S. P., R. C. Jain, "Road Following Using Vanishing Points," Proceedings of IEEE Computer Society Conference on Computer Vision and Pattern Recognition, 1986, pp. 41–46.
16. Nevatia, R., "Machine Perception," Prentice Hall, New Jersey, 1982.

17. Fu, K. S., Gonzalez, R. C., Lee, C. S. G., "Robotics; Control, Sensing, Vision, and Intelligence," McGraw-Hill, 1987.

18. Marr, D., T. Poggio, "A Computational Theory of Human Stereo Vision," Proceedings of the Royal Society, London, B204, 1979, pp. 301–328.

19. Marr, D., "Vision," Freeman and Co., 1982.

20. Pipitone, Frank, T. G. Marshall, "A Wide-field Scanning Triangulation Rangefinder for Machine Vision," The International Journal of Robotics Research, Vol. 2, No. 1, Spring 1983, pp. 39–49.

21. Moravec, H. P., "Obstacle Avoidance and Navigation in the Real World by Seeing Robot Rover," Stanford Artificial Intelligence Laboratory Memo, AIM-340, September 1980.

22. Thompson, J. E., "A 36-Mbit/s Television Coder Employing Psuedorandom Quantization," IEEE Transactions on Communication Technology, COM-19, No. 6, December 1971, pp. 872–879.

23. Goldstein, Gina, "Engineering the Ultimate Image, The Voyager 2 Mission," *Mechanical Engineering*, December 1989, pp. 30–36.

24. Pearson, J. J., R. M. Simonds, "Adaptive, Hybrid, and Multi-Threshold CAQ Algorithms," Proceedings of SPIE Conference on Advanced Image Transmission Technology, Vol. 87, August 1976, pp. 19–23.

25. Arnold, J. F., M. C. Cavenor, "Improvements to the CAQ Bandwidth Compression Scheme," IEEE Transactions on Communications, COM-29, No. 12, December 1981, pp. 1818–1822.

26. Chattergy, R., "Some Heuristics for the Navigation of a Robot," The International Journal of Robotics Research, Vol. 4, No. 1, Spring 1985, pp. 59–66.

27. Wilson, Andrew, Editor, "Robot Vision System Locates Radioactive Pucks," Vision Systems Design, May 2002, pp. 7–8.

28. "Using Vision to Enable Robotic Random Bin Picking," Imaging Technology, June 2008, pp. 84–86.

29. "Creating an Automated Brake Inspection System with Machine Vision," Imaging Technology, June 2008, pp. 88–90.

30. "Vision System Measures Motions of Robots and External Objects," NASA Tech Briefs, November 2008, pp. 24–26.

31. Thilmany, Jean, "Accessible Vision," *Mechanical Engineering*, July 2009, pp. 42–45.

32. "Adaptive Behavior for Mobile Robots," NASA Tech Briefs, August 2009, pp. 52–53.

33. "Imaging System Analyzes Crop Health," Defense Tech Briefs, August 2009, pp. 32–33.

34. "Vision-Based Maneuvering and Manipulation by a Mobile Robot," NASA Tech Briefs, March 2002, pp. 59–60.

35. Ashley, Steven, Associate Editor, "Roving Other Worlds by Remote," *Mechanical Engineering*, July 1997 pp. 74–76.

36. Hallett, Joe, Contributing Editor, "3-D Imaging Guides Surgical Operations," Vision Systems Design, May 2001, pp. 25–29.

37. Niku, S. B., "Active Distance Measurement and Mapping Using Non Stereo Vision Systems," Proceedings of Automation '94 Conference, July 1994, Taipei, Taiwan, R.O.C., Vol. 5, pp. 147–150.

38. Niku, S. B., "Camera Calibration and Resetting with Laser Light," proceedings of the 3rd International Conference on Mechatronics and Machine Vision in Practice, September 1996, Guimaraez, Portugal, Vol. 2, pp. 223–226.

Problems

Note: If you do not have access to an image, simulate the image by creating a file called $I_{m,n}$ where m and n are the row and column indices of the image. Then, using the following image matrix, create an image by substituting numbers 0 and 1 or gray-level numbers in the file. In a binary image 0 represents off, dark or background pixel, while 1 represents on, light, or object pixels. In gray images, each pixel is represented by a corresponding greyness level value. A computer routine can then be written to access this file for image data. The result of each operation can be written to a new file such as $R_{m,n}$, where R represents result of the operation, and m and n are the row and column indices of the resulted file.

Alternately, you may use your own graphics system or any commercially available graphics language to create, access, and represent an image.

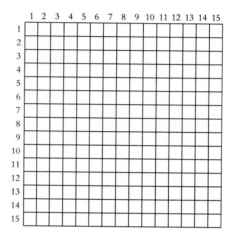

Figure 9.70   A blank image grid.

9.1. Calculate the necessary memory requirement for a still color image from a camera with 10 megapixels at:

- 8-bits per pixel (256 levels)
- 16-bits per pixel (65,536 levels)

9.2. Consider the pixels of an image, with values as shown in Figure P.9.2 , as well as a convolution mask with the given values. Calculate the new values for the given pixels.

4	7	1	8
3	2	6	6
5	3	2	3
8	6	5	9

1	0	1
1	1	1
1	0	1

Figure P.9.2

9.3. Consider the pixels of an image, with values as shown in Figure P.9.3, as well as a convolution mask with the given values. Calculate the new values for the given pixels. Substitute 0 for negative gray levels. What conclusion can you make from the result?

5	5	5	7	7	8
5	5	5	8	8	9
10	10	10	10	10	10
8	8	8	9	9	8
8	8	8	9	9	10

1	0	1
0	-4	0
1	0	1

Figure P.9.3

9.4. Repeat Problem 9.3, but substitute the absolute value of negative gray levels. What conclusion can you make from the result?

9.5. Repeat Problem 9.3, but apply the mask of Figure P.9.5 and compare your results with Problem 9.3. Which one is better?

-1	0	-1
0	4	0
-1	0	-1

Figure P.9.5

9.6. Repeat Problem 9.3, but apply the mask of Figure P.9.6 and compare your results with Problem 9.3. Which one is better?

0	1	0
1	-4	1
0	1	0

Figure P.9.6

9.7. Repeat Problem 9.3, but apply the mask of Figure P.9.7 and compare your results with Problem 9.3. Which one is better?

1	1	1
1	-8	1
1	1	1

Figure P.9.7

9.8. An image is represented by values shown below.

(a) Find the value of pixel 2c when mask 1 is applied.

(b) Find the value of pixel 3b when mask 2 is applied.

(c) Find the values of pixels 2b and 3c when a 3×3 median filter is applied.

(d) Find the area of the major object that results when a threshold of 4.5 is applied based on a +4-connectivity (start at the first on-pixel).

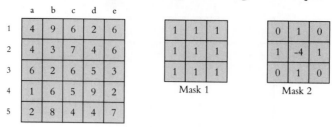

Figure P.9.8

9.9. An image is represented by values shown below.

  (a) Find the value of pixel 3b when mask 1 is applied.

  (b) Find the values of pixels 2b, 2c, 2d when mask 2 is applied.

  (c) Find the value of pixel 3c when a 5 × 5median filter is applied.

  (d) Find the area of the major object that results when a threshold of 4.5 is applied based on a ×4-connectivity (start at the first on-pixel).

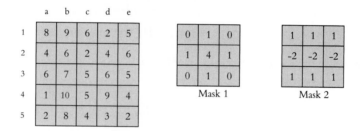

Figure P.9.9

9.10. Write a computer program for the application of a 3×3 averaging convolution mask unto a 15×15 image. Refer to the note on page 415 for more information.

9.11. Write a computer program for the application of a 5×5 averaging convolution mask unto a 15×15 image. Refer to the note on page 415 for more information.

9.12. Write a computer program for the application of a 3×3 high-pass convolution mask unto a 15×15 image for edge detection. Refer to the note on page 415 for more information.

9.13. Write a computer program for the application of an n×n convolution mask unto a k×k image. Refer to the note on page 415 for more information. You should write the routine such that the user can choose the size of the mask and the values of each mask cell individually.

9.14. Write a computer program that will perform the Left-Right search routine for a 15×15 image. Refer to the note on page 415 for more information.

9.15. Using the Left-Right search technique, find the outer edge of the object in Figure P.9.15:

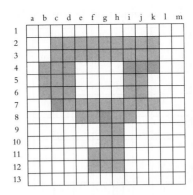

Figure P.9.15

9.16. The x and y coordinates of 5 points are given as (2.5, 0), (4,2), (5,4), (7,6), and (8.5,8). Using the Hough transform, determine which of these points form a line and find its slope and intercept.

9.17. Write a computer program that will perform a region growing operation based on +4 connectivity. The routine should start at the 1,1 corner pixel, search for a nucleus, grow a region with a chosen index number, and after finishing that region, must continue searching for another nuclei until all object pixels have been checked. Refer to the note on page 415 for more information.

9.18. Write a computer program that will perform a region growing operation based on ×4 connectivity. The routine should start at the 1,1 corner pixel, search for a nucleus, grow a region with a chosen index number, and after finishing that region, must continue searching for other nuclei until all object pixels have been checked. Please refer to the note on page 415 for more information.

9.19. Using +4 connectivity logic and starting from 1,a pixel, write the sequence of pixels in correct order that will be detected by a region growing routine for the object in Figure P.9.19:

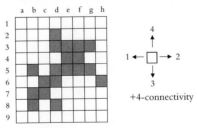

Figure P.9.19

9.20. Using ×4 connectivity logic and starting from 1,a pixel, write the sequence of pixels in correct order that will be detected by a region growing routine for the object in Figure P.9.20:

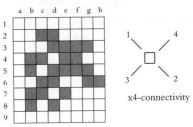

Figure P.9.20

9.21. Find the union between the two objects in Figure P.9.21.

0	0	0	0	0	0	0	0	0	0	0	0	0
0	0	0	0	0	0	0	0	0	0	0	0	0
0	0	0	0	0	0	0	0	0	0	0	0	0
0	0	0	0	0	0	0	0	0	0	0	0	0
0	0	0	0	0	0	0	0	0	0	0	0	0
0	0	0	0	0	0	0	0	0	0	0	0	0
0	0	0	0	1	1	1	1	0	0	0	0	0
0	0	0	0	0	0	0	1	0	0	0	0	0
0	0	0	0	0	0	0	0	0	0	0	0	0
0	0	0	0	0	0	0	0	0	0	0	0	0
0	0	0	0	0	0	0	0	0	0	0	0	0
0	0	0	0	0	0	0	0	0	0	0	0	0
0	0	0	0	0	0	0	0	0	0	0	0	0

0	0	0	0	0	0
0	1	0	0	0	0
0	1	0	0	0	0
0	1	1	1	1	0
0	0	0	0	1	0
0	0	0	0	1	0
0	0	0	0	0	0

Figure P.9.21

9.22. Apply a single-pixel erosion based on 8-connectivity on the image of Figure P.9.22.

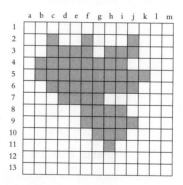

Figure P.9.22

9.23. Apply a one-pixel dilation to the result of Problem 9.22 and compare your result to Figure P.9.22.

9.24. Apply an open operation to Figure P.9.24.

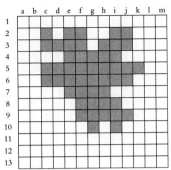

Figure P.9.24

9.25. Apply a close operation to Figure P.9.24.

9.26. Apply a skeletonization operation to Figure P.9.26 .

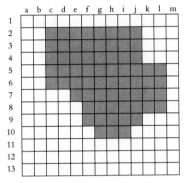

Figure P.9.26

9.27. Write a computer program in which different moments of an object in an image can be calculated. The program should ask you for moment indices. The results may be reported to you in a new file, or may be stored in memory. Refer to the note on page 415 for more information.

9.28. Calculate the $M_{0,2}$ moment for the result of Problem 9.8(d) based on +4-connectivity.

9.29. For the binary image of a key in Figure P.9.29, calculate the following:

- Perimeter, based on the Left-Right search technique.
- Thinness, based on $P^2/_{Area}$.
- Center of gravity.
- Moment $M_{0,1}$ about the origin (pixel 1,1) and about the lowest pixel of a rectangular box around the key (2,2).

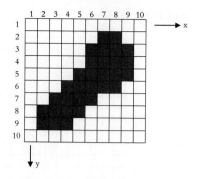

Figure P.9.29

9.30. Using moment equations, calculate $M_{0,2}$ and $M_{2,0}$ about the centroidal axes of the part in Figure P.9.30.

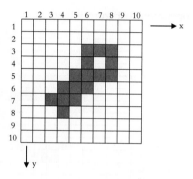

Figure P.9.30

9.31. Using moment equations, calculate $M_{0,2}$ and $M_{2,0}$ about the centroidal axes of the part in Figure P.9.31.

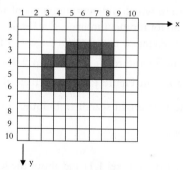

Figure P.9.31

CHAPTER 10

Fuzzy Logic Control

10.1 Introduction

"Tuesday, October 26, 1993 was supposed to be a very warm day in San Luis Obispo, and in fact it turned out to be pretty hot. When the robotics lab was opened in the morning, we found out that the steam line had leaked into the room and much heat and humidity had been released into the environment. When the hydraulic power unit for the robots was turned on, it added even more heat to the lab, raising the temperature even further. Eventually, it got so hot that we had to bring in large fans to cool down the lab a bit to make it a little more comfortable for students."

This true statement is a very good example of what fuzzy logic is about. Let's look at the statement again, noticing the underlined words:

"Tuesday, October 26, 1993 was supposed to be a very warm day in San Luis Obispo, and in fact it turned out to be pretty hot. When the robotics lab was opened in the morning, we found out that the steam line had leaked into the room and much heat and humidity had been released into the environment. When the hydraulic power unit for the robots was turned on, it added even more heat to the lab, raising the temperature even further. Eventually, it got so hot that we had to bring in large fans to cool down the lab a bit to make it a little more comfortable for students."

As you can see, a number of "descriptors" are used in this statement to describe certain conditions that are not very clear. For example, when we state that the day was supposed to be very warm, what do you think it was supposed to be? 85°F? Or maybe 100°F? If you live in San Luis Obispo, even 80°F may be too warm. Then, as you can see, this description of the temperature is, in fact, fuzzy. It is not very clear what the temperature may have been. And the statement continues to be fuzzy. We also don't know exactly what is meant by so hot, or a bit, or large fans. How large? How much cooler did the temperature get when we turned on the fans? Now, read the paragraph once again and see how many other fuzzy descriptions are used in addition to the ones we are discussing.

So, obviously, fuzzy statements are very common in everyday speech, and they are constantly used in all matters of conversation.

What makes this even more (you notice <u>even more</u>?) interesting is that these fuzzy descriptions of events and other phenomena are context dependent. A warm day in Alaska is very different from a warm day in Dallas or Rio de Janeiro, and a warm day in Dallas means something entirely different if it happens in the summer or in winter. It is expected that summer days in Dallas are actually hot. So, an 85°F temperature will feel relatively cool, whereas the same temperature in the middle of January in Dallas will feel much warmer. As another example, consider the description of pain by a patient to a doctor. If a young child has just fallen, and "it really hurts," the understanding of the doctor will be different than someone who has had a serious accident and "it really hurts." The description of intensity of pain (as well as its understanding by someone else) while delivering a baby will also be different than the pain during stitching a wound. So, we must extend the fuzziness of descriptions to the context as well.

We have learned to listen to these fuzzy descriptions and understand the meaning of them, even if not very accurately and even if it is related to the context and the person who is describing it and the conditions under which it is described. We have learned to associate a certain range of meanings to particular descriptions as they relate to other conditions and we are capable of making inferences related to the fuzzy descriptions. In everyday conversations, this seems to work, even if we have to describe ourselves later or if we have to ask more questions to clarify the meanings. However, let's consider three simple examples where we need to have a better way of describing the situation.

First, let's assume there is an "expert system" used to prescribe medicine to patients in remote areas where there is no access to doctors for routine problems, and having a doctor visit a patient on the phone is not practical. The system asks the patient about his or her condition, the symptoms, pain, fever, coughing, and so on, and compares these conditions to a look-up table or bank of possible reasons and from that, diagnoses the cause and prescribes a medicine. Now suppose a patient is describing a sore throat. If the sore throat is "really bad," or if the fever is "about"100°, what is the expert system to conclude?

Second, consider a temperature control thermostat. Suppose we want to control the temperature at 75° such that if the temperature is higher, the air conditioning turns on; if it is cooler, it turns off. To do this, we want to set the thermostat control to 75°. In a simple microprocessor control, we may have a control statement such as:

$$\text{IF TEMPERATURE} \geq 75° \text{ TURN ON A/C} \tag{10.1}$$

This means that as soon as the temperature is 75° or more, the air conditioning will turn on. However, as you notice, the system will not turn on even if temperature is 74.9°. Considering the fuzzy descriptions we saw earlier, is this what we intend?

Third, consider a washing machine. In most machines, except for a simple time-of-wash, there are no other choices the user can make based on how much fabric is washed and how dirty the clothes are. Would it not be better if there were a system where the water is tested for its cleanliness and the washing (scrubbing) mode is adjusted accordingly? In this case, we need to know how to define clean water versus dirty water. What is considered clean (and how clean is clean) and what is dirty (and how dirty)?

10.2 Fuzzy Control: What Is Needed

Let's reconsider Equation (10.1):

$$\text{IF TEMPERATURE} \geq 75°, \text{THEN TURN ON A/C}$$

One way to improve the flexibility of this control statement is to add another statement to it that would turn on the air conditioning at slightly lower temperature but at a slightly lower power setting (assuming it is possible to change the power setting of the air conditioning system), resulting in the following:

$$\text{IF TEMPERATURE} \geq 75°, \text{THEN TURN ON A/C 90\% FULL\_POWER} \qquad (10.2)$$

$$\text{IF TEMPERATURE} \geq 79°, \text{THEN TURN ON A/C FULL\_POWER} \qquad (10.3)$$

Now we have added a bit of flexibility to the system, as it is not just dependent on one single value to operate but will also react differently depending on the temperature. Notice that still for each statement, the controller reacts to one single value. As a result, even at 78.99° the system is at 90% full-power.

There are still two major problems with this type of control statement. One is that if we intend to have control over a very large range of values, hundreds of this type of statements will be necessary to cover small variations to desired values. Imagine that we want to have control over every 0.1°F variation in temperature in a chemical process, which may vary ±10°F. There would be about 200 control statements! Second, even if we do this, and write control statements for all possible variations in a variable, we still cannot relate the statements to everyday spoken words. As a result, the medical expert system of the previous example would not be able to communicate with the patients, and the washing machine would not be able to relate to dirty and clean water.

This is why we need to find a way of systematically defining spoken fuzzy descriptors into useful engineering descriptions that a system can use. This is done using a technique called Fuzzy Inference Control. In the next sections, we will see how fuzzy inferences can be defined (called *Fuzzy Sets* and *Fuzzification*), how a collection of control laws (called Fuzzy Inference Rules Base) can be written, and how to convert the results into a useful engineering output (called *Defuzzification*). The fuzzy control idea started with the publication of a paper by Lotfi Zadeh,[1] and since then, much has been added to the field. Although much more can be discussed about fuzzy logic, we will only discuss some fundamentals of fuzzy logic in this book related to developing a fuzzy logic controller for a device or related to simple machines, including robots. For further reading, please refer to related books and journal articles on this topic.[2–4]

10.3 Crisp Values versus Fuzzy Values

In the aforementioned examples, all values mentioned in the statements are called "crisp" values. A crisp value is a clearly defined value with one interpretation. A crisp value of 75°F means the same in any system, and it is a clearly defined and measurable value. It is also called a singleton value as opposed to a set of values that may be defined by a fuzzy value. In contrast, a fuzzy value is unclear and may be interpreted differently depending on the circumstances.

10.4 Fuzzy Sets: Degrees of Membership and Truth

To be able to use a fuzzy description in a control setting, we define a fuzzy set whose members describe the fuzzy variable at different degrees of membership or truth. Each value in the fuzzy set has a degree of membership within the set, varying from 100% (1) to 0% (0). This means that, in contrast to a crisp value that is the only true value and all other values relative to it are false, a fuzzy set has fuzzy values with different degrees of truth, varying from 100% to 0% true.

To understand this, let's once again consider a washing machine and the following statement:

IF WATER_SAMPLE = CLEAN_WATER, THEN WASH_TIME = 0

If we assume that CLEAN_WATER represents a purely clean water sample, then as a crisp value, when the water sample is purely clean (no other material in the water), the clean-water statement is true; otherwise, for all other samples, even with a slight amount of impurity, the statement is false and the water is not clean. No deviation is allowed in that definition. However, in a fuzzy set defining CLEAN_WATER, a purely clean water sample will have a degree of membership of 100% (or 1) in the set; it is purely clean water. However, water with a slight amount of dirt is still somewhat clean, perhaps 95%. Water with a little more dirt is not as clean, but perhaps 90% clean. So, every value in the set relates to some definition of clean water, with only one single crisp value (called singleton), but also containing countless other clean water possibilities with different degrees of membership and different levels of truth. In this case, a dirty water sample could still be a part of a CLEAN_WATER set, but may have a very low degree of membership. On the other hand, if we also define a fuzzy set called DIRTY_WATER, the purely clean water sample mentioned earlier will have a degree of membership (or truth) of zero in the DIRTY_WATER set, while the dirty water sample will have a 100% degree of membership in that set. The water with a 90% degree of membership in the CLEAN_WATER set may have a 15% degree of membership in the DIRTY_-WATER set as well. So, if we define two sets, a water sample will have two defined values—one in each set, each with a different degree of membership.

Considering a general crisp rule as:

IF RULE THEN CONSEQUENCE   (10.4)

if RULE is 100% true, CONSEQUENCE will be executed.

However, in a fuzzy rule, values are not necessarily 100% true (although they occasionally happen), but have degrees of membership in the set. The corresponding defined membership value is used in the rule to calculate an output. Assuming that two input variables called INPUT1 and INPUT2 are used in a system to control an output variable called OUTPUT, we may write a general set of rules as:

IF INPUT1 = degree-of-membership in INPUT1-SET AND

INPUT2 = degree-of-membership in INPUT2-SET THEN

OUTPUT = degree-of-membership in OUTPUT-SET

In the next sections, we will discuss the process of fuzzification, development of a rule base, and defuzzification.

10.5 Fuzzification

Fuzzification is the process of converting input and output values into their membership functions. The result of fuzzification is a set of graphs or equations that describe the degree of membership of different values in different fuzzy variables.

To fuzzify a variable, its range of possible values is divided into a number of sets, each describing a particular portion of the range. Subsequently, each range is represented by an equation or a graph that describes the degree of truth or membership of each value within the range. The number of sets, the range that each set represents, and the type of representation is arbitrary and a choice of the designer of the system. As we will see later, these can be modified and improved when the system is simulated and analyzed.

A number of possible representations are available for each set. If you create your own fuzzy system, you may use any representation you find appropriate. However, when you use a commercial system, you may be limited to what is available. The following membership functions are common:[5]

- **Gaussian membership function:** As shown in Figure 10.1, this is a natural way to represent a distribution. Generally, more mathematical operations are needed to use the Gaussian distribution; therefore, as we will see next, the Gaussian representation is modified into simpler forms for easier application.
- **Trapezoidal membership function:** Figure 10.2 shows the common trapezoidal membership function used to represent a Gaussian function in a simpler way. Here, the membership function is represented by three simple lines, requiring only four points. Each section is a straight line between successive points, and therefore, the degree of membership for each value of the variable can easily be calculated from the line equations.
- **Triangular membership function:** This is also a very common membership function that simplifies a Gaussian function, requiring only three points. As shown in Figure 10.3, each section is a straight line between successive points. Degrees of membership for each value of the variable are simply calculated from the line equations.

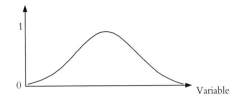

Figure 10.1 A Gaussian membership function.

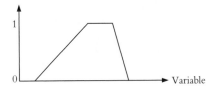

Figure 10.2 A trapezoidal membership function.

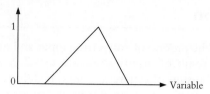

Figure 10.3 A triangular membership function.

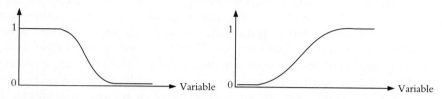

Figure 10.4 Z- and S-shaped membership functions.

- **Z-shaped and S-shaped membership functions:** These second-order functions depicted in Figure 10.4 may be used to represent the upper and lower limits of a variable, where the degrees of membership may remain the same (0 or 1) for a range of values. A trapezoidal membership function with a vertical left or right side may be used as a simple model for S- and Z-shaped functions.

Other membership functions such as a π-shaped function, the product of two sigmoidal functions, and the difference between two sigmoidal functions may also be used.[5]

To see how these membership functions may be used, let's consider a system in which one variable is temperature that may vary between 60°F and 100°F. To define the temperature variable in fuzzy form, we divide the desired range into a number of sets. For the purpose of illustration, let's use triangular and trapezoidal functions and assign corresponding temperature ranges to sets of VERY-HOT, HOT, WARM, and COLD, as shown in Figure 10.5.

Each set contains a range of temperatures, where every temperature has a degree of membership. As we discussed earlier, any temperature, say 78°, has corresponding

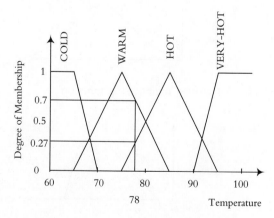

Figure 10.5 Fuzzy sets for a temperature variable.

membership values in different sets. In this case, the values are 0.27 in HOT and 0.7 in WARM. Obviously, the choice of functions, ranges, and number of sets is ours and may be modified as needed. For example, as you should notice, with the choices we made in Figure 10.5, there are gaps between sets, where certain temperatures belong only to one set. We may later change these ranges and close the gaps to improve the response.

The membership functions modeled in this manner are easy to formulate by expressing two points on each segment. All points on the line can then be easily identified. As an example, we may use the following arbitrary syntax to express the membership function for VERY-HOT and HOT:

$$\text{VERY-HOT}: \quad @90,0, \; @95,1, \; @100,1 \qquad (10.5)$$

$$\text{HOT}: \quad @75,0, \; @85,1, \; @95,0 \qquad (10.6)$$

Based on these definitions, all membership values on all sets can be calculated from the limits shown.

10.6 Fuzzy Inference Rule Base

Fuzzy Inference Rule Base is the controller part of the system and is based on truth table logic. The rule base is a collection of rules related to the fuzzy sets, the input variables, and the output variables and is meant to allow the system to decide what to do in each case. It generally takes one of the following forms, depending on the number of input and output variables:

if <condition> then <consequence>
if <condition1 and (or) condition2> then <consequence>
if <condition1 and (or) condition2> then <consequence1 and (or) consequence2>

As an example, for a system where the temperature is one input variable, humidity is the second input variable, and the power setting of the air conditioning system is the output variable, a fuzzy rule may be:

$$\text{IF temperature is HOT \textbf{and} humidity is HUMID then power is HIGH} \qquad (10.7)$$

or $$\text{IF temperature is HOT \textbf{or} humidity is HUMID then power is HIGH} \qquad (10.8)$$

Obviously, these two rules will behave differently. Based on commonly used truth tables, in the first case, both conditions must be true for the consequence, while in the second case, either condition will result in a consequence. However, remembering that these are all rather fuzzy—not crisp—values, they do not result in true or false consequences. Therefore, to evaluate the "and" and the "or" rules, we use the following:

The result of an "and" operation is the minimum of the two values.
The result of an "or" operation is the maximum of the two values.

With this definition, the system can check all the rules for the given inputs and calculate a corresponding output. The logic system that checks the rules and finds the

RULES

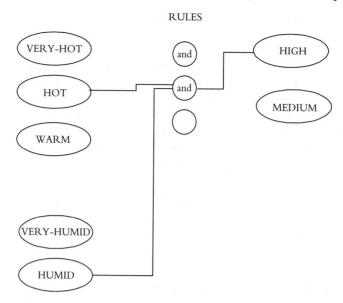

Figure 10.6   Graphical representation of rules.

corresponding output is called a *fuzzy inference engine*. You may write your own fuzzy inference engine or use commercial systems.[5-8]

The total number of rules in a rule base is equal to the product of the numbers of sets of each input variable. For example, if there are three input variables, with m, n, and p fuzzy sets, the total number of rules is $R = m \times n \times p$.

Equations (10.7) or (10.8) can also be demonstrated graphically in order to assist the designer in visualizing the relationships. Figure 10.6 is the graphical representation of the equation. When all the rules are determined, they all may be represented together in a similar manner.

10.7 Defuzzification

Defuzzification is the conversion of a fuzzy output value to an equivalent crisp value for actual use. As the fuzzy rules are evaluated and corresponding values are calculated, the result will be a number related to the corresponding membership values for different output fuzzy sets. As an example, suppose that the output power setting for an air conditioning system is fuzzified into OFF, LOW, MEDIUM, and HIGH. The result of rule base evaluation may be, say, a 25% membership in LOW and a 75% membership in MEDIUM. Defuzzification is the process of converting these values into a single number that can be sent to the air conditioning control system.

A number of different possibilities exist for defuzzification. We will consider two common and useful techniques called Center of Gravity and Mamdani's Inference Method.[6,8,9]

10.7.1 Center of Gravity Method

In this method, the membership value for each output variable is multiplied by the maximum singleton value of the output membership set to get an equivalent value for the output from the membership set in question. These equivalent values for each set are added together and normalized by summation of the output membership values for an equivalent value for the output. The following is a summary of this method:

1. Multiply the membership degrees for each output variable by the singleton value of the output set.
2. Add all of the above together and divide by the summation of output membership degrees.

As an example, suppose the values obtained for the output of the air conditioning system membership sets are 0.4 for LOW and 0.6 for Medium, and further suppose that the singleton value for LOW is 30% and for MEDIUM is 50% of full power. The output value for the air conditioning would then be:

$$Output = \frac{0.4 \times 30\% + 0.6 \times 50\%}{0.4 + 0.6} = 42\%$$

10.7.2 Mamdani's Inference Method

In this method, membership function of each set is truncated at the corresponding membership value, as in Figure 10.7. The resulting membership functions are then added together as an "or" function. This means that all repeated areas are superimposed over each other as one layer only. The result will be a new area that is representative of all areas, once each. The center of gravity of the resulting area will be the equivalent output. Mamdani's method can be summarized as follows:

1. Truncate each output membership function at its corresponding membership value, which is found from the rule base.
2. Add the remaining truncated membership functions with an "or" function in order to consolidate them into one area describing the output.
3. Calculate the center of gravity of the consolidated area as the crisp output value.

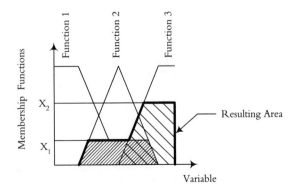

Figure 10.7 Defuzzification based on Mamdani's method.

Through the process of fuzzification, application of the rules base, and defuzzification, an output value is calculated that can be applied to the output. The following examples demonstrate this procedure for the calculation of an output value.

Example 10.1

A cooling system is to be controlled by fuzzy logic. The two inputs are temperature and humidity, and the output is the power setting of the air conditioning system. Design the fuzzy logic control system.

Solution: Based on the discussion above, we will follow the next three steps to design the system.

1. **Fuzzification:** In this part, we will develop the fuzzy sets relating to the two inputs and the output. We will assume that the range of temperatures we are interested in is 60°F to 100°F, the desired range of humidity is 0% to 100%, and the power setting can be 0% to 100%. Figure 10.8 demonstrates the three fuzzy sets for the two inputs and one output. The temperature range is divided into four membership functions: COLD, WARM, HOT, and VERY-HOT. Temperatures below 60°F are all considered to be part of COLD, and temperatures above 100°F are all part of the VERY-HOT set. Similarly, the range of humidity is

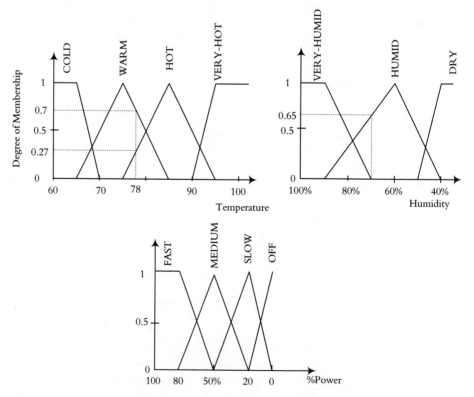

Figure 10.8   Fuzzification of inputs and outputs for Example 10.1.

divided into three membership functions of VERY-HUMID, HUMID, and DRY. All humidity below 40% is considered DRY. The output power setting is divided into 4 membership functions of FAST, MEDIUM, SLOW, and OFF.

Obviously, we could have chosen other ranges for each membership function, divided the ranges differently, decided on different ranges of overlap for the functions, or assigned asymmetrical membership functions. This, as in all design activities, is based on the requirements of the system and the experience of the designer. However, we will study the response of the system later and, if necessary, will make adjustments.

2. **Development of the Rules Base:** Since there are four membership functions for temperature and 3 membership functions for humidity, there will be a total of $4 \times 3 = 12$ rules. We will demonstrate the 12 rules both graphically, in Figure 10.9, and symbolically, as follows. Notice that in each rule (or control law),

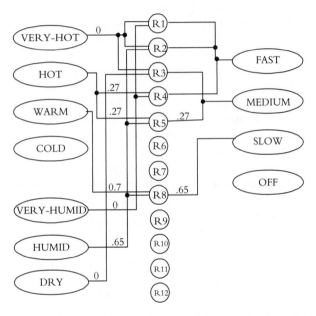

Figure 10.9 Graphical representation of some of the rules in the rules base for Example 10.1. Other rules can be similarly demonstrated, but are not shown for clarity.

Rule 1: If temperature = VERY-HOT and humidity = VERY-HUMID then power = FAST
Rule 2: If temperature = VERY-HOT and humidity = HUMID then power = FAST
Rule 3: If temperature = VERY-HOT and humidity = DRY then power = MEDIUM
Rule 4: If temperature = HOT and humidity = VERY-HUMID then power = FAST
Rule 5: If temperature = HOT and humidity = HUMID then power = MEDIUM
Rule 6: If temperature = HOT and humidity = DRY then power = MEDIUM
Rule 7: If temperature = WARM and humidity = VERY-HUMID then power = MEDIUM
Rule 8: If temperature = WARM and humidity = HUMID then power = SLOW
Rule 9: If temperature = WARM and humidity = DRY then power = SLOW
Rule 10: If temperature = COLD and humidity = VERY-HUMID then power = SLOW
Rule 11: If temperature = COLD and humidity = HUMID then power = OFF
Rule 12: If temperature = COLD and humidity = DRY then power = OFF

one membership function for every input is considered. The consequence for each rule is chosen based on the experience of the designer and the necessities of the design. For example, in Rule 1, if temperature is VERY-HOT and humidity is VERY-HUMID, the consequence, based on experience and the desired result of the rule, may be 100% power setting, or FAST. However, in Rule 10, the consequence for COLD and VERY-HUMID could either be SLOW or OFF. Which one is correct? We cannot really know that until the output of the rule base is simulated and the result is analyzed. If the output is not as desired, the rules (or membership functions) may be adjusted until a satisfactory result is obtained. Also notice that some of the rules could have been based on "and," while others were based on "or."

To understand how the results are found, let's look at some numbers. Suppose the present temperature is 78°F and humidity is 70%. As shown in Figure 10.8, the resulting membership values will be 0.7 WARM, 0.27 HOT, and 0.65 HUMID. All other membership values are zero.

Substituting these values into the corresponding rules (also shown graphically in Figure 10.9) yields output membership values of 0.27 MEDIUM and 0.65 SLOW. Remember that since we are using "and" logic, the minimum value between the two numbers is chosen. So, for example, in Rule 5, the smaller of 0.65 and 0.27 is chosen.

3. **Defuzzification:** Now that we have found the output membership values, we have to defuzzify these values to get a crisp power setting for the system. We calculate the output value based on both the center of gravity method and Mamdani's inference method.

For the center of gravity method, we multiply the output membership values by their corresponding singleton values and then divide by the sum of the membership values to get:

$$\text{Power} = \frac{0.65 \times 20\% + 0.27 \times 50\%}{0.65 + 0.27} = 29\%$$

For Mamdani's inference method, we first truncate the MEDIUM and SLOW functions at 0.27 and 0.65 values, combine the two into a single area, and calculate the center of gravity of the resulted area, as shown in Figure 10.10.

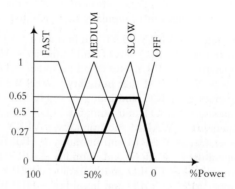

Figure 10.10 Application of Mamdani's inference method.

The center of gravity of the resulting area can be calculated by taking the first moment of the area and dividing it by the total area and is calculated to be at 34% power rating, which is somewhat different from the center of gravity method.   ∎

10.8 Simulation of Fuzzy Logic Controller

So far, we have made a few somewhat arbitrary choices in the number of sets, ranges of variables, and the rules, that can have potentially detrimental effects on the outcome. Consequently, it is necessary that we simulate the system and analyze the results. This is usually done through fuzzy logic programs such as MATLAB's^{TM} Fuzzy Logic toolbox and others. These systems simulate the fuzzy control system by running a fuzzy inference engine and calculating the output for all possible input values and plotting the results. The plot is used to check the rules and the membership functions and to see if they are appropriate and whether modifications are necessary to improve the output. If necessary, the rule base or the fuzzy sets are modified until the output curves are as desired. Figure 10.11 is the 3-D depiction of the output of the air conditioning system of Example 10.1 from MATLAB's toolbox. When a satisfactory system is achieved, the fuzzy program is converted to machine language (or other real-time code) and downloaded into a microprocessor controller. The microprocessor will then run the machine or the system based on the fuzzy control rules. Although the process seems long, it is actually relatively easy to do. And it does add interesting "intelligence" to a machine.

As Figure 10.11 shows, there are certain areas within the output surface that although the input variables change, the output of the system remains flat. In certain systems, this may be desirable. For example, for an automobile's transmission, unless it is continuously variable (CVT), the output can only be a few discrete values (first, second, third, and so on). In that case, it is desirable to have a constant output for a range of inputs. For systems where the output is continuous, a smoothly varying output is more desirable. Therefore, the designer may choose to modify the input and output membership functions and/or the rules to achieve a more continuously varying output surface.

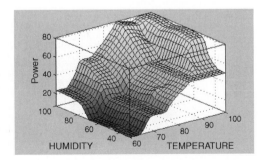

Figure 10.11 The 3-D output result of the air conditioning example generated by MATLAB's Fuzzy Logic toolbox. The graph can be used for modifying and adjusting the fuzzy sets or the rules for best output result.

Example 10.1 continued

To improve the output of Example 10.1, let's modify the inputs by closing the gaps, as shown in Figure 10.12. The result of the simulation is shown, too. The output is much smoother.

Next, we will also try to improve the system by changing the membership functions from triangular and trapezoidal to Gaussian, as shown in Figure 10.13. The output is still smoother and more continuous.

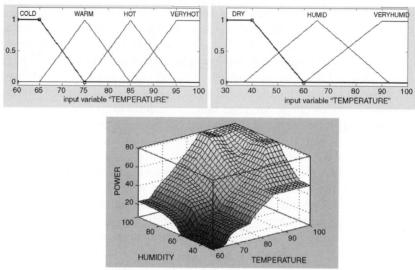

Figure 10.12   The gaps between different sets of input and output membership functions are closed, therefore improving the output of the system.

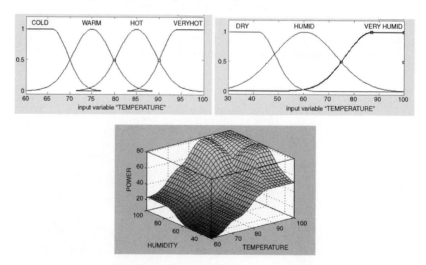

Figure 10.13   The fuzzy sets in input and output membership functions were modified to Gaussian to further improve the output.

10.9 Applications of Fuzzy Logic in Robotics

Fuzzy logic control systems can be used for both controlling robots as well as for adding intelligence to applications where other systems may be inadequate or difficult to use. For example, in one application, fuzzy logic was used to directly control the torque output of a switched reluctance motor by a current modulation scheme.[10] In another application, fuzzy logic was used to provide a force feedback signal to a wheelchair's joystick for a blind user. Therefore, as the wheelchair got closer to an obstacle or a drop-off, it became harder to push the joystick.[11] Although fuzzy logic can be used for controlling robots in lieu of, or in conjunction with, classical control systems, there are many other applications where fuzzy logic may perhaps be more appropriate, if not the only way to control a function. It is for this reason that the discussion on fuzzy logic has been presented here. Through these applications, a robot can become unique, more intelligent, or more useful. As an example, consider a mobile robot designed for rough terrain. A fuzzy logic control system can be used to enhance the robot controller in deciding what action to take depending on the speed of the robot, the terrain, the robot's power, and so on. Or imagine a robot whose end effector must exert a force proportional to two other inputs, say, the size of a part and its weight. In yet another example, suppose that a robot is used to sort a bag of objects based on their colors according to the colors of the rainbow. In these, and countless other similar examples, fuzzy logic may be the best choice to incorporate the intelligence needed to accomplish the task. Additionally, many peripheral devices are integrated with robots or work with a robot through their own controller. In these cases, too, fuzzy logic may be incorporated into the processor for better performance.

Example 10.2

Design a fuzzy logic system for the motion control of a mobile robot on rough terrain.

Solution: We assume that the inputs to the system are the slope of the terrain and the terrain type, while the output is the robot's speed. We assume the slope can range between −30 to +30 degrees, divided into LargeNegative, Negative, Level, Positive, and LargePositive sets. We further assume that the terrain can be VeryRough, Rough, Moderate, and Smooth. The output speed can range between 0–20 miles per hour and is divided into VerySlow, Slow, Medium, Fast, and Very-Fast. Figure 10.14 shows the fuzzy sets describing the above. The rules base, with its 20 rules, is also shown. Figure 10.15 shows the result of the simulation of the rules base by MATLAB. Do you think any of the rules or fuzzy set limits should be modified?

The surface shown in Figure 10.15 is the control surface. It means that for every possible value of the two inputs, there is a corresponding output based on the rules. For example, for the input values of Slope = 20 degrees and Terrain type = 30 (corresponding to half way between Smooth and Moderate), the output speed of the robot is about 11 miles per hour. As shown, the speed of the robot decreases as the slope increases up or down as well as when the terrain becomes increasingly rough.

Here too, by closing the gap and changing the input membership functions to Gaussian, the result may be improved, as shown in Figure 10.16.

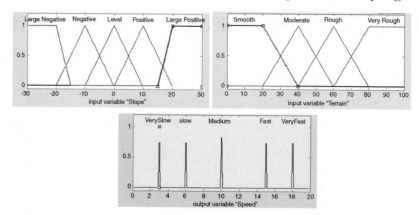

Figure 10.14 The fuzzy sets describing the inputs and the output of Example 10.2.

1. If slope is LargePositive and terrain is VeryRough then speed is VerySlow;
2. If slope is LargePositive and terrain is Rough then speed is Slow;
3. If slope is LargePositive and terrain is Moderate then speed is Medium;
4. If slope is LargePositive and terrain is Smooth then speed is Medium;
5. If slope is Positive and terrain is VeryRough then speed is VerySlow;
6. If slope is Positive and terrain is Rough then speed is Slow;
7. If slope is Positive and terrain is Moderate then speed is Medium;
8. If slope is Positive and terrain is Smooth then speed is Fast;
9. If slope is Level and terrain is VeryRough then speed is Slow;
10. If slope is Level and terrain is Rough then speed is Medium;
11. If slope is Level and terrain is Moderate then speed is Fast;
12. If slope is Level and terrain is Smooth then speed is very-Fast;
13. If slope is Negative and terrain is VeryRough then speed is VerySlow;
14. If slope is Negative and terrain is Rough then speed is Slow;
15. If slope is Negative and terrain is Moderate then speed is Medium;
16. If slope is Negative and terrain is Smooth then speed is Fast;
17. If slope is LargeNegative and terrain is VeryRough then speed is VerySlow;
18. If slope is LargeNegative and terrain is Rough then speed is VerySlow;
19. If slope is LargeNegative and terrain is Moderate then speed is Slow;
20. If slope is LargeNegative and terrain is Smooth then speed is Medium;

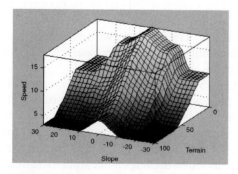

Figure 10.15 The result of the simulation of the system of Example 10.2 with MATLAB software.

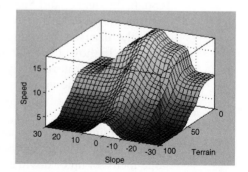

Figure 10.16 The improved result for Example 10.2.

Example 10.3

In a particular application, a robot is used to sort diamonds by weight and by color and determine a price for the diamonds. Design a fuzzy logic system to control the process.

Solution: Diamonds are classified by weight, color (indicated by letters, where A is extremely clear and other letters indicate tints of yellow in the diamond), and purity (the size of inclusions). The clearer the diamond, the smaller the inclusions, and the larger the size, the more expensive the diamond per carat. In this example, we assume we are sorting the diamond by color and size (weight) only. We assume that a vision system can take an image of the diamond and compare its color with a data bank of colors to estimate the color range. We will also assume that using the techniques mentioned in Chapter 9, the vision system can recognize the diamond, its surface can be measured, and its weight can be estimated based on its size. We further assume that the size of the diamonds will be in one of the sets of Small, Medium, Large, and Very-Large, as shown in Figure 10.17. The colors of the diamonds are divided into three color ranges of D, H, and L. The price per carat of the diamonds will be in the ranges of Ten, Fifteen, Twenty, Thirty, Forty,

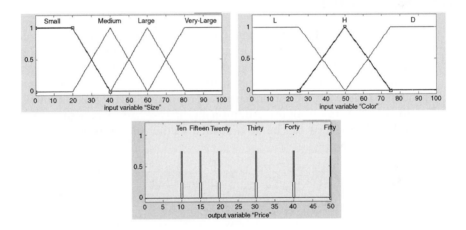

Figure 10.17   Fuzzy sets showing the input and the output of Example 10.3.

1.   If size is Small and color is D then price is Twenty;
2.   If size is Medium and color is D then price is Thirty;
3.   If size is Large and color is D then price is Forty;
4.   If size is Very-Large and color is D then price is Fifty;
5.   If size is Small and color is H then price is Fifteen;
6.   If size is Medium and color is H then price is Twenty;
7.   If size is Large and color is H then price is Thirty;
8.   If size is Very-Large and color is H then price is Forty;
9.   If size is Small and color is L then price is Ten;
10. If size is Medium and color is L then price is Fifteen;
11. If size is Large and color is L then price is Twenty;
12. If size is Very-Large and color is L then price is Thirty;

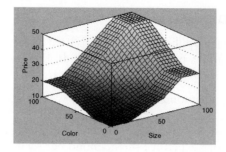

Figure 10.18 The result of the simulation of example 10.3.

and Fifty (all times a normalized base price). The rule base as well as the result of the simulation of this system is shown in Figure 10.18.

As shown, for every color and weight combination, there is a corresponding price. With this fuzzy logic system and only 12 rules, a vision system could estimate the corresponding price of diamonds automatically. ∎

10.10 Design Project

If you have access to a fuzzy logic control simulator, you may want to develop a fuzzy logic control program for a mobile robot, a specific task for a vision system, or other similar applications. The fuzzy logic control program may be written for either controlling the robot motions or it may be written for other purposes. For example, you may write a fuzzy control heuristics program such that the robot will follow a certain path based on fuzzy inputs. Additionally, if you have access to a microprocessor, the developed programs can be downloaded to the microprocessor as part of the control program it runs.

Summary

In this chapter, we discussed how a fuzzy logic control system may be developed, simulated, tested, and used. Fuzzy logic is a very powerful way of including nonexact concepts in everyday systems, including definitions (e.g. temperature, humidity), feelings (e.g. pain, hot, cold), and adjectives (e.g. much, less). Although fuzzy logic systems may be applied to countless different situations, we primarily discussed how they may be used in robotics. Applications in robotics can range from navigation control for mobile robots and telerobotic to expert systems and vision systems. Fuzzy logic systems are generally simulated with a simulator program and, when it is verified that the system behaves as intended, it is used in conjunction with the remaining control programs.

References

1. Zadeh, Lotfi, "Fuzzy Sets," Information and Control, Vol. 8, 1965, pp. 338–353.

2. Cox, Earl, "Fuzzy Logic for Business and Industry," Charles River Media, 1995.

3. McNeill, F.Martin, Ellen Thro, "Fuzzy Logic, a Practical Approach," Academic Press, 1994.

4. Kosko, Bart, "Neural Networks and Fuzzy Systems, a Dynamical Systems Approach to Machine Intelligence," Prentice Hall, 1992.

5. MATLAB Fuzzy Logic Toolbox®.

6. Fuzzy Inference Development Environment (FIDE) User's Manual, Aptronix Inc., San Jose, CA., 1992.

7. FUzzy Design GEnerator (FUDGE), Motorola.

8. Fuzzy Knowledge Builder, McNeill, F. Martin, Ellen Thro.

9. Mamdani, E. H., "Application of Fuzzy Logic to Approximate Reasoning Using Linguistic Synthesis," IEEE Transactions on Computers, Vol. c-26, No. 12, 1977, pp. 1182–1191.

10. Sahoo, N. C., S. K. Panda, P. K. Dash, "A Current Modulation Scheme for Direct Torque Control of Switched Reluctance Motor Using Fuzzy Logic," Mechatronics, The Science of Intelligent Machines, Vol. 10, No. 3, April 2000, pp. 353–370.

11. Sindorf, Brent, S. B. Niku,"Force Feedback Wheelchair Control," masters thesis, *Mechanical Engineering*, Cal Poly, San Luis Obispo, 2005.

Problems

10.1. Develop a fuzzy inference system for a robot, where the force exerted at the hand and the velocity of the hand are the inputs and the % power to the actuators is the output.

10.2. Develop a fuzzy inference system for a washing machine. The inputs are how dirty the fabrics are and how much clothes are being washed, and the output is the wash time.

10.3. Develop a fuzzy inference system for a barbecue. The inputs may be the thickness of the steak and how cooked or rare it is desired to be. The output may be the temperature of the flame and/or the time of cooking.

10.4. Develop a fuzzy inference system for an automatic gearbox. The inputs are the speed of the car and the load on the engine, and the output is the gear ratio of the transmission.

10.5. Develop a fuzzy logic system for a vision system in which the inputs are the intensities of the three colors of red, green, and blue (RGB) in a color image, and the output is the relationship of the combination to the colors of the rainbow.

10.6. Develop a fuzzy inference system for grading a robotics course. The inputs are your effort level in the course and your exam grade, and the output is your letter grade.

Review of Matrix Algebra and Trigonometry

A.1 Matrix Algebra and Notation: A Review

Throughout this book, we use matrices to represent coordinates, frames, objects, and motions. In this Appendix, certain characteristics of matrices that we will need in our calculations are reviewed. You must already have an understanding of matrix algebra to understand the use of matrices. Therefore, only a simple review is presented here.

Matrices

A matrix is a collection of m rows and n columns of values, represented in a bracket. The dimensions of the matrix are $m \times n$, and each element of the matrix is referred to as A_{ij}. A matrix whose number of rows and columns are the same is called a square matrix.

Matrix Transpose

The transpose of a matrix A_{ij}^T is another matrix A_{ji} where elements of each row and column are replaced, as shown:

$$A_{ij} = \begin{bmatrix} a_{11} & a_{12} & a_{13} \\ a_{21} & a_{22} & a_{23} \end{bmatrix} \quad \text{and} \quad A_{ij}^T = A_{ji} = \begin{bmatrix} a_{11} & a_{21} \\ a_{12} & a_{22} \\ a_{13} & a_{23} \end{bmatrix} \quad (A.1)$$

Matrix Multiplication

Matrices can be multiplied by multiplying all the elements of each row by each column and replacing the summation in the corresponding row/column location, as follows:

$$C_{ij} = A_{ik} \times B_{kj} = \begin{bmatrix} d & e & f \\ g & h & l \end{bmatrix} \times \begin{bmatrix} p & s \\ q & t \\ r & w \end{bmatrix} = \begin{bmatrix} dp + eq + fr & ds + et + fw \\ gp + hq + lr & gs + ht + lw \end{bmatrix} \quad (A.2)$$

As you *can* see, the result of an $(m \times n)$ matrix multiplied by an $(n \times p)$ matrix is an $(m \times p)$ matrix. Therefore, the number of columns of the first matrix must be equal to the number of rows of the second matrix. Remember, unlike regular algebra, the order of multiplication of matrices **may not** be changed. In other words, $A \times B \neq B \times A$. This can easily be demonstrated by the fact that if A is a (2×3) matrix and if B is a (3×2) matrix, then $A \times B$ will yield a (2×2) matrix, whereas $B \times A$ will result in a (3×3) matrix, which obviously is different. However, if more than two matrices are to be multiplied, although their order cannot be changed, the result is independent of which pairs of matrices are multiplied first. As a result, the following is true:

$$A \times B \neq B \times A \qquad \text{(A.3)}$$

$$\text{but} \quad A \times B \times C = (A \times B) \times C = A \times (B \times C) \qquad \text{(A.4)}$$

$$(A + B)C = AC + BC \quad \text{and} \quad C(A + B) = CA + CB \qquad \text{(A.5)}$$

Diagonal Matrix

A diagonal matrix is a matrix where all, except the diagonal, elements of the matrix are zero. If all diagonal elements are 1, the matrix is a unit matrix, which effectively acts as a 1; premultiplying or postmultiplying any matrix by a unit matrix will result in the same matrix.

Matrix Addition

Matrix addition can be accomplished by adding each element of one matrix by the corresponding element of the other matrices. Unlike multiplication, addition of matrices is commutative; the order of addition is not important. Obviously, the dimensions of all matrices must be exactly the same for addition. Therefore:

$$A_{ij} + B_{ij} = (A + B)_{ij} \qquad \text{(A.6)}$$

$$A + B + C = B + A + C = C + A + B \qquad \text{(A.7)}$$

Vectors

A vector is a one-dimensional matrix, either a $(1 \times m)$ or an $(n \times 1)$ matrix.

Determinant of a Matrix

The determinant of a matrix can be calculated as follows:

- Pick one row or column.
- Multiply each element in the chosen row or column by the determinant of the matrix that remains after the corresponding row and column of the element are dropped from the matrix, each one with an alternating plus or minus sign.

Example A.1

Calculate the determinant of the following matrix:

$$A = \begin{bmatrix} a & b & c \\ d & e & f \\ g & h & i \end{bmatrix}$$

Solution: First, choose a row or column. In this example, we will pick the first row. The determinant of the matrix is:

$$det(A) = +a(ei - fh) - b(di - fg) + c(dh - eg)$$ ■

Matrix Inversion

This is an important operation in matrix representation of robots. We will use matrix inversions for both inverse kinematics and for differential motions. In this section, two general purpose inversion techniques for square matrices are mentioned.

The inverse of a matrix is another matrix such that if the matrix is multiplied by the inverse, the result will be a unit matrix. In general, a matrix either has a left inverse or a right inverse. If $A \times A^{-1} = I$ (where I is a unit matrix), A^{-1} is called a right inverse. If $A^{-1} \times A = I$, then A^{-1} is called a left inverse. Generally, the left and right inverse matrices are not the same. However, a square matrix will have the same left and right inverse, such that $A \times A^{-1} = A^{-1} \times A = I$. In this case, A^{-1} is simply called an inverse, and it may be premultiplied or postmultiplied by the square matrix, yielding a unit matrix.

Method 1

For square matrices with non-zero determinants only, the inverse of the matrix can be calculated by the following method:

- Calculate the determinant of the matrix.
- Transpose the matrix.
- Replace each element of the transposed matrix by its own minor (this is called an adjoint matrix).
- Divide the adjoint matrix by the determinant to get the inverse.

$$\text{Therefore} \quad A^{-1} = \frac{adj(A)}{det(A)} \tag{A.8}$$

The minor for each element A_{ij} of the matrix is the determinant of the matrix that remains after the row and column of the matrix containing the element are dropped, multiplied by $(-1)^{i+j}$. This creates a sign matrix as shown in Equation A.9. As an example, we can write the following minors for the given matrix in Equation A.9:

$$A = \begin{bmatrix} a & b & c \\ d & e & f \\ g & h & i \end{bmatrix} \quad \text{with} \quad Sign = \begin{bmatrix} + & - & + \\ - & + & - \\ + & - & + \end{bmatrix} \tag{A.9}$$

$$a_{minor} = +(ei - fh)$$
$$\text{and} \quad b_{minor} = -(di - fg)$$
$$\text{and} \quad h_{minor} = -(af - cd)$$

Of course, the minors for a matrix with larger dimensions will be similar but much more involved.

Example A.2

Calculate the inverse of the matrix below.

$$A = \begin{bmatrix} 1 & 0 & 1 \\ 0 & 1 & 4 \\ 5 & -2 & -1 \end{bmatrix}$$

Solution: First, we will calculate the determinant of the matrix:

$$det(A) = 1(-1 + 8) - 0(0 - 20) + 1(0 - 5) = 7 - 5 = 2$$

The transpose of the matrix is: $A^T = \begin{bmatrix} 1 & 0 & 5 \\ 0 & 1 & -2 \\ 1 & 4 & -1 \end{bmatrix}$

The adjoint of the matrix is:

$$A_{adj} = A_{minor}^T = \begin{bmatrix} +(-1+8) & -(0+2) & +(0-1) \\ -(0-20) & +(-1-5) & -(4-0) \\ +(0-5) & -(-2+0) & +(1-0) \end{bmatrix} = \begin{bmatrix} 7 & -2 & -1 \\ 20 & -6 & -4 \\ -5 & 2 & 1 \end{bmatrix}$$

The inverse can be found by dividing the matrix by the determinant as:

$$A^{-1} = \begin{bmatrix} 3.5 & -1 & -0.5 \\ 10 & -3 & -2 \\ -2.5 & 1 & 0.5 \end{bmatrix}$$

To ensure that the result is correct, you may multiply A by A^{-1} as:

$$\begin{bmatrix} 1 & 0 & 1 \\ 0 & 1 & 4 \\ 5 & -2 & -1 \end{bmatrix} \times \begin{bmatrix} 3.5 & -1 & -0.5 \\ 10 & -3 & -2 \\ -2.5 & 1 & 0.5 \end{bmatrix} = \begin{bmatrix} 1 & 0 & 0 \\ 0 & 1 & 0 \\ 0 & 0 & 1 \end{bmatrix}$$

Please verify that $A^{-1} \times A$ will result in a unit matrix as well. ∎

Method 2

In this method, we assume an inverse matrix of the following form exists, such that when multiplied by the given matrix, a unit matrix results as:

$$\begin{bmatrix} a_{11} & a_{12} & . & a_{1i} \\ a_{21} & . & & \\ . & . & & \\ a_{i1} & . & & a_{ii} \end{bmatrix} \times \begin{bmatrix} x_{11} & x_{12} & . & x_{1i} \\ x_{21} & . & & \\ . & . & & \\ x_{i1} & . & & x_{ii} \end{bmatrix} = \begin{bmatrix} 1 & 0 & 0 & 0 \\ 0 & 1 & 0 & 0 \\ 0 & 0 & 1 & 0 \\ 0 & 0 & 0 & 1 \end{bmatrix} \quad (A.10)$$

where the x_{ii} matrix is the inverse of the A matrix that we are looking for. Note that this represents a set of i^2 equations and i^2 unknowns. If you multiply the first matrix A by the first column of the x matrix, the result will be the first column of the unit matrix. Then there will be a set of i equations that you will have to solve for each column.

Example A.3

Find the inverse of the following matrix using method 2.

$$A = \begin{bmatrix} 1 & 0 & 1 \\ 0 & 1 & 4 \\ 5 & -2 & -1 \end{bmatrix}$$

Solution: Based on the above, we write:

$$\begin{bmatrix} 1 & 0 & 1 \\ 0 & 1 & 4 \\ 5 & -2 & -1 \end{bmatrix} \times \begin{bmatrix} x_{11} & x_{12} & x_{13} \\ x_{21} & x_{22} & x_{23} \\ x_{31} & x_{32} & x_{33} \end{bmatrix} = \begin{bmatrix} 1 & 0 & 0 \\ 0 & 1 & 0 \\ 0 & 0 & 1 \end{bmatrix}$$

Multiplying the given matrix by the first column of the inverse matrix and equating it with the corresponding first column of the unit matrix yields the following three equations:

$$\begin{aligned} x_{11} + x_{31} &= 1 \\ x_{21} + 4x_{31} &= 0 \\ 5x_{11} - 2x_{21} - x_{31} &= 0 \end{aligned} \qquad \text{and} \qquad \begin{aligned} x_{11} &= 3.5 \\ x_{12} &= 10 \\ x_{13} &= -2.5 \end{aligned}$$

Similarly, if you multiply the given matrix by the second and then the third columns of the inverse matrix and equate each one with the second or third column of the unit matrix respectively, you will get the remainder of the unknowns. As you can see, the result is exactly the same. Please do this to verify that you get the same results. ∎

Trace

The sum of the diagonal elements of a matrix A is called *traceA*. Therefore,

$$traceA = \sum_{j=1}^{n} a_{jj}$$

Specifically, the trace for the product of a vector of n elements and its transpose is:

$$trace\left[V \times V^T\right] = trace \begin{bmatrix} v_1 \\ v_2 \\ \vdots \\ v_n \end{bmatrix} \begin{bmatrix} v_1 & v_2 & \cdots & v_n \end{bmatrix} = trace \begin{bmatrix} v_1^2 & v_1 v_2 & \cdots & v_1 v_n \\ v_2 v_1 & v_2^2 & & \\ \vdots & & & \\ v_n v_1 & & & v_n^2 \end{bmatrix} = \sum_{j=1}^{n} v_j^2$$

This is used in the calculation of kinetic energy in Chapter 4.

Transpose of Products of Matrices

The following is true:

$$\text{If} \quad [B] \times [C] = [A] \quad \text{then} \quad [C]^T \times [B]^T = [A]^T \quad (A.11)$$

For example, we can see that the following is true:

$$[a \quad b]\begin{bmatrix} c & d \\ e & f \end{bmatrix} = [\, ac + be \quad ad + bf \,]$$

$$\begin{bmatrix} c & e \\ d & f \end{bmatrix}\begin{bmatrix} a \\ b \end{bmatrix} = \begin{bmatrix} ac + be \\ ad + bf \end{bmatrix} = [\, ac + be \quad ad + bf \,]^T$$

A.2 Calculation of An Angle from its Sine, Cosine, or Tangent

There are many instances in robotic analysis when we need to determine the magnitude of an angle from $\sin\theta$, $\cos\theta$, or $\tan\theta$. Although this seems to be a trivial matter, in reality, it is very important because there can be grave ambiguities in the answer, resulting in erroneous values and preventing a robot controller from functioning properly. This is true even with a calculator or a computer. To understand this, let's do a simple test.

Suppose you use your calculator to calculate $\sin 75°$ as 0.966. If you enter the same number into your calculator and calculate the angle from it, you will find the same 75°. However, if you do the same with $\sin 105°$, you will find the same 0.966 as before. As a result, if you calculate the angle again, of course you will get 75° and not 105°. Here lies the basic error. The sine of two angles with equal distance from 90° is always the same, and therefore, the calculator always returns the smaller angle. The same is true for cos and tan of an angle; the cos of the plus or minus of the same angle is the same, while the tan of an angle is the same if 180° is added to it. This is simply demonstrated by the trigonometric relationships, as in Figure A.1.

In order to know the exact magnitude of an angle, it is necessary to determine in what quadrant the angle lies. This will enable us to correctly know what the angle really is. However, to determine the quadrant of an angle, it is necessary to know the signs of both the sin and the cos of the angle. If we know the signs of the sin and the cos of the angle, we can determine what quadrant it is in and, based on that, we can correctly calculate the angle.

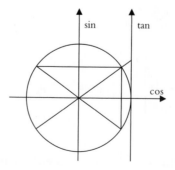

Figure A.1 Trigonometric functions.

In the previous example, if you calculate the values of cos 75° and cos 105°, you will notice that they are respectively 0.259 and –0.259. Considering both the sin and the cos of 75° and 105°, we can easily determine the correct angles. The same principle is true for tan of an angle.

In robotics calculations, we will encounter the same situation, where tan of angles are generally found. If the simple atan (arctan) function of a calculator or computer is used, it may yield an erroneous result. But if both the sin and the cos of the angle are found and used in a function, we can calculate the correct angle. Some languages such as C^{++}, MATLAB®, and FORTRAN computer languages have a function called ATAN2(sin, cos), in which the values of the sin and cos of the angle, entered as arguments, are automatically used to return the value of the angle. In all other situations, either with your calculator or other computer languages, you will have to write such a function. As a result, it is generally necessary to find two equations for each angle, one that yields the sin of the angle, and one that yields the cos of the angle. Based on the sign of the two, we determine the quadrant, and therefore, the correct value of the angle. This will be emphasized throughout this book whenever possible.

The following is a summary of rules for calculating the angles in each quadrant. You may program this into your robotic routines or your calculator for future use.

- If sin is positive and cos is positive, the angle is in quadrant 1, then angle = atan(α).
- If sin is positive and cos is negative, the angle is in quadrant 2, then angle = 180 − atan(α).
- If sin is negative and cos is negative, the angle is in quadrant 3, then angle = 180 + atan(α).
- If sin is negative and cos is positive, the angle is in quadrant 4, then angle = − atan(α).

The program should also check to see if either the sin or the cos are zero. In that case, instead of calculating the tangent, it should directly use the cosine or the sine to calculate the angle to prevent an error.

Problems

A.1. Show that the determinant of a matrix can be calculated by picking any row or column.

A.2. Calculate the determinant of the following (4×4) matrix.

$$A = \begin{bmatrix} 1 & 1 & 0 & 0 \\ 0 & 1 & 2 & 0 \\ 3 & 0 & 1 & 1 \\ 1 & 0 & 0 & 1 \end{bmatrix}$$

A.3. Calculate the inverse of the following matrix using method 1:

$$B = \begin{bmatrix} 1 & 1 & 2 \\ 0 & 1 & 0 \\ 2 & 0 & 3 \end{bmatrix}$$

A.4. Calculate the inverse of the following matrix using method 2:

$$C = \begin{bmatrix} 1 & 0 & 1 \\ 0 & 2 & 1 \\ 3 & 1 & 0 \end{bmatrix}$$

APPENDIX B

Image Acquisition Systems

The following discussion is about analog and digital image acquisition systems. Analog cameras are no longer common; however, there are still analog television sets around, although fast disappearing. Regardless, these systems help in understanding about electronic light intensity sensing, transmission, digitization, and many others. Therefore, a short discussion is presented. The presentation of digital systems is also very basic. There has been a tremendous explosion of data processing capability in digital cameras, from kilo–pixel range to mega-pixels, and from bulky cameras to mini and micro sizes, all in less than a decade. This Appendix only discusses the fundamental ways images are captured.

B.1 Vidicon Camera

A vidicon camera is an analog device that transforms an image into an analog signal. The signal, a variable voltage (or current) versus time, can be stored, digitized, broadcast, or reconstructed into an image. Figure B.1 shows a simple schematic of a vidicon camera. As shown, using a lens, the scene is projected onto a screen made up of two layers, one a transparent metallic film and one a photoconductive mosaic, sensitive to light. The mosaic reacts to the varying intensity of light by varying its resistance. As a result, as the image is projected onto it, the magnitude of the resistance at each location varies with intensity of light. An electron gun generates and projects a continuous cathode beam (a stream of electrons with a negative charge) through two pairs of capacitors (deflectors), perpendicular to each other. Depending on the charge of each pair of capacitors, the electron beam is deflected up or down, and left or right, and is projected onto the photoconductive mosaic. At each instant, as the beam of electrons hits the mosaic, its electrical potential is conducted to the metallic film and can be measured at the output port. The voltage measured at the output is $v = i \times R$, where i is the current (of the beam of electrons), and R is the resistance of the mosaic at the point of interest. Since the

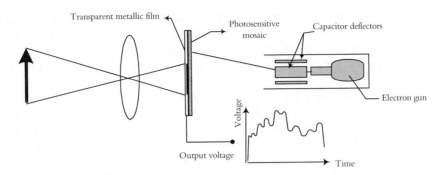

Figure B.1   Schematic of a vidicon camera.

resistance is a measure of the light intensity at that location, the voltage represents the intensity of light at that point.

Now suppose we routinely change the charges in the two capacitors to force the beam to deflect both sideways and up/down and cause it to scan the mosaic (called raster scan). As the beam scans the image, at each instant, the output is proportional to the resistance of the mosaic, or proportional to the intensity of the light on the mosaic. By reading the output voltage continuously, an analog representation of the image can be obtained.

To create moving images in televisions, the image is scanned and reconstructed 30 times a second. Since human eyes possess a temporary hysteresis effect of about 1/10 second, images refreshed at 30 times a second are perceived as continuous, and therefore, moving. The refresh rate of 30 per second and maximum channel bandwidth of 6 MHz yields a resolution of 200,000 "pixels." To increase the resolution and decrease demand on the electronics, the image is divided into two 240-line images, interlaced onto each other. Consequently, a television image is composed of 480 image lines, changing 30 times a second. In order to return the beam back to top another 45 lines are used (for both half-images), creating a total of 525 lines. In countries other than the U.S., this number is different (such as 625). This is shown in Figure B.2.

If the signal is to be broadcast, it is usually frequency-modulated (FM), where the frequency of the carrier signal is a function of the amplitude of the image signal. The broadcast signal is received by a receiver, where it is demodulated back to the original signal, creating a variable voltage with respect to time. To recreate the image, for example, in a television set, this voltage must be converted back to an image. To do this, the voltage is fed into a CRT (cathode ray tube), with an electron gun and similar deflecting capacitors as in a vidicon camera. The intensity of the electron beam in the

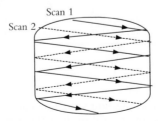

Figure B.2   A Raster scan depiction of a vidicon camera

television is now proportional to the signal voltage and is scanned similar to the camera. But in the television set, the beam is projected onto a phosphorous-based material on the screen that glows proportional to the intensity of the beam, thus recreating the image.

For color images, the projected image is decomposed into three primary colors of red, green, and blue (RGB). The exact same process is repeated for the three images, and three simultaneous signals are produced and broadcast. In the television set, three electron guns regenerate three images in RGB, which are projected onto the screen, except that the screen has three set of small dots (pixels) that react by glowing in RGB colors and are repeated throughout the screen. All color images in any system are divided into RGB images and are dealt with as three separate images.

If the signal is not to be broadcast, it is either recorded for later use, it is digitized (as discussed in Chapter 9), or it is fed into a monitor for closed circuit viewing.

B.2 Digital Camera

A digital camera is based on solid state technology. Similar to other cameras, a set of lenses is used to project the area of interest onto the image area of the camera. The main part of the camera is a solid state silicon wafer image area, which has hundreds of thousands to millions of extremely small photosensitive areas called photosites printed on it. Each small area of the wafer is a picture-cell or pixel. As the image is projected onto the image area, a charge is developed at each pixel location of the wafer proportional to the intensity of light at that location (and therefore, these cameras are called a Charge Coupled Device, or CCD camera or Charge Integrated Device, or CID camera). The collection of charges, if read sequentially, represent the image pixels (see Figure B.3).

The wafer may have millions of pixels on an area with dimensions of a fraction of an inch. Obviously, it is impossible to have direct wire connections to all of these pixels to measure the charge in each one. To read this tremendously large number of pixels, 30 times a second, the charges on each line of pixels are moved to optically isolated shift-registers next to each photosite, moved down to an output line and read.[1,2] The result is that at each 1/30th of a second, the charge in all pixel locations are read sequentially and stored or recorded. The output is a discrete representation of the image, a voltage sampled in time, as shown in Figure B.4(a). Figure B.4(b) is the CCD element of a VHS camera.

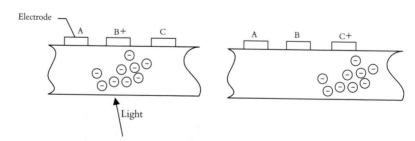

Figure B.3 Image acquisition with a digital camera involves the development of a charge at each pixel location proportional to the light at the pixel. The image is then read by moving the charges to optically isolated shift-registers and reading them at a known rate.

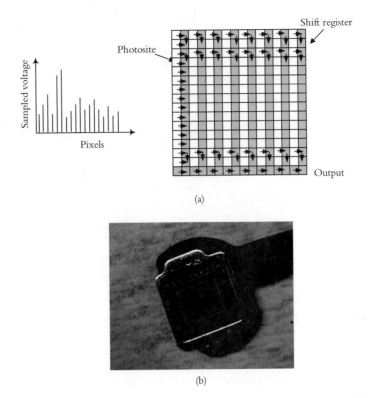

(a)

(b)

Figure B.4 (a) Image data collection model, (b) the CCD element of a VHS camera.

In addition to CCD cameras for visible lights, the same can be done for long wavelength infrared cameras, which yield a television like image of the infrared emissions of the scene.[3]

References

1. Madonick, N., "Improved CCDs for Industrial Video," Machine Design, April 1982, pp. 167–172.

2. Wilson, A., "Solid-State Camera Design and Application," Machine Design, April 1984, pp. 38–46.

3. "A 640 × 486 Long-Wavelength Infrared Camera," NASA Tech Briefs, June 1999 pp. 44–47.

APPENDIX C

Root Locus and Bode Diagram with MATLAB™

MATLAB is a powerful program with many engineering applications. The following is a short tutorial on how to use MATLAB to draw root locus of a characteristic equation and how to use it for design purposes. The tutorial is based on Version 7.6 (R2008a) as available at the time of this writing. No doubt, future versions will have minor differences. Please experiment with the program to discover more options and to better learn the system.

C.1 Root Locus

Start the program and type: **G = zpk ([z$_1$, . . .],[p$_1$, . . .],[k])** where G is an arbitrary name used for the characteristic equation, z_1, p_1, . . . are the zeros and poles, and k is the system gain. Remember, MATLAB is case-sensitive. For example, for the following equation, G will be:

$$GH = \frac{5(s + 0.5)(s + 6)}{s(s + 1)(s + 5)(s + 8)} \qquad G = \text{zpk} ([-0.5, -6], [0, -1, -5, -8], 5)$$

Use k = 1 if the gain is not given. Leave the brackets blank if no zeros or poles exist. MATLAB will respond by rewriting the equation in standard form. Make sure it is entered correctly. Otherwise, correct the equation.

Next, type **rlocus(G)** and enter. MATLAB will draw the root locus in a new window as shown in Figure C.1. You may study the figure, for example, by clicking on any location on the locus. MATLAB will show the root, the gain, %overshoot, damping, and frequency at that location.

Next, type **rltool(G)** and enter. MATLAB creates a single-input-single-output (SISO) GUI with a Control and Estimations Tool Manager window and a SISO design window, showing the root locus.

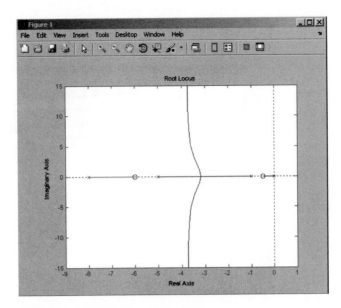

Figure C.1 The root locus drawn by MATLAB.

Select **Analysis Plot** on the Control and estimation Tool Manager window. Select **step** for the plot type for Plot 1 and **All** for the Closed Loop r to y, as shown in Figure C.2.

Next, use the SISO design window to design the system. To do so, right-click on the window. Then select **Design Requirements** and **New**. A New Design Requirement dialog box appears from which you may choose settling time, percent overshoot, damping ratio, natural frequency, or region constraints. Select the design requirement of your choice and enter the desired value and OK. The SISO Design window will add the design requirement to the root locus, indicating where the roots may or may not be

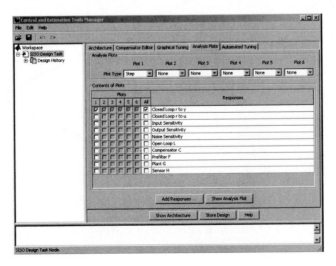

Figure C.2 The Control and Estimation Tools Manager window in MATLAB.

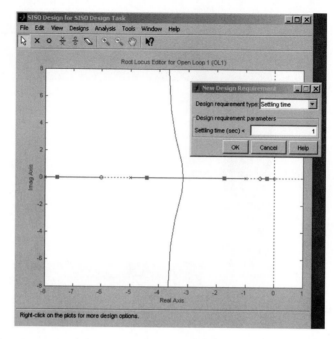

Figure C.3 The SISO Design window of MATLAB.

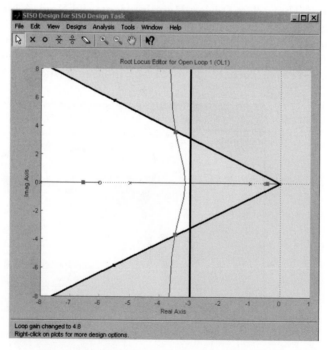

Figure C.4 The design constraints are shown on the root locus as drawn by MATLAB.

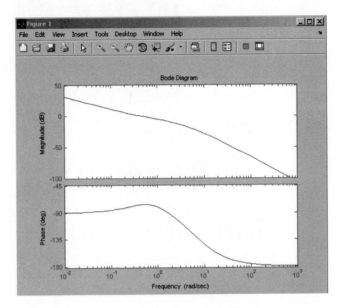

Figure C.5 The Bode diagram as plotted by MATLAB.

located. Later, you may place the pointer on the constraint and drag it to other values, or you may right-click within the constraint area to either delete or edit the constraint. Right-click outside the constraint area for a new design requirement (Figure C.3).

Figure C.4 shows the window for settling time of 1.3 sec and percent overshoot of 5%. All loci beyond the design constraint lines are acceptable. Each location yields a different gain, damping ratio, settling time, and other characteristics. You may choose any appropriate location for the roots depending on your desired characteristics. Figure C.4 shows the location of the roots for the fastest rise time. Clicking on this location will show the damping ratio, the gain, the root locations, and natural frequency.

Please experiment with other capabilities of the system or its features, and refer to online tutorials for more information about the applications and capabilities of MATLAB on this subject.

C.2 Bode Diagram

Similar to the process for root locus, enter your characteristic equations as in section C.1. Then type **bode (G).** The Bode diagram will be displayed, as shown in Figure C.5. You may use the diagram for design purposes in a similar fashion.

APPENDIX D

Simulation of Robots with Commercial Software

At least four commercial software programs are available in the market, that, to different degrees, are capable of simulating robot manipulator motions both kinematically and dynamically. They include SimulationX^{TM}, MapleSim^{TM}, MATLAB^{TM}, and Dymola^{TM}. In these systems, a robot's configuration is entered through modeling the links, the joints, and their relationship to each other. They allow prismatic, revolute, and spherical joints with multiple choices for body (link) types. Later, the axis of rotation (for revolute joints) or linear motion (for prismatic joints) relative to a World coordinate is selected for each joint through which the robot configuration is completed. You may also enter the physical characteristics of each link such as its mass, length, and moments of inertia. Finally, you may enter other information such as initial and final location and velocities and accelerations for each joint.

The user also specifies input profiles for each joint. Through this, the motions of the robot are simulated, animated, and displayed in 3D. You may also specify output values for each joint. For example, you may request the torque on each joint as the motion is completed, and therefore, have an indication of how much torque is needed to maintain the motion as specified. The torque graphs for the entire motion may be used to determine maximum required values from which appropriate actuators may be designed or selected.

A copy of the student version of SimulationX program may be downloaded free of charge from the ITI GmbH website at **www.iti.de**. A step-by-step tutorial available on the same website teaches the user how to make a 3-DOF robot and simulate it.

458

Index

A

Absolute encoder, 324–327
AC motor, 279–280, 285
AC/DC universal motor, 279–280
Acceleration sensor, 331
Accumulator, 273
Accuracy, 321
Acquisition of images, 352
Actuator
  electroactive polymer, 266, 308
  harmonic differential, 311
  hydraulic, 266, 272
  magnetostrictive, 307
  muscle-wire, 266, 307
  piezoelectric, 266, 311
  pneumatic, 266, 278
  shape-memory, 307
ADC, 255, 303–304, 320, 359
Aliasing, 361, 409
AML, 16
Amplitude modulation, 341
Analog image acquisition, 451, 453
Analog to digital converter, 255, 359
Angle criterion, 233
Anthropomorphic, 60
Antistatic foam, 333
Approach, 39
Articulated, 12, 60, 66, 89
Artificial intelligence, 351
Aspect ratio, 398
Asymptote angle, 233
Asymptote center, 233
Asymptotes, 233, 236
ATAN2, 68, 71, 451
Austenitic, 307
Autonomous navigation, 413
Autopass, 17

B

Back emf, 284, 287, 299–300, 306–307
Backlash, 256, 268, 271
Bandwidth, 247
Bifilar stepper motor, 292, 296–297
BigDog robot, 219
Binary image, 353, 365, 378, 389, 398
Binary morphology, 391–396
Biometal, 307, 314
Bitmap, 353
Block diagram, 204, 219
Bode diagram, 247, 459
Break-away point, 234
Breakdown voltage, 299
Break-in point, 234
Brightness, 365
Brushless DC motor, 279–280, 286, 291,
  303, 329
Byte, 303

C

CAD/CAM, 351
Cal Poly, 98, 198, 413
Caliper, 295
Camera
  analog, 451
  digital, 453
  vidicon, 453
Canstack stepper motor, 290–294
Capacitance, 207
Capacitive sensor, 336
Capek, Karel, 4
Cartesian, 12
Cartesian coordinates, 33, 60, 66, 72
Cartesian-space, 179, 195
Cathode-ray tube, 453
CAT-scan, 351

CCD, 453
Center of Gravity, 430–431, 434
Center-tapping, 286, 292-297
Cermet, 324
Characteristic equation, 225, 230–232, 248, 455
Charge-coupled device, 453
Charge-integrated device, 455
Cincinnati Milacron T3, 17, 272
Close operation, 391, 395
Closed-loop transfer function, 217, 233–234, 247–248
CMYK, 351
Coefficient of friction, 269
Color image, 353, 411
Common normal, 75, 82
Commutator, 283, 285, 289
Compensator, 344
  lag, 246–247, 254
  lead, 246–247, 254
Compiler, 16
Complex conjugate, 226, 227
Compliance, 267, 271
Computed tomography, 351, 405
Conductive polymer, 324, 331
Connectivity, 372–374
CONSIGHT, 408
Constant area quantization, 410
Contrast, 365
Convolution, 368–372, 384–375, 380
Coordinates
  articulated, 60, 66
  Cartesian, 33, 60, 66, 72, 179, 195
  cylindrical, 33, 60, 61, 66
  rectangular, 60, 66
  spherical, 33, 60, 64, 66
Coriolis acceleration, 158, 164–166
Coronal plane, 10
Correspondence, 407
Coupling inertia, 158
Crisp value, 425
Critical damping, 241
Critically damped system, 227, 230, 234, 239
Cross correlation, 341
Cross product, 42
CRT, 452
Cylindrical coordinates, 12, 33, 60, 61, 66

D
da Vinci, 23
DAC, 303, 304
DARPA, 28
Data compression, 409–411

DC brushless motor, 279–280
DC motor, 279–280, 283, 286
Deflection, 34
Defuzzification, 425
Degeneracy, 95–96
Degree of membership, 426
Degree of truth, 427
Degrees of freedom, 8–11
Denavit-Hartenberg, 5, 60, 73, 96
Depth analysis, 406, 408
Depth measurement, 351, 406–409
Derivative controller, 254
Derivatives, 128
Detent torque, 289, 298
Determinant, 56
Devol, George, 5
Dexterity, 95–96
Dielectric constant, 338
Dielectric elastomer, 308
Differential
  change, 124–125
  higher-order, 121
  motion, 114–119, 123, 127, 134–135, 147–148
  operator, 123, 125, 131
  rotation, 120–123, 129
  transformation, 123
  translation, 120, 123
Differential dithering, 278
Digital
  camera, 453
  control, 254
  filtering, 254
  image acquisition, 451
Dilation, 391–393, 396
Direct drive DC motor, 279–280
Direct-drive electric motor, 266, 286
Directional cosine, 42
Directional Vector, 37, 40
Discrete Fourier Descriptors, 405
Disk motor, 285
Disparity, 407–408
Displacement sensor, 328
Dot product, 43, 57
Dymola, 459
Dynamic analysis, 147
Dynamics of robots, 178

E
Eddy current, 336, 339
Edge detection, 358, 377–383, 388, 397, 406

Edge detectors
    Left–Right, 382, 397
    Prewitt, 381
    Roberts, 381
    Sobel, 381
Effective moment of inertia, 158
Electric motor, 266, 279, 283
Electroactive polymer actuator, 266, 308
Electromotive force, 339
Encoders, 278, 286, 312, 324–327, 330
    absolute, 324–327
    incremental, 324–326
    velocity sensor, 330
End-effector, 7, 10
Erosion, 391–394
Error signal, 204
Estimator, 253
Euler, 70–72
Expert system, 425

F
Fast Fourier transform, 356
Feature extraction, 406
Feedback gain, 218
Feed-forward loop, 235
Feed-forward transfer function, 217–219
Fill operation, 391, 396
Filters
    Gaussian, 375
    high-pass, 358, 380–383
    low-pass, 357, 375
    median, 377
Final value theorem, 211
Finger-spelling hand, 24, 98
Finite difference, 255, 379
First moment of area, 399, 403
First-order transfer function, 221, 223
Flaw detection, 341
Flexural spring, 334
Flow rate, 274
FM, 452
Force
    analysis, 170–174
    control, 171
    sensing resistor, 332
    sensor, 331
Force-current analogy, 207
Force-voltage analogy, 207
Foreshortening gradient, 408
Forward kinematics, 33, 59, 66, 72–78
4-bar mechanism, 35, 345

Fourier series, 354–357
Fourier transform, 354–358, 361, 374–376
Freewheeling diode, 299
Frequency
    content, 343, 354
    domain, 247, 354, 374, 376
    modulation, 341
    response, 321
    spectrum, 354, 376, 405
Fuzzification, 425, 430, 434
Fuzzy
    control, 425
    descriptors, 425
    inference control, 425–432, 435
    inference engine, 430–431, 434
    inference rule base, 425, 427, 432,
        437, 455
    sets, 425, 429, 433
    value, 425

G
Galvanometer, 333
Gaussian elimination, 94–95, 135
Gaussian filter, 375
GPS, 342
Gradient, 378
Gray image, 398
Gray morphology, 396

H
Halftoning, 410
Hall-effect sensor, 286, 325, 329, 336
Hallback array, 291
Hand frame, 125
Harmonic drive, 309–310
Harmonic linear differential actuator,
    311
Harmonics, 357
H-bridge, 306
Heat dissipation, 205, 280–281,
    286, 289
Heuristics, 377, 412
High-pass filter, 358, 380–383
Histogram, 354, 364–365, 375, 413
Histogram equalization, 365–366
Holding valve, 273
Hollow-rotor motor, 285
Homogeneous matrix, 40, 44, 46, 57
Homogeneous transformation, 44
Hough transform, 385–388, 405
Humerus, 10

Hybrid stepper motor, 290, 294
Hydraulic
    actuator, 266, 267, 272, 276
    power, 267
    ram, 60
Hysteresis, 256

I
Image acquisition, 352, 451
Image(s)
    acquisition, 352
    analysis, 350, 396–409
    averaging, 375–376
    binary, 353, 359, 365, 378,
        389, 398
    color, 353, 411
    digital, 352
    gray, 391, 398
    processing, 350, 363
    sharpening, 383–384
Incremental encoder, 324–326
Independent joint control, 259
Indexer, 289
Inductance, 207
Infrared sensor, 335
Instantaneous center, 345
Integral controller, 254
Intelligent motor controller, 304
Interfacing, 320
Interframe coding, 410–411
Interpreter, 16
Intraframe coding, 409–410
Inverse
    of matrices, 54–59
    Jacobian, 134–141
    Kinematics, 33, 59, 65–66, 71–73, 87–95, 97,
        182, 196, 199
    Laplace transform, 211–215
    Transformation matrix, 54–59

J
Jacobian, 116–118, 128–141, 171–173
Jacobian, inverse, 134–141
Joint variable, 178, 196
Joint-space, 179, 184
JPEG, 411
JPG, 353

K
Kinematic equations, 141, 178
Kinematic position, 147

Kinematics
    forward, 33, 59, 66, 72–78
    inverse, 33, 59, 65–66, 71–73, 87–95, 97
    equation, 128
Kinetic energy, 148, 152, 155, 158–163
Kirchhoff's law, 207

L
Lag compensator, 246–247, 254
Lagrangian, 148, 155, 158,164
Lagrangian mechanics, 148–157
Laplace transform, 208–214, 250, 255, 260
    inverse, 211–215
Laplacian, 379–380
Lapsed time, 339–340
Lead compensator, 246–247, 254
Left-Right edge detector, 382–383, 397
LIDAR, 342
Light sensor, 335
Linear Variable Differential Transformer,
    327
Linearity, 320
Load sensor, 334
Lossless compression, 411
Lossy compression, 411
Low-pass filter, 305, 357, 375
LVDT sensor, 327

M
Magnetic flux density, 282
Magnetostrictive actuator, 307
Magnetostrictive sensor, 328
Magnitude criterion, 233
Mamdani's inference method, 435
Manipulator, 2, 6
Manipulator, parallel, 35
MapleSim, 459
Martensitic, 307
MATLAB, 235, 237, 247, 435–436, 459
Matrices, 443
Matrix
    addition, 444
    adjoint, 56, 447
    algebra, 443
    determinant, 56, 95, 445
    diagonal, 444
    homogeneous, 40, 57
    inversion, 445
    multiplication, 121, 444
    pseudo inertia, 162
    representation, 36

square, 56
trace, 449
transpose, 56, 443
unitary, 57
Max operator, 396
Maxwell stress, 308
Median filter, 377
Membership function, 430, 435
MEMS, 29
Microstepping, 299, 301
MIMO system, 249
Min operator, 396
Mobile robot, 101
Modulus of elasticity, 268
Moment invariant, 401, 403
Moment of inertia, effective, 158
Moments, 398–404
Motors
    AC, 279–280, 285
    AC/DC universal, 279
    bifilar stepper, 292, 296–297
    brushless DC, 279–280, 286, 291, 303, 329
    Canstack stepper, 290–294
    DC, 279–280, 283, 286
    direct drive DC, 279
    direct-drive electric, 266, 286
    disk, 285
    electric, 266, 279, 283
    hollow-rotor, 285
    hybrid stepper, 290, 294
    intelligent controller, 304
    reactance, 284
    reversible AC, 286
    servo, 266, 287, 304
    stepper, 266, 267, 280, 288–303
    three-phase AC, 279
Mount Erebus, 24
multiple-input, multiple output, 249
Muscle-wire actuator, 266, 307

N
NASA, 25
Natural frequency, 247, 321
Navigation, 351, 406
NC machines, 5
Neighborhood averaging, 374–375
Neodymium, 285
Newtonian mechanics, 148–149
Nibble, 303
Noise reduction, 374–377
Nondexterous, 96

Nonlinear control, 256
Nonlinearity, 320
Nutating gear train, 309, 311
Nyquist rate, 409

O
Object recognition, 378, 397–406
Open operation, 391, 395
Open-loop, 34
Open-loop transfer function, 216, 229–231, 247–248
Optical encoder, 286
Orbidrive, 309–310
Overdamped system, 227, 230

P
Parabolic blend, 188–191
Parallel axes theorem, 400
Parallel manipulator, 35
Partial fraction expansion, 211–214, 223
Path planning, 178
Payload, 15
PD Controller, 241
Peak time, 225
Percent overshoot, 225, 236, 239, 243, 246
Phase compensation, 341
Phoneme, 344
Photosite, 453
Phototransistor, 335
PI controller, 239
PID controller, 244, 277
Piezoelectric, 331
    actuator, 266, 311
Pixel, 352, 357, 365, 372, 377, 386, 396, 401–411, 451
Planetary gear train, 309–311
Pleo life form, 315
Pneumatic actuator, 266, 267, 278
Pole placement, 236
Pole/zero placement, 225
Poles, 221, 225, 245
Polynomials, 183–188, 214, 221
Portable Gray Map, 353
Position sensor, 323
Potential energy, 148, 152, 163
Potentiometer, 278, 286, 312, 323–324
Power to weight ratio, 267, 271, 285
Prague, 1
Precision, 15
Predictive coding, 410
Pressure sensor, 331
Prewitt edge detector, 381
Prismatic joint, 11, 73–75, 80–85, 160, 171, 258

Product of inertia, 168, 400
Programming, 93
Proportional controller, 235, 238, 254
Proportional gain, 223, 235, 240
Proportional-derivative controller, 241
Proportional-integral controller, 239
Proximity sensor, 336–339
Pseudo inertia matrix, 162
Pseudorandom Quantization Dithering, 409
Pull-out torque, 298
Pulse Code Modulation, 409
Pulse width modulation, 304–305

Q
Quadrants, 63
Quadratic equation, 182
Quantization, 358–360

R
Radius of curvature, 346
Range array acquisition, 341
Range finders, 338, 339–342, 406, 413
   ultrasonic, 340
   infrared, 341–342
   laser, 342
Rapid prototyping, 276
Raster scan, 452
RCC device, 268, 344–347
Reactance of motors, 284
Reduction gears, 268, 270
Reflectance, 408
Region growing, 378, 388–391, 397, 406
Reliability, 321
Reluctance, 290–299
Remote center compliance device, 268, 344–347
Remote sensing, 351
Repeatability, 15, 321
Residual torque, 289, 298
Resistance, 207
Resolution, 303, 304, 320, 358–360
Resolver, 286, 328
Response time, 321
Reversible AC motor, 286
Revolute joint, 11, 159, 171, 258
RGB, 353, 411
Rigid body, 41, 59
Rise time, 222, 225, 321
Roberts edge detector, 381
Robot
   actuation, 257
   characteristics, 15

   classification, 3
   control, 257
   fixed sequence, 4
   intelligent, 4
   numerical control, 4
   payload, 15
   playback, 4
   precision, 15
   repeatability, 15
   validity, 15
   variability, 15
   variable sequence, 4
   workspace, 15
Roll, pitch, yaw, 66–70, 72
Root locus, 230–254
Rossum's Universal Robots, 4
Rotating frame, 52–54
Rotation, 197
   about an axis, 45
   pure, 46
RPY, 66–70, 72

S
Sagittal plane, 10
Sampling rate, 254, 358, 360
Sampling theorem, 360–363
Scale factor, 37, 57
SCARA robot, 12, 73, 344–347
Second moment of area, 400, 403
Second-order transfer function, 223, 225–226
Segmentation, 378, 388–391, 408
Selective compliance, 345
Sensitivity, 320
Sensors, 204
   acceleration, 331
   antistatic foam, 333
   capacitive, 336
   displacement, 328
   force, 331
   GPS, 342
   Hall-effect, 286, 325, 329, 336
   infrared, 335
   LIDAR, 342
   light, 335
   load, 334
   LVDT, 327
   magnetostrictive, 328
   piezoelectric, 331
   position, 323
   pressure, 331
   proximity, 336–339

Sensors (*continued*)
  range finder, 338–342
  resolver, 328
  sniff, 343
  Strain gauge, 332
  tachometer, 330
  tactile, 335–336
  taste, 343
  torque, 333–334
  touch, 335–336
  velocity, 330
  voice recognition, 343
Servocontroller, 204, 289, 312
Servomechanism, 258
Servomotor, 266, 275, 287, 304
Servovalve, 273, 275, 278
Settling time, 222–225, 236, 246, 321
Shape-memory metal, 307
Simulation, 435
Simulation-X, 459
Single-input, single-output, 249, 259
Singleton value, 425
SISO system, 249, 259, 458
Skeletonization, 391–395, 413
Sniff sensor, 343
Sobel operator, 381
Spatial domain, 354, 368–374
Spherical, 12
Spherical coordinates, 33, 60, 64, 66
Spool valve, 274–275
Spot checking, 341
Square matrix, 56
Stability, 249
Stanford Arm, 85, 96–97, 127, 161
State-space control, 250–253
Static force analysis, 170–174
Static position error coefficient, 229, 239
Static velocity error coefficient, 230
Steady-state
  error, 228–229, 237–240, 244, 246–249, 261
  gain, 224
  value, 224
Stepper driver, 300–302
Stepper motor, 266, 280, 289, 312
Stepper translator, 300
Stereo imaging, 406–409, 413
Stiffness, 267, 271, 347
Strain gauge, 332
Surgical robots, 22–23, 413
System dynamics, 205
System type, 230, 238, 239

T
Tachometer, 289, 330
Tactile sensor, 335–336
Taste sensor, 343
Taylor series, 256, 257
Telerobotics, 4
Template matching, 404
Terfenol-D, 307
Thickening, 391–393
Thinness, 398
Three-phase AC motor, 279–280
Thresholding, 354, 365–367, 378, 389, 406, 413
TIFF, 353
Time constant, 221, 222, 225, 321
Time of flight, 339–342
Time-to-amplitude converter, 341
Torque
  constant, 281–283
  detent, 289, 298
  holding, 298
  pull-out, 298
  residual, 289, 298
  sensor, 171, 333–334
Touch sensor, 335–336
Trace, 162, 165
Trajectory planning, 178–200
Transfer function, 216–219, 242, 260–261
  closed-loop, 217, 232–233, 247–248
  feed-forward, 217–219
  first-order, 221, 223
  open-loop, 216, 229, 231, 247–248
  second-order, 223, 225–226
Transformations, 35
  homogeneous, 44
  combined, 49–52
  relative, 52–54
Translation, 45, 62, 76, 197, 347
Transpose, 443
Triangulation, 339–342, 407
Tunnel diode oscillator, 330, 334

U
Ultrasonic, 336, 351, 405
Underdamped system, 227, 230
Union, 391
United Auto Workers, 29

V
V+, 16–17
VAL , 16–17
Validity, 15

Value
  crisp, 425
  fuzzy, 425
  singleton, 426
Valve
  holding, 273
  servo, 274–278
  spool, 274–275
Variability, 15
Vector(s), 35
  direction, 37
  directional, 40
  in space, 36
Velocity sensor, 330
Vidicon camera, 452
Virtual reality, 329
Virtual work, 171–172
Viscosity, 271
Viscous coefficient of friction, 207

Voice
  coil, 287
  recognition, 343
  synthesis, 344
Voltage divider, 323
Voxel, 352, 372
Voyager 2, 410

W
Wheatstone bridge, 332
Work envelope, 180
Workspace, 95
World War II, 5

Z
Zadeh, Lotfi, 425
Zero-pole cancellation, 245
Zeros, 225, 245
z-transform, 255